Freshwater Red Algae of the World

Dedicated to Paul Silva in celebration of his 80th birthday

Freshwater Red Algae
of the World

Shigeru Kumano

Biopress Limited

© Biopress Ltd., 2002

All rights reserved. No part of this book may be reproduced or transmitted in any form or by any means, electronic or mechanical, including photocopy, recording or any information storage and retrieval system, without permission from the publisher.

ISBN: 0–948737–60–3

PUBLISHED BY:

Biopress Ltd.
The Orchard
Clanage Road
Bristol
BS3 2JX
England

Acknowledgement

Many illustrations and diagrams in this volume have been obtained from the publications. Some of the original figures have been slightly modified. In all cases reference is made to the original publication. The full source can be found in the reference list at the end of each chapter. Permission for the reproduction of this material is gratefully acknowledged.

British Library Cataloguing in Publication Data

A catalogue record for this book is available from the British Library

© Published by Biopress Limited, Bristol, England
Printed in Great Britain by Henry Ling Ltd. at The Dorset Press, Dorchester, Dorset, England

2. *Batrachospermum* **gelatinosum** group

1) *B. theaquum* Skuja ex Entwisle et Foard..p. 103, pl. 59
2) *B. debilis* Entwisle et Foard..p. 105, pl. 60
3) *B. gelatinosum* (Linnaeus) De Candolle..p. 107 (ref. p. 86)
4) *B. campyloclonum* Skuja ex Entwisle et Foard...p. 107, pl. 61

3. *Batrachospermum* **antipodites** group

1) *B. antipodites* Entwisle...p. 109, pls 62, 63
2) *B. discorum* Entwisle et Foard..p. 111, pl. 64
3) *B. kraftii* Entwisle et Foard..p. 113, pl. 65
4) *B. ranuliferum* Entwisle et Foard..p. 115, pl. 66
5) *B. antiquum* Entwisle et Foard..p. 116, pl. 67

4. *Batrachospermum* **prominens** group

1) *B. prominens* Entwisle et Foard..p. 118, pl. 68
2) *B. terawhiticum* Entwisle et Foard..p. 118, pl. 69

5. *Batrachospermum* **wattsii** group

1) *B. wattsii* Entwisle et Foard...p. 121, pl. 70

6. *Batrachospermum* **cayennens**e group...............p. 123 (ref. p. 163)

2. Section *Setacea* De Toni

1) *B. androinvolucrum* Sheath, Vis et Cole...p. 124, pl. 71
2) *B. latericiuum* Entwisle..p. 124, pl. 73
3) *B. diatyches* Entwisle...p. 127, pls 72, 74
4) *B. atrum* (Hudson) Harvey..p. 129, pls 75, 76, 77
5) *B. puiggarianum* Grunow...p. 131, pls 77, 78

3. Section *Turfosa* Sirodot *sensu* Necchi

1) *B. tapirense* Kumano et Phang..p. 135, pl. 79
2) *B. orthostichum* Skuja..p. 136, pl. 80
3) *B. keratophytum* Bory *emend.* Sheath, Vis et Cole..p. 137, pls 81, 82
4) *B. periplocum* (Skuja) Necchi...p. 139, pls 83, 84
5) *B. turfosum* Bory *emend.* Sheath, Vis et Cole..p. 141, pl. 85

4. Section *Virescentia* Sirodot

1) *B. crispatum* Kumano et Ratnasabapathy...........p. 143, pl. 86
2) *B. gombakense* Kumano et Ratnasabapathy...........p. 144, pl. 87
3) *B. vogesiacum* Schultz ex Skuja...........p. 145, pl. 88
4) *B. gulbenkianum* Reis...........p. 147
5) *B. bakarense* Kumano et Ratnasabapathy...........p. 147, pl. 89
6) *B. azeredoi* Reis...........p. 148
7) *B. ferreri* Reis...........p. 149
8) *B. elegans* Sirodot *emend.* Sheath, Vis et Cole...........p. 150, pls 90, 91
9) *B. transtaganum* Reis...........p. 150, pl. 97 (a)
10) *B. helminthosum* Bory *emend.* Sheath, Vis et Cole...........p. 152, pls 92, 93
11) *B. desikacharyi* Sankaran...........p. 154, pl. 94

5. Section *Gonimopropagulum* Sheath et Whittick

1) *B. breutelii* Rabenhorst...........p. 156, pls 95, 96

6. Section *Hybrida* De Toni

1) *B. virgato-decaisneanum* Sirodot...........p. 158, pls 97 (b), 98, 99, 100
2) *B. abilii* Reis...........p. 158

7. Section *Aristata* Skuja

7-1. Subsection *Aristata* Kumano

1) *B. cayennense* Montagne...........p. 163, pls 101 (a), 101 (b), 102
2) *B. longiarticulatum* Necchi...........p. 166, pls 102, 103 (a)
3) *B. turgidum* Kumano...........p. 166, pl. 103 (b)
4) *B. beraense* Kumano...........p. 168, pl. 104 (a)

7-2. Subsection *Macrospora* Kumano

1) *B. macrosporum* Montagne...........p. 168, pls 104 (b), 105
2) *B. equisetifolium* Montagne...........p. 171, pl. 106 (b)
3) *B. hypogynum* Kumano et Ratnasabapathy...........p. 171, pl. 106 (a)

8. Section *Contorta* Skuja

8-1. Subsection *Intorta* Kumano

1) *B. intortum* Jao..p. 174, pls 107 (a), 107 (b)
2) *B. pseudocarpum* Reis..p. 175
3) *B. woitapense* Kumano..p. 176, pl. 108
4) *B. lusitanicum* Reis..p. 177

8-2. Subsection *Torrida* Kumano

1) *B. henriquesianum* Reis..p. 178
2) *B. tortuosum* Kumano var. *tortuosum*..p. 178, pl. 109
3) *B. tortuosum* Kumano var. *majus* Kumano...p. 179, pl. 110
4) *B. torridum* Montagne..p. 180, pl. 111 (a)
5) *B. faroense* Kumano et Bowden-Kerby..p. 181, pls 112 (a), 112 (b)
6) *B. curvatum* Shi...p. 182, pl. 113

8-3. Subsection *Procarpa* Kumano

1) *B. equisetoideum* Kumano et Necchi..p. 187, pl. 114 (b)
2) *B. procarpum* Skuja var. *procarpum*...p. 187, pl. 115 (a)
3) *B. procarpum* Skuja var. *americanum* Sheath *et al*...............................p. 188, pl. 115 (b)

8-4. Subsection *Kushiroense* Kumano

1) *B. spermatiophorum* Vis et Sheath..p. 190, pl. 116
2) *B. kushiroense* Kumano et Ohsaki..p. 190, pl. 117 (a)
3) *B. iriomotense* Kumano..p. 192, pl. 117 (b)
4) *B. louisianae* Skuja...p. 193, pls 111 (b), 111 (c)
5) *B. breviarticulatum* (Necchi et Kumano) Necchi.....................................p. 194, pl. 118 (a)
6) *B. tabagatense* Kumano et Bowden-Kerby..p. 195, pl. 118 (b)
7) *B. guyanense* (Montagne) Kumano..p. 196, pl. 119 (a)
8) *B. nonocense* Kumano et Liao...p. 196, pl. 119 (b)
9) *B. globosporum* Israelson...p. 197, pls 114 (a), 120 (a), 120 (b)
10) *B. nechochoense* Kumano et Bowden-Kerby..p. 199, pl. 121 (a)
11) *B. capense* Starmach ex Necchi et Kumano..p. 201, pl. 122
12) *B. skujanum* Necchi..p. 201, pl. 121 (b)

8-5. Subsection *Ambigua* Kumano

1) *B. gibberosum* (Kumano) Kumano...p. 203, pl. 123 (a)
2) *B. deminutum* Entwisle et Foard..p. 204, pl. 124

3) *B. vittatum* Entwisle et Foard..p. 207, pl. 125

4) *B. hirosei* Kumano et Ratnasabapathy..p. 207, pl. 123 (b)

5) *B. australicum* Entwisle et Foard..p. 209, pl. 126

6) *B. mahlacense* Kumano et Bowden-Kerby...p. 209, pl. 127

7) *B. dasyphillum* Skuja..p.211, pl. 129

8) *B. nodiflorum* Montagne...p. 212, pl. 128

9) *B. gracillimum* W. West et G. S. West *emend.* Necchi...p. 212, pl. 130

10) *B. torsivum* Shi..p. 213, pl. 131

11) *B. iyengarii* Skuja..p. 214, pl. 129

12) *B. tiomanese* Kumano et Ratnasabapathy..p. 214, pl. 133

13) *B. zeylanicum* Skuja..p. 216, pl. 129

14) *B. kylinii* Balakrishnan et Chaugule..p. 217, pl. 135

15) *B. mahabaleshwarensis* Balakrishnan et Chaugule..p. 218, pl. 129

16) *B. omobodense* Kumano et Bowden-Kerby..p. 220, pl. 134

17) *B. ambiguum* Montagne..p. 221, pls 136 (a), 136 (b)

Genus *Sirodotia* Kylin

1) *Sirodotia yutakae* Kumano..p. 224, pls 137 (a), 141 (b)

2) *S. segawae* Kumano..p. 225, pls 137 (b), 141 (b)

3) *S. sinica* Jao...p. 227, pls 138 (a), 138 (b)

4) *S. geobelii* Entwisle et Foard...p. 227, pl. 139

5) *S. suecica* Kylin...p. 230, pls 140, 141 (a), 142

6) *S. gardneri* Skuja ex Flint...p. 234, pl. 145

7) *S. huillensis* (Welwitsch ex W et G. S. West) Skuja...p. 234, pls 143, 144, 145

8) *S. delicatula* Skuja..p. 236, pl. 146

Genus *Tuomeya* Harvey

1) *Tuomeya americana* (Kützing) Papenfuss..p. 238, pls 147, 148, 149, 150

Genus *Nothocladus* Skuja

1) *Nothocladus afroaustralis* Skuja...p. 243, pl. 151

2) *N. lindaueri* Skuja...p. 243, pls 151, 152

3) *N. nodosus* Skuja..p. 246, pls 153, 154

Family **Psilosiphonaceae** Entwisle *et al.*
Genus *Psilosiphon* Entwisle

1) *Psilosiphon scoparium* Entwisle...p. 249, pls 155 (a), 155 (b)

Family **Lemaneaceae** C. Agardh
Genus *Lemanea* Bory

1) *Lemanea borealis* Atkinson..p. 253

2) *L. simplex* Jao..p. 254, pl. 156

3) *L. condensata* Israelson..p. 254

4) *L. fluviatilis* (Linnaeus) C. Agardh...p. 256, pl. 157

5) *L. sudetica* Kützing..p. 256, pl. 160

6) *L. ciliata* (Sirodot) De Toni..p. 260, pl. 160

7) *L. sinica* Jao...p. 261, pl. 161

8) *L. rigida* (Sirodot) De Toni...p. 261, pls 161, 165

9) *L. mamillosa* (Sirodot) De Toni..p. 263, pls 156, 159

10) *L. fucina* (Bory) Atkinson var. *fucina*..p. 263, pls 158, 159

11) *L. fucina* (Bory) Atkinson var. *parva* Vis et Sheath..p. 264

Genus *Paralemanea* (Silva) Vis et Sheath

1) *Paralemanea mexicana* (Kützing) Vis et Sheath...p. 265

2) *P. catenata* (Kützing) Vis et Sheath..p. 265, pls 162, 163

3) *P. annulata* (Kützing) Vis et Sheath..p. 265, pls 163, 164

4) *P. grandis* (Wolle) Kumano...p. 269, pl. 165

Species of the genus *Paralemanea* from Indiana and California by Blum

1) *Paralemanea deamii* Blum..p. 271, pl. 166

2) *P. gardnerii* Blum..p. 271, pls 167, 169

3) *P. parishii* Blum..p. 271, pls 168, 169

4) *P. tulensis* Blum...p. 275, pl. 168

5) *P. californica* Blum..p. 275, pls 167, 169

6) *P. brandegeei* Blum...p. 275, pls 168, 169

Order **Thoreales** Sheath, Müller et Sherwood

Family **Thoreaceae** (Reichenbach) Hassall

Genus ***Thorea*** Bory

1) *Thorea clavata* Seto et Ratnasabapathy..p. 279, pls 170, 171

2) *T. zollingeri* Schmitz *emend.* Sheath, Vis et Cole..p. 280, pl. 171

3) *T. violacea* Bory...p. 280, pl. 171

4) *T. hispida* (Thore) Desvaus *emend.* Sheath *et al*...p. 280, pl. 171

5) *T. conturba* Entwisle et Foard..p. 282, pl. 172

6) *T. okadae* Yamada..p. 282, pls 171, 174

7) *T. bachmannii* Pujals..p. 284, pls 175, 176

8) *T. riekei* Bischoff..p. 284, pl. 173

9) *T. prowsei* Ratnasabapathy et Seto..p. 286, pl. 177

10) *T. gaudichaudii* C. Agardh...p. 286

11) *T. brodensis* Klas...p. 289

Genus ***Nemalionopsis*** Skuja

1) *Nemalionopsis tortuosa* Yoneda et Yagi..p. 290, pls 171, 173, 178

2) *N. shawii* Skuja..p. 292, pl. 171

Order **Hildenbrandiales** Pueschel et Cole

Family **Hildenbrandiaceae** Rosenvinge

Genus ***Hildenbrandia*** Nardo

1) *Hildenbrandia rivularis* (Liebmann) J. Agardh...p. 293, pl. 179

2) *H. angolensis* Welwitsch ex W. West et G. S. West..p. 293, pl. 180

Order **Balliales** Choi, Kraft et Saunders

Family **Ceramiaceae** Choi, Kraft et Saunders

Genus ***Ballia*** Harvey

1) *Ballia pinnulata* Kumano..p. 298, pl. 181 (a)

2) *B. prieurii* Kützing...p. 298, pls 181 (b), 182, 183, 190

Genus ***Ptilothamnion*** Thuret

1) *Ptilothamnion richardsii* Skuja..p. 300, pls 184, 185

Order **Ceramiales** Oltmanns

Family **Delesseriaceae** Bory

Genus ***Caloglossa*** (Harvey) Martens

1) *Caloglossa ogasawaraensis* Okamura.................................p. 304, pls 186 (a), 187, 190

2) *C. beccarii* (Zanardini) De Toni...p. 306, pl. 186 (b)

3) *C. leprieurii* (Montagne) Martens..p. 306, pls 190, 197

4) *C. continua* (Okamura) King et Puttock..p. 307, pls 187, 188

5) *C. saigonensis* (Tanaka et Pham-Hoàng Hô) King et Puttock.............p. 310, pl. 189

Family **Rhodomelaceae** Areschoug

Genus ***Bostrychia*** Montagne nom. cons.

1) *Bostrychia moritziana* (Sonder) J. Agardh....................................p. 312, pls 191, 192, 193

2) *B. radicans* (Montagne) Montagne...p. 315, pls 192, 194

3) *B. simpliciuscula* Harvey ex J. Agardh..p. 318, pl. 196 (a)

4) *B. flagellifera* Post..p. 318, pl. 196 (b)

5) *B. tenella* (Lamouroux) J. Agardh..p. 320, pls 192, 195

6) *B. scorpioides* (Hudson) Montagne..p. 320, pls 197, 198

Genus ***Polysiphonia*** Greville nom. cons.

1) *Polysiphonia subtilissima* Montagne...p. 324, pls 190, 199

Abbreviations used in nomenclature..p. 325
Abbreviations of herbaria compiled in this book...p. 327
Glossary: Phycological terms relating to the freshwater Rhodophyceae..........p. 329
Bibliography...p. 333
Species Index ....p. 367
Addendum..p. 375

Preface

Millions of human beings around the world depend on waters for their livelihood, and barely fifty years ago the world's environment was still largely in balance. The springs and the rivulets, the brooks, the rivers, and the lakes, seem to give life to Nature, and were indeed regarded by our ancestors as living entities themselves.

The environment in which the rivers flow makes a critical difference to the divergence of wildlife that are found there. Today we are told that our planet is in crisis, that we are destroying and polluting our way to global catastrophe. Freshwater pollution has affected the fish and shellfish as well as freshwater algae which inhabit the clean waters.

Many taxa of the freshwater Rhodophyceae mainly inhabit unpolluted clean inland waters, so that they can be used as an indicator of clean waters. Many phycologists have reported the fact that human activities have an influence on their survival and their extinction would cause a crisis. In other words, the extinction of the freshwater Rhodophyceae means extinction of clean waters because of pollution of water by human activities.

In such circumstances, many people have become interested in the freshwater Rhodophyceae as an indicator for quality of water. However, the identification of this group is very difficult, because comprehensive books on the taxonomy of this group have not been published until now. It is very important to compile the hitherto described taxa so that many more workers can be assisted in the study of the freshwater Rhodophyceae. Although many books have been published, they have treated the freshwater Rhodophyceae as merely a part of the freshwater or marine algal taxa.

Investigations concerning the freshwater Rhodophyceae have been carried out by many phycologists in different fields and regions in the world. The system of classification of the species of the freshwater Rhodophyceae has been revised according to the progress of phycology. In the near future, a revision of the classification system of the freshwater Rhodophyceae will be made because much evidence has been reported by many authors. In such circumstances, it is very difficult to compile the taxa without helpful advice given by several authorities and experts on the freshwater Rhodophyceae.

Acknowledgements

The author wishes to express his sincere thanks to Dr Paul C. Silva, Dr Richard L. Moe (Herbarium, Department of Botany, University of California, Berkeley, California, USA) and Dr Tadao Yoshida (Professor Emeritus of Hokkaido University, Sapporo, Japan), whose helpful review in rewriting the manuscript in English and critical reading of the manuscript has enhanced the quality of the final presentation. The advice and suggestions of Dr Mitsuo Chihara (Professor Emeritus of Tsukuba University, Tsukuba, Japan), Dr Hiroshi Kawai (Kobe University, Kobe, Japan) and Dr Takao Yamagishi (Japanese Institute of Plankton, Kanagawa, Japan) are gratefully acknowledged.

The author extends his thanks to the following phycologists: Dr Orlando Necchi, Jr. (UNESP, Departmento Botanica, Campus S. Jose Rio Preto, Brasil) for help with the families Audouinellaceae, Lemaneaceae and the Brazilian taxa of freshwater Rhodophyceae, Dr Timothy J. Entwisle (Royal Botanical Gardens, Sydney, New South Wales, Australia) for assistance with the Australian taxa of freshwater Rhodophyceae, Dr Ryozo Seto (Nishinomiya, Japan) for families Compsopogonaceae and Thoreaceae, Dr Jiroh Tanaka (Tokyo University of Fisheries, Tokyo, Japan), Dr Mitsunobu Kamiya (Kobe University, Kobe, Japan) for *Caloglossa* and *Bostrychia* and Dr Shi Zhixin (Institute of Hydrobiology, Academia Sinica, Wuhan, China) for the recent information of the Chinese taxa of freshwater Rhodophyceae.

Thanks are also given to the many colleagues who generously permitted reproduction of their published material especially the original figures and photomicrographs of taxa of freshwater Rhodophyceae: to Biopress Ltd., especially Professor Frank E. Round and Mrs Gillian Lockett for supporting the development of this publication.

Most of all thanks to my family, especially my wife, Kazuko, for their constant support and moral encouragement.

Biopress Limited wish to acknowledge with thanks the full permission received from Uchida Rokakuho Publishing Company Limited of Japan to publish the English version of their publication *Taxonomy of Freshwater Rhodophyta* by Shigeru Kumano, published in 2000, without restriction or provisos, including the reproduction of all illustrations published therein.

The taxa compiled in this book

This book deals with the taxa of the freshwater Rhodophyceae including the brackish species reported from the freshwater habitats. Most taxa compiled in this book are placed according to the existing systems, however, some are placed in the tentatively proposed new systems for regrouping of the genus *Batrachospermum* from Australia and New Zealand (Entwisle & Foard 1997).

The author has tried to compile not only the recognized and recently described taxa, but also the historically important taxa. The initial drawings and photomicrographs made by the original authors are brought together in this book. In addition, wherever possible the classical material has been supplemented by the authors' own observations.

In the text, the journals for basionyms and synonyms are referred to by codes which refer to the bibliography, where the full references are given. The bibliography includes floral references of the freshwater and marine Rhodophyceae.

Information concerning the type specimen such as the specimen number, the herbarium housing the type specimen, abbreviations of herbaria (Index Herbaria ver. 8, 1990) cited in this book are also listed. The type localities are noted as precisely as possible and described in detail in the languages used by the original authors.

Short history of studies on the freshwater Rhodophyceae

Linnaeus (1753) divided the Phanerogams into twenty three classes, while he classified all algae together with fungi, mosses and Pteridophytes into the 24th class, the Cryptogamia. He accepted fourteen genera of algae, *Jungermannia, Targionia, Marchantia, Blasia, Riccia, Anthoceros, Lichen, Chara, Tremella, Fucus, Ulva, Conferva, Byssus* and *Spongia*. Among them, only four genera, *Conferva, Ulva, Fucus* and *Chara* contained organisms now regarded as algae. In addition, two other genera, *Byssus* and *Tremella*, contained a few species of algae. Linnaeus treated the species of red algae under the genera, *Conferva, Ulva* and *Fucus*.

Roth (1797) accepted ten genera of algae, *Fucus, Ceramium, Batrachospermum, Conferva, Mertensia, Hydrodictyon, Ulva, Rivularia, Linkia* and *Tremella*. The third genus *Batrachospermum* was established with the description of its characters by Roth who reconfirmed *Batrachospermum dichotomum, Conferva gelatinosa* in detail in 1800.

Vaucher (1803) divided *Conferva* into several genera such as *Ectosperma, Conjugata, Hydrodictyon, Polyspermum, Prolifera* and *Batrachospermum*.

On the basis of colour, Lamouroux (1813) was the first to segregate the algae now placed in the Rhodophyceae from other algae of similar external morphology. He established the category Floridées from which the name of the currently accepted major groups of red algae, the Florideophycidae, was derived.

Agardh (1824) arranged the algae into five well-defined orders, Diatomae, Nostochinae, Confervoideae, Ulvaceae and Florideae. Nostochinae included many cyanophycean species, the Confervoideae comprised the filamentous green algae. He emphasized the importance of the cystocarps in Rhodophycean taxonomy; however, some species of Rhodophycean algae were placed in Confervoideae, in which he accepted ten families, *viz.*, Funginae, Lichenoideae, Byssoideae, Leptomistae, Batrachospermeae, Oscillatorinae, Converveae, Characeae, Ceramieae and Ectocarpeae.

On the basis of their pigmentation, Harvey (1836) divided the all algal genera into four major divisions, *viz.*, Rhodospermeae (red algae), Melanospermeae (brown algae), Chlorospermeae (green algae) and Diatomaceae. In general his assignment of genera was reasonably accurate by modern standards, however, the considerable colour variation in many genera of red algae produced some problems. The genera *Porphyra* and *Bangia* and most freshwater red algae were placed in Chlorospermeae (green algae) rather than in Rhodospermeae (red algae), because of the frequent greenish colour of fronds of the above-mentioned red algae.

Kützing (1843) divided the algae as a whole into two classes, *viz*. class Isocarpeae including two tribes, tribe Gymnospermeae and tribe Angiospermeae, and class Heterocarpeae which compose the red algae. Freshwater red algae, however, were retained as the second order Cryptospermeae of the tribe Gymnospermeae in the class Isocarpeae. Order Cryptospermeae was composed of five families, *viz*. Lemanieae, Chaetophoreae, Batrachospermeae, Liagoreae and Mesogloeaceae.

Bornet & Thuret (1866, 1867) were the first to describe clearly the female organ in a number of the red algae. They determined the nature of the female organ, which Nägeli (1861) had observed but misinterpreted. They also observed several red algae including the genera *Ceramium, Dudresnaya* and *Nemalion*, and that the antheridia (spermatia) fused with trichogynes, thereby effecting fertilization. They also recorded cystocarp formation as a consequence of fertilization. This was followed by Schmitz's (1883) observation that in some red algae, sporogenous filaments (gonimoblast filaments) develop directly from the fertilized carpogonium, whereas in others, sporogenous filaments (connecting filaments) first fuse with other cells in the female gametophyte before cutting off the gonimoblast. Schmitz named these cells "auxiliary cells".

Rabenhorst (1868) first removed the family Batrachospermaceae together with several other freshwater species from green algae to Rhodophyceae. He divided the algae into four classes, *viz.*, Phycochromophyceae (blue-greens), Chlorophyllophyceae (greens), Melanophyceae (browns) and Rhodophyceae (reds). The last class, Rhodophyceae, comprised five freshwater families, *viz.*, Porphyraceae including the genera *Porphyridium, Porphyra* and *Bangia*, Chantransiaceae including *Chantransia*, Batrachospermaceae including the genera *Batrachospermum* and *Thorea*, Hildenbrandiaceae including the genus *Hildenbrandia* and Lemaneaceae including the genera *Lemaea* and *Compsopogon*.

Schmitz (1883) was the first to observe that in certain red algae the fertilized carpogonium produced the filaments (connecting filaments) that fused with a neighbouring cell. He termed this the auxiliary cell, from which the gonimoblast (carposporophyte) developed. The present period in the classification of the red algae was ushered in by Schmitz's findings. On the basis of the fundamental differences in the ontogeny of the gonimoblast (carposporophyte), especially with regard to the development and the function of the auxiliary cell, Schmitz (1893) later proposed a regrouping of the red algae along lines that portrayed a much more natural arrangement than had been possible previously. He divided the Florideae into four orders; (1) Nemaliales (as Nemalionales), characterized by a gonimoblast developing directly from the zygote (auxiliary cell absent), (2) Gigartinales, characterized by paired carpogonial branches and auxiliary cells (procarpic) and inwardly developing gonimoblast, (3) Rhodymeniales, characterized by paired carpogonial branches and auxiliary cells (procarpic) and outwardly developing gonimoblast, and (4) Cryptonemiales, characterized by carpogonial branches and auxiliary cells scattered in the thallus (non-procarpic) and either outwardly and inwardly developing gonimoblast.

Cooke (1882–1884) included four families in the freshwater Rhodophyceae, *viz.*, family Porphyraceae (genus *Bangia*), family Chantransiaceae (genus *Chantransia*), family Batrachospermaceae (genera *Batrachospermum* and *Thorea*) and family Lemaneaceae (genus *Lemanea*).

Taxonomic studies on the family Batrachospermaceae have made remarkable progress since the magnificent monograph entitled *Les Batrachospermes* was published by Sirodot (1884). Based on the colour of fronds, the shape of the trichogyne and the location of the gonimoblast (carposporophyte), Sirodot grouped the species of the genus *Batrachospermum* into six sections, *viz.*, Moniliformes, Helminthoides, Setaces, Turficoles, Verts and Hybride. Thus, Sirodot gave the foundation for modern taxonomy of the family Batrachospermaceae.

Oltmanns (1904) established the order Ceramiales for the order Rhodymeniales of Schmitz's system, in which the auxiliary cell is formed after the fertilization of the carpogonium, and then, Kylin (1923) established the order Gelidiales for the family Gelidiaceae placed in the order Nemalionales.

Pascher (1925) had made remarkable progress when he published the magnificent monograph entitled *Die Süsswasser-Flora, Deutschlands, Österreichs und der Schweiz*, in which he compiled the hitherto-known species of freshwater algae from Europe – the species of Heterokontae, Phaeophyta, Rhodophyta, Charophyta are included in the above-mentioned series, Heft 11.

Pascher & Schiller (1925) divided the freshwater Rhodophyceae into Bangiales including the family Bangiaceae (genera *Bangia, Phragmonema, Asterocytis, Chroothece, Porphyridium, Vanhoeffenia*) and Florideae including the family Thoreaceae (genus *Thorea*), the family Helminthocladiaceae (genera *Chantransia, Batrachospermum, Sirodotia*), the family Lemaneaceae (genus *Lemanea*), the family Ceramiaceae (genus *Ceramium*) and the genus *Hildenbrandia*.

The morphology and the mode of the reproduction of many British freshwater algae were reported in *A treatise on the British algae* by West & Fritsch (1927). Of the British genera of red algae, the genera *Chantransia* and *Batrachospermum* belonged to the family Helminthocladiaceae and the genus *Lemanea* to the family Lemaneaceae, while they stated that *Thorea, Compsopogon* and *Hildenbrandia* were best regarded as genera *incertae sedis*. In this book it is most appropriate to consider the genera alone, without reference to the families in which they may belong.

In Smith's (1933, 1950) book *The freshwater algae of the United States*, the morphology and the reproduction of many genera and species were discussed. He divided the freshwater Nemalionales into four families, *viz.*, Chantransiaceae, Batrachospermaceae, Thoreaceae and Lemaneaceae.

Skuja (1931 etc.) was without any doubt one of the most remarkable phycologists working with freshwater red algae, for example, he wrote many papers under the title *Untersuchungen über die Rhodophyceen des Süsswassers. I–XII*, and established two new genera, *Nothocladus* and *Nemalionopsis*, and many new species of freshwater red algae.

Post (1936 etc.) carried out the systematic and the phytogeographical studies of the *Bostrychia-Caloglossa* associations in the subtropical and tropical regions.

Israelson (1938) reported many freshwater red algae from Sweden. Jao (1941) dealt with the freshwater red algae collected from different localities in China, including eight genera, *viz., Bangia, Compsopogon, Audouinella, Batrachospermum, Sirodotia, Lemanea, Hildenbrandia* and *Caloglossa*. Flint (1948 etc.) described many species of freshwater red algae from the genera *Batrachospermum, Sirodotia* and *Tuomeya* from North America.

Prescott (1951) wrote a history of phycology and published *The algae of the Western Great Lakes Area*, in which he reported ten species of freshwater red algae.

Tiffany and Britton (1951) also reported two species of the genus *Batrachospermum* in their *Algae of Illinois*.

Krishnamurthy (1961) published his noteworthy work on the morphology and taxonomy of the genus *Compsopogon*. Chihara (1977 etc.) reported some Japanese species of the genera *Compsopogon* and *Compsopogonopsis*.

The *Lemanea*-type of life history was described by Magne (1967) and Balakrishnan & Chaugule (1980a, 1980b) reported that the genus *Batrachospermum* also had the *Lemanea*-type of life history.

Bourrelly also made remarkable progress with his books entitled *Les Algues d'Eau Douce*, "Tome I: *Les Algues vertes*" (1966), "Tome II; *Les Algues jaunes et brunes*" (1968), "Tome III: *Algues rouges et Algues bleues, les Eugléniens, Peridiniens et Cryptomonadines*" (1970), in which he compiled the hitherto-known species of freshwater red algae from all over the world. In the third volume, the morphology and taxonomy of the freshwater red algae were globally reviewed and Bourrelley (1970) divided the freshwater Rhodophyceae into the subclass Bangiophycidées and the subclass Florideophycidées. Subclass Bangiophycidées included four orders, Porphyridiales (family Porphyridiacées), Goniotrichales (families Goniotrichacées, Phragmonematacées), Bangiales (families Bangiacées, Boldiacées), Compsopogonales (family Compsopogonacées). Subclass Florideophycidées included orders, Acrochaetiales (families Audouinellacées, Acrochaetiacées), Nemalionales (families Batrachospermacées, Lemaneacées, Thoreacées), Cryptonemiales (family Hildenbrandiacées), Ceramiales (families Ceramiacées, Rhodomelacées, Delesseriacées).

On the other hand, Garbary & Gabrielson (1990) considered that because inter- as well as infra-ordinal relationships among taxa previously assigned to the Bangiophyceae (Bangiophycidae) are currently unresolved, the recognition of one class, Rhodophyceae, appears to be the best taxonomic procedure. The present author follows this opinion and does not recognize the subclasses, Bangiophycidae and Florideophycidae.

Reis (1970 etc.) described numerous species of the genus *Batrachospermum* mainly from Portugal whilst Mori (1975) described the Japanese taxa of the genus and discussed their phylogenetic relations.

Starmach (1977) wrote the book entitled *Phaeophyta* (*Brunatnice*) *and Rhodophyta* (*Krasnorosty*), in which he reviewed and illustrated many freshwater taxa reported from all over the world. Although written in Polish, this monograph is considered to be very important and convenient in classifying the freshwater Rhodophyta and Phaeophyta of the world. He recognized the division Rhodophyta which was divided into two classes, *viz*. Bangiophyceae and Florideophyceae.

Kumano (1977 etc.) dealt with collections of the freshwater Rhodophyceae from Japan, Malaysia, India, the Philippines, Micronesia and Papua New Guinea, and cooperative studies were carried out with many phycologists such as Hirose, Ohsaki, Ratnasabapathy, Yadava, Phang, Liao, Seto, Borden-Kerby and Johnstone.

Pueschel & Cole (1982) found that there were differences in the morphology and number of cap layers that overlay red algal pit plugs and that these differences appeared to be useful in distinguishing among taxa at higher ranks. They proposed two new orders, Hildenbrandiales for the monotypic family Hildenbrandiaceae, and Batrachospermales for families Batrachospermaceae, Lemaneaceae and Thoreaceae. Thus, the order Batrachospermales is characterized by the combination of lack of tetrasporic meiosis, heterotrichy, determinate lateral branches, discoid or laminate chloroplasts and pit plugs with inner cap layer, and dome-shaped outer cap layer.

Sheath (1984) in a review *Biology of the Freshwater Rhodophyceae*, listed all species and twenty eight genera of the freshwater Rhodophyceae distinguished in a key. Since, Sheath and his co-workers (1983 etc.), Vis *et al.* (1996 etc.) have continued work on many studies on morphology, taxonomy, distribution and the ecology of the freshwater Rhodophyceae from North America.

Entwisle & Kraft (1984 etc.), Entwisle (1989 etc.) and Entwisle & Foard (1997 etc.) have surveyed the freshwater Rhodophyceae of Australia and New Zealand and discussed their distribution in Australia and New Zealand.

Kumano & Necchi (1985 etc.), Necchi (1986 etc.) and Necchi & Zucchi (1997 etc.) have dealt with the taxonomy of the freshwater Rhodophyceae and considered their distribution mainly in Brazil in South America.

Blum (1993b, 1997 ["1994"]) described 6 new species of the genus *Lemanea* from Indiana and California in USA in North America.

King & Puttock studied the morphology and taxonomy of the genera *Bostrychia* and *Stictosiphonia* (1989) and the genus *Caloglossa* (1994).

Some phycologists such as Shi, Wie, Li, Hua, Xie and Ling described some taxa of genera *Compsopogon* (1998) and *Batrachospermum* (1994, 1996) from China.

Lee, Y. P. & Yoshida (1997) proposed a new taxonomic system for the family Acrochaetiaceae, in which there is no consensus of the generic concept (cf. Papenfuss, 1945, 1947, Woelkerling 1971, Lee, Y. P. 1980, Lee, Y. P. & Lee, I. K. 1988).

The Rubisco spacer, rbcL gene and 18S rRNA gene sequences showed that the North American and European freshwater populations of the genus *Bangia* were nearly identical but appeared to differ from their marine counterparts (Müller *et al.* 1998). Based on molecular and morphological data, phylogenetic and biogeographical studies were made on *Batrachospermum gelatinosum* in North America (Vis & Sheath 1997) and on *Sirodotia* (Vis & Sheath 1998, 1999). Phylogeny of Batrachospermales (Vis *et al.* 1998), Thoreales (Sheath *et al.* 2000, Hanyuda *et al.* 2001) and Balliales (Choi *et al.* 2000). Compsopogonales (Rintoul, T. L. Sheath, R. G. & Vis, M. L., 1998, 1999) was inferred from rbc L gene and 18S rRNA gene sequences. These regions of the Rubisco gene have been used successively for phylogenetic and biogeographical studies at the rank of genera, species and populations of *Porphyra* (Brodie *et al.* 1996), *Bostrychia* (Zuccarelo & West 1997), *Porphyra* and *Bangia* (Brodie *et al.* 1998) and *Caloglossa* (Kamiya *et al.* 1998, 1999, 2000).

Many books concerning phycology, especially the taxonomy of algae, have been published by Japanese authors, for example, *Shinsen Nihon Shokubutsu Zusetsu, Katoh Inka-bu* (Matsumura & Miyoshi, ed. 1901), *Icons of Japanese Algae* (Okamura 1907–1909), *Nihon Kaiso-shi* (Okamura 1936), *Inka Shokubutsu Zukan* (Asahina, ed. 1939), *General Phycology* (Hirose 1965), *Illustrations of the Japanese Freshwater Algae* (Hirose & Yamagishi, ed. 1977). With regard to freshwater algae, *Photomicrographs of the Freshwater Algae*, Vols 1–20 (Yamagishi & Akiyama, ed. 1984–1997) was published, and then *An Illustrated Atlas of the Life History of Algae, Vol. 2, Brown and red Algae* (Hori, ed. 1993) and *Biology of Algal Diversity* (Chihara, ed. 1897) appeared, providing excellent guide books on the life history and phylogeny of algae. *Marine Algae of Japan* (Yoshida 1998) included the description of all Japanese taxa reported up to 1997. Taxa were compiled at the rank of genera in *Introduction to Freshwater Algae* (Yamagishi 1999).

The present book draws together all the data on the freshwater red algae occurring worldwide into a comprehensive, taxonomic and illustrated account in English and is based on the published Japanese version *Taxonomy of Freshwater Rhodophyta* (Kumano 2000).

Key to the genera of the freshwater Rhodophyceae compiled in this book

Bangiophycideae

1. Unicells or loose colonies..2
1. Filamentous or saccate..5
2. Chloroplast blue, parietal laminate..3
2. Chloroplast red, axial stellate..*Porphyridium*
3. Asexual reproduction by cell division..............................*Cyanidioschyzon*
3. Asexual reproduction by autospores..4
4. 4 autospores formed..*Cyanidium*
4. 4–32 autospores formed...*Galdieria*
5. Pseudofilaments containing a loose series of cells in a gelatinous matrix........6
5. True filaments or saccate tissue...7
6. One axial chloroplast with a pyrenoid..............................*Chroodactylon*
6. Several parietal chloroplasts without pyrenoid....................*Kyliniella*
7. Monostromatic sac...*Boldia*
7. Filamentous..8
8. Unbranched filaments, uniseriate to multiseriate...................*Bangia*
8. Branched filaments, axial filament with cortication..9
9. Rhizoids confined to base, coarse filaments...................*Compsopogon*
9. Rhizoids throughout contributed to the cortication, fine filaments.........*Compsopogonopsis*

Nemaliophycideae

1. Thallus crustose...*Hildenbrandia*
1. Thallus not crustose..2
2. Uniseriate filaments without cortication..3
2. Multiseriate axis or uniseriate main axis with cortication.................4
3. Lateral branches with limited growth..5
3. Lateral branches with irregular and unlimited growth.......................6
4. Main axis monopodial...*Ballia*
4. Main axis sympodial...*Ptilothamnion*
5. Small epiphytes without rhizoids...*Balbiania*
5. Mostly free-living with rhizoids...*Audouinella*

6. Uniseriate axis..7
6. Multiseriate axis..8
7. Without cortication, trichogyne linear..*Rhododraparnaldia*
7. With cortication, trichogyne thicker..9
8. Pseudoparenchyma with flattened leaf-like sections.................................*Caloglossa*
8. Compacted filaments in cylinders...10
9. Sporangia at base of lateral branches, laterals not contained in a common mucilage, thereby having a hairy appearance...*Thorea*
9. Sporangia at apex of lateral branches, laterals in a common mucilage.............*Nemalionopsis*
10. Filaments not compacted into pseudoparenchymas..11
10. Pseudoparenchymatous...12
11. Carposporophyte a branched filaments creeping along the main axis, carpogonial base with a profusion...*Sirodotia*
11. Carposporophyte a distinct, compact mass of filaments, carpogonial base symmetrical..*Batrachospermum*
12. Lateral branches enclosed in a common mucilage but without a distinct cortex....*Nothocladus*
12. Carified pseudoparenchyma with a distinct cortex...13
13. Nodes not obvious, frequent branching...14
13. Nodes obvious in macroscopic view as a series of swelling, few branches............16
14. Axial cell remarkable, spermatangial papillae present..15
14. Axial cell not remarkable, spermatangial papillae absent...........................*Psiosiphon*
15. Ray cell T- or Y-shaped..*Lemanea*
15. Ray cell simple shaped..*Paralemanea*
16. Whorled laterals obvious at the node...*Tuomeya*
16. Lateral branches not obvious at the node..17
17. Trichoblast absent..*Bostrychia*
17. Trichoblast present...*Polysiphonia*

Class **Rhodophyceae** Ruprecht in Middendorff (1851 : 205)

The class Rhodophyceae differs from all other algal classes in having an alternation of three phases, namely, a sexual phase (gametophyte), a spore-producing phase (tetrasporophyte), and an additional phase representing amplification of zygotic tissue (carposporophyte).

Sexual reproduction is oogamous, involving non-motile male reproductive cells termed spermatia produced in reproductive structures called spermatangia and specialized female reproductive cells termed carpogonia. The carpogonium typically is an elongate cell, relatively enlarged at the base and extending distally into a process called the trichogyne, to which spermatia may become attached. Upon fertilization of the carpogonium by the spermatium, the carpogonium may either directly or indirectly produce a diploid phase called the carposporophyte, with which the term gonimoblast is essentially synonymous. The carposporophyte may consist of sterile gonimoblast filaments bearing carposporangia, or may be entirely converted to carposporangia.

In a typical life history in the Rhodophyceae the carpospore is released and germinates into a diploid phase termed the tetrasporophyte, which eventually becomes reproductively mature and produces tetrasporangia. These tetrasporangia constitute the site of meiosis, the four haploid nuclei being incorporated into four products termed tetraspores.

Of the three different life history phases, the gametophyte and the tetrasporophyte, are independent and free-living, whereas the carposporophyte is essentially parasitic in that it is attached and dependent on the gametophyte, from which it arises by sexual reproduction. In the *Lemanea*-type life history reported for *Lemanea* (Magne 1967) and *Batrachospermum* (Balakrishnan & Chaugule 1980a, 1980b), the carpospores germinate into small diploid plants called the chantransia phase. Meiosis takes place in an apical cell of the chantransia phase. After each of two nuclear divisions a small portion of cytoplasm with one of the nuclei is cut off by a wall and the rest degenerates, with only one nucleus surviving. The apical cell, which becomes haploid without changing its outer appearance, then gives rise to a new haploid gametophyte.

Pit plugs, which are structures unique to most of the Rhodophyceae, are intercellular connections consisting of a core, a cap membrane enveloping the core and a cap layer, are classified as follows: type 1: core only, type 2: core enveloped by cap membrane only, type 3: core enveloped by a cap layer, type 4: core enveloped by a cap layer and a cap membrane, type 5: core enveloped by inner and outer cap layers with intercalary cap membrane, type 6: core enveloped by inner and outer dome-shaped cap layers with intercalary cap membrane. Pit plug structure is a useful character at ordinal rank.

More than 200 taxa of Rhodophyceae are found in freshwater habitat. In freshwater habitat they live in moving water, attached to rocks or other substrata. Sexual reproduction has not been reported from most freshwater red algae, but asexual cells such as endospores, monospores and gemmae are common.

Porphyridiales

Order **Porphyridiales** Kylin (1937:122)

Type: family Porphyridiaceae Kylin ex Skuja

Plants unicellular, free living, or embedded in a common mucilaginous matrix or sheath to form irregularly shaped colonies or uniseriate or multiseriate pseudo-filaments without rhizoids. Cells containing a single stellate chloroplast with axial pyrenoid or many parietal chloroplasts without pyrenoids. Asexual reproduction by binary division or autospores (endospores). Sexual reproduction unknown.

Family **Porphyridiaceae** Kylin ex Skuja (1939:38)

Type: genus *Porphyridium* Nägeli, nom. cons.

Plant unicellular, with one stellate chloroplast usually with a central pyrenoid. Asexual reproduction by binary division. Without pit-plug.

Genus *Porphyridium* Nägeli (1849:71, 138. pl. 41, fig. 9), nom. cons.

Type: *Porphyridium cruentum* (Gray) Nägeli

1) ***Porphyridium purpureum*** (Bory) Drew et Ross (1965:93)
Basionym: *Phytoconis purpurea* Bory (1997:55),
Synonym: *Porphyridium cruentum* (Gray) Nägeli (1849:71, 138, pl. 41, fig. 9)

plate 1 (a), figs 1–4 (Hara 1993)

Cells blood-red or brown-red in colour, 7–12(–15) µm in diameter, containing one stellate chloroplast with a central pyrenoid. Colonies large, diffuse, forming red or brown patches on wet ground. Occurring in eutrophic places. Asexual reproduction by cell division; single chloroplast stellate or lobe-like with a pyrenoid.
Type Locality:
Holotype:
Distribution: Known worldwide from shady and eutrophic places.

Family **Cyanidiaceae** Geitler (1933:624) *emend.* Chapman (1974:673)

Type: genus *Cyanidium* Geitler

Plants unicellular, with each cell containing one or more parietal chloroplasts without a pyrenoid. Asexual reproduction by autospores (endospores).
Note: Ott *et al.* (1994) pointed out that the presence of only one mitochondrion, a chloroplast with a club-like nucleoid, the absence of vacuoles, and trienoic fatty acids (Merola *et al.* 1981) and the utilization of nitrates, are best to be considered only of secondary, taxonomic importance and, in any event, are not significant bases for generic and familial distinctions. However, *Cyanidium* and *Galdieria* can be clearly distinguished with electron microscopy (Nagashima & Fukuda 1981, Nagashima 1993a,b,c, 1995, 1997) and with molecular phylogenetical analyses (Kondoh *et al.* 1998, Yokoyama *et al.* 1998). The author prefers to maintain *Cyanidium* and *Galdieria* as separate genera.

Key to the genera of the family **Cyanidiaceae**

1. Asexual reproduction by cell division...*Cyanidioschyzon*
1. Asexual reproduction by autospores...2
2. Four autospores formed...*Cyanidium*
2. Four to thirty two autospores formed..*Galdieria*

Genus *Cyanidioschyzon* De Luca *et al.* (1978:43) in NCU–3 (1993:297)

Type: *Cyanidioschyzon merolae* De Luca *et al.*

Cell containing one large ovoidal chloroplast without a pyrenoid. Asexual reproduction by binary cell division.

1) **Cyanidioschyzon merolae** De Luca, Taddei et Varano (1978:37).

plate 1 (b), figs 1–3 (Nagashima 1993a)

Cells forming powdery covering on stones in acidic hot springs. Cells blue-green in colour, club-shaped, 3–4 µm long, 1–2 µm in diameter, containing one elongate chloroplast without a pyrenoid. Cell wall absent or very thin. In the process of cell division, the chloroplast dividing longitudinally, with the cell becoming first V-shaped then linear. Cell division immediately following nuclear division. Asexual reproduction occurring only by means of binary cell division.
Type Locality:
Holotype:
Distribution: Occurs worldwide on stones in acidic hot spa. In Japan, Kusatsu Spa in Gunma, Noboribetsu in Hokkaido.

Genus *Cyanidium* Geitler (1933:624) in NCU–3 (1993:297)

Type: *Cyanidium caldarium* (Tilden) Geitler

Synonym: *Rhodococcus* Hirose (1958:347),

non *Rhodococcus* Hansgirg (1885:19, pl. 41, figs 14–16).

Cell containing a large ovoidal chloroplast without a pyrenoid. Asexual reproduction by means of endospores.
Note: See above under Cyanidiaceae for the justification for recognition of this genus.

1) **Cyanidium caldarium** (Tilden) Geitler (1933:622)
Basionym: *Protococcus botryoides* f. *caldaria* Tilden (1898:94, pl. VII, fig. 18);
Synonym: *Rhodococcus caldarius* (Tilden) Hirose (1958:351, figs 1, 4–5)

plate 2 (b), figs 1–4 (Nagashima 1993c)

Cells forming powdery covering on stones in acidic hot springs. Cells blue-green in colour, spherical or rounded-pyramidal, 4–6 µm in diameter, containing one large ovoidal chloroplast without a pyrenoid. Autosporangium with 4 autospores; produced by mature cell, with first chloroplast division, then nuclear division preceding cytokinesis and autospore wall formation.

Porphyridiales

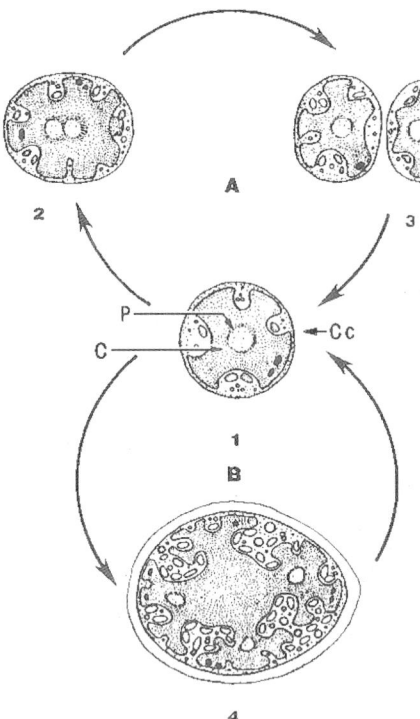

Plate 1 (a)
Porphyridium purpureum, figs 1–4 (Hara 1993)
A. asexual reproduction (cell division), B. asexual reproduction (formation of a multinucleate giant cell).
1. a vegetative cell, 2. a cell initiating chloroplast division, 3. a dividing cell, 4. a giant cell.
(C: chloroplast, Cc: cell envelope, P: pyrenoid)

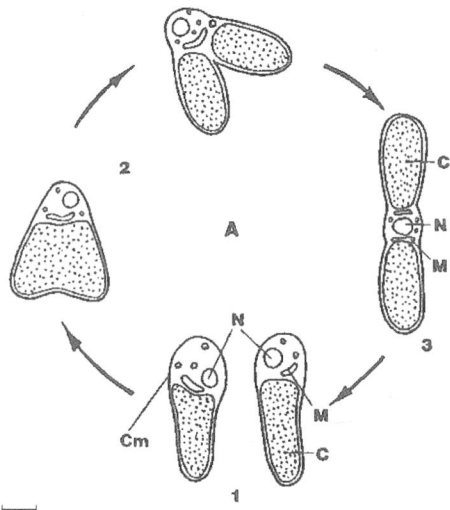

Plate 1 (b)
Cyanidioschyzon merolae, figs 1–3
(Nagashima 1993a)
A. life history, B. photomicrographs showing stages of life history in YR–89.
1. a vegetative cell, 2. the first stage of division, 3. the later stage in division.
(C: chloroplast, Cm: cell membrane, N: cell nucleus, M: mitochondrion)

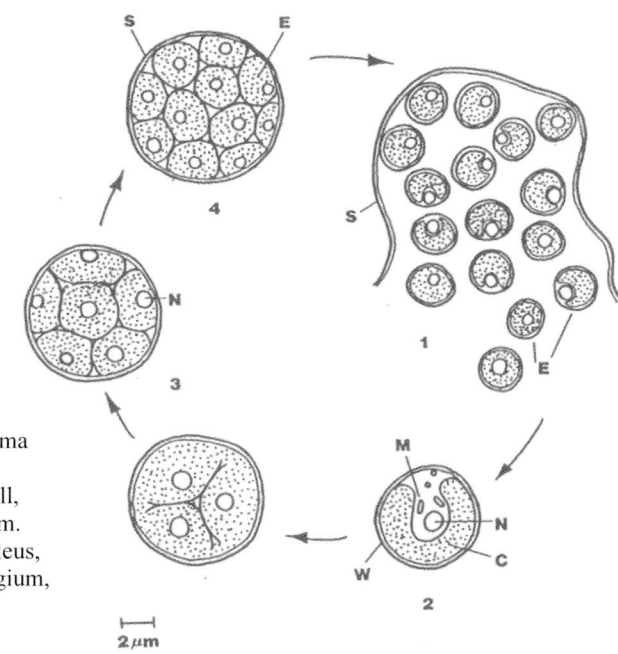

Plate 2 (a)
Galdieria sulphuraria, figs 1–4 (Nagashima 1993b)
1. an endospore release, 2. a vegetative cell, 3. an endospore formation, 4. a sporangium.
(C: chloroplast, E: endospore, N: cell nucleus, M: mitochondrion, S: wall of endosporangium, W: cell wall)

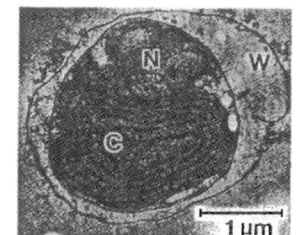

Plate 2 (b)
Cyanidium caldarium, figs 1–4 (Nagashima 1993c)
A. life history, B. electron photomicrograph of stage 2.
1. an endospore release, 2. a vegetative cell, 3. an endospore formation. 4. a sporangium.
(C: chloroplast, E: endospore, N: nucleus, M: mitochondrion, S: wall of endosporangium, W: cell wall)

Porphyridiales

Autospores 2–3 µm in diameter, trihedrally arranged, liberated following dissolution of autosporangium wall.
   Type Locality: West America (Tilden 1898)
   Holotype:
   Distribution: Known worldwide from stones in acidic hot spa such as in Italy in Europe and Japan, Eastern Asia. In Japan; Kusatsu Spa in Gunma, Noboribetsu in Hokkaido.

Genus ***Galdieria*** Merola in Merola *et al.* (1982:193) in NCU–3 (1993:447)

Type: *Galdieria sulphuraria* (Galdieri) Merola

Cell containing a parietal, lobed, laminate chloroplast without a pyrenoid. Asexual reproduction by means of autospores.
   Note: See above under Cyanidiaceae for the justification for recognition of this genus.

1) ***Galdieria sulphuraria*** (Galdieri) Merola in Merola *et al.* (1982:193)
Basionym: *Pleurococcus sulphurarius* Galdieri (1899:162, figs A–I),
Synonym:*Cyanidium sulphuraria* (Galdieri) Ott in Ott *et al.* (1994:149).
Misapplied name: *Chroococcidiopsis thermalis* Geitler var. *nipponica sensu* Negoro (1943:310, fig. 9).

plate 2 (a), figs 1–4 (Nagashima 1993b)

Cells forming powdery covering on stones in acidic hot springs. Cells blue-green in colour, spherical, 7–14 µm in diameter, containing one parietal, lobed, laminate chloroplast without a pyrenoid. Autosporangium with 4–16 (–32) autospores; produced by mature cell, with first chloroplast division, then nuclear division preceding cytokinesis and autospore wall formation. Autospores 2–5 µm in diameter, trihedrally arranged, liberated following dissolution of autosporangium wall.
   Type Locality: Italy, Garderi.
   Holotype:
   Distribution: Known worldwide from stones in acidic hot spa such as in Italy in Europe and Japan in Eastern Asia. In Japan; Kusatsu Spa in Gunma, Noboribetsu in Hokkaido.

Order **Goniotrichales** Skuja (1939:31)

Type: family Goniotrichaceae G. M. Smith

Plants forming uniseriate or parenchymatous filaments with diffuse growth. Asexual reproduction by gonidia or by walled reproductive cells arising from ordinary vegetative cells, either directly or after successive binary fission into smaller cells.

Family **Goniotrichaceae** G. M. Smith (1933:121–122)

Type: genus *Goniotrichum* Kützing

Cells arranged in more or less distinct filaments, each containing one stellate chloroplast with a central pyrenoid. Asexual reproduction by means of naked monospores.

Genus ***Chroodactylon*** Hansgirg (1885:14)

Type: *Chroodactylon wolleanum* Hansgirg

Synonym: genus *Asterocytis* (Hansgirg) Gobi et Schmitz in Engler et Prantl (1896:314, pl. 42, figs 1 a 3, pl. 45, figs 1–2)

Plants irregularly branched or unbranched filaments consisting of elongate-elliptical to flat-spherical cells surrounded by a thick, gelatinous matrix. Cells containing one stellate chloroplast with a pyrenoid. Asexual reproduction by means of monospores.

1) ***Chroodactylon ramosum*** (Thwaites) Hansgirg (1885:19)
Basionym: *Hormospora ramosa* Thwaites in Harvey (1048:pl.213),
Synonym: *Asterocytis ramosa* (Thwaites) Gobi (1879:85)

plate 3, figs 1–6 (Watanabe 1994)

Plants filamentous or palmelloid, growing on submerged substratum, up to several mm high. Filaments composed of uniseriate cells and gelatinous sheaths, with 'false' branches, 9–15 µm in diameter, up to 20 µm at the basal portion, branches dichotomous and lateral. Cells ellipsoidal or irregular in shape at the basal portion of the filaments and in palmelloid plants. Cells containing a single, stellate blue-coloured chloroplast with a central pyrenoid.
Type Locality: in a salt-water lake near Wareham, Dorset, U.K.
Holotype:
Distribution: Known from stones or epiphytically on aquatic plants in Europe and in Eastern Asia. In Japan in Akan-ko Lake in Hokkaido.
Note: Vis & Sheath (1993) stated that Sheath & Hymes (1980) misapplied *Chroodactylon ornatum* as this species. Entwisle & Kraft (1983) regarded this species as a synonym of *Chroodactylon ornatum*.

Goniotrichiales

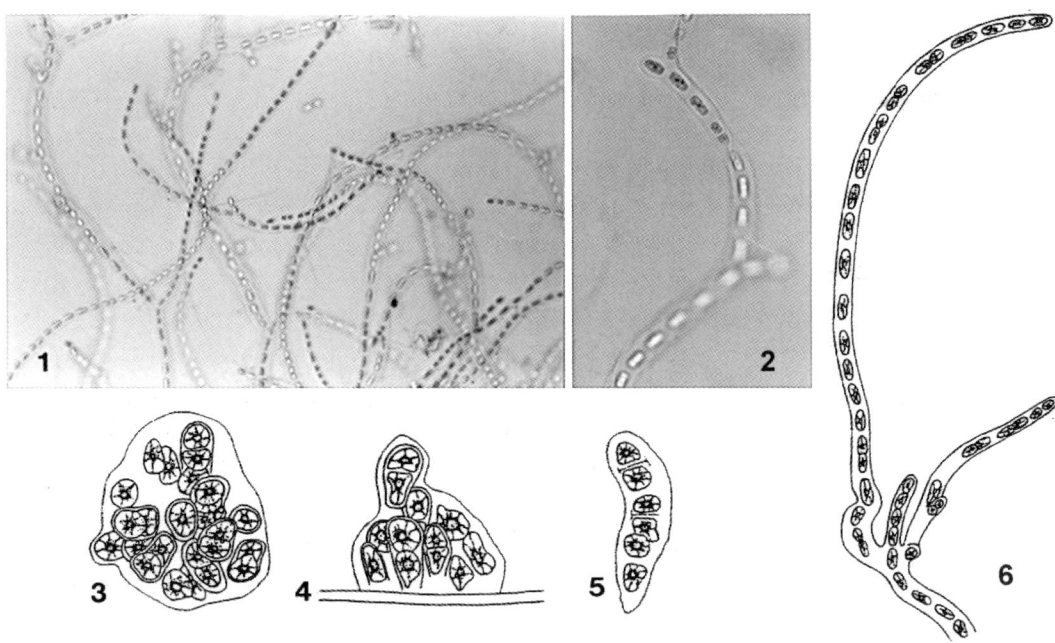

Plate 3
Chroodactylon ramosum, figs 1–6 (Watanabe 1994)
1–2, 6. filaments composed of uniseriate cells and gelatinous sheaths with false branches, 3–5. cells containing a single stellate chloroplast with a central pyrenoid, ellipsoidal or irregular in shape at the basal portion of the filaments and in the palmelloidal plant.

2) ***Chroodactylon ornatum*** (C. Agardh) Basson (1979:67)
Basionym: *Conferva ornata* C. Agardh (1824:104)

plate 4, figs 1–3 (Vis & Sheath 1993)

Pseudofilaments with variable number of false branches (0–6) composed of rectangular to ellipsoidal cells loosely arranged in a linear fashion within a broad gelatinous matrix. Cells with an axial, blue-coloured chloroplast containing a prominent central pyrenoid. Cell diameter 5.8–11.6 µm, cell length 7.1–16.6 µm and filaments length 24–1240 µm.
Type Locality: Lake Maelaren, Taneberg, Sweden in Europe.
Holotype: PC herb. Thuret no. 69.
Distribution: Known from Europe, Australia, North America. In North America, occasional component of the epiphyton of *Cladophora* and *Rhizoclonium* in warm, alkaline streams (Vis & Sheath 1993). This species has been found only on *Cladophora glomerata* in Ontario, it has been found on *Rhizoclonium hookeri* and various other plants in Manitoba (Sheath & Hymes 1980, Daily 1943) in Canada. It has also been reported from New York and Arizona (Vis & Sheath 1993) in USA. In Australia, it Known from Yarra River, Warburton, Spitters Creek, Mornington Peninsula, Victoria (Entwisle & Kraft 1984).
Note: Sheath & Hymes (1980) reported this species as *Chroodactylon ramosum.* Vis & Sheath (1993) regarded *Asterocytis smaragdina* (Reinsch) Forti as a synonym of this species.

Chroodactylon

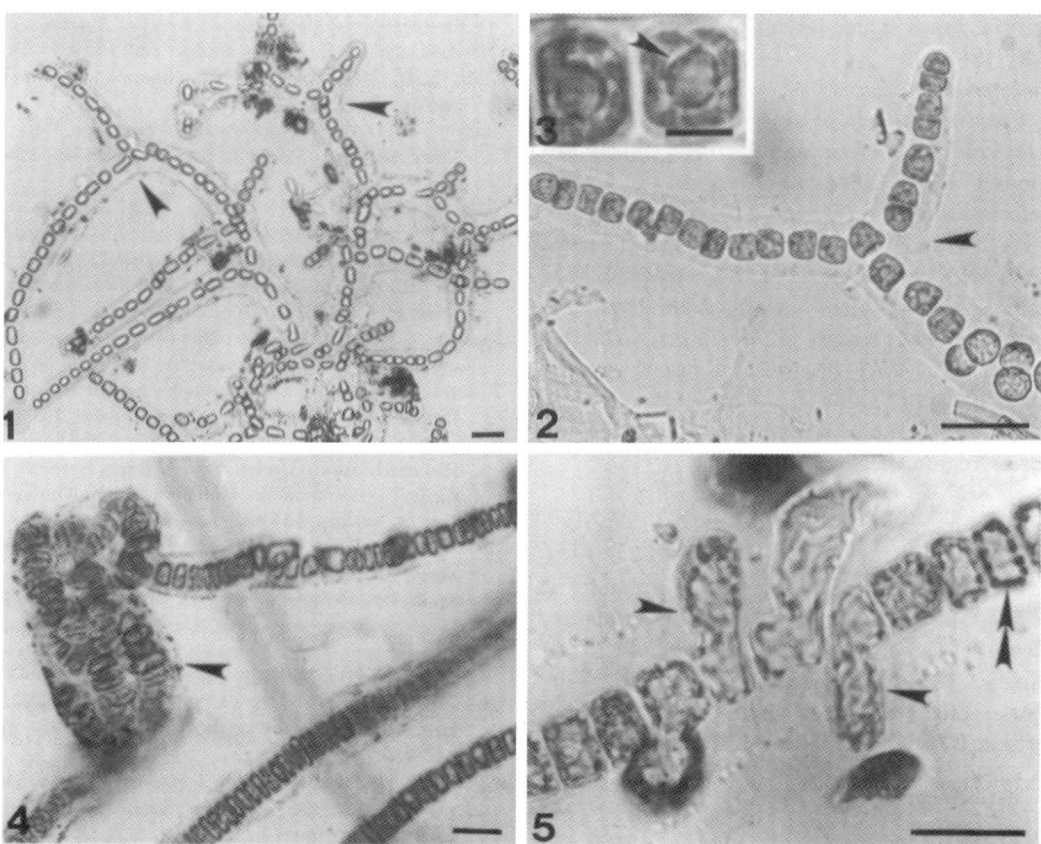

Plate 4
Chroodactylon ornatum, figs 1–3 (Vis & Sheath 1993)
1. a complex of pseudo-filaments with false branches (arrowheads), 2. a linear arrangement of cells in a broad gelatinous matrix with a false branch (arrowhead), 3. two cells showing an axial, stellate chloroplast with a prominent, central pyrenoid (arrowhead).
Scale bars = 20 µm (figs 1–2); 5 µm (fig. 3).

Kyliniella latvica, figs 4–5 (Vis & Sheath 1993) 4. pseudo-filaments arising from discoidal base (arrowhead), 5. densely packed cells with parietal, discoidal chloroplasts (double arrowhead), some cells produce rhizoidal outgrowths (arrowheads).
Scale bar = 20 µm.

Goniotrichiales

Family **Phragmonemataceae** Skuja (1939:23)

Type: genus *Phragmonema* Zopf

Plants forming filamentous colonies with true and false branching; one or more parietal chloroplasts without pyrenoids. Asexual reproduction by means of small spores (microgonidia).

Genus *Kyliniella* Skuja (1926:4)

Type: *Kyliniella latvica* Skuja

Plants filamentous, growing out from a pseudoparenchymatous base.

1) ***Kyliniella latvica*** Skuja (1926:4, pl. 42, figs 4–6, pl. 43, pl. 70, fig. 1)

plate 4, figs 4–5 (Vis & Sheath 1993)

Pseudofilaments arising from a discoid base composed of rectangular cells tightly arranged in a linear fashion in a broad gelatinous matrix. Rhizoidal outgrowths arising from attachment cells. Cells with several parietal, blue-coloured, discoid chloroplasts. Cells 7.4–14.8 µm in diameter, 4.9–17.3 µm long, rhizoid 17.3–24.7 µm long and filament 12.4–32.1 µm long.

Type Locality: Lake Usma, Latvia in Europe. Epiphytic on *Phragmites*.
Holotype: RIG.
Distribution: Worldwide, it has been reported from Latvia, Austria, France in Europe, but it is infrequently collected (Bourrelly 1985). In North America, from the littoral zone of only two streams in the deciduous forest region in Rhode Island (Sheath & Burkholder 1985, Sheath & Cole 1992, Vis & Sheath 1993) and in New Hampshire (Flint 1953), USA.

Order **Compsopogonales** Skuja (1939:23)

Type: family Compsopogonaceae Schmitz

Composed of uniseriate filaments becoming corticated with small cells surrounding the central axial row. Young cells containing a single parietal chloroplast, which later becomes fragmented into many small, discoid chloroplasts. Monospores formed via unequal division of the parental wall by an oblique or curving wall. The Compsopogonales including the Boldiaceae, Compsopogonaceae and Erythrotrichiaceae was supported as a valid entity through the phylogenetic analysis of the rbcL gene and 18S rRNA gene sequences, (Rintoul, Sheath & Vis 1998, 1999).

Family **Compsopogonaceae** Schmitz in Engler et Prantl (1897)

Type: genus *Compsopogon* Montagne

The family Compsopogonaceae includes two genera, *Compsopogon* and *Compsopogonopsis*. Understanding of their relationship, and of the relationship of the Compsopogonaceae to other families would benefit from detailed studies of pit connection type.

Note: Rintoul, Sheath & Vis (1998) stated that in comparisons of both genes of the genus *Compsopogon* and genus *Compsopogonopsis*, sequence divergence was very low, and this results suggests that these genera should be synonymised.

Vis, Sheath & Cole (1992) recognized two species of the genus *Compsopogon*, *C. coeruleus* (Balbis) Montagne, and *C. prolificus* Yadava et Kumano, and one species of the genus *Compsopogonopsis*, *Cs. leptoclados* (Montagne) Krishnamurthy, based on an examination of type specimens and other collections. Variation of quantitative characters were studied within and among six populations of *C. coeruleus* in Brazil (Necchi & Pascoaloto 1995). The species recognized by Vis, Sheath & Cole (1992), and the other species discussed by Seto & Kumano (1993) are included here.

Key to the genera of the family **Compsopogonaceae**

1. Rhizoidal filaments throughout the plants, composing the cortex...................*Compsopogonopsis*
1. Rhizoidal filaments restricted to the basal part of the plants............................*Compsopogon*

Genus *Compsopogonopsis* Krishnamurthy (1962:219)

Type: *Compsopogonopsis leptoclados* (Montagne) Krishnamurthy

Plants filamentous, cylindrical, profusely branched; the cortical cells initiated by means of rhizoidal filaments arising from uniseriate filaments; fully developed cortex consisting of one layer generally, occasionally two layers of cells; cortical cells almost globular; reproduction by monospores produced from uniseriate axial cells as well as from outermost cells of the cortex. Sexual reproduction unknown.

Compsopogonales

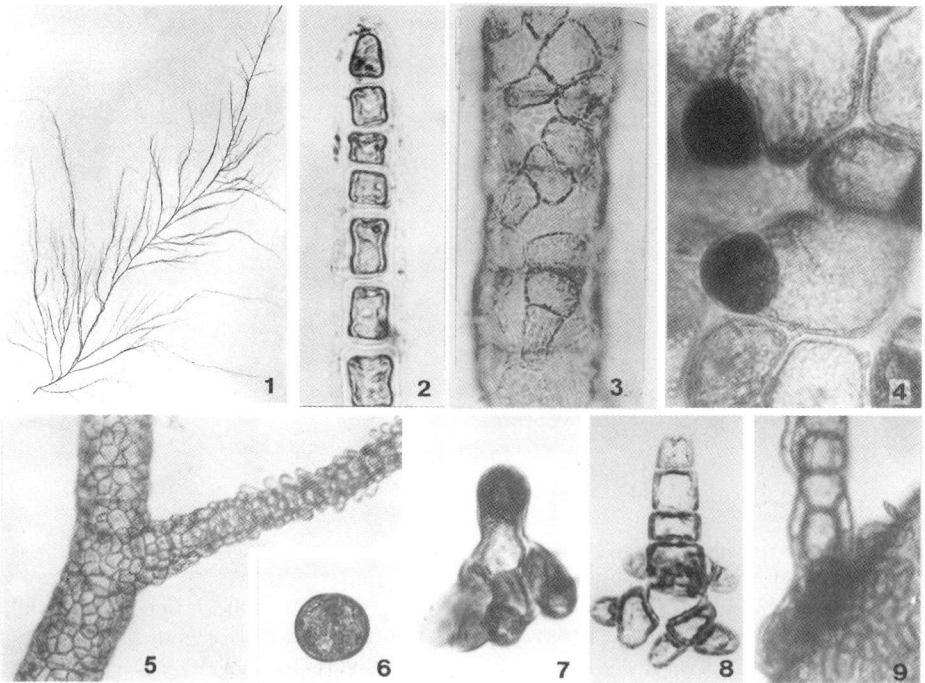

Plate 5 (a)
Compsopogonopsis japonica, figs 1–9 (Nakamura 1984e)
1. a habit of mature plant, 2. an apical portion of a young plant showing uniseriate axial cells, 3. the development of cortical initials arising as rhizoid-like protuberances, 4. the surface view of a mature plant showing monosporangia, 5. a mature plant showing fully developed cortical layers, 6, a liberated monospore, 7–9. the successive stages in development of a basal disc and erect filaments from germinating monospores.

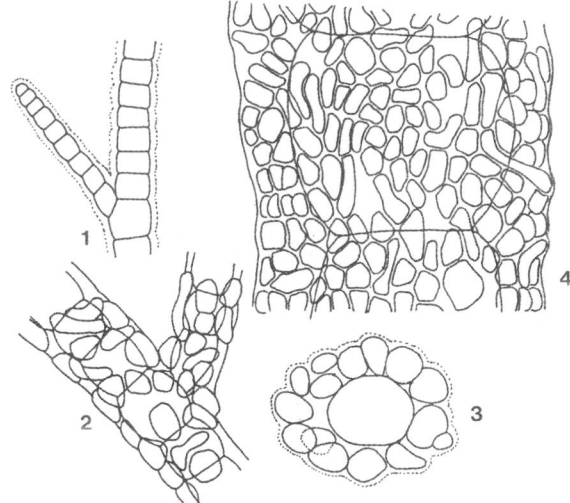

Plate 5 (b)
Compsopogonopsis leptoclados as *Compsopogon leptoclados*, figs 1–4 (Bourrelly 1970)
1. an apical portion of a plant showing uniseriate axial cells, 2. the development of cortical initials arising as rhizoidal cells and fully developed cortical layers.

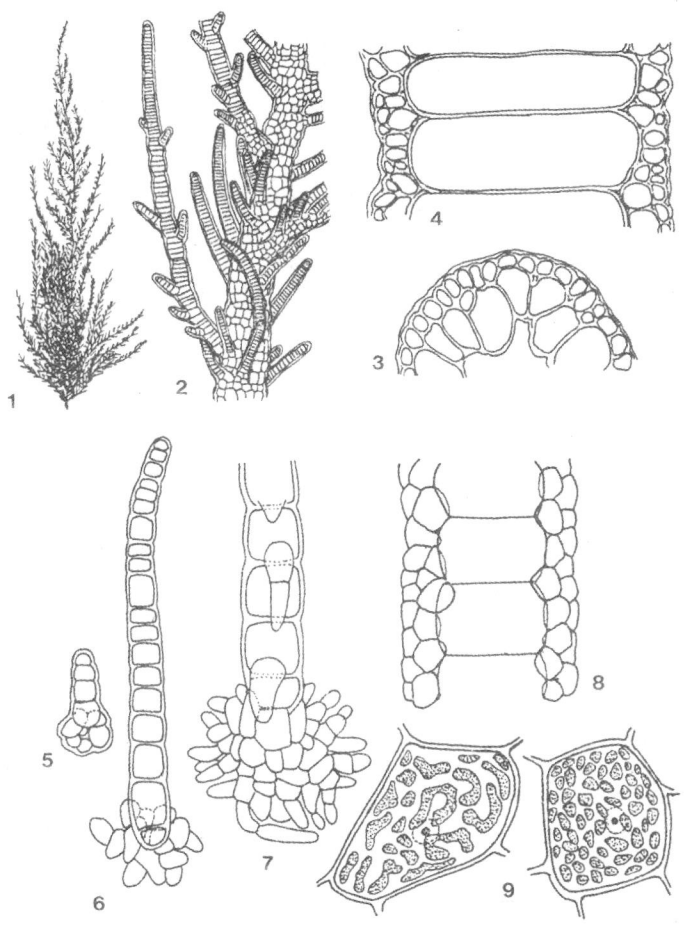

Plate 6
Compsopogonopsis fruticosa as *Compsopogon fruticosus,* figs 1–4 (original author, Jao 1941)
1. the habit of a single plant, 2. the portion of a plant, showing type of ramification and constrictions of main filaments and branches, 3. the cross sectional view of a fully developed plant, 4. the longitudinal section of a well developed plant, showing axial and cortical cells.

Compsopogon hookeri, figs 5–9 (Friedrich 1966)
5–6. the successive stages in development of an erect filament and basal disc showing uniseriate axial cells, 7. the basal portion of an erect plant showing basal disc and rhizoidal cells at the early stage in development, 8. the longitudinal section of a plant showing axial cells and two layers of cortical cells, 9. chloroplasts in irregular in shape.

Key to the species of the genus *Compsopogonopsis*

1. Outermost cortical cells up to ca. 30 μm in size and with
   one cortical layer...1) *Compsopogonopsis leptoclados*
1. Outermost cortical cells up to ca. 50 μm in size and with two cortical layers............................2
2. Rhizoidal filaments initiated from the segments on both sides of
   the axial cells...2) *Cs. japonica*
2. Rhizoidal filaments initiated from tubular outgrowths on
   the axial cells...3) *Cs. fruticosa*

Compsopogonales

1) ***Compsopogonopsis japonica*** Chihara (1976:289, fig. 1, 2 A–I)

plate 5 (a), figs 1–9 (Nakamura 1984e)

Plant heterotrichous, consisting of prostrate system and erect filaments, dark blue-green. Prostrate system a polymorphous disk or holdfast with the form depending on the substratum; 1 or 2 erect filaments arising from prostrate system. Erect filament 0.5–1 mm in diameter, 30–40 cm long, well branched. Cortex formed by means of elongation of rhizoidal protuberances from axial cells. Cortex 1–2 cell-layered; cortical cells polygonal, 25–40 μm in diameter, 30–50 μm long. Monosporangia produced from cortical cells by unequal divisions, spherical, 17–23 μm in diameter.

Type Locality: near Sakai along Tone River in Gunma, Japan in Eastern Asia. In a small man-made pond.

Holotype: TNS Al–24051.

Distribution: Known from the type locality and in a creek, Saitama Mound, Gyoda in Saitama and a small spring, Huga-ka, near Kapira, Ishigaki-jima in Okinawa, Japan in Eastern Asia.

2) ***Compsopogonopsis leptoclados*** (Montagne) Krishnamurthy (1962:219)
Basionym: *Compsopogon leptoclados* Montagne (1850:298).

plate 5 (b), figs 1–4 (Bourrelly 1970) as *Compsopogon leptoclados* Montagne

Rhizoidal filaments present throughout the plants, giving rise to the cortex. The main branches 130–400 μm in diameter. The fully developed cortex consisting of one layer of cells 4–15 μm wide, 9–31 μm long. Monosporangia 12–16 μm in diameter.

Type Locality: near Cayenne, French Guiana in South America. In a freshwater stream.

Holotype: (L 940.28.440), PC Leprieur No 1098.

Distribution: Known from the type locality and Maui Island, Hawaii Island, Oahu Island, Kauai Island, Hawaii Islands in Pacific Ocean; Guadeloupe, Lesser Antilles, Caribbean Sea in Central America.

3) ***Compsopogonopsis fruticosa*** (Jao) Seto (1987:265)
Basionym: *Compsopogon fruticosus* Jao (1941:248, pl. II, figs 10–14)

plate 6, figs 1–4, (original author, Jao 1941) as *Compsopogon fruticosus*

Plants very densely bushy, elongate-pyramidal in outline, up to 15 cm high, deep blue-green, very abundantly branched, here and there distinctly constricted, capillary-fine in young branches, irregularly thickened in adult portions up to 200–500 μm in diameter. Branches and branchlets irregularly arranged. Rhizoidal cortical filaments present throughout the plants, giving rise to thecortex. Axial cells very short, 3–5 times shorter than diameter, disc-shaped, cortical cells 3–5 angular. The fully developed cortex consisting of two layers of cells; the outermost cortical cells 12–35 μm wide, 14–55 μm long. Monosporangia 15–22 μm in diameter.

Type Locality: Kan-tung-tze, about five miles southeast of Pehpei in Szechwan, China in Eastern Asia. On the concrete wall of a mill dam, in fast-running water emanating from a limestone cave.

Holotype: SC 1145.

Distribution: Known from the type locality only.

Note: According to Seto (1987), examination of an isotype specimen of *Compsopogon fruticosus* shows that the initials of cortical cells are produced from the lower part of the axial cells, and cut off from the tubular outgrowths by oblique or horizontal walls in a mode similar to that observed in *Compsopogonopsis leptoclados*.

Genus *Compsopogon* Montagne in Bory et Durieaux (1846:152)

Type: *Compsopogon coeruleus* (Balbis ex C. Agardh) Montagne

Plants filamentous, cylindrical, often profusely branched; juvenile erect axes uniseriate; mature axes becoming multiseriate and consisting of cortex and axial cells. Cortex consisting of one or more layers of cells. Axial cells linear, large. Asexual reproduction mostly by monospores, involving the unequal division of the parental wall by an oblique or curving wall, sometimes by microspores produced from uniseriate axial cells as well as from the outermost cells of the cortex. Sexual reproduction unknown.

Key to the species of the genus *Compsopogon*

1. Erect filaments with many short spinous branchlets.....................1) *Compsopogon aeruginosus*
1. Erect filaments without many short spinous branchlets..2
2. Mature filaments with knot-like structures...2) *C. prolificus*
2. Mature filaments without knot-like structures..3
3. Monospores < 20 μm in diameter..3) *C. chalybeus*
3. Monospores > 20 μm in diameter..4
4. Cortex consisting of up to 5 layers of cells...4) *C. corticrassus*
4. Cortex consisting of fewer than 3 layers of cells..5
5. Cortex consisting of 2–3 layers of cells..8
5. Cortex consisting of 1– (rarely 2–) layers of cells...6
6. Main axis < 200 μm in diameter..5) *C. tenellus*
6. Main axis > 200 μm in diameter...7
7. Plant 2 cm high...6) *C. minutus*
7. Plant 15–20 cm high..7) *C. sparsus*
8. Erect filaments sparsely branched..8) *C. hookeri*
8. Erect filaments well branched..9) *C. coeruleus*

1a) ***Compsopogon aeruginosus*** (J. Agardh) Kützing (1849:433)
Basionym: *Pericystis aeruginosa* J. Agardh (1847:6).

plate 7 (a), figs 1–7 (Nakamura 1984a)

Plants heterotrichous, consisting of prostrate system and erect filaments, dark blue-green. Prostrate system a polymorphous disk or holdfast 180–390 μm in diameter, depending on the substratum. Erect filament 0.7 mm in diameter, 25–50 cm long, well branched; with many short spinous branchlets, 130–550 μm in diameter, 80–300 μm in diameter. Cortical layers 50–150 μm thick, 1–2 cell-layered; cortical cells 12–30 μm in diameter, 16–40 μm long. Monosporangia spherical, 10–15 μm in diameter.

Compsopogonales

Type Locality: near Havana, Cuba in Caribbean Sea in Central America. In running waters.
Holotype: (L 940.284.410), LD herb Agardh 12408–9
Distribution: Known from the type locality and in streams, Israel in Europe; India in Asia; in Eastern Asia; the Philippines, Indonesia in Southeast Asia; North America; Central America; South America. In Japan, Matsue, Hirata along Shinji-ko Lake in Shimane.

1b) ***Compsopogon aeruginosus*** (J. Agardh) Kützing var. *catenatum* Yadava et Pandey (1980:20, figs 1–9).

plate 8, figs 1–4 (Seto, Yadava & Kumano 1991)

Plants heterotrichous, dark bluish violet, consisting of disc-shaped holdfast and erect filaments of about 1 mm in diameter, up to 40 cm long, nodulated wavy margins and furrows, richly branched. Branches arising at 30° to 60°. Axial cells in mature filaments 25–36 μm in diameter, 40–58 μm long. Lateral branches alternate, cells 20–38 μm in diameter, 6–18 μm long. Spinous branchlets present on main filament. Peripheral cells 8–12 μm in diameter, 10–18 μm long. Chloroplasts parietal and plate-shape. Both monosporangia- and microsporangia present. Monosporangia initiated from the cortical or peripheral cells, up to 13 μm in diameter. Microsporangia occurring in groups, up to 10 μm in diameter. Plants anchored by a disc-shaped attachment.
Type Locality: Dorania River, Bareilly, Rohilkhand Division, Uttar Pradesh, India in Asia.
Holotype: DCP 37, Botany Department, Allahabad University, Allahabad, India.
Distribution: Known from the type locality, Nakatia River and Khannaut River near Shahjahanpur, Rohilkhand Division, Uttar Pradesh, India in Asia.

2) ***Compsopogon prolificus*** Yadava et Kumano (1985:19, figs 1–7).

plate 7 (b), figs 1–7 (original authors, Yadava & Kumano 1985, Kumano 1997k)

Plants heterotrichous, consisting of prostrate system and one or two erect filaments, dark blue-green. Prostrate systems polymorphous, holdfast disc-shaped and tubular rhizoids depending on the substratum. Erect filament 200 μm in diameter, 42 cm long, coarsely and profusely branched. Branches arising at an angle of 30–70 degrees to main branches. Axial cells in mature filaments 44–50 μm in diameter, 58–70 μm long. Cortical cells 6–16 μm in diameter, 16–30 μm long, one layered in young filaments, while about three layered in mature ones. Monosporangia produced from cortical cells by unequal divisions; monosporangia spherical, 15–20 μm in diameter. Microsporangia produced from cortical cells, in groups of 8–16; microspores spherical, 10–18 μm in diameter. Knot-like structures in mature filaments produced for vegetative propagation.
Type Locality: Allahabad, Uttar Pradesh, India in Asia. Grows epiphytically on the filaments of *Cladophora* sp. on bricks and also in muddy beds in the water channel from a tube-well.
Holotype: DCP no. 38.
Distribution: Known from the type locality and Kido-gawa River in Chiba, Japan in Eastern Asia.

Compsopogon

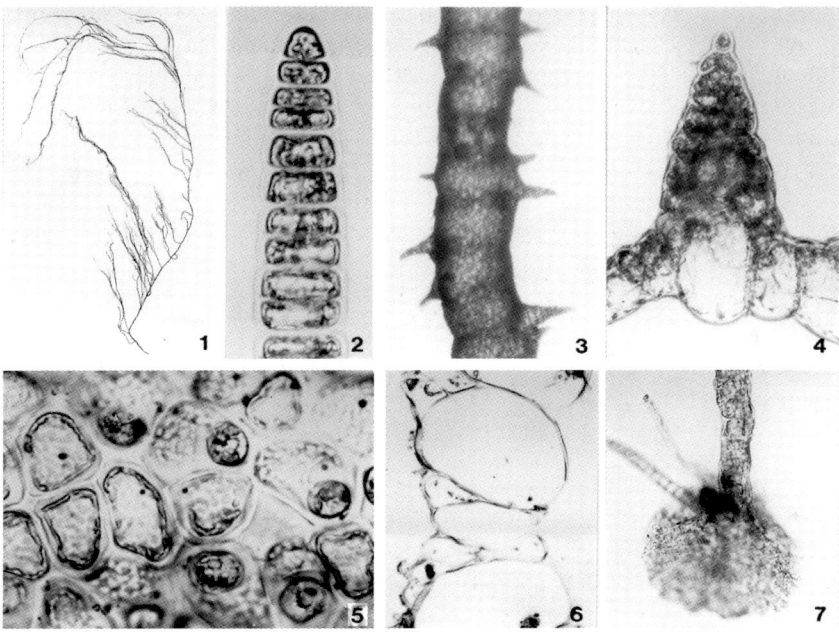

Plate 7 (a)
Compsopogon aeruginosus, figs 1–7 (Nakamura 1984a)
1. the habit of a plant showing ramifications, 2. the apical portion of a young plant showing uniseriate axial cells, 3. a mature plant with many spinous short branchlets, 4. the cross sectional view of a spinous short branchlet, 5. the surface view of cortical cells and monosporangia, 6. the cross sectional view of monosporangia, 7. three erect filaments arising from a basal small disc.

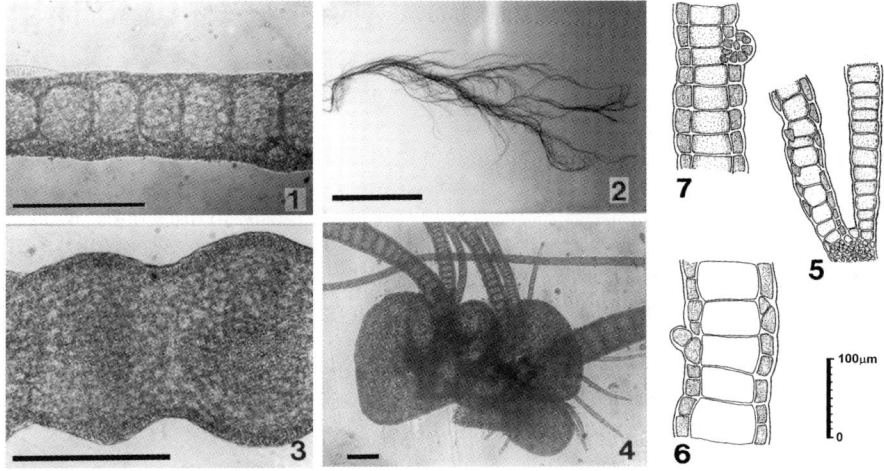

Plate 7 (b)
Compsopogon prolificus, figs 1–7 (original authors, Yadava & Kumano 1985, Kumano 1997k)
1. an erect filament showing axial cells and cortical cells of 2–3 layers, 2. the habit of whole plants, 3, a main branch showing nodulation, 4. knot-like structures of vegetative discs, 5. two erect filaments arising from a small disc type of a prostrate system, 6. a multiseriate plant consisting of axial cells and cortical cells in single layer, 7. microsporangia forming in cortical layer.
Scale bars = 100 μm (figs 1, 3–4); 10 cm (fig. 2).

Compsopogonales

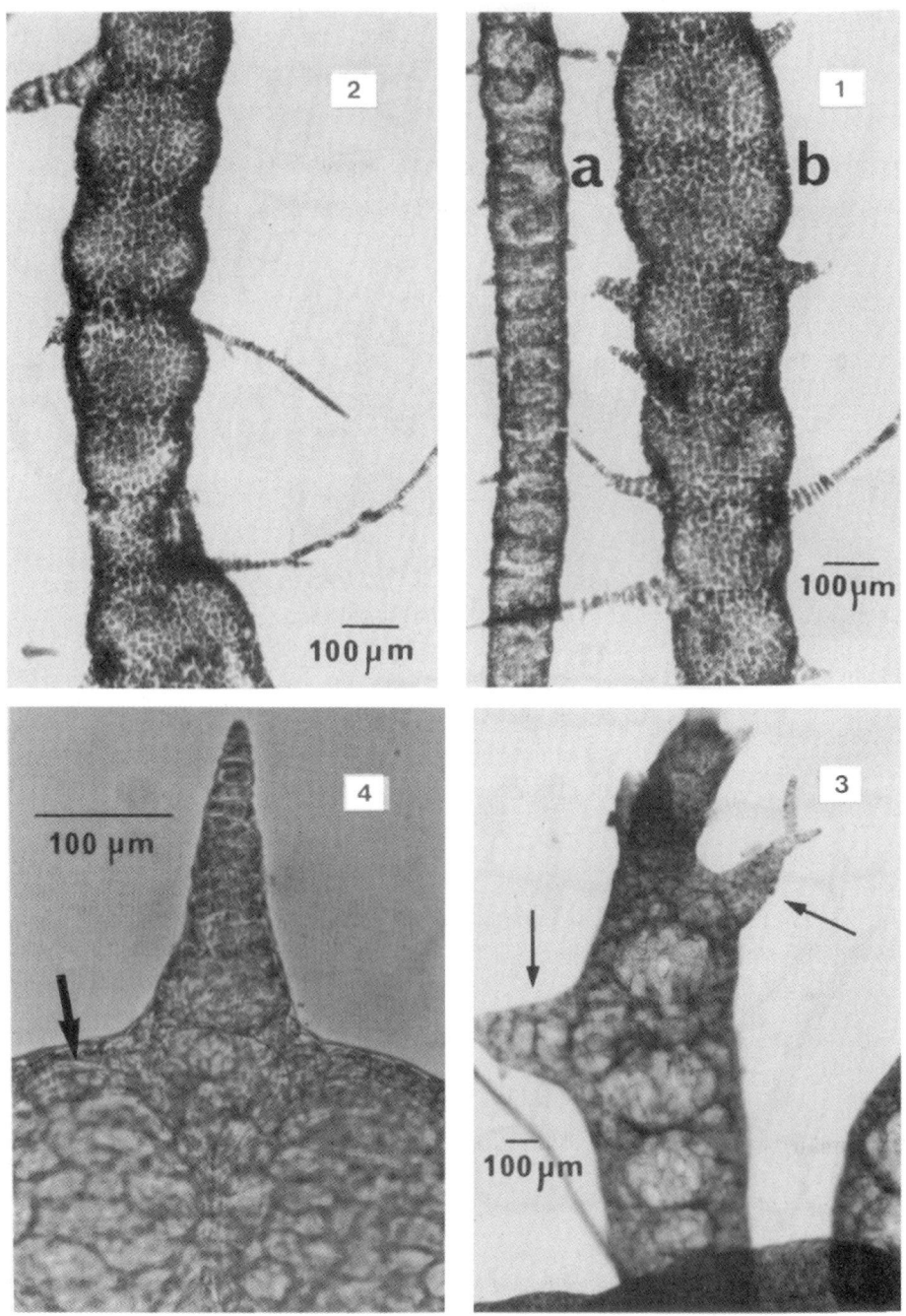

Plate 8
Compsopogon aeruginosus var. *catenatum*, figs 1–4 (Seto, Yadava & Kumano 1991)
1. various developmental stages of short spinous branchlets, a: a corticated plant with many spinous short branchlets, b: a nodulated plant with many well-developed spinous short branchlets, 2. a plant with two uniseriate filamentous branchlets and three short spinous branchlets, 3. two spinous short branchlets (arrow) newly formed by the division of axial cells on the old spinous branchlets, 4. the development of a short branchlet formed by the division of an axial cell (arrow).

3) ***Compsopogon chalybeus*** Kützing (1849:432, 1857: tab. 89, fig. II)
Synonym: *Compsopogon corinaldii* (Meneghini) Kützing (1857:35, tab. 88, fig. I a, a'–b, b');
Lemanea corinaldii Meneghini (1841:186)

plate 11, figs 1–2 (Kützing 1849) as *Compsopogon chalybeus*,
figs 6–8 (original author, Kützing 1849) as *Compsopogon corinaldii*

Plants branching from the base, erect filaments constricted at transverse walls, in portions producing cortex 100–250 μm in diameter, cells of the cortex 12–19 μm in diameter, 16–28 μm long, monosporangia 12–16 μm in diameter.
Type Locality: Cayenne, French Guiana in South America.
Holotype: (L 940.284.409), PC herb. Leprieur 828.
Distribution: Known from a tropical aquarium at Catania, Sicily and Ukraine in Europe; India in Asia; Florida, USA in North America. Known from rivers and streams.

4) ***Compsopogon corticrassus*** Chihara et Nakamura (1980:136, figs 1, 2 A–J, 3 A–L, 4 A–C)

plate 9, figs 1–8 (Nakamura 1994b)

Plants heterotrichous, consisting of many erect filaments arising from a prostrate system, dark blue-green. Prostrate systems polymorphous, holdfast conical or hemispherical, 170–250 μm high, 300–400 μm in diameter, depending on the substratum. Erect filaments 2–3 mm in diameter, 50–80 cm long, sparsely branched. Cortical layers 300–380 μm thick, 3–5 cell-layered, cortical cells 7–20 μm in diameter, 14–36 μm long. Medullar cells of main axes fragmented in older plants. Monosporangia produced from cortical cells by unequal divisions; spherical, 16–22 μm in diameter.
Type Locality: Minuma-yohsui, Gyoda in Saitama, Japan in Eastern Asia. Growing on the concrete bed in a riffle.
Holotype: TNS Al–35602.
Distribution: Known from the type locality and in Yase-gawa River, Ohta in Gunma, Kanzaki-gawa River, Shido-gawa River, Kashima-gawa River, Oba-gawa River in Chiba, Yunoo Falls, Egawa-bashi along Sendai-gawa River in Kagoshima, Japan in Eastern Asia.

5) ***Compsopogon tenellus*** Ling et Xie in Xie & Ling (1998a:81, pl. 1, figs 1–7)

plate 13 (a), figs 1–7 (original authors, Xie & Ling 1998)

Plants small, 1–2 cm high and 90–120 μm in diameter, densely caespitose, green, generally slightly attenuated at the base with basal rhizoids, usually little branched with branchlets generally slightly attenuated at their apex; axial cells depressed-spherical, 70–90 μm in diameter, 30–60 μm long, cortical cells polygonal or ovoidal, 1–layered, 10–25 μm in diameter. Monosporangia rarely produced from cortical cells of proximal portion of plants, pyramidal-ovoidal.
Type Locality: Known from Taiyuan, Shanxi, China in Eastern Asia.
Holotype: SAS 92086.
Distribution: Known from the type locality only.

Compsopogonales

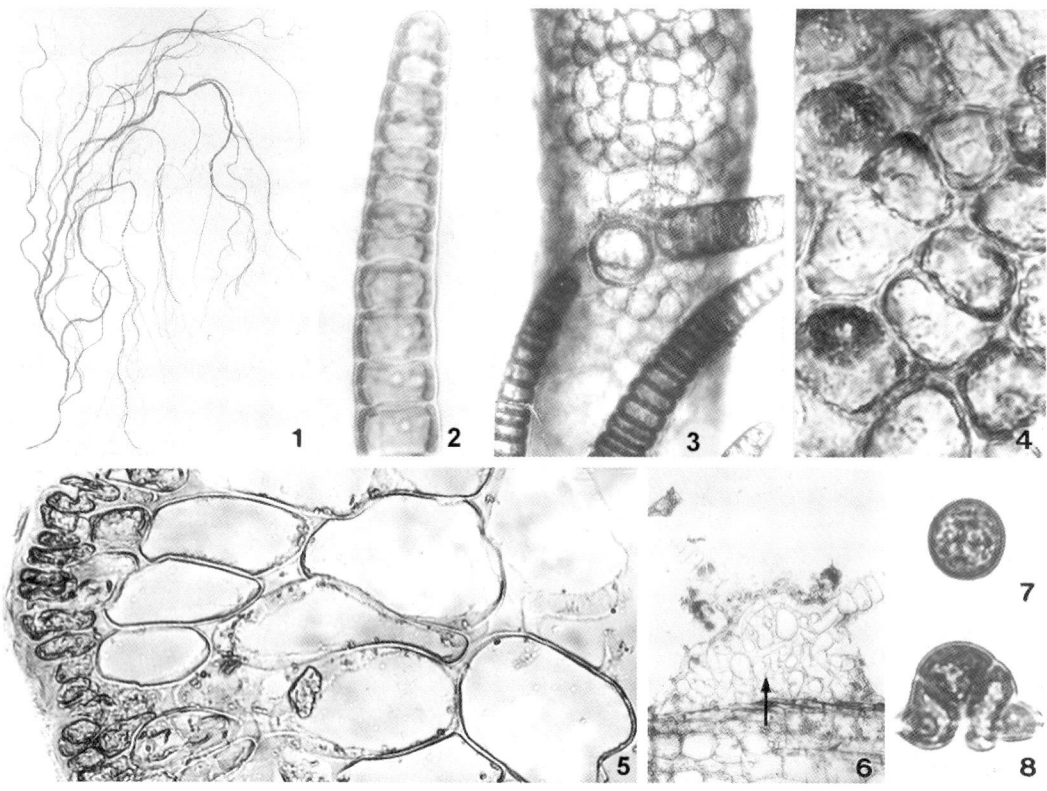

Plate 9
Compsopogon corticrassus, figs 1–8 (Nakamura 1994b)
1. the habit of plant showing ramification, 2. an apical part of a young plant showing uniseriate axial cells, 3. a mature plant with fully developed cortical layers, 4. the surface view of a mature plant showing cortical cells and monosporangia, 5–6. the cross or longitudinal section of a plant showing several layers of cortical cells, 7–8. a monospore and its germination.

6) ***Compsopogon minutus*** Jao (1941:249, pl. I, figs 2–8)

plate 12, figs 1–5 (original author, Jao 1941)

Plants minute, up to 2 cm high and 300 μm in diameter, densely pulvinate, olive-green, usually very abundantly branched from the base, branches and branchlets lateral, not dichotomous, rectangular patent, generally slightly attenuated at their base and apex, branchlets very rare, axial cells depressed-globose, usually 2 times shorter than their diameter, cortical cells 3- or 5-angular, usually 1–layered or very rarely partly 2–layered.
  Type Locality: Hua-kai-shan, Kiangtsin, Szechwan, China in Eastern Asia. On submerged roots and leaves in a mountain stream.
  Holotype: SC 1010.
  Distribution: Known from the type locality only.

Compsopogon

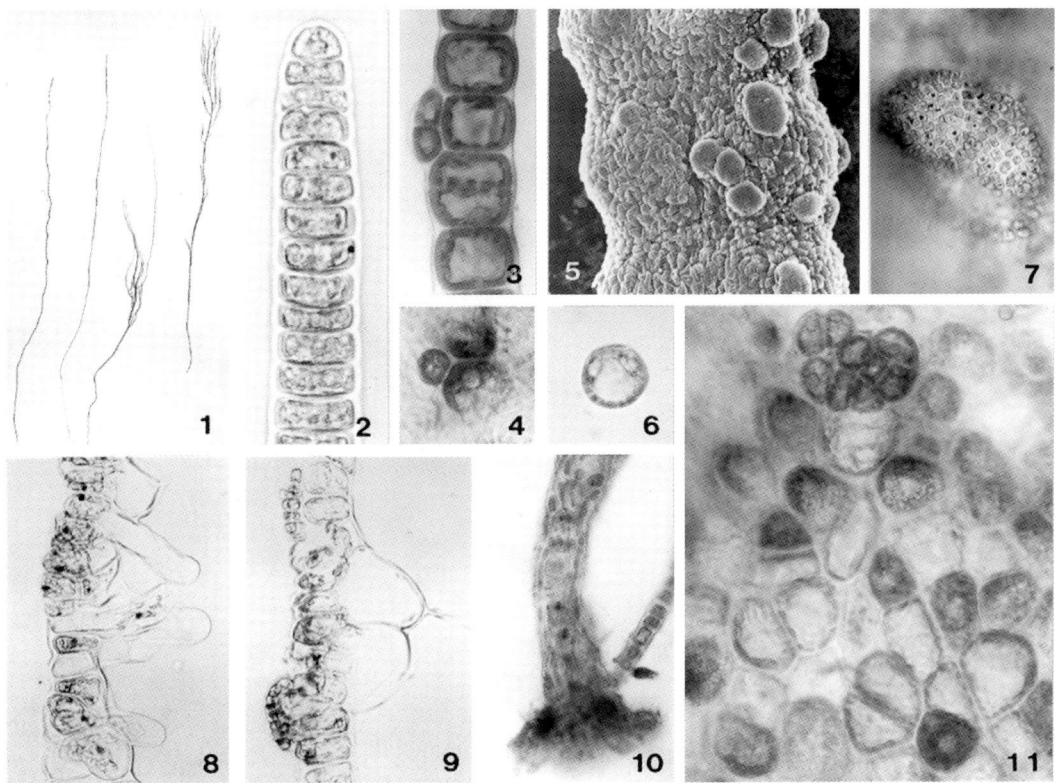

Plate 10
Compsopogon hookeri, figs 1–11 (Nakamura 1984c)
1. the habit of a plant showing ramification, 2. an apical portion of a young plant showing uniseriate axial cells, 3. the longitudinal view of microsporangia development, 4. the surface view of microsporangia development, 5. the surface view of a plant showing microsporangial sori, 6. a monospore, 7. spherical microsporangial sori, 8–9. the longitudinal section of a plant showing 1–2 layers of cortical cells, inner cell of which elongated inward, 10. the lower part of a plant showing small disc covered by rhizoidal filaments. 11. the surface view of a plant showing cortical cells and monosporangia.

7) ***Compsopogon sparsus*** Xie et Ling (1998a:81, pl. 2, figs 1–9)

plate 13 (b), figs 1–9 (original authors, Xie & Ling 1998a)

Plants consisting of simple rhizoidal holdfast pad and erect filaments, 15–20 cm high and 300 μm in diameter, loosely caespitose, blue green, distinctly constricted, generally attenuated toward basal rhizoids, abundantly branched proximal, sparsely branched distally. Branchlets alternate, departing at a right angle from the main filaments, gradually attenuated towards apex, abruptly constricted at the apex. Axial cells depressed spherical, 240–250 μm in diameter, ca. 200 μm long, cortical cells polygonal-ovoidal, 1-layered, 20–40 μm in diameter, 20–30 μm long. Monosporangia unknown.
Type Locality: in Bose, Guangxi, China in Eastern Asia.
Holotype: GX 89035.
Distribution: Known from the type locality only.

Compsopogonales

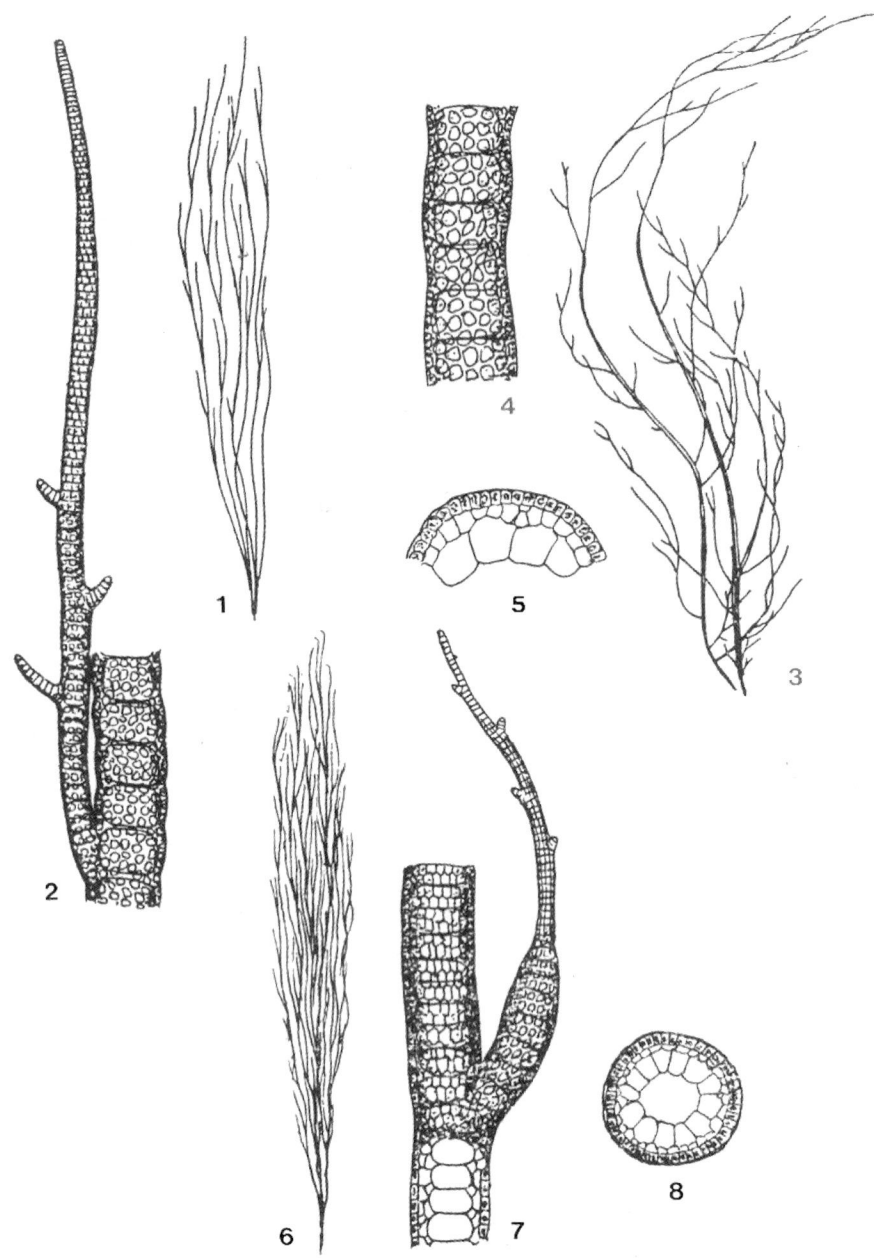

Plate 11
Compsopogon chalybeus, figs 1–2 (original author, Kützing 1849)
1. the habit of plants showing ramification, 2. an apical portion of branch.

Compsopogon coeruleus, figs 3–5 (Kützing 1849)
3. the habit of plants showing ramification, 4. a part of a plant showing axial cells and cortical cells, 5. the cross sectional view of cortical layer.

Compsopogon chalybeus as *C. corinaldi,* figs 6–8 (original author, Kützing 1857)
6. the habit of plants showing ramification, 7. an apical portion of branch and longitudinal section showing axial cells and cortical layer, 8. the cross sectional view of cortical layer.

8) ***Compsopogon hookeri*** Montagne (1846: 157, pl. 46, figs 3, 10; pl. 47, fig. 1)

plate 6, figs 5–9 (Friedrich 1966)
plate 10, figs 1–11 (Nakamura 1984c)

Plants heterotrichous, consisting of prostrate system and erect filaments, dark blue-green. Prostrate system a polymorphous disk, holdfast 140–250 µm in diameter, covered with descendant rhizoidal outgrowths from cortical cells, depending on the substratum. Erect filaments 0.5–2 mm in diameter, 20–50 cm long, sparsely branched. Cortical layers 130–180 µm thick, 1–2 cell-layered; cortical cells 10–21 µm in diameter, 15–35 µm long, rhizoidal extensions produced inwardly from cortical cells into medulla. Monosporangia produced from cortical cells by unequal divisions; spherical, 13–19 µm in diameter. Microsporangia in groups on main axis and branches, spherical, 8–12 µm in diameter.

Type Locality: near Madras, India in Asia.
Holotype: BM (K)
Distribution: Known from River Erth (affluent of Rhone), France in Europe; Japan in Eastern Asia. In Japan, Kito-gawa River in Chiba, Matsue, Hirata and Komatsu along Shinji-ko Lake in Shimane.

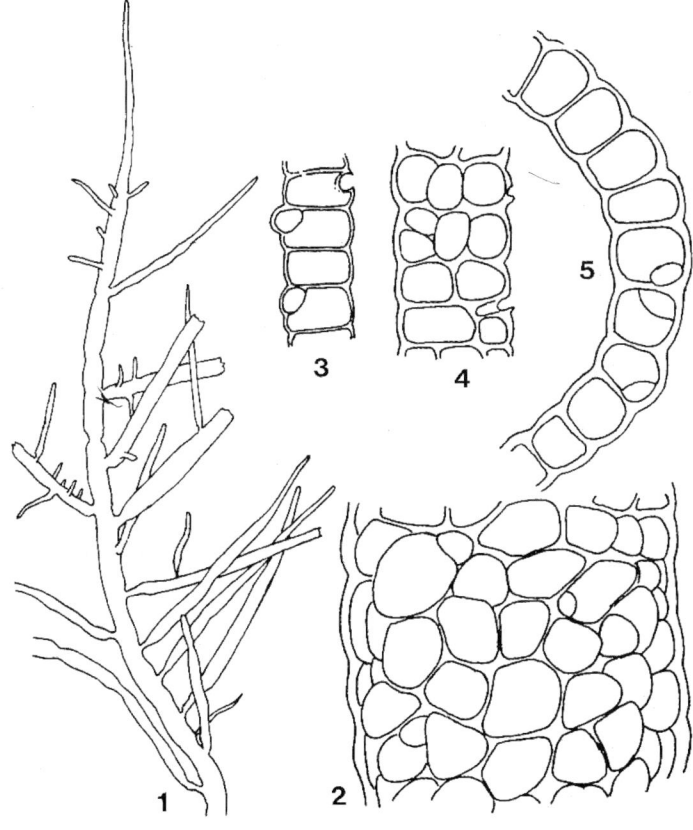

Plate 12
Compsopogon minutus, figs 1–5 (original author, Jao 1941)
1. the habit of plant, 2. the surface view of a fully developed plant showing cortical cells, monosporangia, 3–4. an upper portion of a branch, 5. the cross sectional view of a well-developed plant.

Compsopogonales

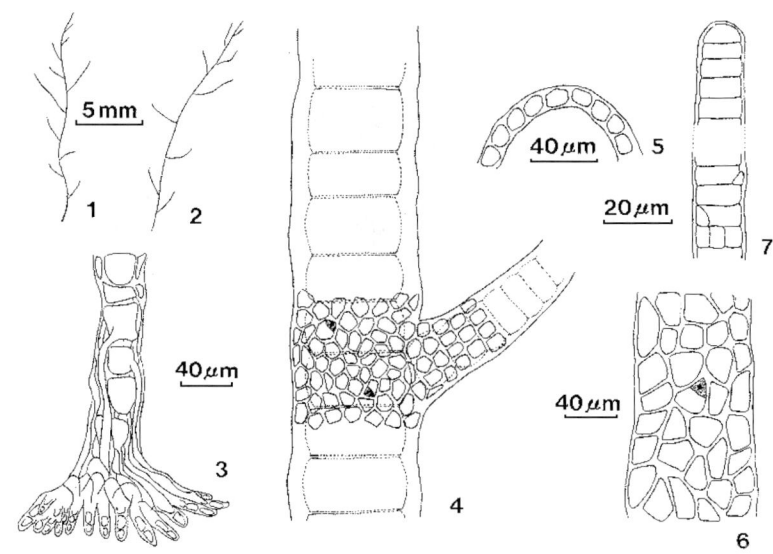

Plate 13 (a)
Compsopogon tenellus, figs 1–7 (original authors, Xie & Ling 1998)
1–2. the habit of plants showing ramification, 3. the basal portion of a plant showing rhizoidal filaments, 4. a part of a plant showing axial cells and monosporangia developed from cortical cells, 5. the cross sectional view of one layered cortical cells, 6. the surface view of cortical layer, 7. an apical portion of branch.

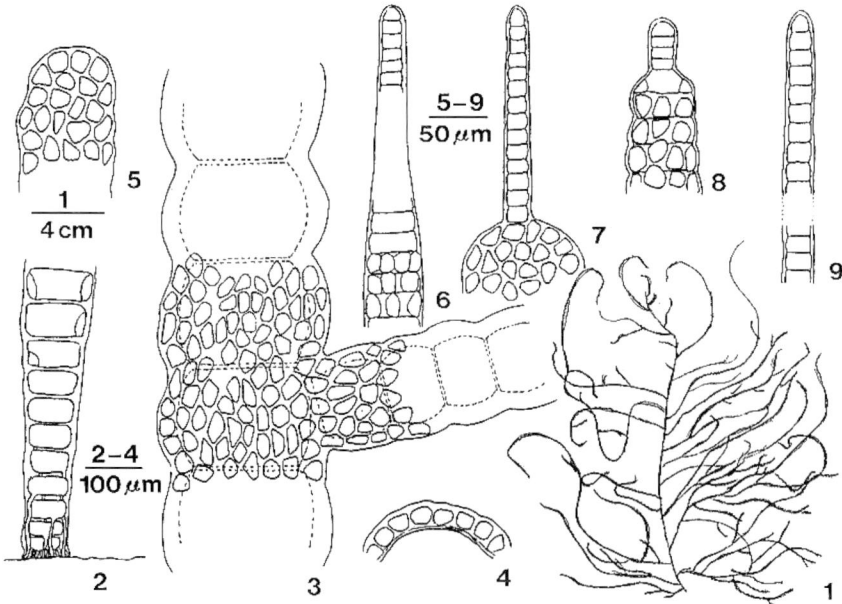

Plate 13 (b)
Compsopogon sparsus, figs 1–9 (original authors, Xie & Ling 1998a)
1. the habit of a plant showing ramification, 2. the basal portion of plants showing rhizoidal filaments, 3. a part of a constricted plant showing axial cells and cortical cells, 4. the cross sectional view of one layered cortical cells, 5–9. the apical portion of plants showing rounded and abruptly constricted monosiphonous apices.

Compsopogon

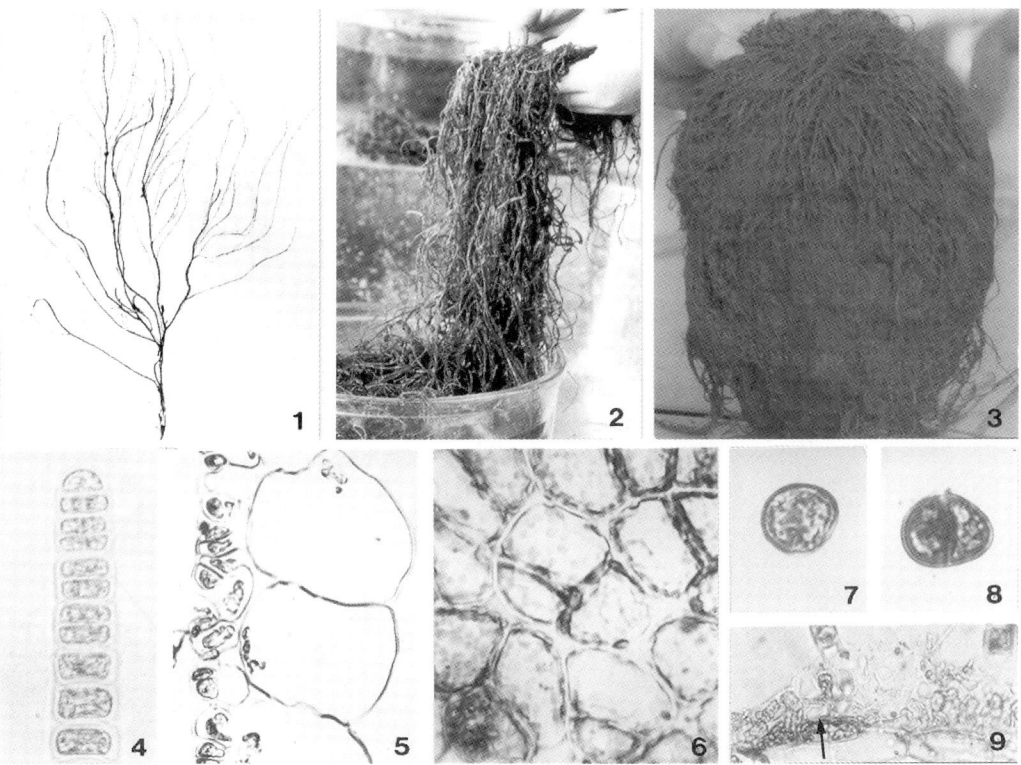

Plate 14
Compsopogon coeruleus as *C. oishii*, figs 1–9 (Nakamura 1994d)
1. the habit of a plant showing ramification, 2–3. vegetative plants growing in natural habitats, 4. an apical portion of a plant showing uniseriate axial cells, 5. the cross section of a plant showing 2–layered cortical cells and monosporangia, 6. the surface view of a plant showing cortical cells and monosporangia, 7–8. germination of monospore.

9) ***Compsopogon coeruleus*** (Balbis ex C. Agardh) Montagne (1846:154)
Basionym: *Conferva coerulea* Balbis ex C. Agardh (1824:122),
Synonym: *Compsopogon oishii* Okamura (1915:128, pl. 132–133);
Compsopogon helwanii El-Gamal et Salah El-Din (1999:38, figs A–J)

plate 11, figs 3–5 (Kützing 1849) as *C. coeruleus*,
plate 14, figs 1–9 (Nakamura 1994d) as *C. oishii*

Plants heterotrichous, consisting of prostrate system and erect filaments, dark blue-green. Prostrate system polymorphous, with disc-shaped holdfast and flat-conical rhizoids, 300–500 μm in diameter, depending on the substratum; 1 to several erect filaments arising from prostrate system. Erect filament 1–3 mm in diameter, 10–80 cm long, coarsely and profusely branched. Cortical layers 130–180 μm thick, 2 or 3 cells thick in mature thalli; cortical cells 10–27 μm in diameter, 15–36 μm long. Monosporangia produced from cortical cells by unequal divisions; monosporangia spherical, 15–20 μm in diameter.
Type Locality: Puerto Rico, Great Antilles in Caribbean Sea in Central America.
Holotype: L 940284413.
Isotype: PC herb. Balbisii

Compsopogonales

Distribution: Known from a freshwater canal in Malta, central Mediterranean in Europe; on *Vallisneria spiralis* and other aquatic plants in a well with connections to slow-running springs and in streams near Yangso, Kwangsi, China in Eastern Asia; Central America; on aquatic plants in streams and rivers, Queensland, New South Wales in Australia and tropical regions. In Central America, in a small stream with swift currents in a village of Yaguaracual, between Cuman·and Puerto La Cruz, eastern Venezuela.

[*C. oishii.* Type Locality: Growing on leaves and stems of *Vallisneria*, on gravel and on wood in a spring brook, Yanokuchi, Tama-gawa River in Tokyo, Japan in Eastern Asia. Lectotype: SAP herb. Okamura by Oishi]

C. oishii, known from Japan, southern regions such as in Sesekimoi, Akan-ko Lake in Hokkaido, in Kawakami-yu-numa Lake in Fukushima, Bizen-bori-gawa River in Saitama, Yaba-gawa River in Gunma, Kido-gawa River in Chiba, in Aichi, Ohomi-hachiman in Shiga, Aboshi, Kobe in Hyogo, Aya-kawa River in Kagawa, Takabaru in Miyazaki and in Kagoshima.

[*C. helwanii.* Type Locality: Attached to stones, pipes, floating objects such as twinges and plastic bags in a natural, underground spring, Helwan, located about 25 km south of Cairo in Egypt. Holotype No. 1011 by El-Gamal & Salah El-Din]

Order **Bangiales** Schmitz in Engler (1892:15).

Type: family Bangiaceae Engler, nom. cons.

Plants of the order Bangiales are unicellular or branched or unbranched filaments, or uniseriate or multiseriate or expanded blades either one or two cells in thickness. For freshwater members, sexual reproduction unknown. Asexual reproduction by means of monosporangia. Pit-plug is type 3, consisting of a core enveloped by a cap layer.

Family **Bangiaceae** Engler (1892:12), nom. cons.

Type: genus *Bangia* Lyngbye

The suggestion of Garbary *et al.* (1980) that the order Bangiales comprises the single family Bangiaceae has been accepted. Family Bangiaceae, with two genera (*Bangia* and *Porphyra*), forms a well-circumscribed family based on life history and developmental features.

Genus *Bangia* Lyngbye (1819:82)

Type: *Bangia fuscopurpurea* (Dillwyn) Lyngbye

One species of the genus *Bangia* is reported from freshwater. The filaments of *Bangia* are unbranched and grow attached to rock and woodwork. Asexual reproduction by means of monosporangia is reported for Japanese freshwater species.

Note: Geesink (1973) concluded that the freshwater *B. atropurpurea* and the marine *B. fuscopurpurea* were conspecific and should be synonymyzed under the older epithet *B. atropurpurea*. However, the analyses of RuBisCo spacer, rbcL gene and 18S rRNA gene sequences showed that the North American and European freshwater specimens are nearly identical and appear to be considerably different from marine specimens (Müller, Sheath, Vis, Crease & Cole 1998, Müller, Gutell & Sheath 1998). Sequence analyses reveal a strong similarity between *Bangia* isolates in eastern Australia over a wide geographical range, but taxonomy of *B. atropurpurea* may need to be re-examined in light of sequence differences between these and northern hemisphere isolates of this species (Woolcott & King 1998).

1) *Bangia atropurpurea* (Roth) C. Agardh (1824: 76)
Basionym: *Conferva atropurpurea* Roth (1806: 208, pl. 6)

plate 15, figs 1–8 (Kumano 1998a)

Plants erect, filamentous, unbranched, caespitose, dark purple tinged with brown, 2–4.5 cm high, 20–98 µm in diameter; cells 8–20 µm long, 18–20 µm in diameter in sterile plants, 5–15 µm long, 35–50 µm in diameter in mature plants; 50–98 µm in diameter in monosporangial plants. Fastened to the substratum by means of rhizoids which grow downwards from the lower cells through the common outer wall. Filaments consisting initially of a single row of cylindrical cells, 6 µm in diameter, 8 µm long when young, later disc-shaped, 6 µm in diameter, 20 µm long. Plants soon becoming polysiphonous by longitudinal cell divisions: cells cylindrical ca. 6 µm in diameter, ca. 10 µm long, containing a stellate chloroplast with a central pyrenoid. Asexual

Bangiales

reproduction by means of monosporangia; all cells in mature plant divided into monosporangia; monospore when liberated naked and amoeboid with a stellate chloroplast, then spherical, ca. 10 μm in diameter rounding up within in a day or two, secreting a cell wall, and by cell division developing into a new plant. Sexual reproduction unknown.

Type Locality: Bremen, Germany in Europe.

Holotype: probably destroyed (Koster 1969)

Distribution: Known from Belgium, Netherlands in Europe; Chialing River near Pehpei, Szechwan, China; Amahata-gawa River in Yamanashi, Japan in Eastern Asia; in all of Great Lakes except Lake Superior, Canada and USA in North America. On rocks in mountain streams.

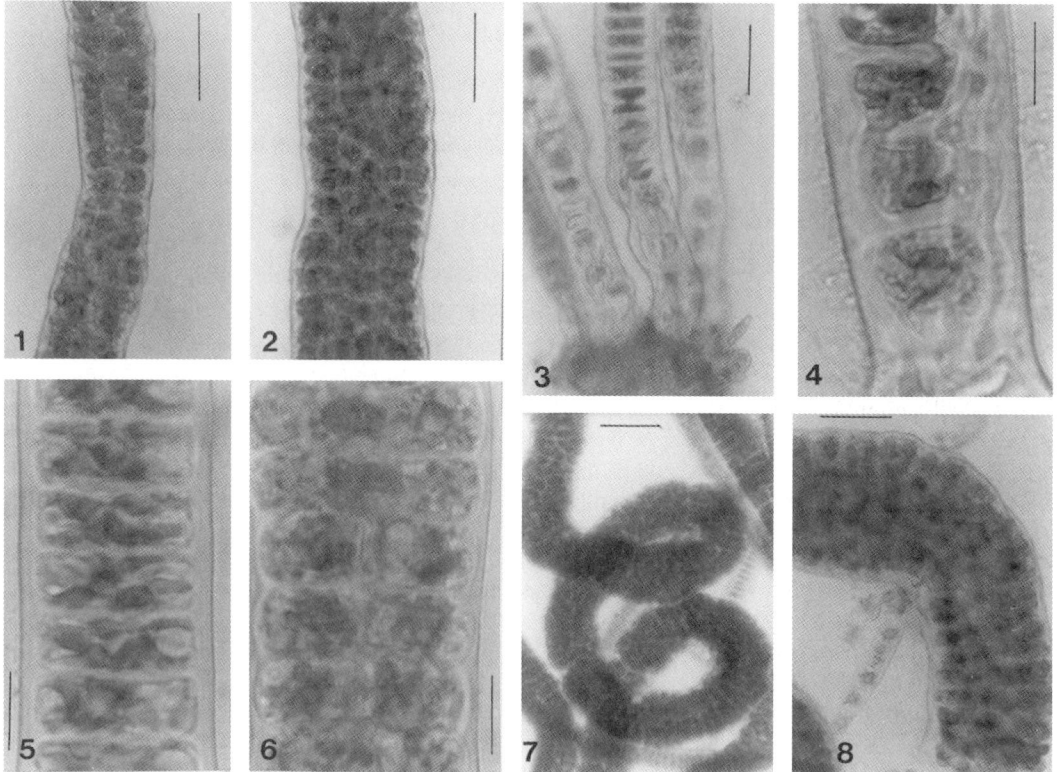

Plate 15
Bangia atropurpurea, figs 1–8 (Kumano 1998a)
1–2, 6. mature filaments, 5. sterile filaments, 2, 7–8. fertilized filaments with monosporangia, 3–4. a rhizoidal portion of a plant.

Family **Boldiaceae** Herndon (1964:575) in NCU–3 (1993:139)

Type: genus *Boldia* Herndon

Plants saccate, as in genus *Enteromorpha* of Chlorophyceae; reproduction by means of monospores formed on separate filaments.

Genus *Boldia* Herndon (1964:576)

Type: *Boldia erythrosiphon* Herndon

1) ***Boldia erythrosiphon*** Herndon (1964:576, pl. 45, fig. 4; pl. 46, figs 1–2; pl. 50, fig. 1) *emend* Howard et Parker (1980:413, figs 1–26).
Synonym: *Boldia angustata* Deason et Nichols (1970:40, pl. 1; figs 1–6, 9–11; pl. 2, figs 12–17).

plate 16, figs 1–4 (Sheath 1984)

Plants 0.2–12.5(–20) cm high, 0.1–1.3(–1.5) cm in diameter, consisting of a pink to reddish tan, monostromatic sac or tube occurring singly or in clusters arising from a microscopic disc of cells. Mature plants containing isodiametric or rectangular cells, 5–8 µm in diameter, 5–13 µm long, attached to the basal disc by elongate rhizoidal cells as described by Herndon (1964) and Nichols (1964). Secondary cell filaments produced as branching outgrowths from the vegetative cells. These dividing and elongating between and above the plane of the isodiametric cells. The secondary cell filaments eventually producing small monosporangia, 1.4–5 µm in diameter. Young plants arising from the basal cushion of cells and composed of two ranks of isodiametric cells, appearing to divide by a prominent apical cell which is not present in mature plants. Each vegetative cell containing a few discoidal or ribbon-shaped chloroplasts which are the *Polysiphonia* type (Hara & Chihara 1974).
Type Locality: Brg Walker Creek, Giles, Virginia, USA in North America.
Holotype: US
Distribution: Known from Ontario streams in Canada (Sheath 1984). This species is reported from Missouri, south-eastern USA (Prescott 1978) in North America. In quiet pools at the sides of rivers as well as in riffles on the downstream sides of rocks, and may be a co-dominant with *Lemanea fucina*.
Note: *Boldia angustata* Deason & Nichols (1970) was reduced to synonymy with the earlier described *B. erythrosiphon* Herndon, because of significant overlap between the specific characteristics by Howard et Parker (1980).

Bangiales

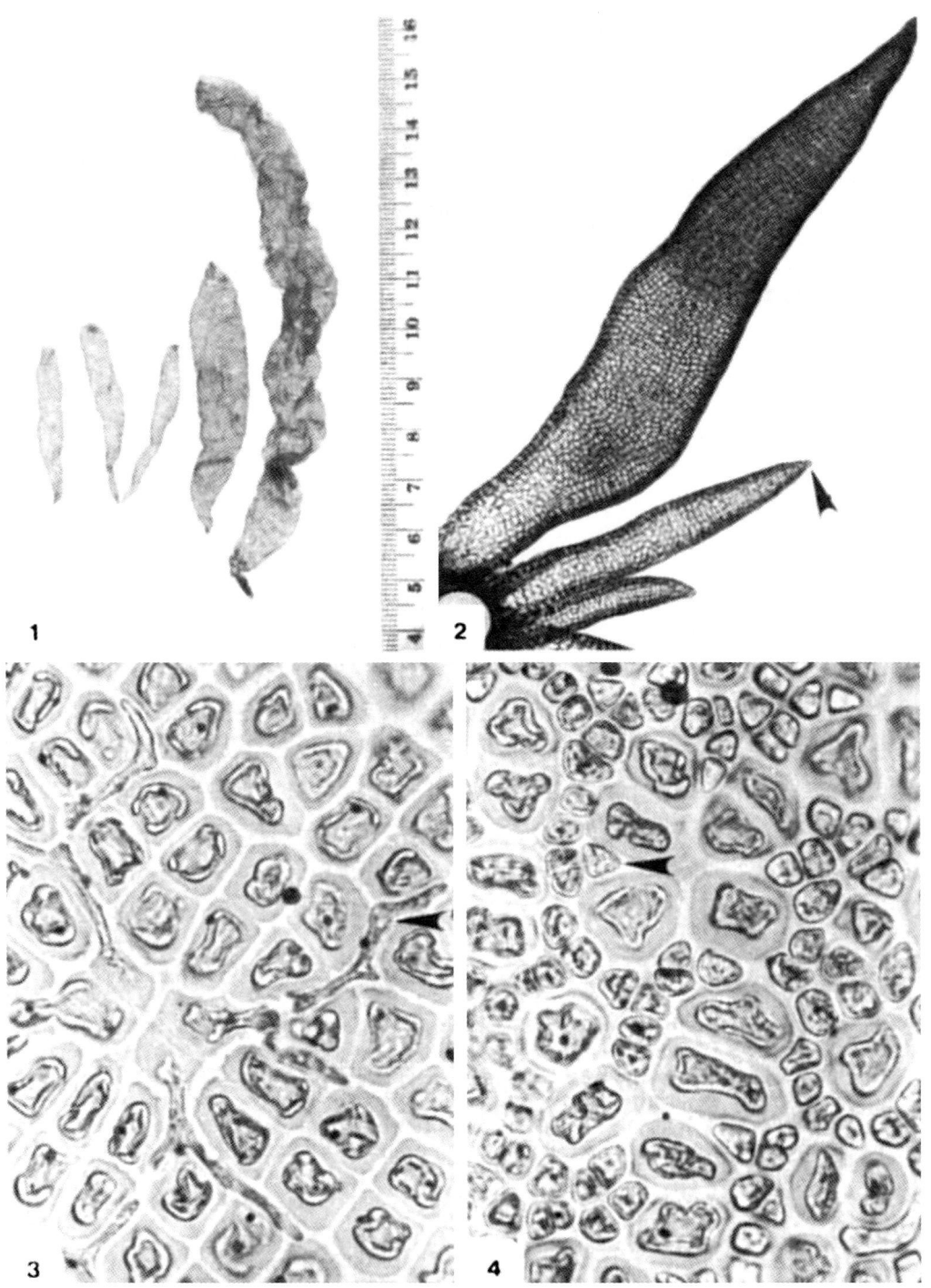

Plate 16
Boldia erythrosiphon, figs 1–4 (Sheath 1984)
1. the macroscopic view showing saccate plant, 2. young plants with prominent apical cells (arrow), 3. the surface of a plant containing large primary cells forming secondary cell filaments (arrow), 4. the plant surface with a small monosporangium (arrow) developed between and above the large primary cells.

Order **Acrochaetiales** Feldmann (1953:12)

Type: family Acrochaetiaceae Fritsch ex Taylor, nom. cons.

Plants composed of monosiphonous, branched, heterotrichous, filaments with apical growth. Erect filaments growing upwards although those of several gametophytic plants with a single cell base developing obliquely or parallel to the surface of host. Chloroplasts various in shape: laminate, stellate, discoid, ribbon-shaped, and parietal or axial in position. Laminate chloroplasts tending to include pyrenoids at the margin; stellate chloroplasts having a central pyrenoid. Pit plug with core enveloped by inner and outer cap layers (type 5), with intercalary cap membrane; some taxa with outer dome-shaped cap layers (type 6).

Asexual reproduction by means of monospores or by cruciately divided tetraspores. Sexuality is monoecious or dioecious. Simple carpogonium terminal on vegetative branchlets or 2–3 celled carpogonium-bearing branch. Gonimoblast filaments derived directly from fertilized carpogonium or from divided carpogonium, then developing into carposporophyte.

Family **Acrochaetiaceae** Fritsch ex Taylor (1957:209–210), nom. cons.

Type: genus *Acrochaetium* Nägeli

Synonym: family Rhodochortaceae Nasr (1947:92)

family Audouinellaceae Feldmann (1962:220)

There is no consensus concerning generic concepts in the Acrochaetiaceae (cf. Papenfuss 1945, 1947, Feldmann 1962, Woelkerling 1971, Y. P. Lee 1980, Y. P. Lee & I. K. Lee 1988). The taxonomic system proposed by Y.P. Lee & I. K. Lee (1988) and Y. P. Lee & Yoshida (1997) is adopted in this book.

Key to the genera of the family **Acrochaetiaceae**
(* not compiled in this book)

1. Monosporangia formed on both gametophyte and sporophyte...2
1. Monosporangia not formed on either gametophyte and sporophyte....................*Rhodochorton**
2. Chloroplasts laminate or ribbon-shaped...*Audouinella*
2. Chloroplasts stellate..*Acrochaetium*

Genus *Audouinella* Bory (1823a:340), orth. cons.

Type: *Audouinella miniata* Bory (1823a:341) in NCU–3 (1993:102) [= *Auduinella*]

Plants consisting of monosiphonous erect filaments and multicellular prostrate rhizoids (or a single, unicellular rhizoid) attached to the substratum. Chloroplasts laminate or ribbon-shaped, with or without pyrenoids. Monosporangia formed on both gametophytes and sporophytes. Sexual reproduction known except few species such as *A. hermannii*. After fertilization carpogonium undivided or divided. Carposporophyte composed of new gonimoblast filaments. Carpospores terminal.

Acrochaetiales

Note: Garbary (1987) treated the genus *Balbiania* as a synonym of the genus *Audouinella* and some freshwater species were transferred from the genus *Acrochaetium* to the genus *Audouinella*. However, Sheath & Müller (1998) separated the above mentioned two genera and proposed a new order, the Balbianiales.

Skuja (1934) proposed criteria to distinguish freshwater *Audouinella* from chantransia phases of the genus *Batrachospermum*: true species of *Audouinella* having a reddish colour, with the *Audouinella*-phases of *Batrachospermum* having a bluish colour.

Necchi & Zucchi (1997) found that in some American populations, plants initially identifiable as *Audouinella macrospora* ultimately produced juvenile *Batrachospermum* gametophytes which produced carpospores developing into plants resembling *Audouinella*. This result raises the possibility that most bluish freshwater plants such as *A. macrospora* are chantransia-phases of *Batrachospermum*, and that most reddish ones are true species of *Audouinella*.

Of the six species recognized in recent monographic treatments of Western Hemisphere species (Necchi, Sheath & Cole 1993b, 1993c, Necchi & Zucchi 1995) four are reddish, *A. eugenea* (Skuja) Jao, *A. hermannii* (Roth) Duby, *A. tenella* (Skuja) Papenfuss and *A. meiospora* (Skuja) Garbary, and two are bluish, *A. macrospora* (Wood) Sheath et Burkholder and *A. pygmaea* (Kütz.) Weber-van Bosse.

The revised descriptions given by Necchi *et al.* (1993) and Necchi & Zucchi (1995) are supplementary to the descriptions by original authors. A tentative key to the recognized species of the genus *Audouinella* and the uncertain blue taxa is shown here.

Key to the bright red or violet plants of the genus *Audouinella*

1. Cells < 7 μm in diameter..2
1. Cells > 7 μm in diameter..3
2. Cells 4–7 μm in diameter, with tetrasporangia..1) *A. amahatana*
2. Cells 3–3.5 μm in diameter, without tetrasporangia......................................2) *A. tenella*
3. Monosporangia < 16 μm in diameter..4
3. Monosporangia > 16 μm in diameter..3) *A. eugenea*
4. Monosporangia 8.5–11 μm in diameter..4) *A. hermannii*
5. Monosporangia 10–14 μm in diameter..5) *A. lanosa*

1) ***Audouinella amahatana*** (Kumano) Garbary (1987:19)
Basionym: *Acrochaetium amahatanum* Kumano (1978:105, figs 1–6)

plate 17, figs 1–6 (original author, Kumano 1978)

Plants minute, up to 500 μm high, heterotrichous, composed of prostrate and erect systems; ramification alternate, rarely unilateral; terminal hairs absent; cells of prostrate filaments fusiform, 4–6 μm in diameter and 5–12 μm long; cells of erect filaments cylindrical, 4–7 μm in diameter and 5–15 μm long; chloroplasts single, parietal, irregularly lobed, without pyrenoids. Asexual reproduction by means of monospores and tetraspores; monosporangia solitary or in groups on short lateral branchlets of erect filaments, obovoidal, 5–8 μm in diameter and 7–10 μm long; tetrasporangia solitary or in groups, mixed with monosporangia on short lateral branchlets of erect filaments, obovoidal, 7–9 μm in diameter and 11–14 μm long, cruciately divided. Sexual reproduction unknown (Kumano 1978).

Type Locality: Okusawa-tani, a tributary of Amahata-gawa River in Yamanashi, Japan in Eastern Asia. Epiphytic on *Bangia atropurpurea* (Roth) Ag. and submerged moss in a mountain stream.
Holotype: Kobe University, Kumano 1/XI, 1973.
Distribution: Known from the type locality only.

2) ***Audouinella tenella*** (Skuja) Papenfuss (1945:326)
Basionym: *Chantransia tenella* Skuja (1934:177)

> plate 18, figs 1–2 (original author, Skuja 1934) as *Chantransia tenella*
> plate 24, figs 1 (Necchi, Sheath & Cole 1993b) as *Audouinella tenella*
> plate 25, figs 11–12 (Necchi & Zucchi 1995) as *Audouinella tenella*

Plants macroscopic, up to 10 mm long, and composed of > 50 cells, reddish; basal portion consisting of an irregular prostrate system of densely aggregated filaments; lateral branches developing at angle ≥ 25°; vegetative cells of main branches cylindrical, 4.5–6.0 μm in diameter, 14–25.5 μm long; tetrasporangia spherical (undivided) to obovoidal (after first division), single or in couple, 7.0–10.5 μm in diameter, 8.0–12.0 μm long. Carpogonium and carposporophytes not observed (Necchi & Zucchi 1995).
Type Locality: Mt. Tamalpais, Marin, California, USA in North America.
Isotype: UC 395493.
Distribution: Known from the type locality and from Pres. Figueiredo, on the road between Manaus and Caracarai (Route BR–174) 115 km, Amazonas, Brazil in South America.

3a) ***Audouinella eugenea*** (Skuja) Jao (1940:362)
Basionym: *Chantransia eugenea* Skuja (1934a:177, pl. 1, figs 3–5)

> plate 18, figs 3–4 (original author, Skuja 1934a) as *Chantransia eugenea*,
> plate 24, figs 2 (Necchi, Sheath & Cole 1993b) as *Audouinella eugenea*

Plants 1–3 mm high, red-violet, solitary or velvet-like cushion forming velvety aggregations developing from prostrate pseudoparenchymatous basal system. Plants monosiphonous, without cortication, monopodial, generally alternately to oppositely branched. Branchlets short to moderately long, straight, solitary; cells 7–10 μm in diameter, 4–8 times longer than diameter, cylindrical, cross wall not constricted. Terminal hairs absent. Cell wall hyaline, up to 1 μm thick. Chloroplasts spiral-band-shaped with more or less lobed margins, without pyrenoids, several in each cell. Monosporangia numerous, mostly only on inner side of short lateral branchlet, often in rows, sessile or on short, 1–2 celled short branchlets, single or grouped, compressed-ovoidal, 14–15 μm in diameter, 16–19 μm long (Skuja 1934, Starmach 1976).
Type Locality: Punjab, India in Asia.
Holotype:
Distribution: Known from India in Asia, tropical North America. Grows in pond on aquatic plants.

Acrochaetiales

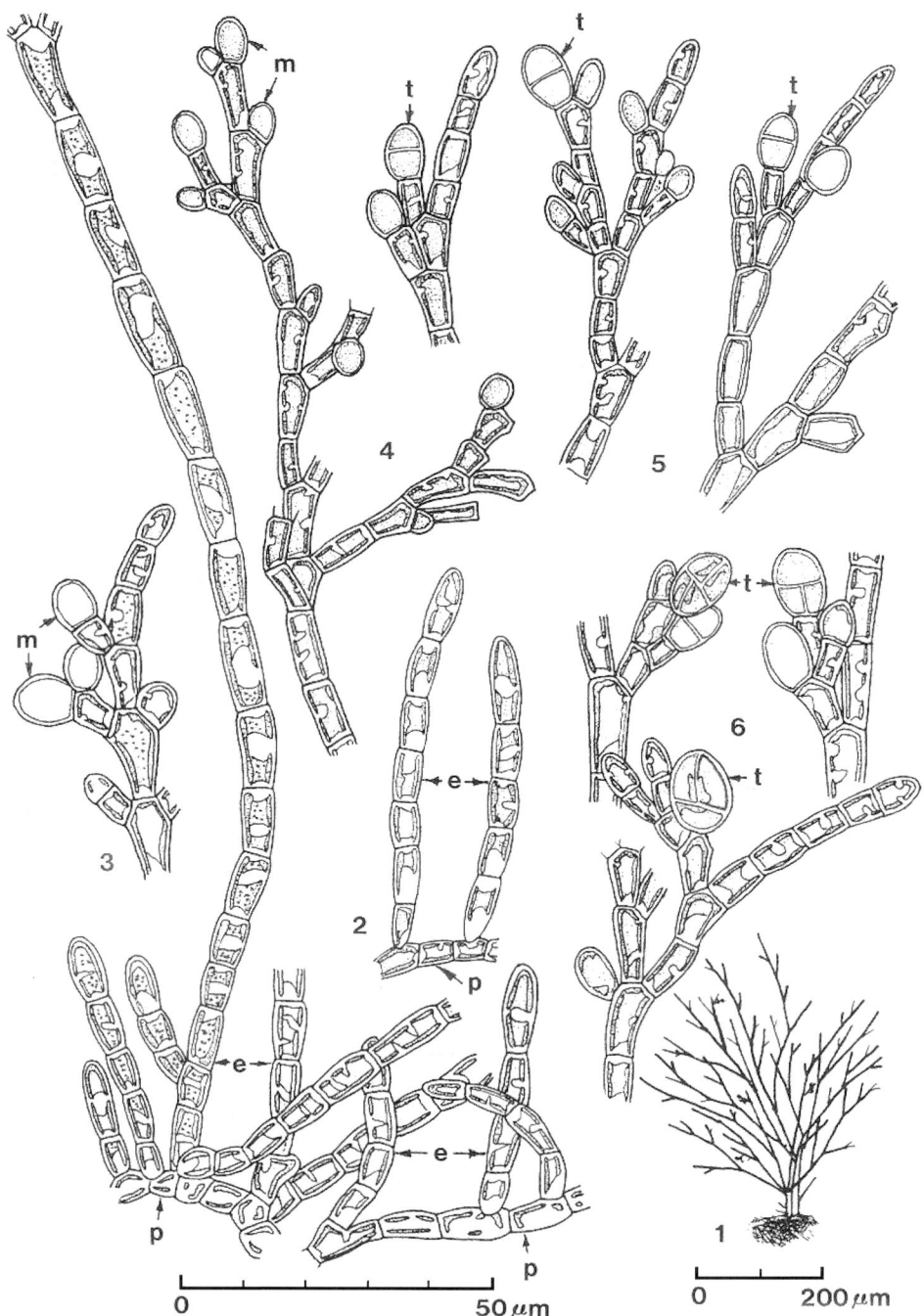

Plate 17
Audouinella amahatana as *Acrochaetium amahatanum*, figs 1–6 (original author, Kumano 1978)
1. the heterotrichous habit of a plant, 2. prostrate and erect systems, each cell of which containing a single parietal irregularly lobed chloroplast, 3–4. monosporangia, 5. young tetrasporangia on short lateral branches of erect filaments, the first division of tetraspore already occurs, 6. a mature tetraspore cruciately divided, the second division of tetraspore is perpendicular to the first, each spore contains a single parietal chloroplast.
p: prostrate system, e: erect system, m: monosporangium, t: teterasporangium.

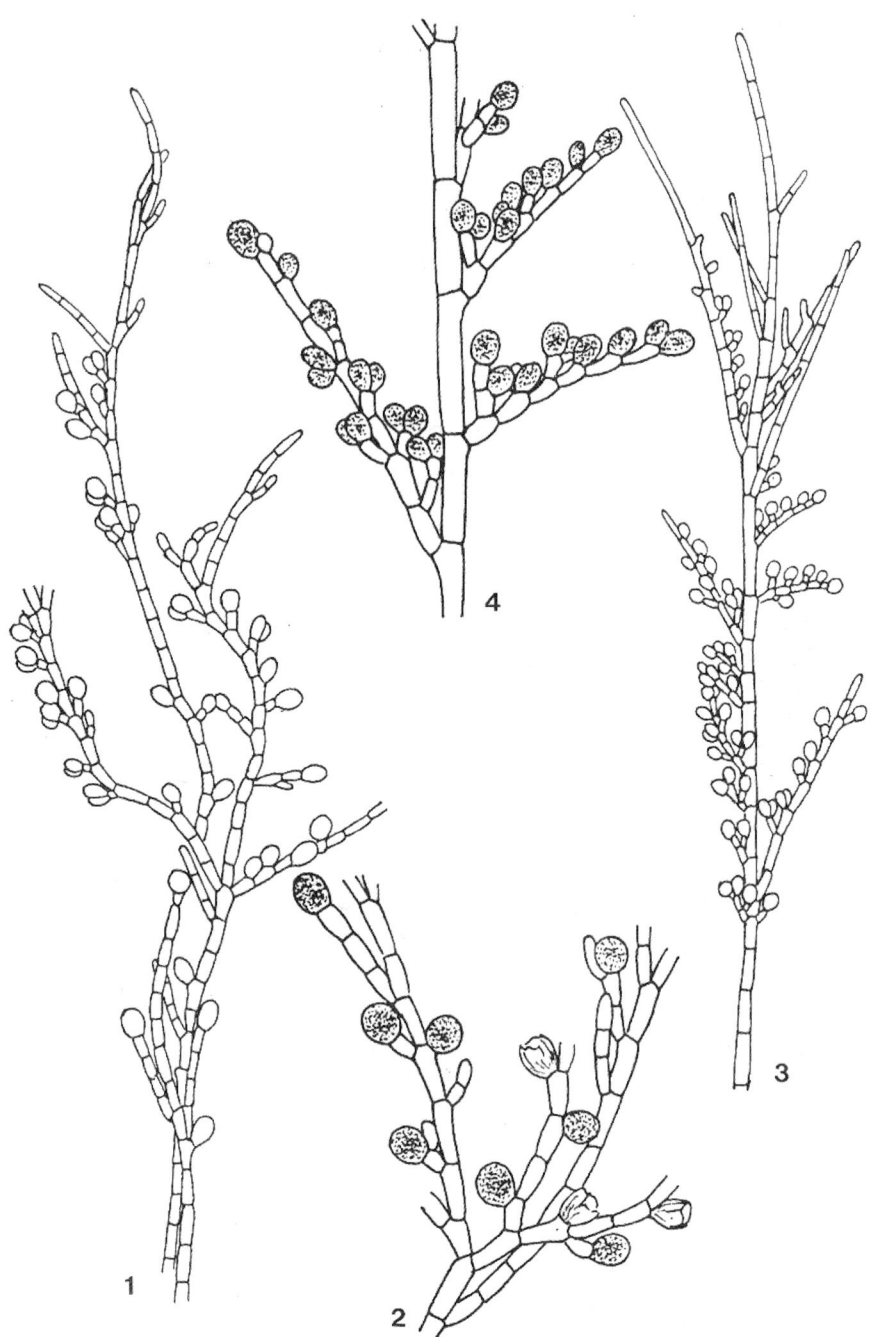

Plate 18
Audouinella tenella as *Chantransia tenella*, figs 1–2, (original author, Skuja 1934a)
1. the portion of a plant showing the habit of short lateral branchlets bearing monosporangia, 2. a filament with short lateral branchlets bearing monosporangia.

Audouinella eugenea as *Chantransia eugenea*, figs 3–4 (original author, Skuja 1938a)
3. a portion of a plant showing the habit of short lateral branchlets bearing monosporangia, 4. a filament with branchlets bearing monosporangia.

Acrochaetiales

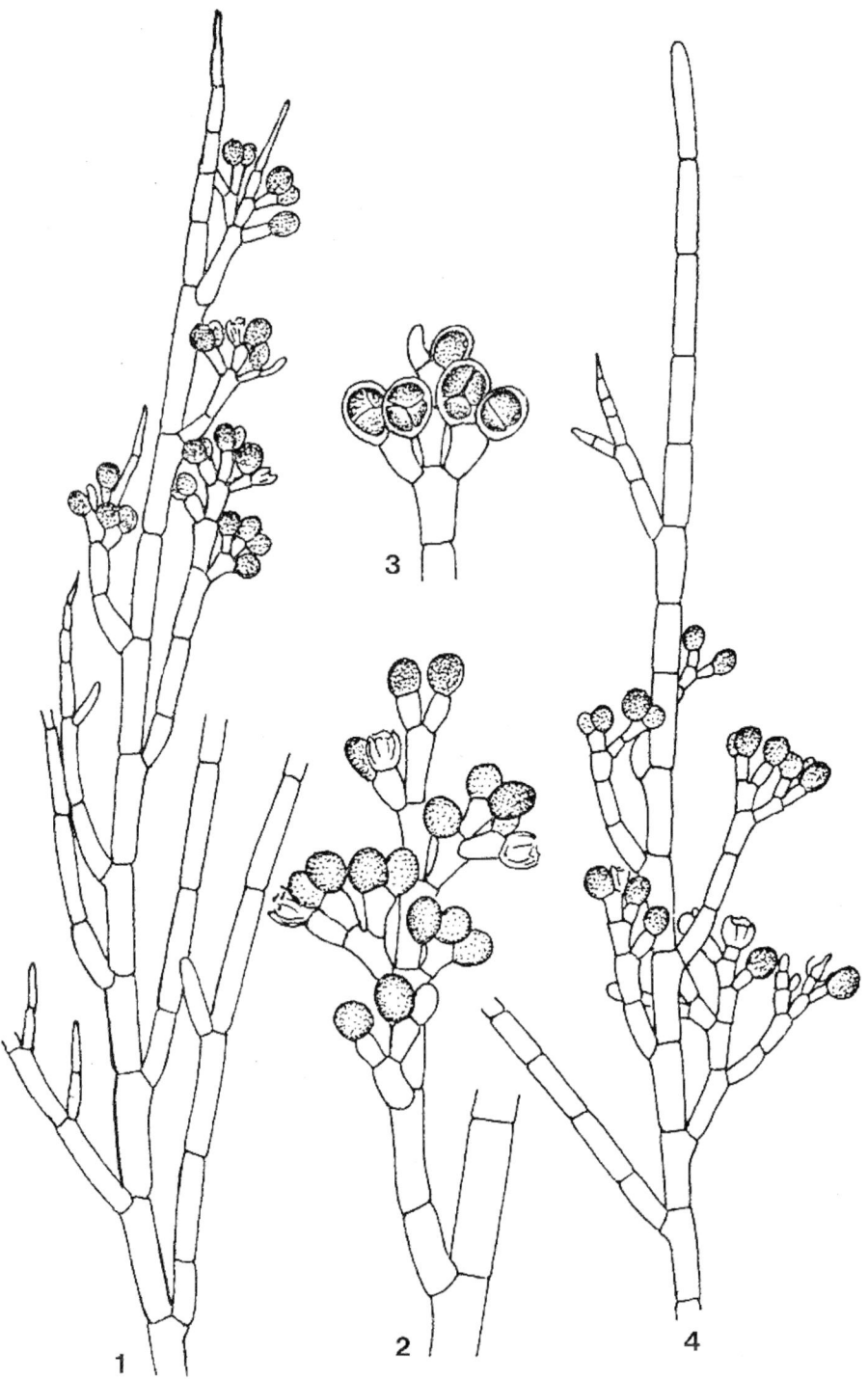

Plate 19
Audouinella hermannii as *Chantransia hermannii*, figs 1–4 (Starmach 1977)
1. a filament bearing monosporangia, 2. monosporangia in clusters on short lateral branchlet, 3. tetrasporangia, 4. a filament with short lateral branchlets bearing monosporangia.

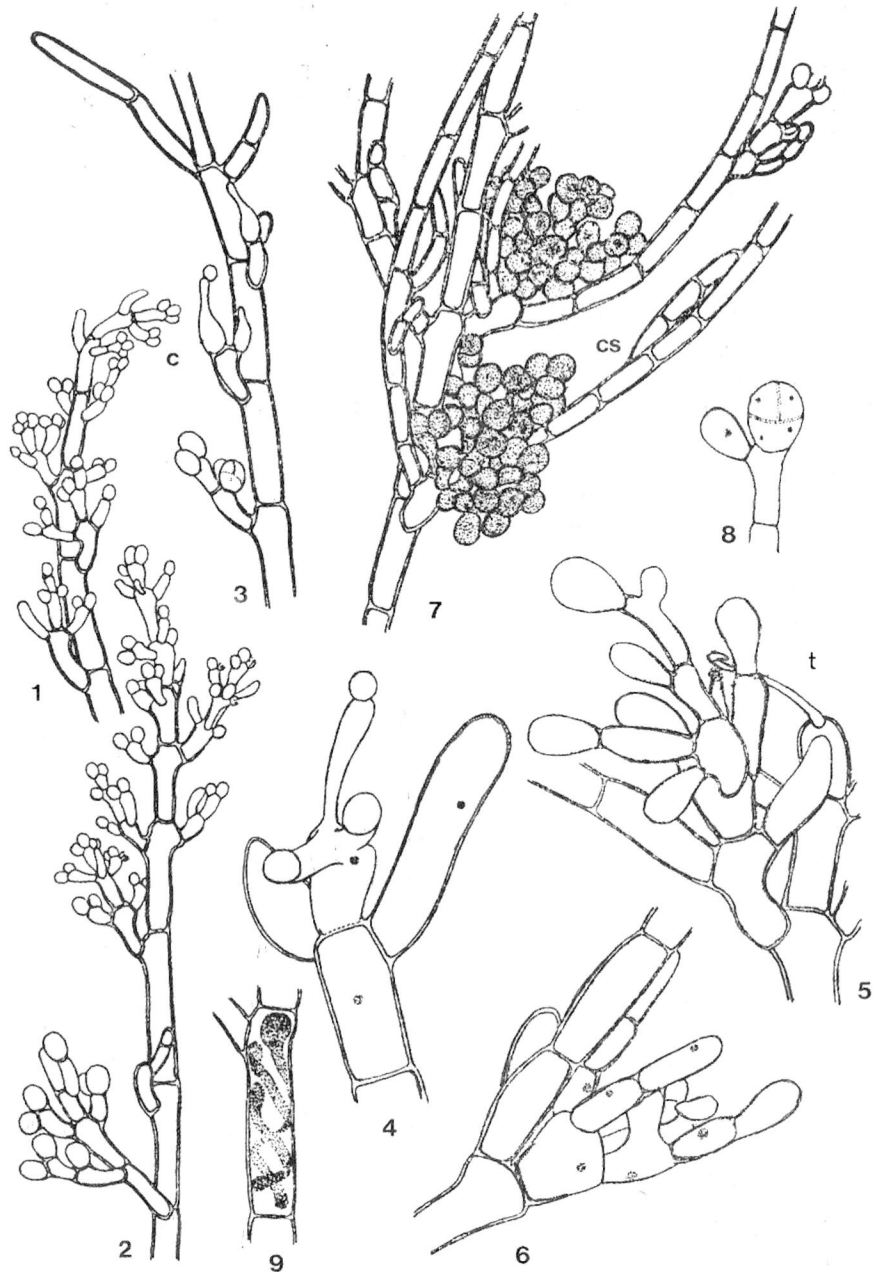

Plate 20
Audouinella hermannii as *Rhodochorton violaceum* (Kützing) Drew, figs 1–9 (Drew 1935)
1. an apical portion of a male plant, 2. a filament bearing both spermatangia and monosporangia, 3. a female plant with carpogonia and also monosporangia, 4. a mature sessile carpogonium with two spermatia adhering to a trichogyne, 5. the fertilized carpogonium, an apical portion of the basal portion has begun to swell, a trichogyne beginning to decay, 6. spermatia adhering to a trichogyne, two gonimoblast filaments developing at an apical end of a basal portion of a carpogonium, two branches have started to grow out from the cell below carpogonium, 7. the later stage in the development of a carposporophyte, showing three gonimoblast filaments, the shrinking trichogyne, 8. monosporangium (tetrasporangium), 9. spiral ribbon-shaped chloroplasts, c: carpogonium, cs: carposporangium, t: trichogyne.

Acrochaetiales

3b) *Audouinella eugenea* (Skuja) Jao var. *secundata* Jao (1941:254, pl. 3, figs 21–22)

Plants caespitose, usually hemispherical in outline, up to 3 mm high, dark steel-blue, composed of prostrate and erect filaments; prostrate filaments composed of short cells, irregularly branched, laterally united at the central portion into a pseudoparenchyma; erect filaments elongate, abundantly and alternately branched; branches erect-patent, slightly attenuated towards the end, terminal hairs absent; cells cylindrical, 9–12 μm in diameter, 30–63 μm long, each containing 3–5, slightly lobed and laminate chloroplasts; pyrenoid absent; monosporangium-bearing branchlets 1–5 (usually 2–3) cells long, scattered or opposite, very rarely verticillate; monosporangia sessile or with 1–celled short lateral branchlets, 1–2 on each cell of monosporangium-bearing short lateral branchlets, secund or terminal, ovate, 9–14 μm in diameter, 16–20 μm long (Jao 1941).

Type Locality: Chin-liang Temple, Sin-tien-tze, Pa-hsien, Szechwan, China in Eastern Asia. Growing with *Audouinella cylindrica* Jao on submerged root and rocks in a slowly-running mountain stream emanating from a limestone cave.

Holotype: SK 9.

Distribution: Known from the type locality only.

Note: The Chinese plants are undoubtedly closely allied to *A. eugenea* (Skuja) Jao, but are distinguished from the latter chiefly by the plants being regularly hemispherical in outline, by having unilaterally arranged and narrower monosporangia, and by having monosporangia restricted to branchlets that are oppositely branched

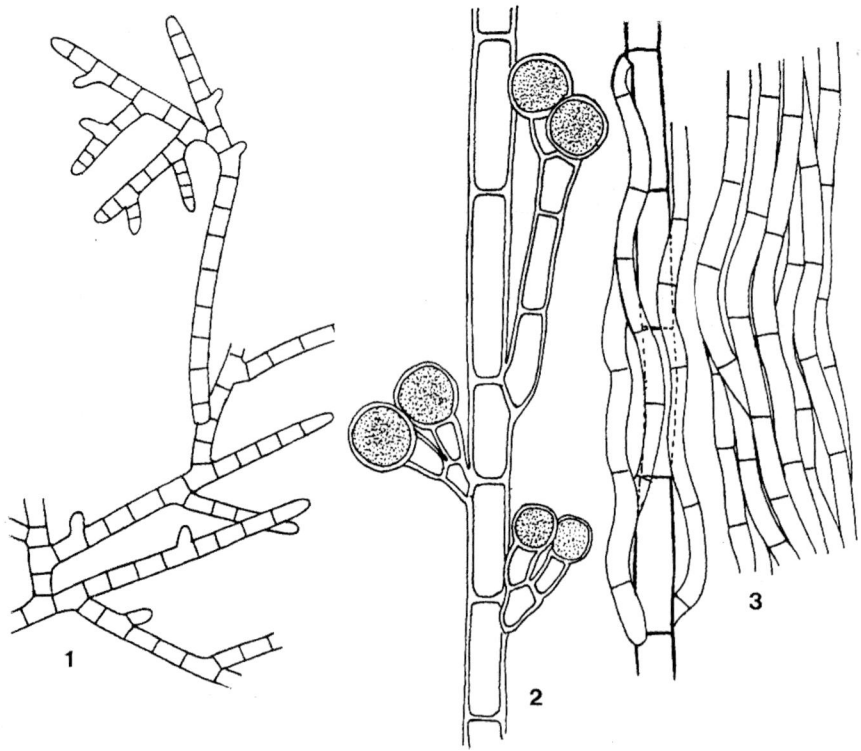

Plate 21
Audouinella serpens as *Pseudochantransia serpens*, fig. 1 (original author, Israelson 1942)
1. a portion of a plant, showing the habit of ramification.

Audouinella macrospora as *Chantransia macrospora*, figs 2–4 (Starmach 1977)
2. a filament with short lateral branchlets bearing monosporangia, 3–4. rhizoidal filaments.

4) *Audouinella hermannii* (Roth) Duby in De Candolle (1830:972)
Basionym: *Conferva hermannii* Roth (1806:180),
Synonym: *Chantransia hermannii* (Roth) Desv. (1809: 310); *Ch. violacea* Kützing (1845:231); *Audouinella miniata* Bory (1823a:341), *A. violacea* (Kützing) Hamel (1925:46); *Rhodochorton violaceum* (Kützing) Drew (1935: 439, figs 1–19)

> plate 19, figs 1–4 (Starmach 1977) as *Chantransia hermannii*
> plate 20, figs 1–9 (Drew 1935) as *Rhodochorton violaceum* (Kützing) Drew
> plate 24, figs 3–7 (Necchi, Sheath & Cole 1993b) as *Audouinella hermannii*
> plate 25, fig. 1 (Necchi & Zucchi 1995) as *Audouinella hermannii*
> plate 197, figs 1–2 (Entwisle & Kraft 1984)

Plants macroscopic, up to 15 mm long, and composed of > 50 cells, reddish; basal portion consisting of an irregular prostrate system of densely aggregated filaments; lateral branches developing at angles ≥ 25°; vegetative cells of main branches cylindrical, 11.0–16.0 m in diameter, 35.0–67.0 μm long; monosporangia obovoidal to subspherical, 8.5–11.0 m in diameter, 9.0–12.0 μm long; carpogonium, tetrasporangia and carposporophytes not observed. (Necchi & Zucchi 1992)

Plants 1–3(–5) mm high, red sometimes with a brownish tinge, solitary or in dense groups. Plants composed of a prostrate system of irregularly branched, interweaving threads and of erect, branched filaments forming penicilliform plants. Ramification alternate or sometimes opposite, typically regularly monopodial, sometimes more irregular but always abundant. Branches and branchlets mostly ramifying at an angle of about 45 degrees. Main filaments of cylindrical cells, (7–)9–12(–15) μm long, 4–7 μm in diameter, in sexual plants thinner, 7–10 μm in diameter, 2.5–5 times longer than broad. Cells with a central nucleus and parietal chloroplasts, often dissected into spirally arranged ribbons. Walls 1.7–3 μm thick, in old specimens often stratified. Terminal hairs absent to abundant, varying in length. Monosporangia always (?) present, often abundant, 1–2(–3) together terminating short lateral branchlets of 2–4 cells, usually equally distributed among extensive parts of the branches, ellipsoidal, 7–10 μm in diameter and 8–13(–16) μm long, smaller in sexual specimens than in asexual ones. Tetrasporangia single, terminating the short lateral branchlets, ellipsoidal, 9–10 μm in diameter and 10–13 μm long. Polyoecious but with bisexual plants being much rarer than unisexual ones. Spermatangia 1–4 together terminating branchlets, of very varying abundance, ellipsoidal or almost cylindrical, 3.5–4 μm in diameter and 4.5–5.5 μm long. Carpogonia sessile or usually terminating branchlets, bottle-shaped, 4–6 μm in diameter at the basal portion and 9–13.5 μm long. Trichogynes slightly broader at the distal end and before fertilization distinguished from the terminal hairs by the absence of a wall at the basal end. Gonimoblasts more or less regularly globular, about 50–70 μm in diameter. Carposporangia ellipsoidal, inversely ovoid or pear-shaped, 9–11 μm in diameter and 10–15 μm long (as *Rhodochorton violaceum*, Drew 1935).

Type Locality: near Bremen, Germany in Europe.
Neotype: B 28258 designated by Necchi, Sheath & Cole 1993b.
Distribution: Known from the type locality and in streams, usually together with *Lemanea* and *Batrachospermum* in River Goyt in UK, Belgium and Sweden in Europe; Australia; British Columbia, Canada (Hymes & Cole 1983); Connecticut (Hylander 1928, Wolle 1887), USA in North America; São Roque de Minas, Mexico (Montejano-Zurita DATE) in Central America; São Roque de Minas, Serra da Canastra National Park, Minas Gerais, Brazil in South America (Necchi & Zucchi 1995).

Acrochaetiales

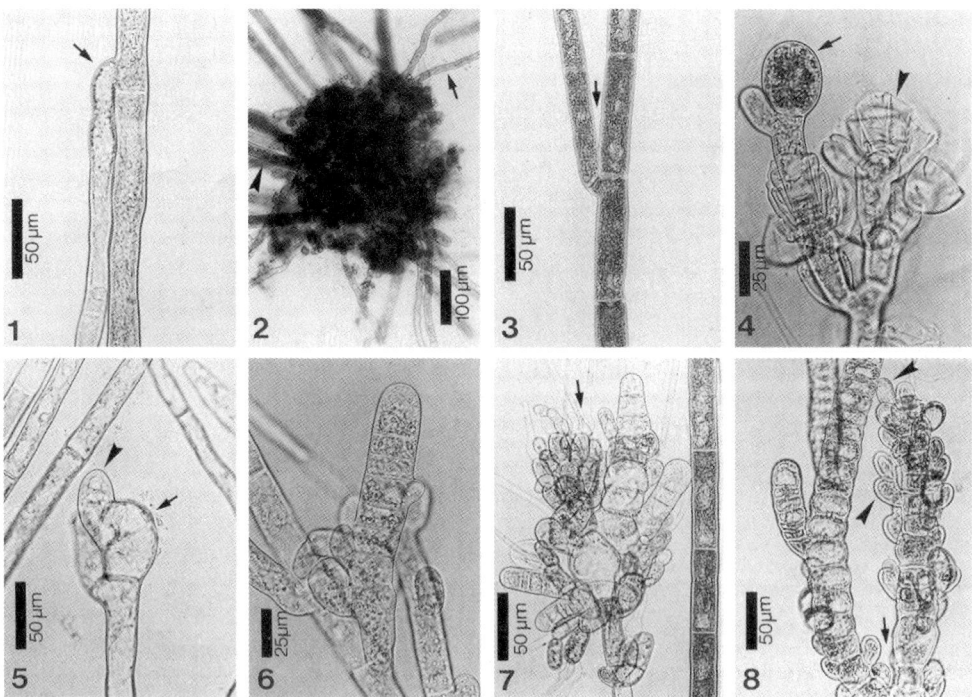

Plate 22
Audouinella macrospora as chantransia phase of the genus *Batrachospermum*, figs 1–8 (Necchi & Zucchi 1997)
1. basal portion of a filament showing well-developed rhizoid (arrow), 2. erect filaments (arrowhead) arise from basal system consisting of a prostrate mass with sparse rhizoids (arrow), 3. filament with a narrow branching angle (arrow), 4. monosporangial short lateral branchlets with a mature monosporangium (arrow) and an empty, discharged monosporangium (arrowhead), 5. an apical cell of a *Audouinella* filaments after division, showing a larger cell (arrow) and a smaller elimination cell (arrowhead), 6. early stage in the production of a juvenile *Batrachospermum* gametophyte from cells of the *Audouinella* filaments, 7. later stage in the production of a juvenile *Batrachospermum* gametophyte showing hairs (arrow), 8. juvenile gametophyte with a branch (arrow) and young fascicle cells (arrowheads)

5) ***Audouinella lanosa*** Jao (1941:256, pl. 3, figs 16–17)

plate 28, figs 2–3 (original author, Jao 1941)

Plants up to 3 mm high, densely pulvinate, widely expanded, woolly, dark purple, composed of prostrate and erect filaments; branches elongate, lower ones longer than upper ones, erect-patent; cells cylindrical, 9–12 μm in diameter, 42–72 μm long; terminal cells slightly attenuated, with an obtuse apex, sometimes piliferous; hairs usually terminating the monosporangium-bearing and the vegetative ramuli, colourless, elongate, up to 350 μm long; short monosporangium-bearing lateral branchlet, 1–5 cells long, alternate or scattered; monosporangia sessile or with a unicellular stalk, opposite or secund, usually seriate, obovoidal, 10–14 μm in diameter, 14–17 μm long (Jao 1941).
Type Locality: Kan-tung-tze, about five miles southeast from Pehpei, Szechwan, China in Eastern Asia. On the concrete wall of a mill dam.
Holotype: SC 1146.
Distribution: Known from the type locality only.

The following gray-blue (steel-coloured) or violet-brown colour (not distinctly red in colour) plants may not represent *Audouinella* but the chantransia phase of the *Batrachospermum*.

1. Lateral branches protruding almost at angle of 90 degree..................................1) *A. serpens*
1. Lateral branches departing at an acute angle...2
2. Monosporangia reaching over 30 μm in diameter...2) *A. macrospora*
2. Monosporangia smaller...3
3. Main branching opposite or alternate, in one plane.............................3) *A. subtilis*
3. Main branching in various planes...4
4. Main filaments up to 15 μm in diameter...5
4. Main filaments up to 10 μm in diameter...6
5. Cells 1–3 times as long as wide..4) *A. pygmaea*
5. Cells 3–7 times as long as wide..5) *A. chalybaea*
6. Plants up to 2 mm high, cells 8–10 μm in diameter, 16–42 μm long.........6) *A. sinensis*
6. Plants about 0.5 mm high, cells 4–7 μm in diameter, 11.7–13.5 μm long, monosporangia, 7–9 μm in diameter, 13–18 μm long. Growing epiphytically on *Vallisneria*...7) *A. cylindrica*
6. Plants small, cells 4–9 μm in diameter, 2–4 times as long, monosporangia 8–12.6 μm in diameter, 11.7–14.4 μm long...................8) *A. glomerata*

1) ***Audouinella serpens*** (Israelson) Sheath (1984:90) in Kumano, *comb. nov.*
Basionym: *Pseudochantransia serpens* Israelson (1942: 59, pl. 1b)

plate 21, fig. 1 (original author, Israelson 1942) as *Pseudochantransia serpens*

Plants bluish green, a loose, cobweb-like coat on mud, consisting of non-adherent filaments up to 5 mm long, straight, very irregularly branched, ramifying at an angle of almost 90 degree. Cells cylindrical, 16–18 μm in diameter and 30–60(–90) μm long. Terminal cells with rounded apex. Terminal hairs absent; old portions of the filaments disintegrating successively. Monosporangia and methods of propagation unknown. Cells with a single, central nucleus and parietal chloroplasts, these often dissected into irregularly spiral-shaped ribbons.
Type Locality: Jämtland, paroec. Hällesjö, in lac Våntjärn, Sweden in Europe. Known from the littoral zone of lakes.
Holotype: Mus. Bot. Upsaliens UPS.
Distribution: Known from the type locality only.

2) ***Audouinella macrospora*** (Wood) Sheath et Burkholder (1985:111)
Basionym: *Chantransia macrospora* Wood in Wolle (1887:59, pl. 59, fig. 1)

plate 21, figs 2–3 (Starmach 1977) as *Chantransia macrospora*
plate 22, figs 1–8 (Necchi & Zucchi 1997) as chantransia phase of *Batrachospermum*
plate 23, figs 3–4 (Necchi, Sheath & Cole 1993c) as *Audouinella macrospora*
plate 25, figs 2–3 (Necchi & Zucchi 1995) as *Audouinella macrospora*

Plants macroscopic, up to 30 mm long, and composed of > 50 cells, bluish; basal portion consisting of well-developed rhizoids; lateral branches developing at angles of < 25°; vegetative cells of main branches cylindrical, 10.5–36.0 μm in diameter, 33.5–108.5 μm long; monosporangia spherical to obovoidal, single or in pairs, 15.5–28.5 μm in diameter, 18.5–38.0 μm long; carpogonia, tetrasporangia and carposporophytes not observed (Necchi & Zucchi 1995).

Acrochaetiales

Type Locality: Aiken, South Carolina, USA in North America.
Lectotype: PH designated by Necchi, Sheath & Cole 1993c.
Distribution: Known epiphytically from about 13 localities in USA in North America (Necchi, Sheath & Cole 1993c); Costa Rica in Central America; about 27 localities in Brazil in South America (Necchi & Zucchi 1995).

Note: Necchi & Zucchi (1997) found that in some North American populations, plants initially identifiable as *Audouinella macrospora* ultimately produced juvenile *Batrachospermum* gametophytes, therefore, those were interpreted as chantransia-phase of *Batrachospermum*. This result raises the possibility that some bluish freshwater plants such as *A. macrospora* are the chantransia phase of the genus *Batrachospermum*.

3) ***Audouinella subtilis*** Jao (1941:253)
Synonym: *Chantransia subtilis* Möbius (1894:309) non *C. subtilis* (C. Agardh) Steudel

Plants 1–3 mm high, caespitose, gray-blue-green or olive-green, main filaments branch in one plane, opposite or alternate, constricted at cross walls; cells 6–10 µm in diameter; monosporangia on short lateral branches, and monosporangia of a different form.
Type Locality: Burpengary, Brisbane in Australia. Growing epiphytically on *Nitella*.
Holotype:
Distribution: Known from the type locality only.

4) ***Audouinella pygmaea*** (Kützing) Weber-van Bosse (1921:191)
Basionym: *Chantransia pygmaea* Kützing (1843:285)
Synonym: *Pseudochantransia pygmaea* (Kützing) Brand (1909:118); *Chantransia leibleinii* Kützing (1845:229); *Pseudochantransia leibleinii* (Kützing) Israelson (1942: 58, Tab. 1a); *Audouinella leibleinii* (Kützing) Palmer in Hirsch et Palmer (1958:378)

plate 23, figs 1–2 (Necchi, Sheath & Cole 1993c) as *Audouinella pygmaea*
plate 25, figs 8–10 (Necchi & Zucchi 1995) as *Audouinella pygmaea*
plate 26, fig. 2 (Israelson 1942);
figs 1, 3–4 (original author, Kützing 1845) as *Chantransia leibleinii*
plate 27, fig. 1 (Starmach 1977) as *Chantransia pygmaea*;
fig. 2 (original author, Kützing 1843) as *Chantransia pygmaea*

Plants macroscopic, up to 20 mm long, and composed of > 50 cells, bluish, basal portion consisting of an irregular prostrate system of densely aggregated filaments; lateral branches developing at angles < 25°; vegetative cells of main branches cylindrical, 5.5–15.5 µm in diameter and 14.0–66.5 µm long; monosporangia obovoidal to subspherical, 6.0–15.0 µm in diameter and 8.0–25.0 µm long. Carpogonia, tetrasporangia and carposporophytes not observed. (Necchi & Zucchi 1995, as *pygmaea*).
Type Locality: Schleusingen am Thübingen Walde
Lectotype: L herb. Kützing, designated by Necchi, Sheath & Cole 1993c.
Distribution: Known from Sweden in Europe and North America in streams.
Note: Israelon (1942) stated that metamorphose into *Batrachospermum* phase are not frequently observed and have not been observed for *Pseudochantransia leibleinii* (Kützing) Israelson.

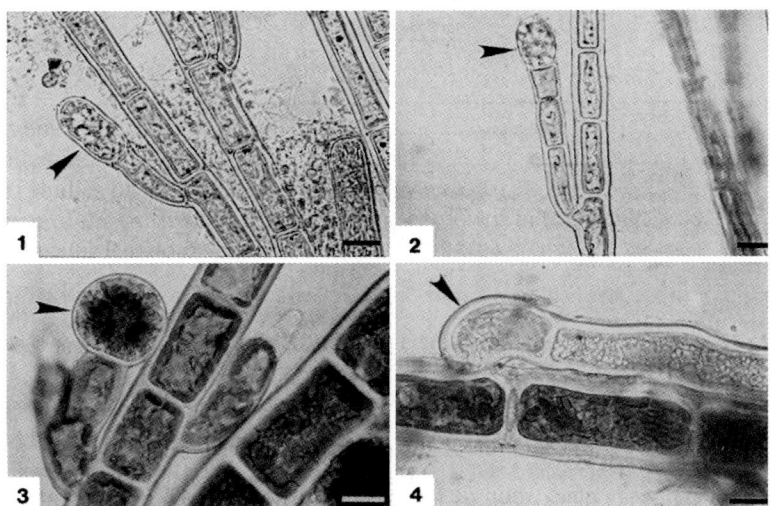

Plate 23
Audouinella pygmaea, figs 1–2 (Necchi, Sheath & Cole 1993c)
1. a plant with ellipsoidal monosporangium (arrowhead), 2. a plant with ovoidal monosporangium (arrowhead).

Audouinella macrospora, figs 3–4 (Necchi, Sheath & Cole 1993c)
3. a plant with spherical monosporangium (arrowhead), 4. basal portion with distinctive, colourless rhizoid (arrowhead). Scale bar = 10 µm (figs 1–4).

5) ***Audouinella chalybea*** (Roth) Bory (1823a:340)
Basionym: *Conferva chalybea* Roth (1806: 286, Tab. 8, fig. 2),
Synomyn:*Trentepohlia pulchella* ß *chalybea* C. Agardh (1824:37); *Trentepohlia aeruginosa* C. Agardh (1824:38); *Chantransia chalybea* (Lyngb.) Fries (1825:338); *Pseudochantransia chalybaea* (Roth) Brand (1909:118)

plate 26, figs 5–6 (Kützing 1845) as *Chantransia chalybea*

Plants penicilliform, solitary or united into extensive carpeting, 2–5 mm high, bright green with a bluish tinge or grayish green, in dull light sometimes darker and more brownish. Ramification more or less abundant, irregular, alternate or opposite, sometimes unilateral in series; branches mostly addressed and frequently reaching the height of the main filaments. Main filaments more or less inconspicuous; cells (6–)8–12(–13) µm in diameter and 30–80 µm long, 3–7 times longer than broad. Terminal cells with rounded, rarely accumulate apex. Cells uninucleate, with parietal chloroplasts, often dissected into spirally arranged ribbons; wall thin, 0.8–1.3(–1.8) µm. Terminal hairs generally absent, rarely abundant. Monosporangia abundant or sparse, sometimes absent, 1–2 together terminating short lateral branchlets consisting generally of 4–7 cells, unequally distributed often restricted to certain portion of thallus, almost globular-ellipsoidal, (7.5–)8–11 µm in diameter and 9–12(–13) µm long. (after Israelson 1942 as *Pseudochantransia chalybea*)

Type Locality: near Leesum, Bremen, Germany in Europe.
Holotype:
Distribution: Known from the type locality and in middle Sweden, Belgium in Europe; in Chien-yuen-shan, Pehpei, Szechwan, China in East Asia. In springs and ß-mesosaprobic streams.
Note: Israelon (1942) stated that metamorphose into *Batrachospermum* phase occur, not only rarely.

Acrochaetiales

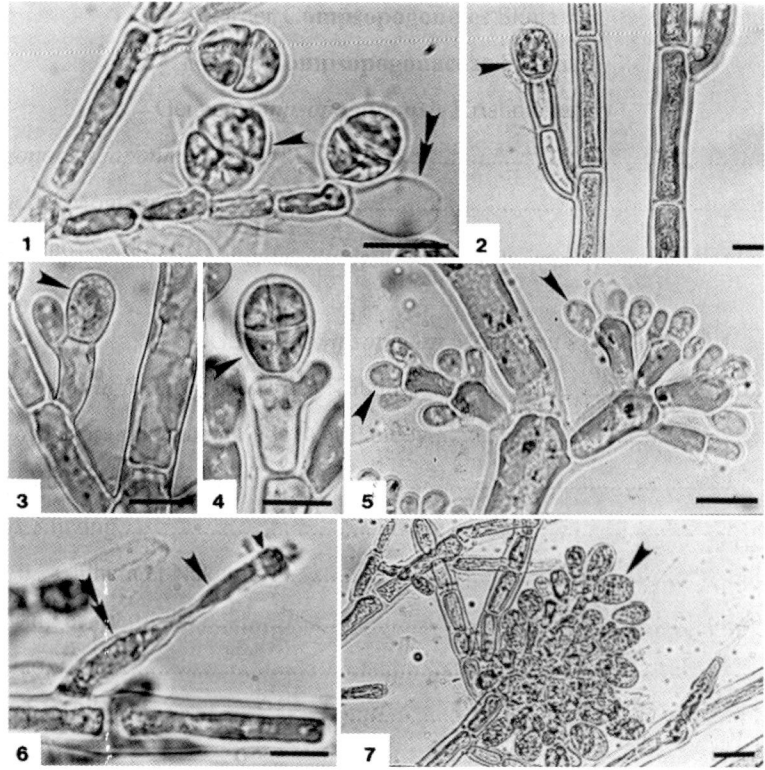

Plate 24

Audouinella tenella, fig. 1 (Necchi, Sheath & Cole 1993b)
1. plant with tetrasporangia (arrowhead) and empty tetrasporangium (double arrowhead)

Audouinella eugenea, fig. 2 (Necchi, Sheath & Cole 1993b)
2. plant with empty tetrasporangium (arrowhead)

Audouinella hermannii, figs 3–7 (Necchi, Sheath & Cole 1993b)
3. monosporangium (arrowhead), 4. tetrasporangium (arrowhead), 5. spermatangia (arrowheads), 6. a fertilized carpogonium with a slender trichogyne (arrowhead), swollen base (double arrowhead) and attached a spermatium (small arrowhead), 7. a carposporophyte with carposporangia (arrowhead). Scale bar = 10 µm.

6) ***Audouinella sinensis*** Jao (1940:241, pl. 1, figs 5–6)

plate 27, figs 3–4 (original author, Jao 1940)

Plants up to 2 mm high, densely caespitose or pulvinate, dark olive, composed of prostrate and erect filaments; prostrate filaments short, irregularly branched, more or less constricted at the cross walls, erect filaments elongate, alternately or irregularly branched, lower branches much longer than upper ones; ramuli usually very short, secund or scattered; cells cylindrical, 8–10 µm in diameter, 16–42 µm long, 2–4 times longer than diameter, not constricted at the joints; cell wall thin; terminal hairs absent; chloroplasts parietal, 2–4 in each cell, laminate with an irregularly undulate margin, pyrenoids absent; monosporangia with a one-celled short lateral branchlet, 2–3 on each short lateral branchlet, very rarely sessile, opposite or rarely secund or verticillate on each cell of a short monosporangium-bearing lateral branchlet, seriate, ovate, 7–8 µm in diameter, 9–13.5 µm long.

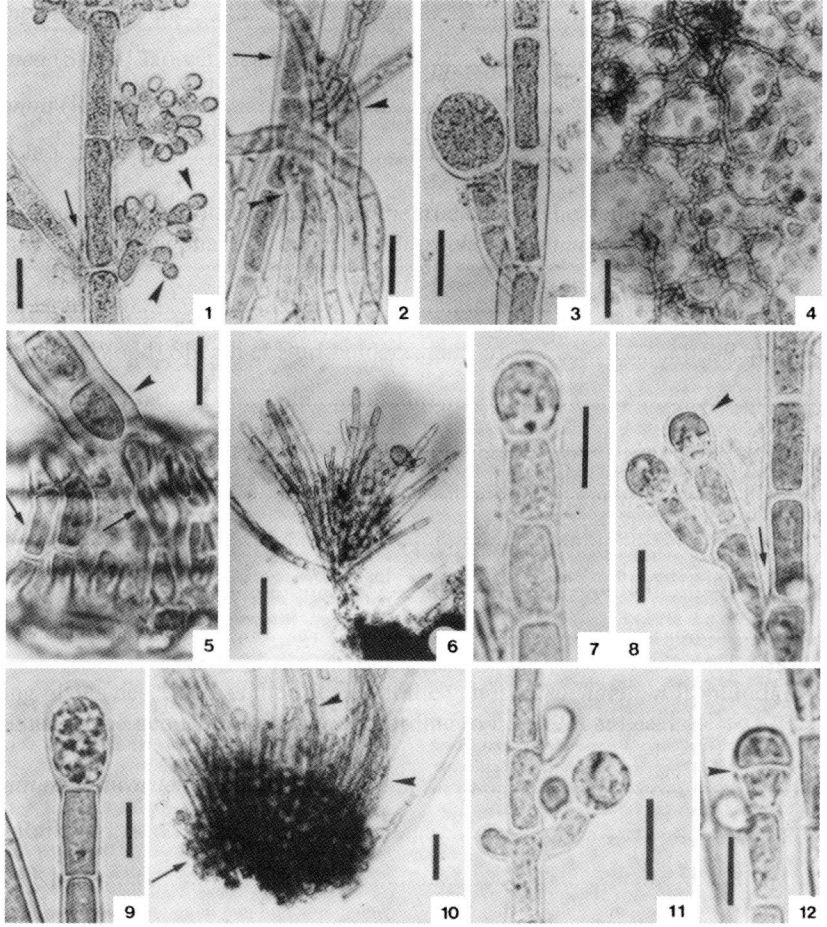

Plate 25
Audouinella hermannii, fig. 1 (Necchi & Zucchi 1995)
1. filament showing branch with angle (25° (arrow) and monosporangia in short lateral branchlets (arrowheads)

Audouinella macrospora, figs 2–3 (Necchi & Zucchi 1995)
2. basal portion of a filaments (arrow) with a rhizoid (arrowhead) and a bundle of rhizoids (double arrowhead), 3. filaments with a monosporangial short lateral branchlet and a subspherical monosporangium.

Balbiania meiospora as *Audouinella meiospora,* figs 4–7 (Necchi & Zucchi 1995)
4. basal system consisting of creeping filaments epiphytic on *Compsopogonopsis leptoclados,* 5. creeping filaments (arrows) developing epiphytically around a filament of *A. macrospora* with an erect filaments (arrowhead), 6. a tuft of erect filaments, 7. a monosporangial short lateral branchlet with a monosporangium.

Audouinella pygmaea, figs 8–10 (Necchi & Zucchi 1995)
8. filaments showing branch angle < 25° (arrow) and monosporangia (arrowhead), 9. monosporangial short lateral branchlets with an ellipsoidal monosporangium, 10. basal system consisting of an irregular prostrate mass (arrow) with erect filaments (arrowhead).

Audouinella tenella, figs 11–12 (Necchi & Zucchi 1995)
11. filaments with a sporangial short lateral branchlet and an undivided spherical tetrasporangium, 12. a sporangial short lateral branchlet with a tetrasporangium after first division (arrowhead).
Scale bars = 50 μm (figs 5, 4, 6, 10); 20 μm (figs 1–3); 10 μm (figs 5, 7–9, 11–12).

Acrochaetiales

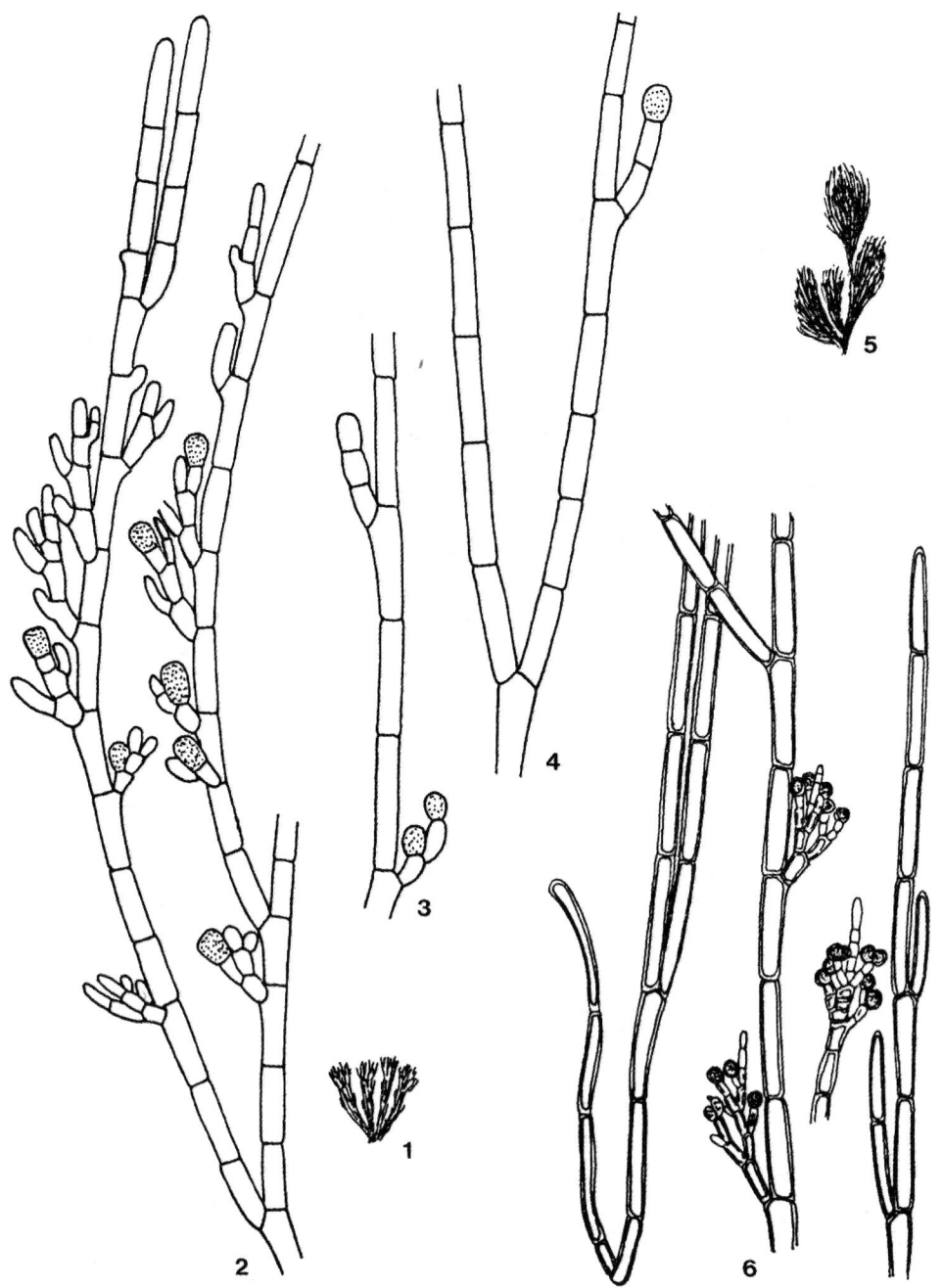

Plate 26
Audouinella leibeinii as ~~*Chantransia leibleinii*~~, figs 1, 3–4 (original author, Kützing 1845), fig. 2 (Israelson 1942)
1. the habit of a plant showing ramification, 2. filaments with normal short lateral branchlets bearing monosporangia, 3–4. filaments with abnormal short lateral branchlets bearing a few monosporangia.

Audouinella chalybea as *Chantransia chalybea*, figs 5–6 (Kützing 1845)
5. the habit of a plant showing ramification, 6. filaments with short lateral branchlets bearing monosporangia in clusters.

Audouinella

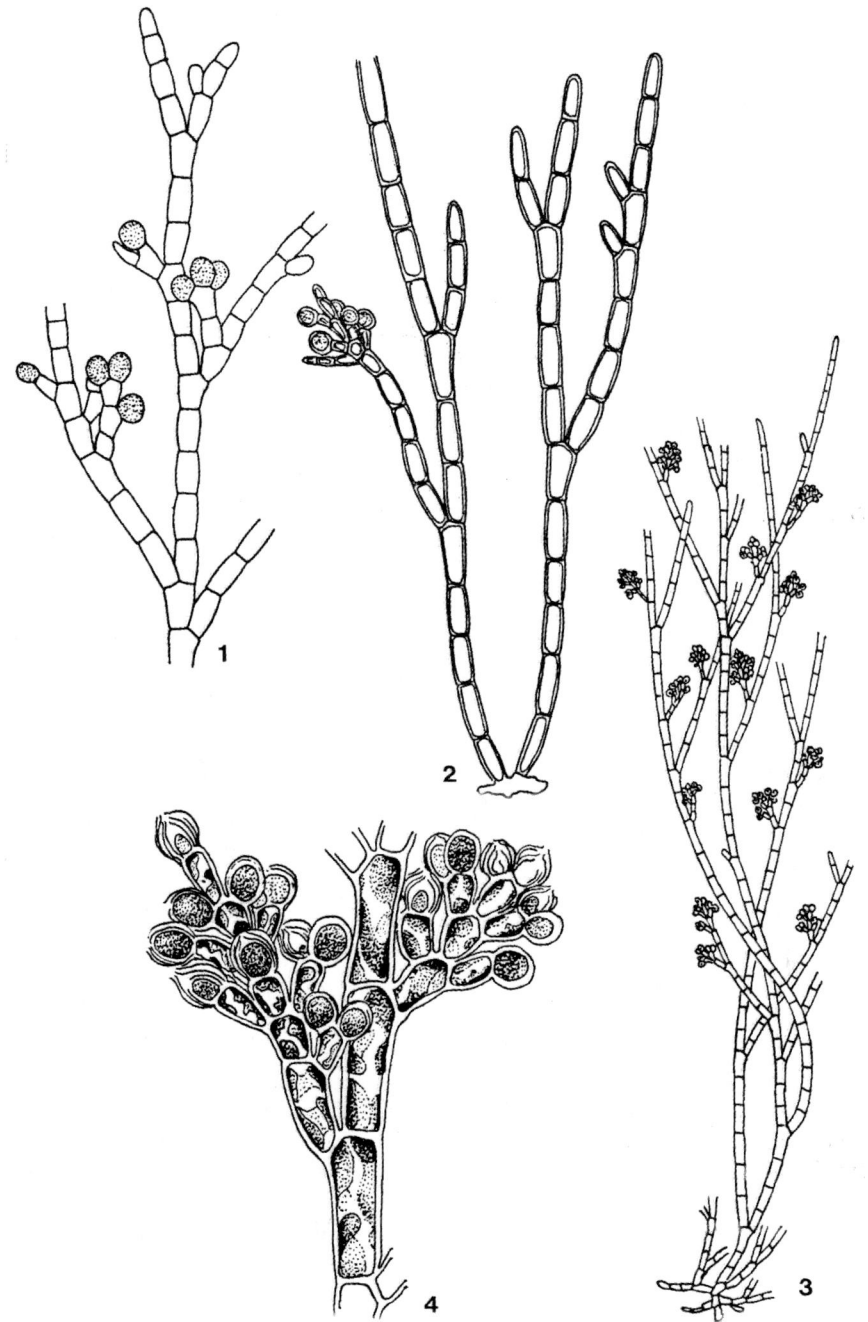

Plate 27
Audouinella pygmaea as *Chantransia pygmaea*, fig. 1 (Starmach 1977), fig. 2 (original author, Kützing 1843)
1–2. filaments with short lateral branchlets bearing monosporangia.

Audouinella sinensis, figs 3–4 (original author, Jao 1940)
3. a portion of a plant showing the habit of growth and short lateral branchlets bearing monosporangia,
4. a filament with short lateral branchlets bearing monosporangia in clusters.

Acrochaetiales

Type Locality: Nanyoh, Hunan, China in Eastern Asia. On rocks in a pool, well shaded by trees.
Holotype: HN121:B, Nov., 1937.
Distribution: Known from the type locality only.

7) *Audouinella cylindrica* Jao (1941:252, pl. 3, fig. 15)

plate 28, fig. 1 (original author, Jao 1941)

Plants very small, up to 0.5 mm tall, hemispherical in outline, blue-green, epiphytic on aquatic plants such as *Vallisneria spiralis*. Plants composed of prostrate and erect filaments; prostrate filaments irregularly branched; erect filaments radiately arranged, alternately branched distally; branches erect, usually much elongated; apical cells round or very rarely slightly attenuated; terminal hairs absent; cells cylindrical, 4–7 µm in diameter, 11.7–13.5 µm long; chloroplasts laminate, with an irregularly sublobed margin, 2–4 in each cell; pyrenoids absent; monosporangium-bearing laterals usually many on the upper parts of the branches, scattered or very rarely opposite; monosporangia cylindrical-ovate 7–9 µm in diameter, 13–18 µm long; sessile or on 1–celled short lateral branchlet, 1 or 2 on each cell of short lateral branchlet, usually terminal (Jao 1941).
Type Locality: near Yangso, Kwangsi, China in Eastern Asia. Epiphytic on the leaves and stems of *Vallisneria spiralis* in streams.
Holotype: KS9:B, Feb. 1938.
Distribution: Known from the type locality only.

8) *Audouinella glomerata* Jao (1941:254, pl. 3, figs 18–20)

plate 28, figs 4–5 (original author, Jao 1941)

Plants minute, dark steel-blue, formed by prostrate and erect filaments; prostrate filaments laterally branched, composed of short, more or less tumid cells, 1.5–2 times longer than diameter; erect filaments numerous, densely pulvinate, with few branches or none; very rarely with ramuli; branches elongate; cells cylindrical, 7–9 µm in diameter, 16–36 µm long; ends of primary filaments and branches gradually attenuated into false terminal hairs; true hyaline terminal hairs absent; chloroplasts elliptic-discoid, with irregular, slightly lobed margin, 2–4 in each cell; monosporangium-bearing short lateral branchlets, usually 1–3, rarely more than 3 cells long, scattered, opposite, or whorled, cells shorter than those of primary filaments and branches; monosporangia sessile or with 1–celled stalk, (1–)2–4 on each cell of monosporangium-bearing short lateral branchlets, usually corymbose-glomerately arranged, ovate, 8–12.6 µm in diameter, 11.7–14.4 µm long.
Type Locality: near North Hot Spring Park, Pehpei, Szechwan, China in Eastern Asia. On rocks and submerged plants in a mountain stream.
Holotype: SC 1080:B.
Distribution: Known from the type locality only.
Note: This species should be compared with *A. hermannii* (Roth) Duby, but differs from the latter in having less branched filaments, vegetative branchlets nearly wanting, corymbose-glomerately arranged monosporangia, and especially in elongate false terminal hairs (Jao 1941).

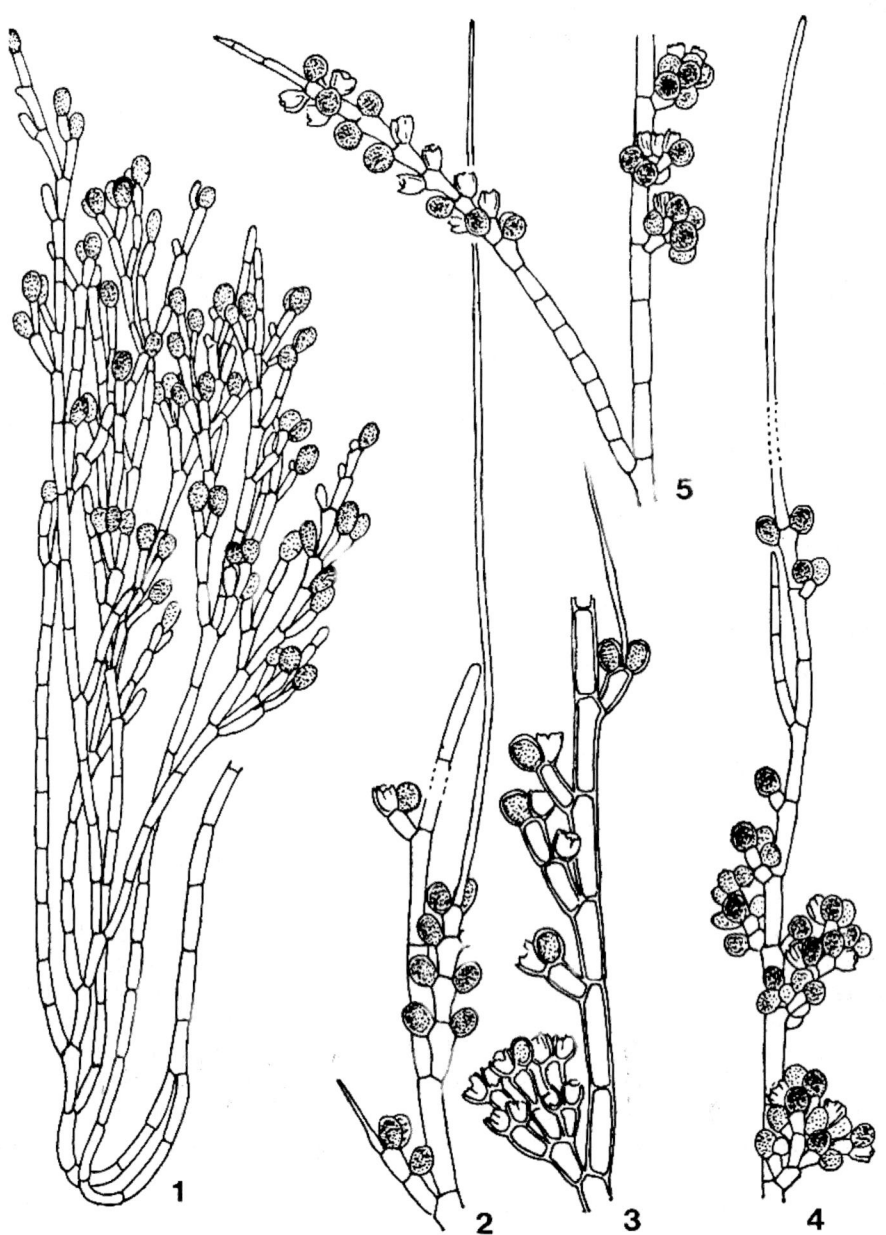

Plate 28
Audouinella cylindrica, fig. 1 (original author, Jao 1941)
1. an upper portion of a filament showing ramification and short lateral branchlets bearing monosporangia.

Audouinella lanosa, figs 2–3 (original author, Jao 1941)
2–3. a portion of a filament showing ramification, short lateral branchlets bearing monosporangia and hairs.

Audouinella glomerata, figs 4–5 (original author, Jao 1941)
4. a portion of a filament showing glomerately arranged monosporangia and false hairs, 5. a portion of filament showing three typical short lateral branchlets and an abnormal short lateral branchlet.

Acrochaetiales

The following species of *Acrochaetium* described from India and Japan may be referable to *Audouinella* according to the taxonomic system proposed by Y. P. Lee & I. K. Lee (1988), but original material has not been available for study.

1. Cells with pyrenoid...1) *Acrochaetium godwardense*
1. Cells without pyrenoid..2
2. Monosporangia terminal, 14.5–15 µm in diameter...3) *A. indica*
2. Monosporangia terminal, 6–12 µm in diameter...4) *A. sarmae*

1) ***Acrochaetium godwardense*** Patel (1970:33, figs 1–12)

plate 29, figs 1–12 (original author, Patel 1970)

Plants heterotrichous, in dense clusters, up to 4.5 mm long, attached to substratum by prostrate system. Branches alternate, rarely opposite or secund, with cylindrical cells. Main filaments 13.0–15.0 µm in diameter, 3–5 times as long as diameter. Cells of branches 7.5–11.3 µm in diameter, cell wall 1.9–3.77 µm thick. Cells containing a nucleus and single lobed parietal chloroplast with mostly one or two pyrenoids. Terminal hairs absent. Reproduction only by monospores. Monosporangia single or paired on branched or unbranched short lateral branchlets, oblong, 16.0–17.0 µm in diameter, 18.0–24.3 µm long. Liberation of monospores through apical dissolution of the sporangial wall. Monosporangia spherical or hemispherical and thin walled, 13.0–16.0 µm in diameter, 1.–0–20.0 µm long.

Type Locality: Balaram River, near Palanpur, Gujarat State, India in Asia. Attached to walls along the banks.

Holotype: Patel 645, Sardar Patel University.

Distribution: Known from the type locality only.

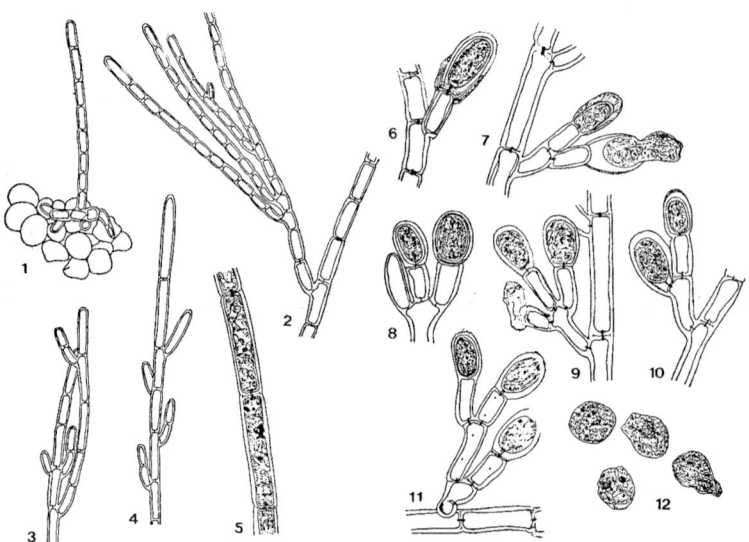

Plate 29
Acrochaetium godwardense, figs 1–12 (original author, Patel 1970)
1. young plant developed from monospore on *Compsopogon* showing the habit of the prostrate and erect filament, 2–4. showing the habit of ramification, 5. portion of the filaments showing detailed cell structure, 6–11. showing monosporangia on different types of short lateral branchlets, 7. showing liberation of monospore, 12. monospores.

2) ***Acrochaetium indica*** Raikwar (1962:102, figs 1–18)

plate 30, figs 1–10 (original author, Raikwar 1962)

Plants heterotrichous and densely tufted, attached to substratum by prostrate system of filaments 10.8 μm in diameter, erect filaments consisting of long cells, 7.2–7.6 μm in diameter, 54–72 μm long. Cells containing one or occasionally more nuclei and a parietal and lobed chloroplast without pyrenoids; terminal hairs absent. Reproduction by monospores. Monosporangia of two types; terminal and oblong, 14.5–15 μm in diameter, 20.5–26 μm long, and lateral and obovoidal, 21.6–23.4 μm in diameter, monospores formed singly in sporangia and liberated at maturity by apical dissolution of sporangium wall.

Type Locality: Anarkali Tal, Bahraich District, Uttar Pradesh, India in Asia. On dead aquatic plants.

Holotype: Banaras Hindo University.

Distribution: Known from the type locality only.

3) ***Acrochaetium sarmae*** Khan (1970:250, figs 1–6)

plate 31, figs 1–6 (original author, Khan 1970)

Plant heterotrichous and densely tufted, attached to substratum by prostrate system of document creeping filaments, 10–12 μm in diameter, erect filaments consisting of long cells, 10–12 μm in diameter, 40–60 μm long. Cells containing a single nucleus and a single, lobed, parietal chloroplast without a pyrenoid. Monosporangia of two types; terminal and oblong, 6–12 μm in diameter, 9–18 μlong, and lateral and obovoidal, 10–12 μm in diameter, monospores formed singly in sporangia and liberated at maturity by apical dissolution of sporangial wall. Reproduction only by monospores.

Type Locality: on moss in outlet of Dehradum Water Works, Dehradum, India in Asia.

Holotype: D. A. V. Post graduate College.

Distribution: Known from the type locality only.

Acrochaetiales

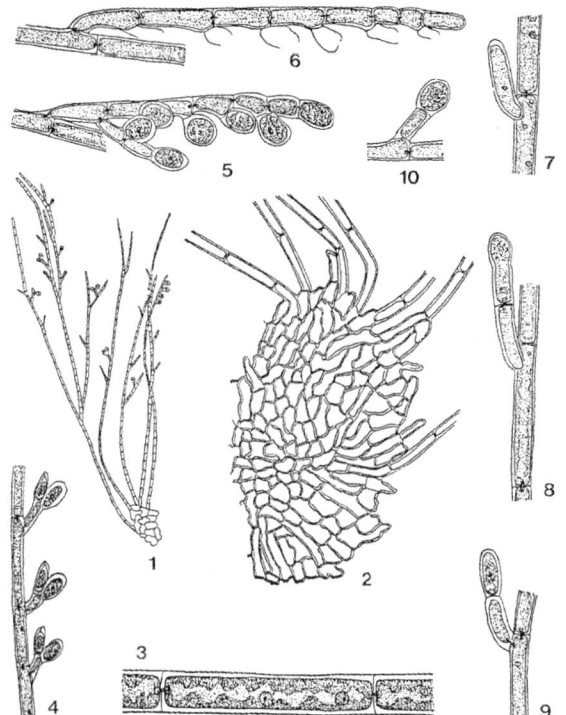

Plate 30

Acrochaetium indica, figs 1–10 (original author, Raikwar 1962)
1. habit of a mature plant, 2. prostrate system, 3. a vegetative cell showing chloroplasts, 4. monosporangia terminal on short lateral branchlets of limited growth, 5. a short lateral branchlet showing further branching and bearing lateral and sessile monosporangia, 6. a short lateral branchlet of unlimited growth showing dehisced lateral and sessile monosporangia, 7–10. a monosporangium terminal on different stages in development of a short lateral branchlet of limited growth.

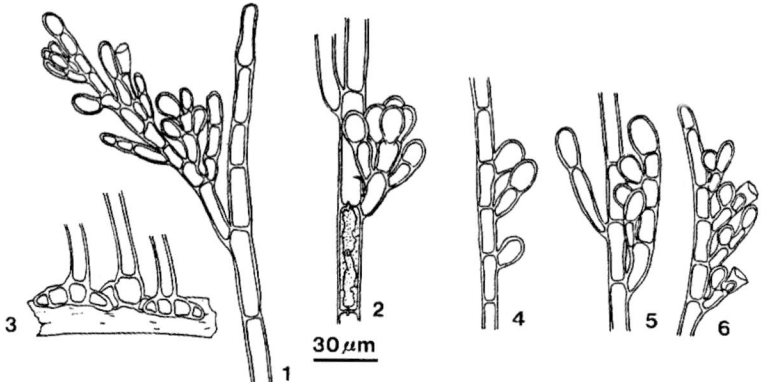

Plate 31

Acrochaetium sarmae, figs 1–6 (original author, Kahn 1970)
1. document prostrate system, 2. a short lateral branchlet showing ramification with monosporangia in cluster, 3. a short lateral branchlet with monosporangia in cluster, basal cell of a short lateral branchlet showing nature of chloroplast, 4. a short lateral branchlet of unlimited growth bearing sessile obovoidal as well as terminal ellipsoidal monosporangia, 5. monosporangia terminal on a short lateral branchlet of limited growth, 6. short lateral branchlets of limited growth showing regeneration of new monosporangia within apical dehisced terminal monosporangia.

Order **Balbianiales** Sheath & Müller (1999:863)

Type family Balbianiaceae Sheath & Müller

Monoscious filaments; spermatangia, together with similar cells filled with starch, borne on elongate differentiated cells. Pit plugs with 2 cap layers of which the outer is dome-shaped, lackinga membrane.

Note: Garbary (1987) treated the genus *Balbiania* as a synonym of the genus *Audouinella* and some freshwater species were transferred from the genus *Acrochaetium* to the genus *Audouinella*. Harper & Saunders (1998) stated that although the genus *Balbiania* and the genus *Rhododraparnaldia* group together, and this lineage warrants ordinal status among neighboring lineages, the phylogenetic affinities of this new order relative to the order seven higher-level lineage discussed herein remain equivocal. Based on the analyses in terms of the morphology, ultrastructure, and the rbcL gene, 18S r RNA gene as well as the first internal transcribe spacer region (ITSI) of the RNA genes, it is clear that the genus Balbiania is a valid taxon and that it is phylogenetically associated with *Rhododraparnaldia*. A new order, the Balbianiales, was proposed (Sheath & Müller (1999). A single family is recognized, the Balbianiaceae

Family **Balbianiaceae** Sheath & Müller (1999:863)

Type genus *Balbiania* Sirodot

The Balbianiaceae has the same characteristics as the order.

Genus ***Balbiania*** Sirodot (1876:149, Tab. 13–15)

Type *Balbiania investiens* (Lenormand ex Kützing) Sirodot

Epiphytic on *Batrachospermum* in European streams, gametophyte with few branches that are equal diameter, diploid phase a tetrasporophyte.

Key to the species of the genus ***Balbiania***

1. Monosporangia 8–11 μm long..1) *B. meiospora*
1. Monosporangia 15–18 μm long..2) *B. investiens*

1) ***Balbiania meiospora*** Skuja (1944:10, tab. 1, figs 1–11)
Basionym: *Audouinella meiospora* (Skuja) Garbary (1987:112)

plate 32, figs 1–3 (original author, Skuja 1944)
plate 25, figs 4–7 (Necchi & Zucchi 1995) as *Audouinella meiospora*

Plants small, monosiphonous throughout, epiphytic on the gelatinous matrix among the whorls of another living red alga, *Nothocladus lindaueri*. Plants at first a low cushion, later developing into a limited red-violet cover, consisting of a prostrate, branched basal portion and an erect branched monopodial shoots. Cells of prostrate portion 1.5–5 μm in diameter, 25 μm long, more or less cylindrical. Erect shoots up to 350 μm high, laterally branched, with cells 4–7 μm in diameter, 8–16 μm long. Chloroplasts parietal, red-violet, irregularly disc-shaped, ribbon-shaped,

Balbianiales

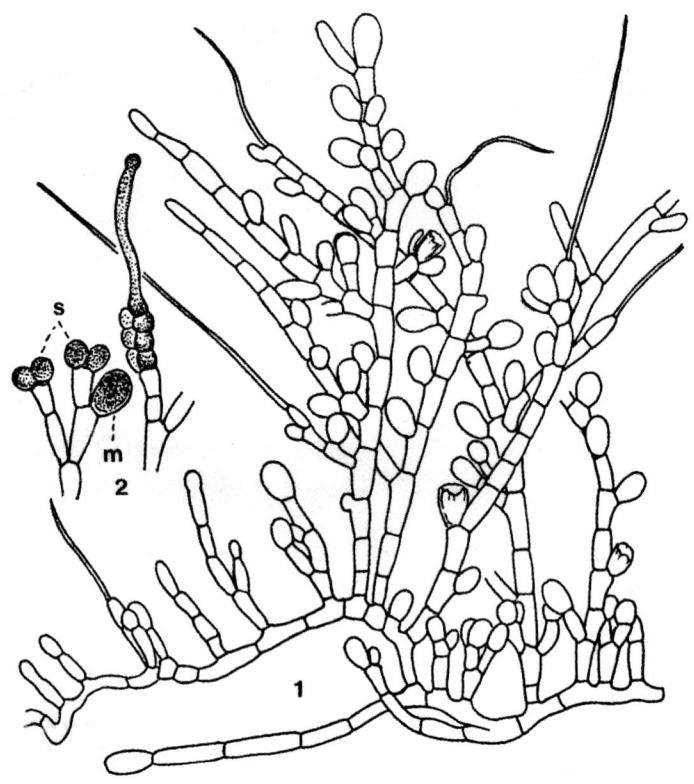

Plate 32
Balbiania meiospora, figs 1–3 (original author, Skuja 1944)
1. a portion of a plant creeping along the cortical portion of *Nothocladus lindaueri*, 2. short lateral branchlets bearing both spermatangia and monosporangium, 3. the formation of a carpogonium. s: spermatangia, m: monosporangium.

and more or less spirally curled, without pyrenoid. Cell wall thin and colourless. Monosporangia, spermatangia and carpogonia on the same plant. Monosporangia ellipsoidal to obovoidal, 6–8 µm in diameter, 8–11 µm long, terminal or lateral, 1–2 celled short lateral branchlets. 1–3 spermatangia terminal on short lateral branchlets, spherical, 3–5 µm in diameter. Carpogonium was shown (pl. 13, fig. 3) but carposporophyte and carposporangia unknown (Skuja 1944).
    Type Locality: River Waitangi near Russel, North Island in New Zealand.
    Holotype:
    Distribution: Known from *Nothocladus lindaueri* in the type locality in New Zealand, and on *Compsopogonopsis leptoclados* in Itanhaém, Rio Branco, 16 km from Route SP–55, São Paulo, Brazil in South America.

2) ***Balbiania investiens*** (Lenormand ex Kützing) Sirodot (1876: 146, Tab. 13–15)
Basionym: *Chantransia investiens* Lenormand ex Kützing (1849:431)
Synonym: *Rhodochorton investiens* (Lenormand) Swale et Belcher (1963:288, figs 1–26); *Batrachospermum rubrum* Hassall (1845: 113, Tab. 15, figs 2–3); *Audouinella investiens* (Sirodot) Garbary (1987:122).

plate 33, figs 1–4 (original author, Sirodot 1876)

Balbiania

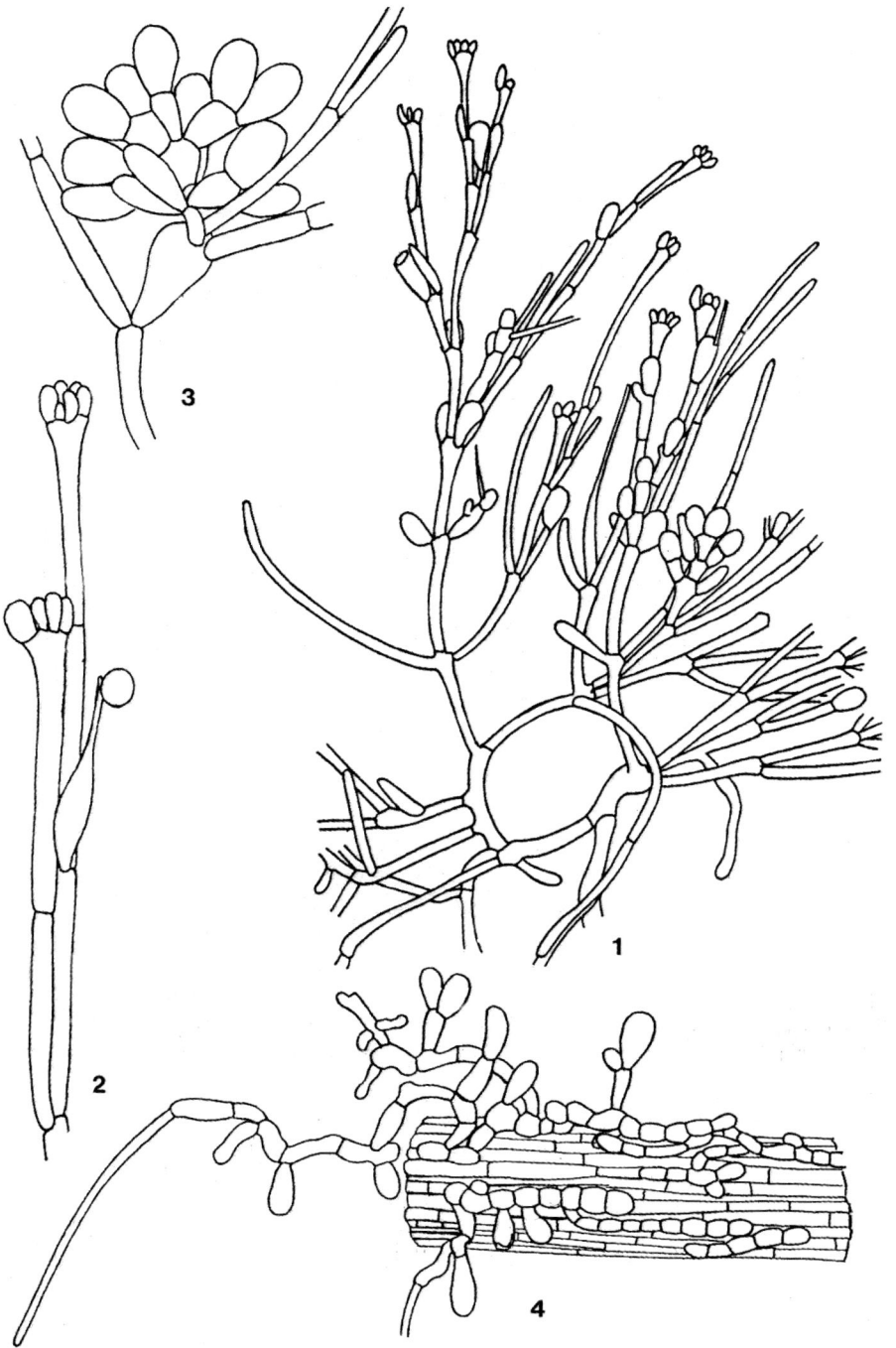

Plate 33
Balbiania investiens, figs 1–4 (original author, Sirodot 1876)
1. a portion of a plant, showing ramification, spermatangia and young carposporophytes, 2. spermatangia terminal on short lateral branchlets, laterally issued carpogonium and an attached spermatium with a trichogyne, 3. a young carposporophyte with carposporangia terminal on gonimoblast filaments, 4. a portion of a plant creeping along the cortical portion of *Batrachospermum gelatinosum* and *B. helminthosum*.

Balbianiales

Plants red-purple, creeping, monosiphonous, epiphytic on *Batrachospermum* plants. Cells 10–15 times as long as diameter, terminal hairs often developed, however, easily detached. Chloroplasts ribbon-shaped and spirally arranged. Asexual reproduction usually by means of monospores, monosporangia terminal on short lateral branchlets, spherical or ovoidal, 7–9 µm in diameter, 15–18 µm long. After liberation, monospores germinating to develop directly into a slender rhizoid and a runner around the plants of *Batrachospermum*. Sexual reproduction occurring in summer. Spermatangia 2–5, in cluster, terminal on short lateral branchlets. Carpogonium terminal on short 2–3 celled lateral branches, ovoidal with a long slender trichogyne. After fertilization, carpogonium rounding off, dividing transversely into two cells, dividing further and producing 4–6 celled short gonimoblast filaments. The terminal cell of such a gonimoblast filament producing a single, naked carpospore. Carpospores germinating on plants of *Batrachospermum* into a new type of plant, different from gametophyte and not producing monospores or carpospores (Starmach 1976).

Type Locality: Penzance, Cornwall, UK.
Holotype:
Distribution: Known from the type locality and Princetown in Devon, Ambleside, Blake Beek in Westmorland, England; Calvados, Ille-et-Vilaine, Finistère in France, near São João da Madeira and near Ovar in Portugal; in the Harz mountains in Germany in Europe. Epiphytic on plants of *Batrachospermum gelatinosum* and *B. helminthosum*.

Genus *Rhododraparnaldia* Sheath, Whittick et Cole (1994:1)

Type: *Rhododraparnaldia oregonica* Sheath *et al.*

Epilithic and thus far knopwn from streams in Oregon, USA. Carpogonium typically acrochaetialean, with a swollen, cylindrical base and thin trichogyne, but carpospores germinating into a diploid phase, a chantransia phase. Gametophyte developing directly from this phase. Pit plugs with two cap layers, the outer typically domed.

Note: Generic name denoting the abrupt change in diameter between cells of the main axis and lateral branches, as the Chlorophyte genus *Draparnaldia* and the crimson colour (Sheath, Whittick et Cole 1994). Since the alga has characteristics of both the Acrochaetiales and the Batrachospermales its classification is uncertain. Using RuBisCo, rbcL gene and 18S rRNA gene sequences, Vis *et al.* (1998) found that *Rhododraparnaldia oregonica* consistently occurs on an early branch within the Acrochaetiales-Palmariales clade and does not appear to be a member of the Batrachospermales.

1) ***Rhododraparnaldia oregonica*** Sheath, Whittick et Cole (1994:1)

plate 34, figs 1–10 (original authors, Sheath, Whittick & Cole 1994)

Plants monoecious, consisting of reddish, uniseriate filaments with dense determinate laterals in the upper two-thirds and little branching at the base. Attached by rhizoidal outgrowths from the lowermost cell. Axial cells barrel-shaped, 17.3–30.1 µm in diameter, 15.1–38.7 µm long. Determinate laterals arising both oppositely and alternately from the mid-portion of axial cells. Lateral branch cells, 4.3–8.5 µm in diameter, considerably smaller than the main axis cells. Spermatangia complex, with two or three cells formed at the tips of colourless stalks, 24.2–43.7 µm long. Axial cell spherical, like a typical spermatangium, with the outer one or two cells more cylindrical. Carpogonium formed on the same branches as spermatangial complex. Carpogonium base cylindrical, 5.2–7.9 µm in diameter, 19.4–29.3 µm long; trichogyne thin and

Rhododraparnaldia

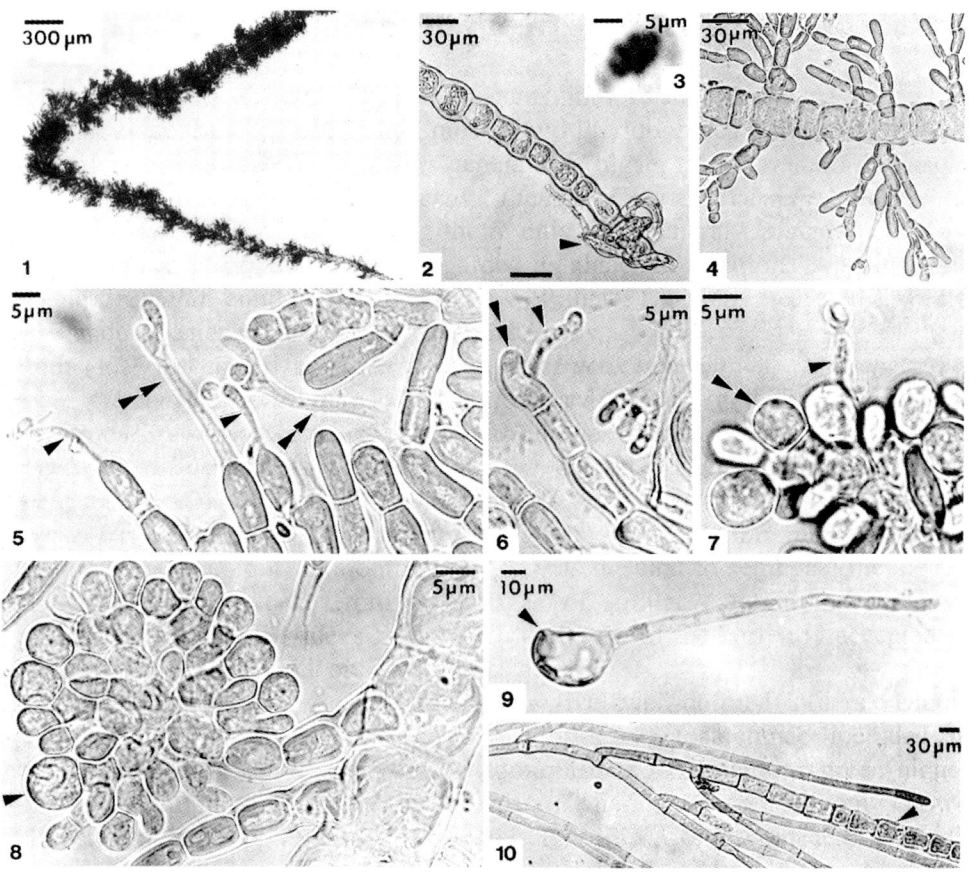

Plate 34
Rhododraparnaldia oregonica, figs 1–10 (original authors, Sheath, Whittick & Cole 1994)
1. plants showing dense lateral branches in the upper portions and few laterals at the base, 2. a base of a plant with rhizoids arising from the basal cell (arrow), 3. chromosomes from a fascicles cell, n=c. 7, 4. an apical portion of plants with determinate lateral branches, both opposite and alternate, arising from large barrel-shaped axial cells, 5. carpogonia with slender trichogynes (arrows) to which are an attached spherical spermatium, spermatangia terminal on long stalks (double arrows), 6. a fertilized trichogyne (arrow) with a carpogonium base forming a shoulder (double arrow), 7. a young carposporophyte with a trichogyne (arrow) still evident and several carposporangia (double arrow), 8. a mature carposporophyte with large, spherical carposporangia (arrow), 9. a germinating carpospore (arrow) forming chantransia phase, 10. chantransia phase forming a gametophyte initial (arrow).

slightly inflated at the tip. Both spherical and cylindrical spermatia seen attached to the trichogyne tip. Carpogonial base enlarging in early post-fertilization development, forming a shoulder around the trichogyne. Carposporophytes consisting of short gonimoblast filaments and numerous, spherical carposporangia, 5.5–7.9 µm in diameter, 7.4–10.8 µm long. Carpospores germinating into an chantransia phase.

   Type Locality: 2.3 km north-east of Blue River Reservoir, Willamette National Forest, Oregon, USA in North America.
   Holotype: UBC A80770.
   Distribution: Known from two streams of Coastal Range and Cascades from Oregon, USA in North America.

Order **Batrachospermales** Pueschel et Cole (1982:717)

Type: family Batrachospermaceae C. Agardh

Plants with a *Lemanea*-type life history and pit connections having two pit plug cap layers, with an enlarged outer layer (Pueschel & Cole 1982).

Carpospores in *Lemanea* (Magne 1967) and *Batrachospermum* (Balakrishnan & Chaugule 1980a, 1980b), germinating into small plants called the chantransia phase with meiosis taking place in an apical cell. After each of two nuclear divisions a small portion of cytoplasm with one of the nuclei is cut off by a wall and then the rest degenerates. Thus only one nucleus survives. The apical cell, which has become haploid without changing its outer appearance, then gives rise to a new gametophyte.

Based on rbcL gene and 18S rRNA gene sequences, Vis, Saunders, Sheath, Dunse & Entwisle (1998) stated as follows. The genus *Batrachospermum* appears to comprise many morphologically similar but distantly related taxa, which will need further investigation to resolve their taxonomic status. Moreover, preliminary analysis of small subunit of rRNA gene of *Ballia prieurii,* has close relationship to the Batrachospermales rather than to the Ceramiales (Necchi, personal communication). The genera *Tuomeya, Sirodotia* and *Nothocladus* are retained at the generic level until further data are obtained. *Psilosiphon scoparium* was not closely allied with the taxa of the Lemaneaceae, lending support to the new family proposed family Psilosiphonaceae (Sheath, Müller, Vis & Entwisle 1996). Sequence data from the remaining taxa of the Lemaneaceae support the concept of a derived monophyletic clade. *Thorea violacea* was not closely related to the other taxa of the Batrachospermales in all trees and hence the Thoreaceae does not appear to be a natural grouping within the Batrachospermales. Sheath *et al.* (2000) and Hanyuda *et al.* (2001) proposed the order Thoreales, although their proposals are invalid at present.

This order includes three families, Batrachospermaceae, Psilosiphonaceae and Lemaneaceae.

Key to the families of the order **Batrachospermales**

1. Plants having a main axis composed of axial cell-row and whorls of short branchlets, more or less mucilaginous, often with a bead-like appearance..family Batrachospermaceae
1. Plants consisting of tubular tufts with regularly placed swellings (nodes); carposporophytes developed within the tubular thallus; spermatangia in form of papillae produced at the nodes..2
2. Outer cortex composed of cells nearly uniform in size in distinct rows: medullar filaments densely interwoven; reproduction by adventitious plantlets derived from cortex...family Psilosiphonaceae
2. Outer cortex composed of cells increasing significantly in size from outer to inner layers and not in distinct rows; interwoven medullar filaments absent; sexual reproduction with spermatangial clusters and carpogonia, no adventitious plants...…..family Lemaneaceae

Family **Batrachospermaceae** C. Agardh (1824: p. XXIII)

as family Batrachospermeae

Type: genus *Batrachospermum* Roth

Plants uniaxial and irregularly branched. Carpogonium-bearing branches arising from periaxial cells and or cells of fascicles or cortical filaments. Carpogonia symmetrical or asymmetrical with elongate conical or club-shaped trichogynes. Carposporophytes definite or indefinite in shape. Gonimoblast filaments of a radially branched determinate type and of a diffuse indeterminate type, developed from the fertilized carpogonia or fused cell. The family Batrachospermaceae includes four genera, *Batrachospermum, Sirodotia, Nothocladus* and *Tuomeya.*

Key to the genera of the family **Batrachospermaceae**

1. Carposporophytes condensed type..2
1. Carposporophytes diffuse type..3
2. Carposporophytes spherical, gonimoblast filaments radially branched determinate type, arising directly from fertilized carpogonium..........genus *Batrachospermum*
2. Carposporophyte spherical, gonimoblast filaments radially branched determinate type, arising from dense mass of fusion cells (placenta), growing radially, having carposporangia on apices arising...............................genus *Tuomeya*
3. Gonimoblast filaments indeterminately creeping along cortical filaments, forming carposporangia in the axial parts..genus *Sirodotia*
3. Gonimoblast filament arising from fertilized carpogonium, indeterminate creeping among outer cortical filaments, forming carposporangia on the surface of thallus..genus *Nothocladus*

Genus *Batrachospermum* Roth (1797:36)

Type: *Batrachospermum gelatinosum* (Linnaeus) DeCandolle

Plants irregularly branched. Carpogonium-bearing branches arising from periaxial cells and or cells of fascicles. Carpogonium-bearing branches of *B. involutum* (section *Batrachospermum*) are similar in ultrastructure to nearby fascicles, having uninucleate cells with abundant starch granules and several parietal, well-developed chloroplasts (Sheath & Müller 1997). The short, carpogonium-bearing branch cells of *B. helminthosum* (section *Virescentia*), by contrast, lacking visible starch and having reduced-type chloroplasts with few thylakoids. Breakdown of cross walls among cells of the carpogonium-bearing branch common in *B. helminthosum* but absent in *B. involutum.* Carpogonia symmetrical or somewhat asymmetrical with elongate conical or club-shaped trichogynes. Carposporophytes definite in shape. Gonimoblast filaments generally of a radially branched determinate type but of a diffuse indeterminate type in section *Turfosa*, developed directly from the fertilized carpogonia.

Notes: Because of evidence accumulating from a variety of sources, the sectional classification system of the genus *Batrachospermum* will require revision in the near future.

According to Kumano (1993), *B. nova-guineense* of the section *Batrachospermum* and *B. bakarense* of the section *Virescentia* with slightly curved carpogonium-bearing branches represent connecting links between the section *Batrachospermum,* the section *Virescentia* and the section *Contorta.* In the early stage of its development, *B. cylindrocellulare* with a relatively long

Batrachospermales

carpogonium-bearing branch closely resembles *B. cayennense* of the section *Aristata*. Some taxa with two types of a determinate and an indeterminate gonimoblast filaments such as *B. tapirense* of the section *Turfosa sensu* Necchi (1990a) are considered as intermediate links between *Batrachospermum* and *Sirodotia*.

The shape of the carpogonium, as used in part by Sheath, Vis & Cole (1994a, 1994c) to separate section *Virescentia* from section *Turfosa*, is of dubious taxonomic value at the sectional level. The distinctions between section *Aristata* and *Batrachospermum* have been questioned by Entwisle (1995) and Sheath, Vis & Cole (1994b). Taxa described from Australia and New Zealand by Entwisle & Foard (1997) are intermediate between section *Batrachospermum* and sections *Turfosa* and *Virescentia*. Australia and New Zealand species of section *Hybrida* may be considered with those of section *Contorta*.

For example, the presence of both indeterminate and determinate gonimoblast filaments in *B. terawhiticum* confirms that the existing system is unworkable. The shape of the carpogonium, as used in part by Sheath, Vis & Cole (1994c) to separate section *Virescentia* from section *Turfosa*, is also of dubious taxonomic value at the sectional level.

According to Entwisle & Foard (1997), taxa such as *B. prominens, B. terawhiticum* and *B. imparum* are intermediate between section *Batrachospermum* and sections *Turfosa* and *Virescentia*, blurring the distinction between 'sessile' or 'axial' and 'pedicellate' carposporophytes.

The distinctions between section *Aristata* and section *Batrachospermum* had been questioned by Entwisle (1995) and Sheath, Vis & Cole (1994b), and are further weakened by the taxa described by Entwisle & Foard (1997).

Sheath, Vis & Cole (1994b) found the length of the carpogonium-bearing branch, the traditional distinguishing character of *Aristata*, to be a poor sectional character. A re-examination by Entwisle & Foard (1997) of an historically important specimen of *B. sporulans* (synonym of *B. skujae sensu* Vis et al. 1995) from France revealed carpogonial branches of 15–18 modified cells, not 5–10 unmodified cells as reported by Vis et al. (1995). This suggested that the specimen examined is part of the *B. cayennense* group ('section *Aristata*').

No sectional classification is entirely satisfying. In this book, the classification of Kumano (1993) is followed. Australian–New Zealand species are placed in the groups proposed by Entwisle & Foard (1997) next to the section *Batrachospermum*.

The sequence data from rbcL gene and 18S rDNA of the representative taxa of six sections of the genus *Batrachospermum* were used to construct phylegenetic hypotheses (Vis, Saunders, Sheath, Dunse and Entwisle 1998).

The close affinity of *B. lousianae* (section *Contorta*) and *B. virgato-deaisneanum* (section *Hybrida*) in the molecular analysis suggests that the separation of these two sections should be reconsidered after more taxa have been analysed.

The results show the close affinity of *B. gelatinosum* and *B. spermato-involucrum* and a similar association of *B. boryanum* and *B. involutum*. However, these taxa clearly do not form a monophyetic clade.

The molecular data lend support to taxonomic schema of Kumano (1993), in which section *Helminthoidea* containing *B. boryanum* is separated from section *Batrachospermum* with *B. gelatinisum*.

B. atrum (section *Setacea*) and *B. helminthosum* (section *Virescentia*) are on divergent branches of the molecular trees. This result supports the hypothesis that section *Setacea* is not made up of merely small whored members of section *Virescentia*, as suggested by Necchi and Entwisle (1990), but constitutes a distinct section as proposed by Sheath et al. (1993).

Like *B. atrum* and *B. helminthosum, B. macrosporum* (section *Aristata*) and *B. turfosum* (section *Turfosa*) appear to be distinct lineages, based on molecular data.

Key to the subgenera and the sections of the genus *Batrachospermum*

1. Carposporophyte consisting of a one celled zygote...................I. subgenus *Acarposporophytum*
1. Carposporophyte multicellular..II. subgenus *Batrachospermum* – 2
2. Carposporophyte consisting of both indeterminate and
   determinate gonimoblast filaments..3. section *Turfosa*
2. Carposporophyte consisting of only determinate gonimoblast filaments....................................3
3. Carpogonium-bearing branches spirally twisted..8. section *Contorta*
3. Carpogonium-bearing branches not spirally twisted...4
4. Carpogonium-bearing branches very long (more than 12 cells).................7. section *Aristata*
4. Carpogonium-bearing branches shorter (2–12 cells...5
5. Whorls reduced..2. section *Setace*
5. Whorls developed normally...6
6. Carposporophyte consisting of gonimoblast filaments
   and propagules..5. section *Gonimopropagulum*
6. Carposporophyte consisting of gonimoblast filaments and carposporangia..............................7
7. Carposporophytes large, 1 or 2, situated in the middle of whorls..8
7. Carposporophytes small, numerous, distributed in whorls..............1. section *Batrachospermum*
8. Trichogyne club-shaped, long, with distinct stalk......................................4. section *Virescentia*
8. Trichogyne short ellipsoidal; plants intense green...6. section *Hybrida*

I. Subgenus *Acarposporophytum* Necchi (1987:446)

Type: *Batrachospermum brasiliense* Necchi

Gonimoblast filaments and carposporangia absent. Filaments of the chantransia phase developing directly from the fertilized carpogonium into the gametophytes. The carposporophyte reduced to a one-celled zygote.

1) ***Batrachospermum brasiliense*** Necchi (1987:442, figs 2–28)

                plate 35 (a), figs 1–6 (Kumano 1996f)
                figs 7–8 (original author, Necchi 1987)

Plants monoecious, moderately mucilaginous, 3–6 cm high, 450 µm in diameter; whorls barrel-shaped or compressed, contiguous; fascicles 2–4, primary fascicles di- or trichotomously branched, 6–10 cell-storeys, proximal cells cylindrical, 6–12 µm in diameter, 40–80 µm long, distal cells obovoidal or spherical, 8–13 µm in diameter, 14–20 µm long, terminal hairs numerous and long; cortical filaments well-developed; secondary fascicles numerous, covering internodes; spermatangia spherical or obovoidal, terminal on fascicles, 6–7 µm in diameter; carpogonium-bearing branches arising from periaxial cells, short, 15–30 µm long, consisting of 3–6 disc- or barrel-shaped cells; carpogonium 30–45 µm long, trichogyne club-shaped or ovoid, indistinctly stalked; upper involucral filaments in rosette, lower involucral filaments numerous, short; carposporophytes one-celled zygotes; filaments of chantransia phase developing directly from one-celled zygotes.
Type Locality: River Claro, Boracéia Biological Station, Biritiba Mirim, Saõ Paulo, Brazil in South America.
Holotype: SP 187180.
Distribution: Known from the type locality and in several localities such as River Negro, Saõ Paulo, Brazil in South America.

Batrachospermales

II. Subgenus *Batrachospermum*

Type: *Batrachospermum gelatinosum* (Linnaeus) DeCandolle

Carposporophyte, gonimoblast filaments and carposporangia present. Filaments of chantransia phase developed from the germination of carpospores.

1. Section *Batrachospermum*

Type: *B. gelatinosum* (Linnaeus) DeCandolle

Synonym: section *Moniliformia* Sirodot (1873), section "*Moniliformes*" Sirodot, section *Helminthoidea* Sirodot ex De Toni (1897), section *Carpocontorta* Sheath *et al.* (1986).

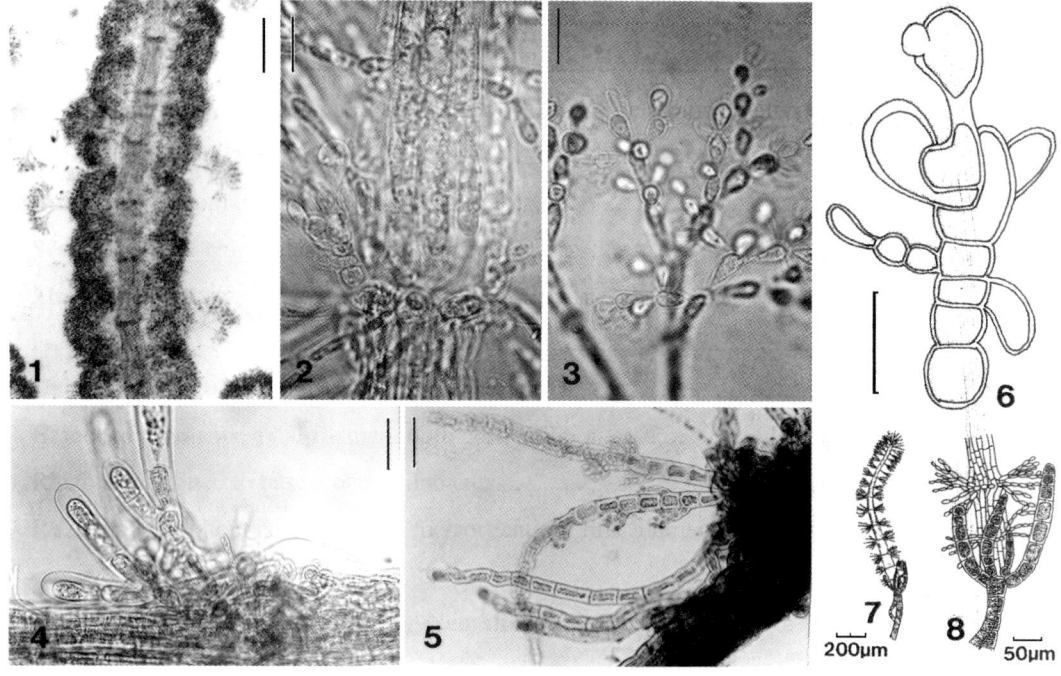

Plate 35 (a)
Batrachospermum brasiliense, figs 1–6 (Kumano 1996f), figs 7–8 (original author, Necchi 1987)
1. a main axis with barrel-shaped, compressed, confluent whorls, 2, 6. a carpogonium with an ovoidal trichogyne and a carpogonium-bearing branch having rosette-like upper involucral filaments, arising from a periaxial cell, 3. spermatangia terminal on fascicle, 4–5. filaments of chantransia phase arising from one celled carposporophyte, 7–8. gametophyte arising from chantransia phase.
Scale bars = 200 μm (fig. 1); 100 μm (fig. 5); 40 μm (figs 2, 4); 20 μm (fig. 3).

1. *Batrachospermum*

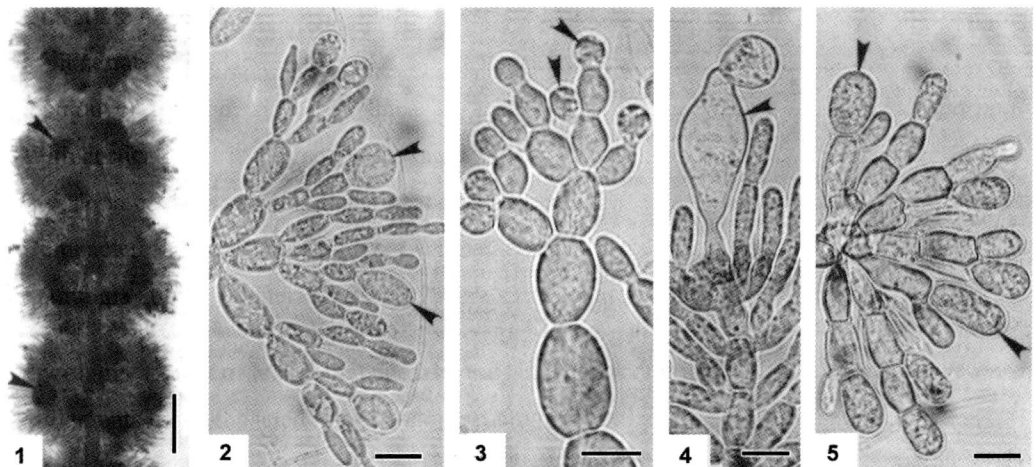

Plate 35 (b)
Batrachospermum skujae, figs 1–5 (Vis, Sheath & Cole 1996b)
1. main axis with confluent, barrel-shaped whorls containing spherical carposporophytes (arrowheads), 2. monosporangia (arrowheads) terminal on fascicle, 3. spermatangia (arrowheads) terminal on fascicle, 4. fertilized carpogonium with inflated, club-shaped, lanceolate trichogyne (arrowhead), 5. carposporophyte with short, dense, 3–4 celled gonimoblast filaments having terminal carposporangia (arrow heads). Scale bar = 10 μm (figs 2–5); 300 μm (fig 1).

Using multivariate morphometric and image analysis, Vis, Sheath & Entwisle (1995), and Vis & Sheath (1996) recognized the following species and gave them revised descriptions (Vis, Sheath & Entwisle 1995).

Key to the species of the section *Batrachospermum*

1. Monosporangia present..2
1. Monosporangia absent...3
2. Plants monoecious...1) *B. skujae*
2. Plants dioecious...2) *B. lochmodes*
3. Spermatangia present on involucral filaments of the carpogonium-bearing branch......................4
3. Spermatangia absent on involucral filaments of the carpogonium-bearing branch......................7
4. Fascicle tips involute; rhizoidal outgrowths from mid-fascicle cells present........3) *B. involutum*
4. Fascicle tips noninvolute; rhizoidal outgrowths from mid-fascicle cells absent..........................5
5. Main axis with heterocortication composed of bulbous and cylindrical cells........4) *B. confusum*
5. Main axis with cortication composed of cylindrical cells only...6
6. Fascicles distinctly curled, carpogonium < 23 μm long..5) *B. pulchrum*
6. Fascicles straight, carpogonium > 39 μm long......................................6) *B. spermatoinvolucrum*
7. Carpogonium contorted..8
7. Carpogonium symmetrical...11
8. Carpogonium with forked trichogyne...7) *B. trichofurcatum*
8. Carpogonium with unforked trichogyne..9
9. Twisting or contortion of trichogyne restricted to the tip, monoecious.......8) *B. trichocontortum*
9. Bends or contortions at various point along the trichogyne, dioecious.................................10
10. Trichogyne with numerous twists and contortions..................................9) *B. carpocontortum*
10. Trichogyne curved or bent with no twisting..10) *B. szechwanense*

Batrachospermales

11. Plants monoecious...12
11. Plants dioecious..17
12. Plants having spherical or pear-shaped whorls...11) *B. heteromorphum*
12. Plants having spherical whorls only...13
13. Fascicle cells cylindrical...12) *B. cylindrocellulare*
13. Fascicle cells obovoidal or ellipsoidal..14
14. Main axis with cortication composed of cylindrical cells..15
14. Main axis with heterocortication composed of bulbous and cylindrical cells......................16
15. Trichogyne enlarged, subpyriform when mature..13) *B. sinense*
15. Trichogyne not enlarged, mostly urn-shaped...14) *B. gelatinosum*
16. Carpogonium ≤ 41 μm long, trichogyne ≤ 10 μm in diameter.........................15) *B. anatinum*
16. Carpogonium ≥ 51 μm long, trichogyne ≥ 11 μm in diameter.........................16) *B. fluitans*
17. Carpogonium-bearing branch with involucral filaments
    on one side only..17) *B. nova-guineense*
17. Carpogonium-bearing branch with involucral filaments
    not confined to one side...18
18. Main axis with cortication composed of cylindrical cells..19
18. Main axis with heterocortication composed of bulbous and cylindrical cells......................20
19. Carpogonium-bearing branch arising from cells of fascicles.........................18) *B. arcuatum*
19. Carpogonium-bearing branch arising from cells of fascicles and
    also cortical cells...19) *B. longipedicellatum*
20. Carpogonium-bearing branch having involucral filaments with
    apical carpogonia present...20) *B. carpoinvolucrum*
20. Carpogonium-bearing branch having involucral filaments with
    apical carpogonia absent...21
21. Main axis with distinct whorls and few secondary fascicles............................21) *B. boryanum*
21. Main axis with distinct whorls and abundant
    secondary fascicle...22) *B. heterocorticum*

1) ***Batrachospermum skujae*** Geitler (1944:127, figs 1–5) *emend.* Vis *et al.* (1995:52)
Synonym: *B. sporulans* Sirodot (1884:216, tab. 11), nom. illeg.

plate 35 (b), figs 1–5 (Vis, Sheath & Cole 1996b)
plate 36, figs 1–5 (original author, Sirodot 1884) as *B. sporulans*

Plants monoecious; whorls confluent or distant, barrel-shaped, 554–1003 μm in diameter with 1–6 carposporophytes scattered within the whorl; main axis with cortication consisting of cylindrical cells only; carposporophytes spherical, pedicellate, 50–152 μm in diameter, gonimoblast filaments of 2–4 cylindrical cells, carpogonia 33–59 μm long with lanceolate or indented lanceolate trichogynes 7–13 μm in diameter; carpogonium-bearing branch undifferentiated, 5–10 cells long; carposporangia obovoidal, 9–19 μm long, 8–14 μm in diameter; monosporangia 7–10 μm long, 7–10 μm in diameter.

Type Locality: In einen Quellbach südlich von Gresten, Niederdonau.
Holotype: WU
For *B. sporulans* Sirodot, PC, Fontaine et doué de Bas-Champs, Betton, France.
Distribution: Known from Portugal, Poland, Crimea, Latvia and Sweden in Europe; Japan in Eastern Asia; and western coniferous forest and tundra biomes in North America. In Japan, Ichinomiya north of Toyokawa, Mukouyama near Toyohashi in Aichi and Notojima in Hiroshima (as *B. sporulans*, Mori 1975).

2) ***Batrachospermum lochmodes*** Skuja (1938:620, pl. 33, figs 1–11).

plate 36, figs 6–9 (original author, Skuja 1938)

Plants dioecious, moderately mucilaginous, more than 3 cm high, ca. 500 μm in diameter, short branches coming out very abundantly at right angles, dark blackish brown to red. Whorls ellipsoidal or spherical, more or less separated. Cortical filaments absent or sparse. Terminal hairs rare, short, ca. 30 μm long. Carpogonium-bearing branches consisting of 13 cells, arising from periaxial cells. Spermatangia unknown. Carpogonia 50 μm long, trichogyne initially obconical, becoming club-shaped, lanceolate, 14 μm in diameter, 41 μm long. Monosporangia ovoid or obovoidal, 8–9.5 μm in diameter, 12–16 μm long, terminal on primary, secondary fascicles, and carpogonium-bearing branches.

Type Locality: Seraju Spring Area in Dijeng Plateau, altitude about 2000 m, Middle Java, Indonesia in Southeast Asia.

Holotype:

Distribution: Known from the type locality only.

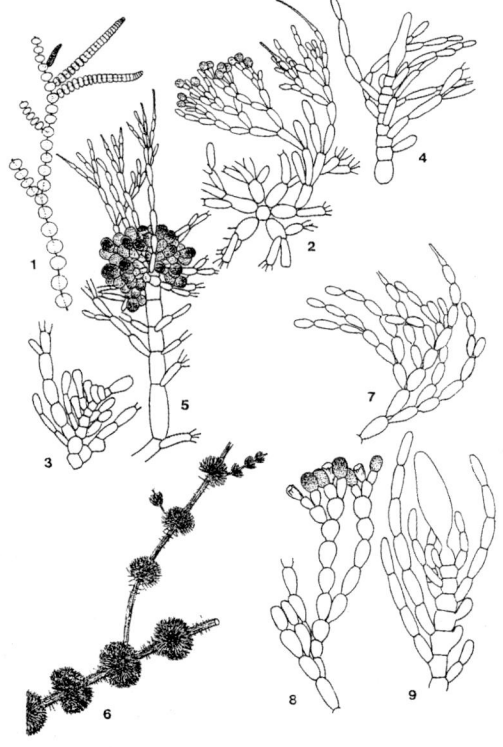

Plate 36

Batrachospermum skujae as *B. sporulans*, figs 1–5 (original author, Sirodot 1884)
1. a main axis with spherical and separated whorls, 2. 6. fascicles arising from a periaxial cell with terminal monosporangia, 3–4. relatively short differentiated carpogonium-bearing branches composed of barrel-shaped cells and a carpogonium with a lanceolate trichogyne, 5. a stalked carposporophyte with short gonimoblast filaments having apical carposporangia.

Batrachospermum lochmodes, figs 6–9 (original author, Skuja 1938)
6. a main axis with spherical, separated whorls, 7. primary fascicles, 8. monosporangia terminated on primary fascicles, 9. a relatively differentiated short carpogonium bearing branch composed of barrel-shaped cells and a carpogonium with a lanceolate trichogyne.

Batrachospermales

3) ***Batrachospermum involutum*** Vis et Sheath (1996:128, figs 25–34)

> plate 37, figs 1–10 (original author, Vis & Sheath 1996)

Plants monoecious, with indistinct whorls, 627–1405 µm in diameter, fascicles 8–12 cell-storeys. Paired fascicle tips turned inwards towards each other. Rhizoidal outgrowths from mid-fascicle cells. Carpogonium-bearing branches undifferentiated, consisting of 3–8 cells with long and short involucral filaments, sometimes terminal with more than one carpogonium. Carpogonium 7–10 µm in diameter, 30–54 µm long, trichogyne typically lanceolate, cylindrical, sometimes curved or branched. Spermatangia observed only terminal on one-celled involucral filaments arising from subtending cell of carpogonium. Carposporophytes spherical, 1–3 per whorl, exserted or within whorl. Mature carposporophytes 112–237 µm in diameter. Carposporangia 7–11 µm in diameter, 12–18 µm long, terminal on 3–4 celled gonimoblast filaments.

Type Locality: a spring-fed river, San Marcos River, San Marcos, Texas, USA in North America.
Holotype: UBC A81614.
Distribution: Known from the type locality only.

4) ***Batrachospermum confusum*** (Bory) Hassall (1845:105) *emend.* Vis *et al.* (1995:54)

The synonymy proposed by Vis *et al.* (1995) is as follows:
Basionym: *B. ludibundum* var. *confusum* ('*Batrachosperma ludibunda* [alpha] *confusa*') Bory (1808:320, pl. 29, fig. 3).
Synonym: *B. alpestre* Shuttleworth ex Hassal (1845:111, pl. XIV, fig. 2); *B. setigerum* Rabenhorst (1859, no 854); *B. helminthosum* Sirodot (1884:240, pls. XXVI–XXVIII), nom. illeg. (non *B. helminthosum* Bory 1808); *B. crouanianum* Sirodot (1884:244, pls. XXIV, XXV); *B. distensum* Kylin (1912: 26, fig. 9 a–g); *B. fruticulosum* Drew (1946:340, pl. viii); *B. confusum* f. *spermatogloberatum* Reis (1962: 62, pl. II).

> plate 38, figs 1–5 (Kumano 1996i)
> figs 6–7 (original author, Sirodot 1884) as *B. crouanianum*
> plate 39, figs 1–7 (original author, Sirodot 1884) as *B. crouanianum*
> figs 8–12 (original author, Kylin 1912) as *B. distensum*
> plate 56, figs 8–13 (Vis, Sheath & Cole 1996b)

Plants monoecious; main axis with inflated irregular cortication; confluent, whorls spherical or barrel-shaped, 412–955 µm in diameter, with 1–19 carposporophytes scattered within the whorl; carposporophytes spherical, pedicellate, 56–112 µm in diameter, gonimoblast filaments of 2–4 cylindrical cells; carpogonia 14–34 µm long with club-shaped, or ellipsoidal trichogynes 5–11 µm in diameter, carpogonium-bearing branch undifferentiated, 4–9 cells long, with involucral filaments having apical spermatangia; carposporangia obovoidal, 8–14 µm long, 6–11 µm in diameter.

Type Locality: near Fougeres, Brittany, France in Europe.
Lectotype: PC herb Thuret, designated by Vis *et al.* 1995.
Distribution: Known from streams in Portugal, UK, France, Poland, Sweden in Europe; in North America, in western coniferous, boreal, hemlock-hardwood and deciduous forest.

1. *Batrachospermum*

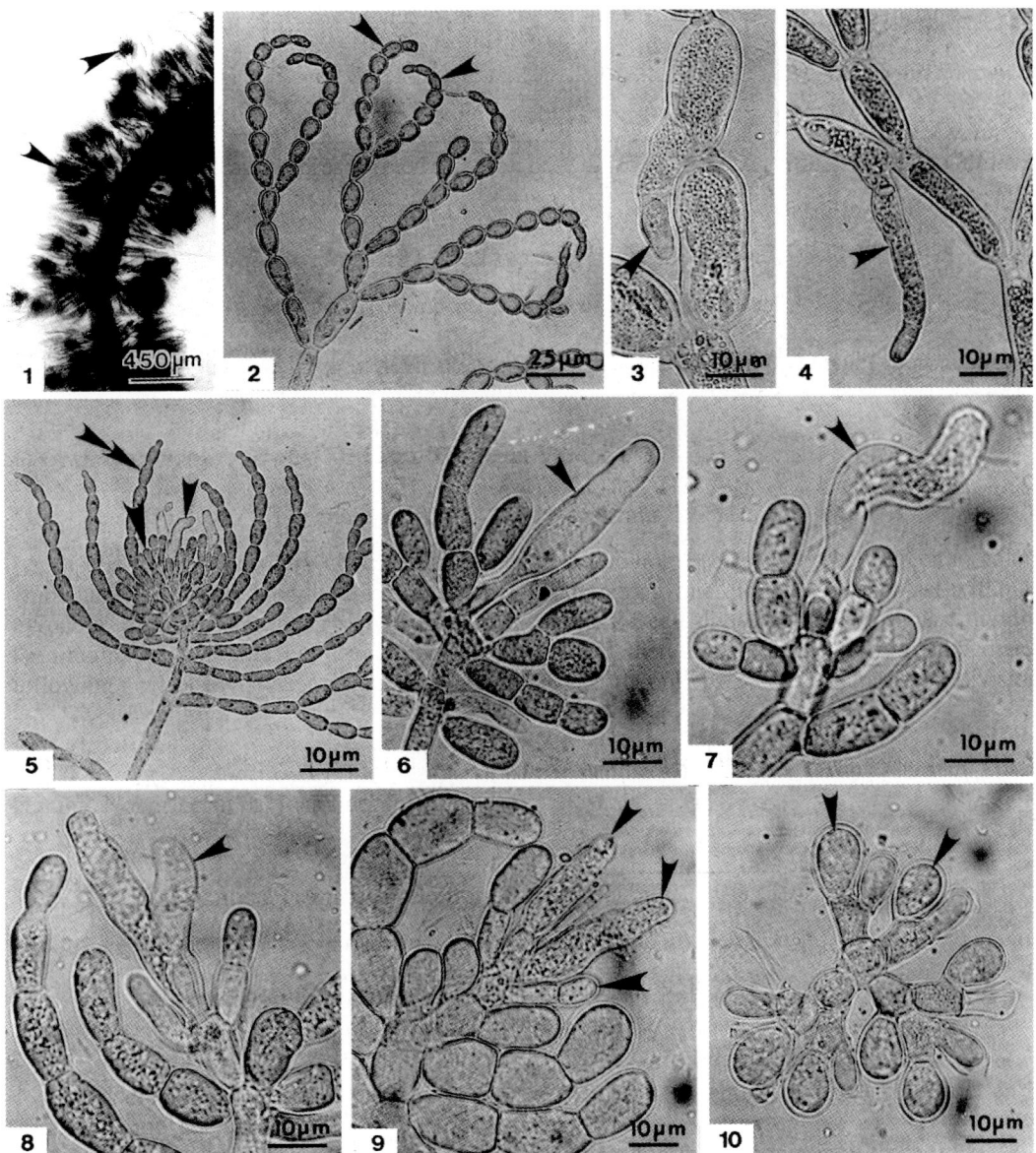

Plate 37
Batrachospermum involutum, figs 1–10 (original authors, Vis & Sheath 1996)
1. a main axis with indistinct whorls and small spherical to exerted carposporophytes (arrows), 2. fascicles with dichotomous involute tips (arrows), 3. intercalarly cells of fascicle with a developing cortical outgrowth (arrow), 4. intercalary cells of fascicle producing a relatively long cortical outgrowth (arrow), 5. a carpogonium-bearing branch with three carpogonia (arrow), short involucral filaments (large arrow) and long involucral filaments (double arrow), 6. a carpogonium with a lanceolate, cylindrical trichogyne (arrow), 7. a carpogonium with a curved lanceolate trichogyne (arrow), 8. a carpogonium with a branched, lanceolate trichogyne (arrow), 9. a carpogonium-bearing branch with two carpogonia (arrows) and a one-celled involucral filament having a spermatangium (large arrow) at its tip, 10. a carposporophyte with short gonimoblast filaments having apical carposporangia (arrows).

Batrachospermales

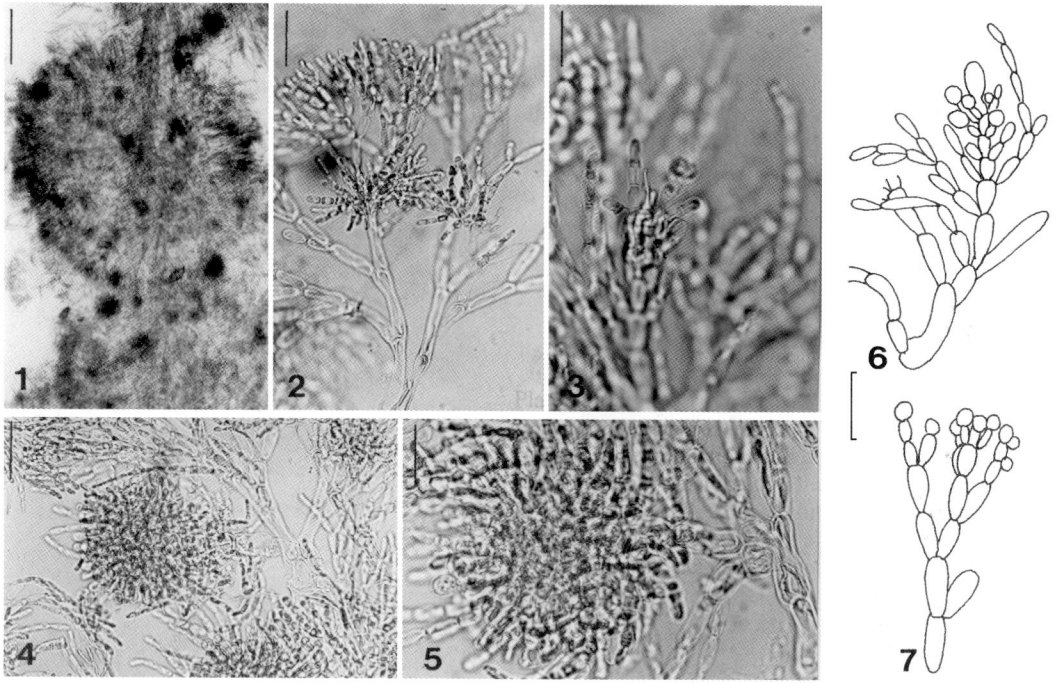

Plate 38

Batrachospermum confusum as *B. crouanianum*, figs 1–5 (Kumano 1996i), figs 6–7 (original author, Sirodot 1884)
1. a main axis with spherical, ellipsoidal, barrel- or disc-shaped, generally confluent, sometimes separated whorls having numerous small spherical carposporophytes, 2–3, 6. a carpogonium and an undifferentiated carpogonium-bearing branch with many involucral filaments, 4–5. carposporophytes, 7. spermatangia terminal on fascicle. Scale bars = 200 µm (fig. 1); 40 µm (figs 2, 4); 20 µm (figs 3, 5–7).

5) ***Batrachospermum pulchrum*** Sirodot (1884:225) *emend.* Vis *et al.* (1995:54)

plate 40, figs 7–12 (Vis, Sheath & Cole 1996b)

Plants monoecious, whorls distant, spherical, 307–485 µm in diameter, composed of few curly fascicles; main axis with cortication consisting of cylindrical cells only; carposporophytes 1–2 per whorl, spherical, pedicellate, 59–108 µm in diameter, gonimoblast filaments of 2–3 cylindrical cells; carpogonia 23–32 µm long with lanceolate trichogyne, 6–10 µm in diameter; carpogonium-bearing branch undifferentiated, 4–8 cells long; with involucral filaments having apical spermatangia; carposporangia obovoidal, 8–17 µm long, 7–14 µm in diameter.
    Type Locality: R. des Ecrevisses, Matouba, Guadeloupe, Lesser Antilles, Caribbean Sea in Central America.
    Holotype: PC herb. Thuret.
    Distribution: Restricted to tropical islands of Martinique and Grenada, Lesser Antillesin in Caribbean Sea in Central America.

1. *Batrachospermum*

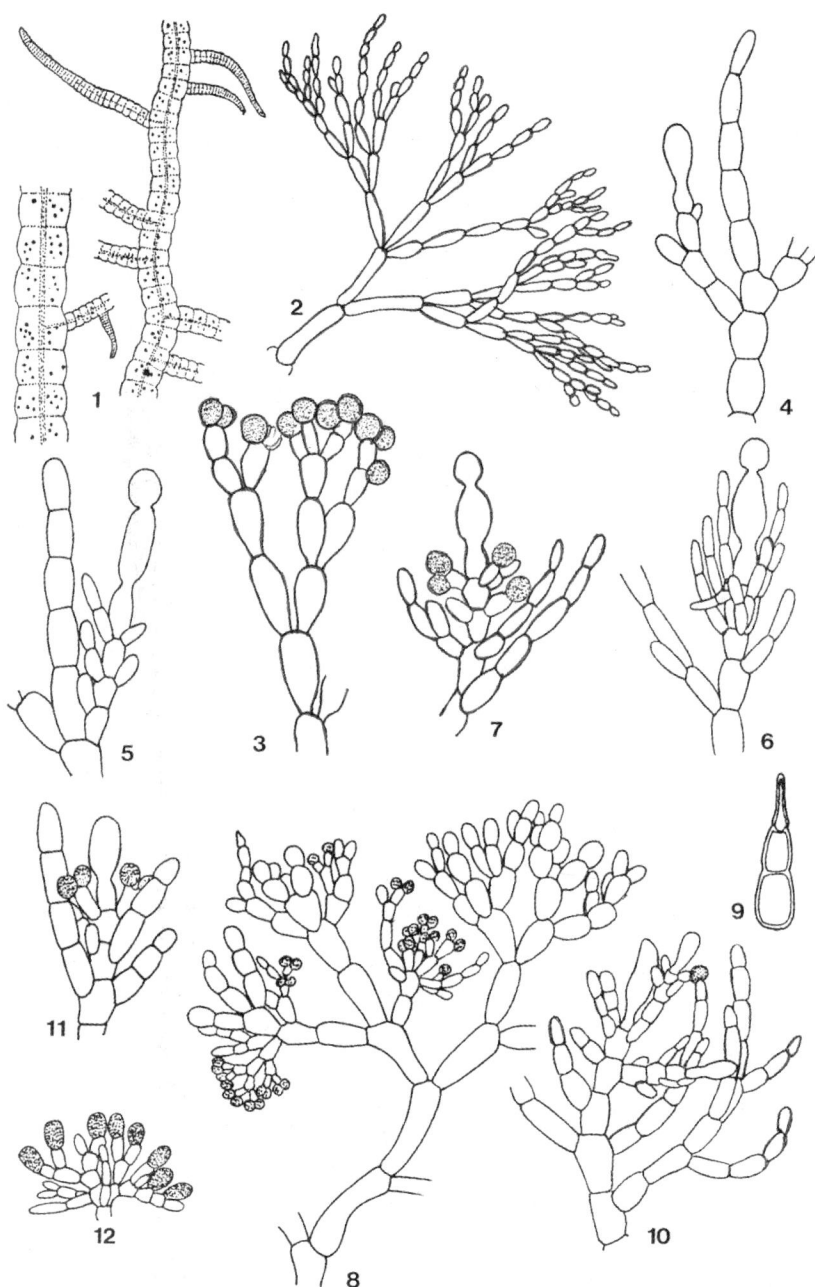

Plate 39

Batrachospermum confusum as *B. crouanianum*, figs 1–7 (original author, Sirodot 1884)
1. a main axis with indistinct whorls having numerous small spherical carposporophytes within whorls, 2. a fascicle, 3. spermatangia terminated on fascicle, 4–6. an undifferentiated carpogonium-bearing branch with carpogonia, short involucral filaments, 7. spermatangia on short involucral filaments.

Batrachospermum confusum as *B. distensum*, figs 8–12 (original author, Kylin 1912)
8. spermatangia on lateral branchlets of fascicles, 9. a terminal hair, 10–11. a carpogonium-bearing branch with carpogonia, and spermatangia terminal on short involucral filaments, 12. carposporangia terminal on a short gonimoblast filament.

Batrachospermales

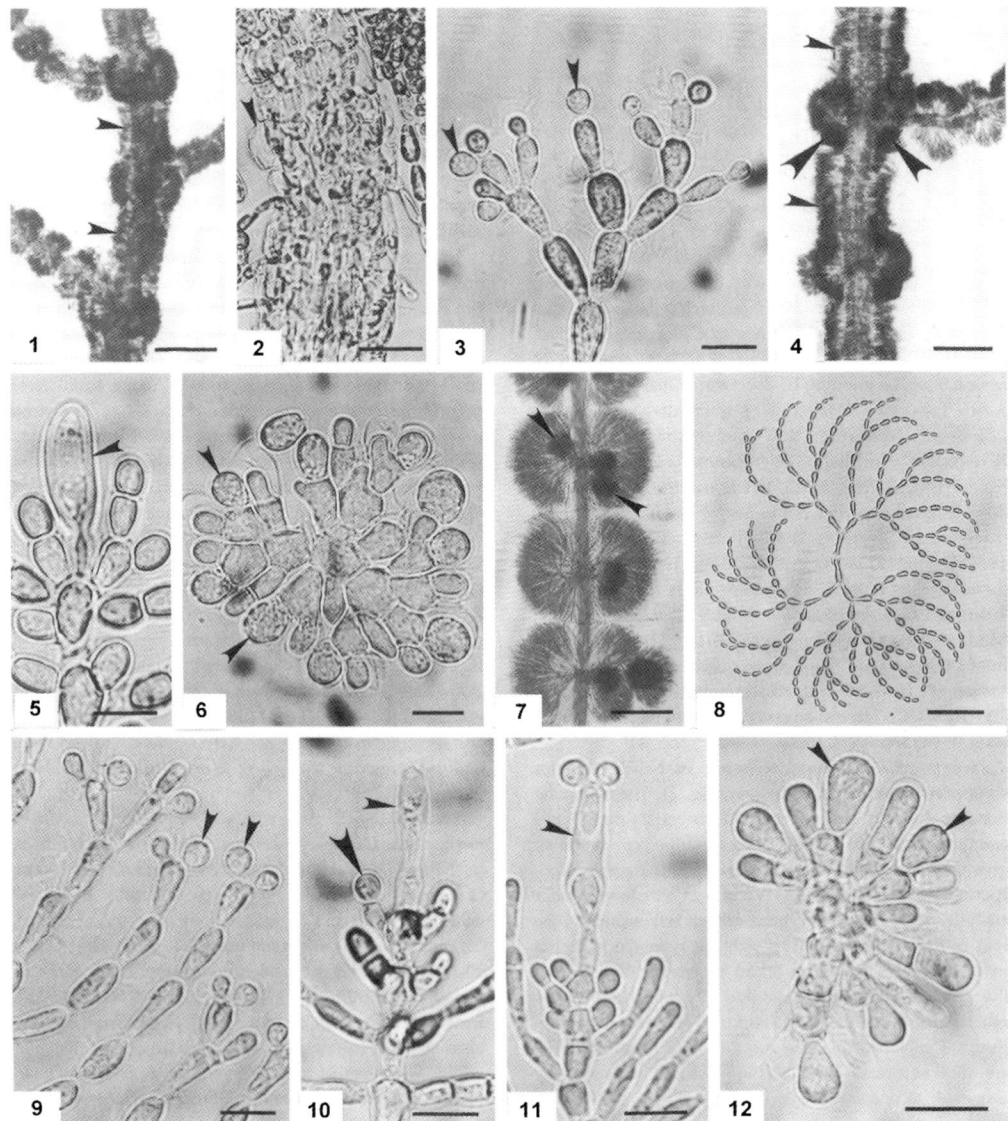

Plate 40

Batrachospermum heterocorticum, figs 1–6 (Vis, Sheath & Cole 1996b)
1. male plant main axis with barrel-shaped whorls and abundant secondary fascicle growth (arrowheads), 2. cortication with bulbous cells (arrowhead), 3. spermatangia (arrowheads) terminal on fascicle, 4. female plant with abundant secondary fascicle growth (small arrowheads) and barrel-shaped whorls containing spherical carposporophytes (large arrowheads), 5. carpogonium with lanceolate trichogyne (arrowhead), 6. carposporophyte with short, dense 2–3 celled gonimoblast filaments having terminal carposporangia (arrowhead).

Batrachospermum pulchrum, figs 7–12 (Vis, Sheath & Cole 1996b)
7. main axis with barrel-shaped whorls containing spherical carposporophytes (arrowhead), 8. fascicles curved in on one direction, 9. spermatangia (arrowheads) terminal on fascicle, 10. carpogonium with lanceolate trichogyne (small arrowhead), involucral filaments with terminal spermatangia (large arrowheads), 11. fertilized carpogonium with lanceolate trichogyne (arrowhead), 12. carposporophyte with short, 2–3 celled gonimoblast filaments having terminal carposporangia (arrowheads).
Scale bars = 350 µm (figs 1–4); 200 µm (fig. 7); 150 µm (fig. 8); 100 µm (fig. 12); 10 µm (figs 5–6, 9–11).

1. *Batrachospermum*

6) ***Batrachospermum spermatoinvolucrum*** Vis et Sheath (1996:124, figs 2–8)

plate 41, figs 1–7 (original authors, Vis & Sheath 1996)

Plants monoecious, whorls confluent, spherical; mature whorls with carposporophytes 239–846 μm in diameter, fascicles consisting of 5–18 cell-storeys; cortical filament cells cylindrical. Spermatangia terminal on vegetative fascicles and on involucral filaments of carpogonium-bearing branch. Carpogonia variable in size and shape, 8–17 μm in diameter, 40–86 μm long. Carposporophytes spherical, 1–10 per whorl and at various distances from axis, 44–158 μm in diameter; carposporangia terminal on 2–4-celled gonimoblast filaments, 6–14 μm in diameter, 6–16 μm long.

Type Locality: Nain, Labrador, Newfoundland, Canada in North America.
Holotype: UBC A81615.
Distribution: Known from 19 northern localities in North America.
Note: Vis & Sheath (1998) reduced this species to a form of *B. gelatinosum*, because the rbcL gene sequence from third population of this species differed by only one nucleotide substitution from the *B. geletinosum* population.

7) ***Batrachospermum trichofurcatum*** Sheath et Vis in Vis & Sheath (1996:128, figs 9–15)

plate 41, figs 8–14 (original authors, Vis & Sheath 1996)

Plants monoecious, whorl confluent, spherical, mature whorls with carposporophytes 448–1220 μm in diameter, fascicles consisting of 10–14 cell-storeys. Spermatangia in clusters terminal on fascicles. Carpogonia 5–9 μm in diameter, 18–27 μm long; carpogonium-bearing branch undifferentiated, consisting of 3–8 cells. Immature trichogyne with slight protrusion; mature trichogyne forked, sometimes with second-order furcations. Carposporophytes spherical, 1–8 per whorl, at various distances from axis, 78–156 μm in diameter; carposporangia terminal on 3–4-celled gonimoblast filaments, 6–9 μm in diameter, 8–12 μm long.

Type Locality: Route 101 in Redwood National Park at Prairie Creek Park, California, USA in North America.
Holotype: UBC A81617.
Distribution: Known from the type locality only.

8) ***Batrachospermum trichocontortum*** Sheath et Vis in Vis & Sheath (1996:131, figs 35–44)

plate 42, figs 1–10 (original authors, Vis & Sheath 1996)

Plants dioecious, having spherical whorls separated by abundant secondary fascicles. Mid-fascicle cells forming rhizoidal outgrowths which can be quite long. Spermatangia in dense clusters terminal on fascicles. Mature female whorls relatively large, 1110–1970 μm in diameter, composed of 12–18 cell-storeys. Carpogonium-bearing branches undifferentiated, consisting of 4–8 cells, with long and short involucral filaments. Carpogonium 7–10 μm in diameter, 23–39 μm long. Trichogyne tip undulated, capitate or helically twisted. Carposporophytes spherical, 1–4 per whorl, at various distances from axis. Mature carposporophytes spherical, 135–304 μm in diameter. Carposporangia 5–9 μm in diameter, 9–14 μm long, terminal on 2–4-celled gonimoblast filaments.

Type Locality: Route 121, 5 km north of Route 76 and outskirts of Newberry, South Carolina, USA in North America.
Holotype: UBC A81616.
Distribution: Known from the type locality only.

Batrachospermales

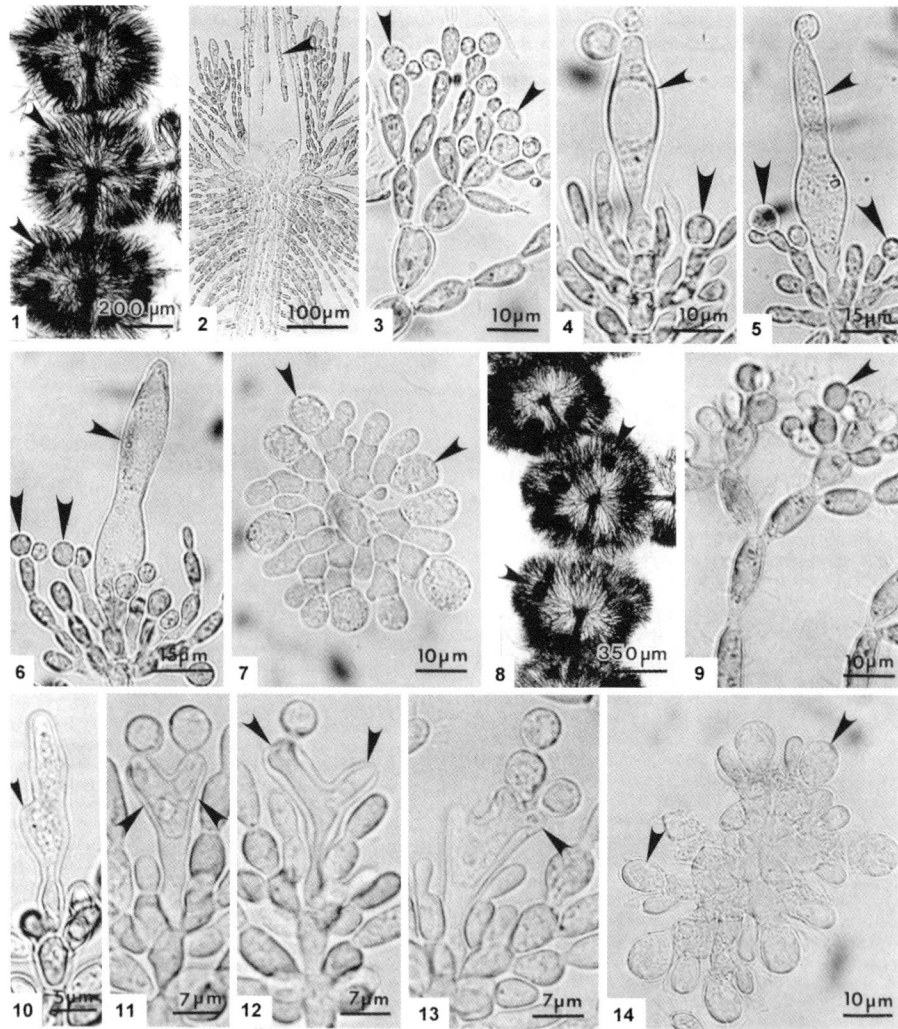

Plate 41
Batrachospermum spermatoinvolucrum, figs 1–7 (original authors, Vis & Sheath 1996)
1. a main axis with confluent, globose whorls having numerous spherical carposporophytes (arrows) at various distances from the axis, 2. axial cells covered by cortical filaments composed of regular, cylindrical cells (arrow), 3. abundant spermatangia (arrows) terminal on fascicles, 4. a fertilized carpogonium with a club-shaped, lanceolate trichogyne (small arrow) and a carpogonium-bearing branch with involucral filaments some of which have apical spermatangia (large arrow), 5. a fertilized carpogonium with an elongate trichogyne (small arrow) and at base surrounded by involucral filaments with terminal spermatangia (large arrows), 6. a carpogonium with an elongate, concave, lanceolate trichogyne (small arrow) and at base surrounded by dense involucral filaments with terminal spermatangia (large arrows), 7. carposporophytes with short, dense 2–3 celled gonimoblast filaments having terminal carposporangia (arrows).

Batrachospermum trichofurcatum, figs 8–14 (original authors, Vis & Sheath 1996)
8. a main axis with spherical whorls containing spherical carposporophytes (arrows) at various distances from the axis, 9. a fascicle terminated abundant spermatangia (arrow), 10. a carpogonium with a trichogyne forming a slight protrusion (arrow), 11. a carpogonium with a forked trichogyne (arrows) having a spermatium attached to each fork, 12. a fertilized carpogonium with a well-forked trichogyne (arrows), 13. a carpogonium with a forked trichogyne and one forked again (arrow), 14. a carposporophyte composed of dense 2–3 celled gonimoblast filaments terminated with carposporangia (arrows).

1. *Batrachospermum*

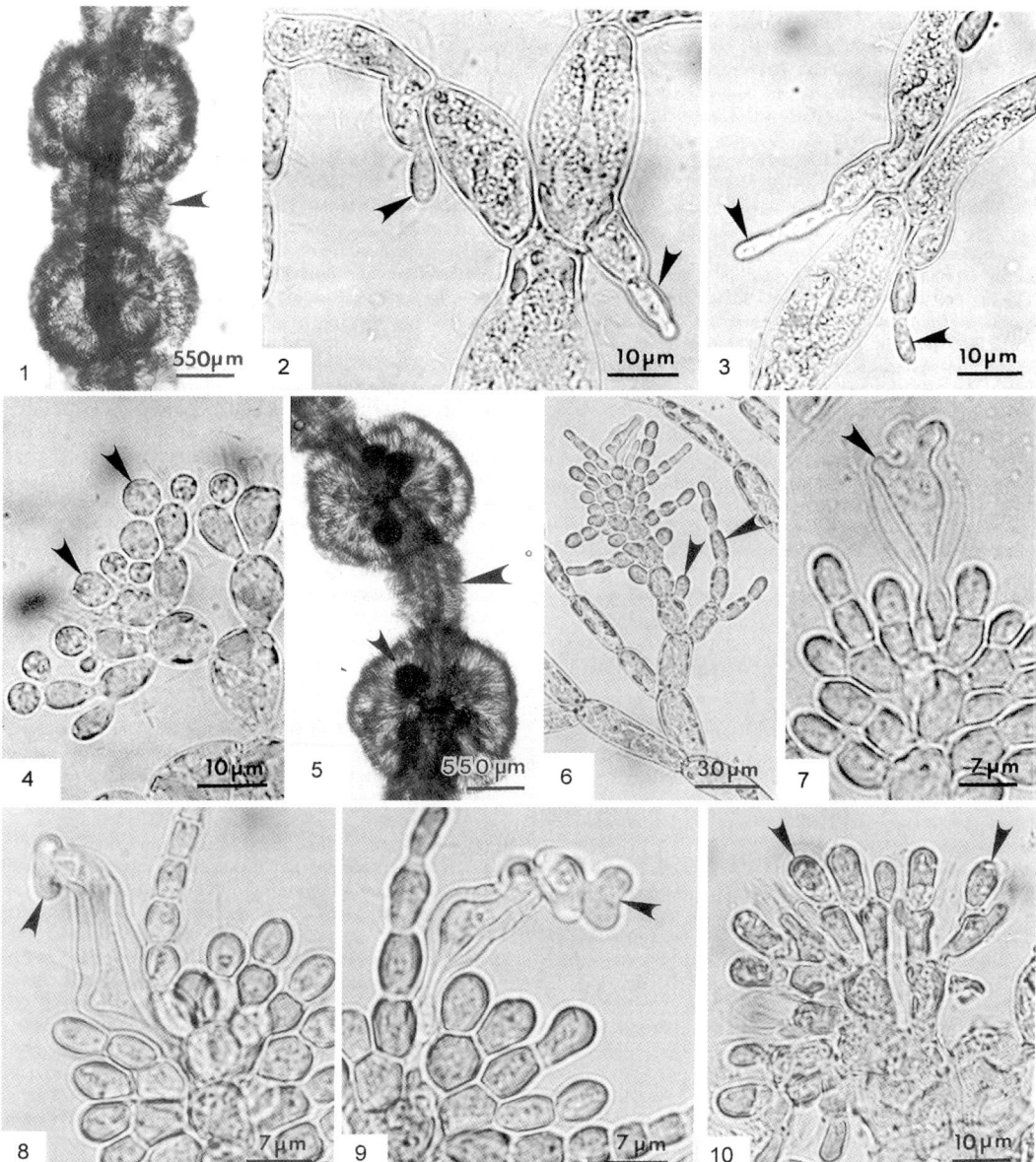

Plate 42
Batrachospermum trichocontortum, figs 1–10 (original authors, Vis & Sheath 1996)
1. a male plant, spherical whorls separated by dense secondary fascicle growth (arrow), 2. intercalary cells of fascicle with cortical outgrowths (arrows), 3. intercalary cells of fascicle with well-developed cortical outgrowths (arrows), 4. abundant spermatangia (arrows) terminal on fascicles, 5. a female plant, a main axis with globose whorls containing spherical carposporophytes (small arrow) separated by dense fascicle growth (large arrow), 6. a carpogonium on a seven-celled undifferentiated branch with dense involucral filaments of various length (arrows), 7. a carpogonium with an undulate trichogyne apex (arrow) and at base surrounded by dense involucral filaments, 8. a carpogonium with a capitate trichogyne apex (arrow) and at base surrounded by dense involucral filaments, 9. a carpogonium with a helically twisted trichogyne apex (arrow) and at base surrounded by dense involucral filaments, 10. a carposporophyte with short gonimoblast filaments having apical carposporangia (arrows).

Batrachospermales

9) ***Batrachospermum carpocontortum*** Sheath, Morison, Cole et Vanalstyne (1986:325, figs 2–15)

plate 43, figs 1–6 (original authors, Sheath *et al.* 1986)

Plants dioecious, 3–8 cm high, olive-green to brown. Whorls barrel-shaped and contiguous, 736–1660 µm in diameter for female; 486–960 µm in diameter for male. Fascicles dichotomous or trichotomous, consisting of 5–7 cell-storeys. Cortical filaments producing abundant secondary fascicles. Proximal cells cylindrical, 5.3–9.0 µm in diameter, terminal hairs rare. Spermatangia terminal on fascicles, spherical to ovoid, 3.9–6.0 µm in diameter. Carpogonium-bearing branches arising from proximal cells; fascicles, straight, composed of 3–5 cells. Carpogonia 4.4–8.1 µm in diameter, 22.6–36.6 µm long; trichogyne sessile, with protrusions and bends. Involucral branches long, extending well beyond the trichogyne. Carposporophytes generally spherical, single or in pairs, in mid to outer portion of whorls, 100–249 µm in diameter. Carposporangia terminal on short lateral branches, 7.1–10.3 µm in diameter.

Type Locality: Cascade River, a small tributary of the Snake River, 5.9 km east of Rockport, on the south side of Route 20, Washington, USA in North America.

Holotype: WA 109, UBC A67724.

Distribution: Known from the type locality and the Skagit River, Washington, USA in North America.

10) ***Batrachospermum szechwanense*** Jao (1941:264, tab. IV, figs 25–27)

plate 44 (a), figs 1–2 (original author, Jao 1941)

Plants dioecious, very mucilaginous, brownish violet, up to 4 cm high, up to 1000 µm in diameter, alternately or irregularly and richly branched. Whorls well-developed, depressed contiguous. Terminal hairs very short and rare. Cortical filaments absent. Spermatangia spherical, terminal or lateral on fascicles, 5.5–8.0 µm in diameter. Carpogonium-bearing branches arising from periaxial cells, consisting of 4–8 cells, usually with secondary carpogonium-bearing branches. Carpogonium with trichogyne ellipsoidal, more or less curved when mature. Carposporophytes unknown.

Type Locality: Kan-tung-tze about five miles northeast of Pehpei, Szechwan, China in Eastern Asia. On the concrete walls of a mill dam, in fast-running water emanating from a limestone cave.

Holotype: SC 1150.

Distribution: Known from the type locality only.

Note: According to Jao (1941) this species differs from *B. boryanum* Sirodot, chiefly in that male and female plants are similar in appearance, cortical filaments are entirely wanting and the trichogyne is curved.

11) ***Batrachospermum heteromorphum*** Shi, Hu et Kumano (1993:295, figs 1–21)
non *B. helminthosum* Bory var. *heteromorphum* Reis (1972:209)

plate 44 (b), figs 1–2 (Kumano 1997e)
figs 3–10 (original authors, Shi *et al.* 1993)

Plants monoecious, caespitose, mucilaginous, 5–9 cm high, 300–600 µm in diameter, densely and irregularly branched; whorls of two types: subspherical whorls closely touching each other, 900–1050 µm in diameter), with primary fascicles dichotomously branched, 11–16 cell-storeys, cells oblong-ovoid and with secondary fascicles rare, unbranched or dichotomously

1. *Batrachospermum*

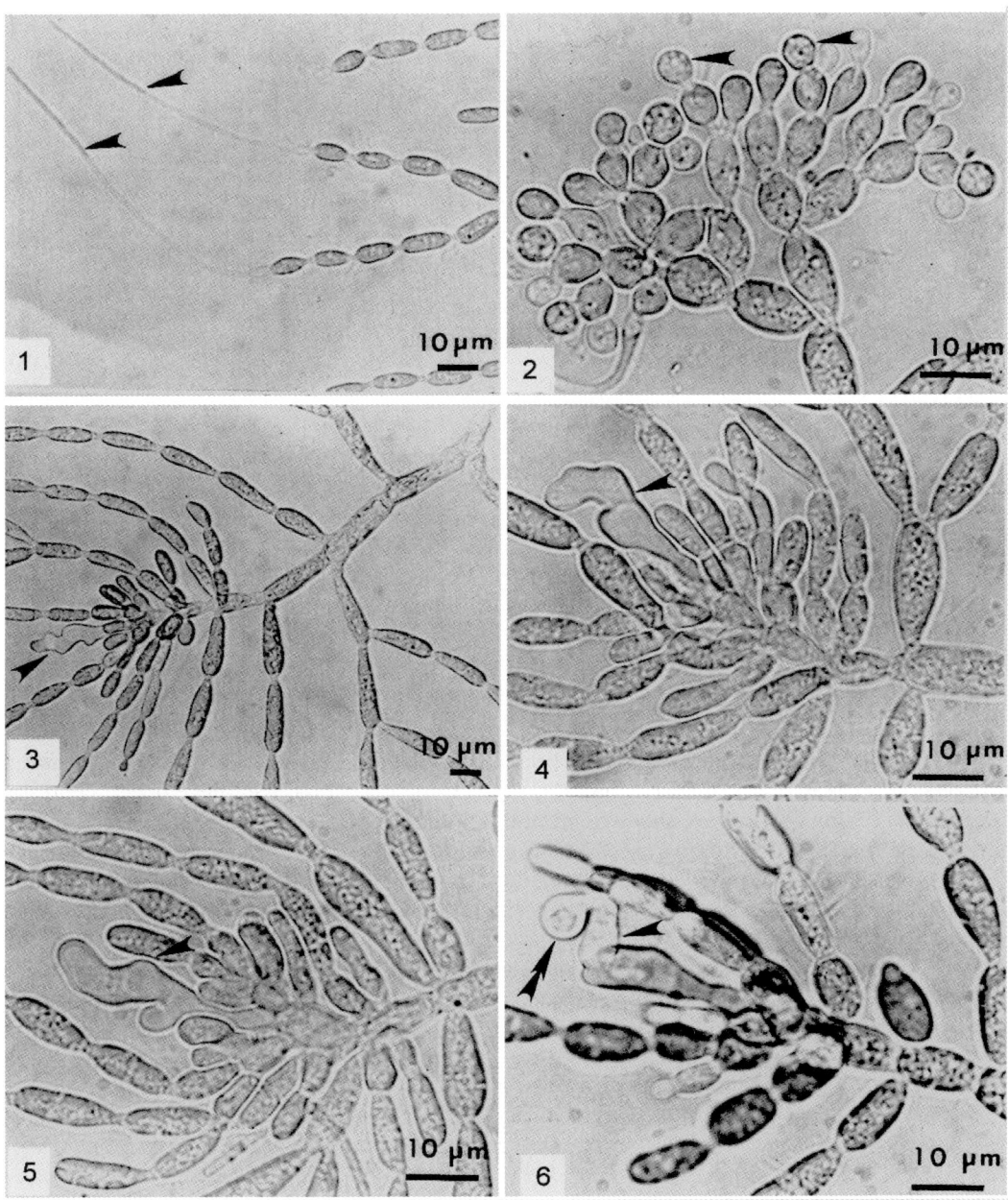

Plate 43
Batrachospermum carpocontortum, figs 1–6 (original authors, Sheath *et al.* 1986)
1. terminal hairs (arrows), 2. spermatangia terminal on fascicles (arrows), 3. a six-celled undifferentiated carpogonium-bearing branch with a mature carpogonium (arrow) and long involucral filaments, 4–5. a mature carpogonium (arrows) exhibiting protrusions and bends, a trichogyne base is partly covered the adjacent fascicle cell, giving an artificial impression of a pedicellate attachment, 6. a fertilized carpogonium (arrow) exhibiting bending and with an attached spermatium (double arrow).

Batrachospermales

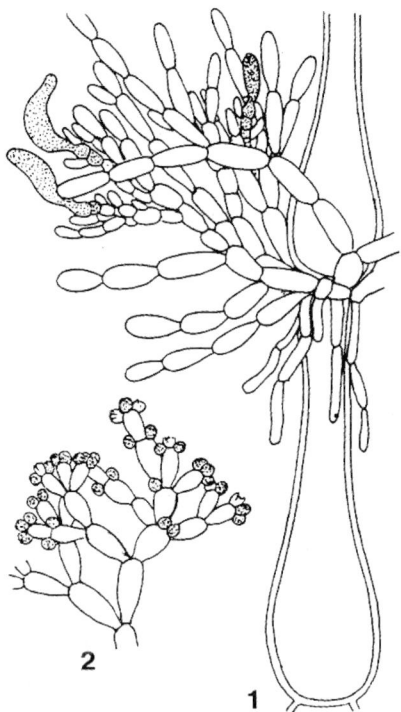

Plate 44 (a)
Batrachospermum szechwanense, figs 1–2 (original author, Jao 1941)
1. undifferentiated carpogonium-bearing branches with three carpogonia, 2. spermatangia terminal on fascicle.

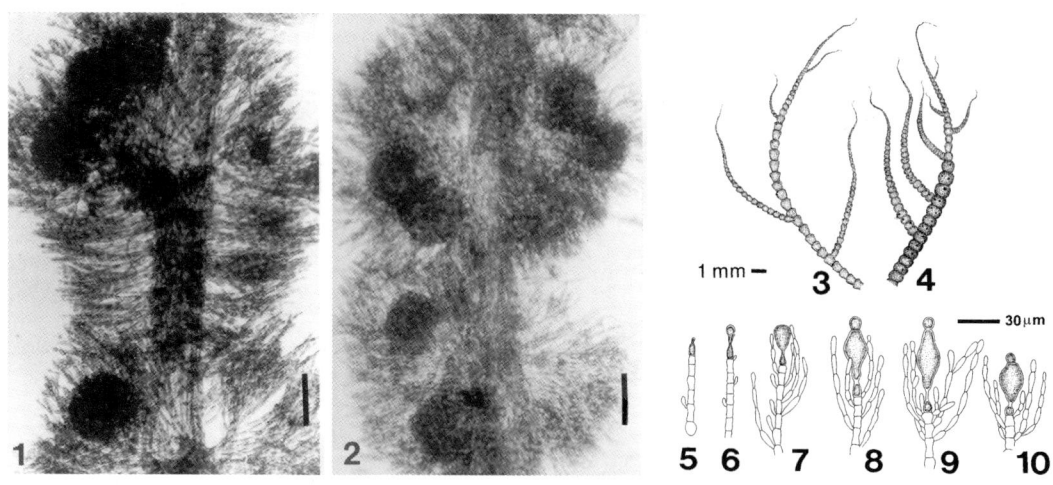

Plate 44 (b)
Batrachospermum heteromorphum, figs 1–2, (Kumano 1997e), figs 3–10 (original authors, Shi *et al.* 1993)
1, 3. obconical or pear-shaped whorls with one or several carposporophytes, 2, 4. hemispherical whorls with several carposporophytes, 5. a carpogonium with a rod-shaped initial of trichogyne, 6–7. young carpogonium with spatula trichogyne, 8–9. a fertilized carpogonia with an angled obovoidal or lanceolate trichogyne, 10. a fertilized carpogonium with an ellipsoidal trichogyne.
Scale bars = 1 mm (figs 3–4); 30 µm (figs 5–10); 10 µm (figs 1–2).

branched, 4–10 cell-storeys and obconical or pear-shaped whorls usually touching each other, 600–800 μm in diameter with primary fascicles dichotomously branched, 12–16 cell-storeys, proximal cells cylindrical or lanceolate, distal cells lanceolate or obovoidal, and with secondary fascicles numerous, unbranched or dichotomously branched densely covering all internodes, 4–14 cell-storeys, terminal hairs rare and short; spermatangia spherical or subspherical, 6–8 μm in diameter, terminal on fascicles; carpogonium-bearing branches straight, consisting of 5–12 cylindrical or barrel-shaped cells, arising from periaxial cells; carpogonia 5–7.5 μm in diameter at the base, 15–20 μm in diameter at the apex, 30–52 μm long, young trichogynes spathulate, mature ones angled-obovoidal, obtrullate, lanceolate, rarely ellipsoidal, indistinctly stalked; carposporophytes single or several, spherical or hemispherical, 70–160 μm in diameter, 80–175 μm high; carposporangia obovoidal, 7–10 μm in diameter, 13–18 μm long.

Type Locality: Dabie Mountain, Hubei, China in Eastern Asia. On rocks and stones in a mountain stream.
Holotype: HP 541.
Distribution: Known from the type locality only.

12) ***Batrachospermum cylindrocellulare*** Kumano (1978:100, fig. 3)

plate 45, figs 1–5 (Kumano 1996k)
figs 6–9 (original author, Kumano 1978)

Plants monoecious, very mucilaginous, deep bluish green, 2–7 cm high, 300 μm in diameter, more or less abundantly and irregularly branched; whorls spherical or ellipsoidal, more or less separated; primary fascicles sparsely branched, 7–11 cell-storeys, cells cylindrical, 4–7 μm in diameter, 7–15 μm long, terminal hairs absent; secondary fascicles numerous, sparsely branched; spermatangia spherical, terminal on laterals or shortened branchlets, 4–6 μm in diameter; carpogonium-bearing branches arising from periaxial cells, 25–40 μm long, consisting of 6–8 barrel-shaped cells; carpogonia 4–5 μm in diameter at the base, 6–9 μm in diameter at the apex, 17–25 μm long, trichogyne ovoid or urn-shaped, distinctly stalked; carposporophytes 2–3, centrally inserted in whorls, spherical, or hemispherical; carposporangia ellipsoidal or ovoid, 8–11 μm in diameter, 13–15 μm long.

Type Locality: Tasek Bera, Fort Iskander, Pahang, Malaysia in Southeast Asia. On submerged leaves of *Pandanus heriocopus* and *Cryptocoryne griffithii* in narrow rivulets covered by swamp forest.
Holotype: Kobe University, Kumano 16/IV 1971.
Distribution: Known from the type locality only.

13) ***Batrachospermum sinense*** Jao (1941:263, tab. IV, figs 28–30)

plate 46, figs 6–8 (original author, Jao 1941)

Plants monoecious, very mucilaginous, violet green, up to 10 cm high, 750 μm in diameter, abundantly, alternately partly unilaterally branched. Whorls spherical or depressed. Primary fascicles abundantly branched, 7–10 cell-storeys, cells elongate, terminal hairs few. Secondary fascicles sometimes very few. Spermatangia spherical, 6–8 μm in diameter. Carpogonium-bearing branches arising from periaxial cells, consisting of 4–8 cells, with involucral filaments. Carpogonia up to 24 μm in diameter, 45 μm long, trichogyne obovoidal, when young, becoming elongate pear-shaped. Carposporophytes 1–2, large, inserted in middle of whorls, subspherical, up to 200 μm in diameter. Carposporangia subovoidal, 10–11 μm in diameter, 24–27 μm long.

Batrachospermales

Type Locality: about four miles south of Hwang-kuo-shu, Pehpei, Szechwan, China in Eastern Asia. On rocks in a mountain stream.
Holotype: SC 1148.
Distribution: Known from the type locality only.
Note: Trichogynes of this species are initially cuneate, but become round or obovoidal, sometimes inflated like a balloon, and carpogonium-bearing branches are composed of barrel-shaped cells and provided with many elongated laterals. Jao (1941) assigned this species to the section *Turfosa*. However, it resembles more closely species of the section *Batrachospermum*, in which these characters are also observed (Kumano 1984b).

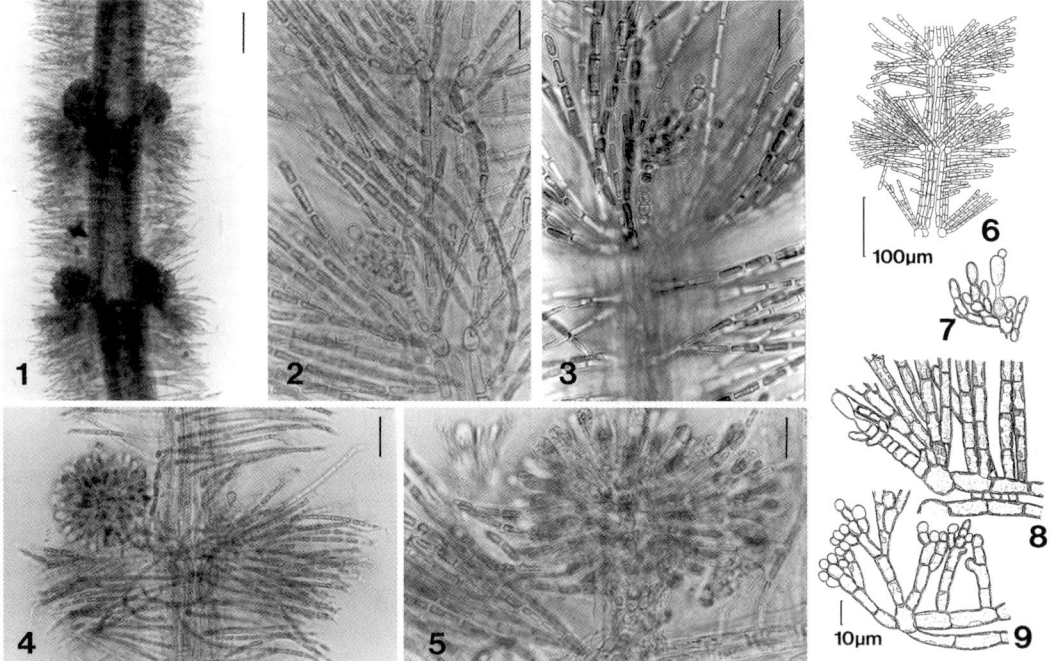

Plate 45
Batrachospermum cylindrocellulare, figs 1–5 (Kumano 1996k), figs 6–9 (original author, Kumano 1978)
1. a main axis with spherical, ellipsoidal, more or less separated whorls having axial carposporophytes, 2, 6. whorls showing primary fascicles composed of cylindrical cells, a differentiated carpogonium-bearing branch consisting of barrel-shaped cells, 3,9. spermatangia in clusters terminal on shortened branchlets, 7–8. a carpogonium-bearing branch and fertilized carpogonium, 4–5, a young and a mature carposporophyte terminated with carposporangia.
Scale bars = 100 μm (figs 1, 6); 40 μm (figs 2–4); 20 μm (fig. 5); 10 μm (figs 7–9).

14) ***Batrachospermum gelatinosum*** (Linnaeus) De Candolle (1801:21) *emend.* Vis *et al.* (1995:52)
Basionym: *Conferva gelatinosa* Linnaeus (1753:1166).
The synonymy proposed by Vis *et al.* (1995) is as follows;
Synonym: *B. ludibundum* var. *pulcherrimum* Bory (1808:323); *B. ludibundum* var. *caerulescens* Bory (1808:324); *B. ludibundum* var. *stagnale* Bory (1808:325); *B. hybridum* Bory (1823:222); *B. moniliforme* f. *lipsiensis* Rabenhorst (1868:405); *B. moniliforme* var. *pisanum* Arcangeli (1882:156); *B. moniliforme* var. *chlorosum* Sirodot (1884:211, pl. I, fig. 3); *B. moniliforme typicum* Sirodot (1884:211, pl. III, fig. 1), nom. illeg.; *B. moniliforme* var. *rubescens* Sirodot

1. *Batrachospermum*

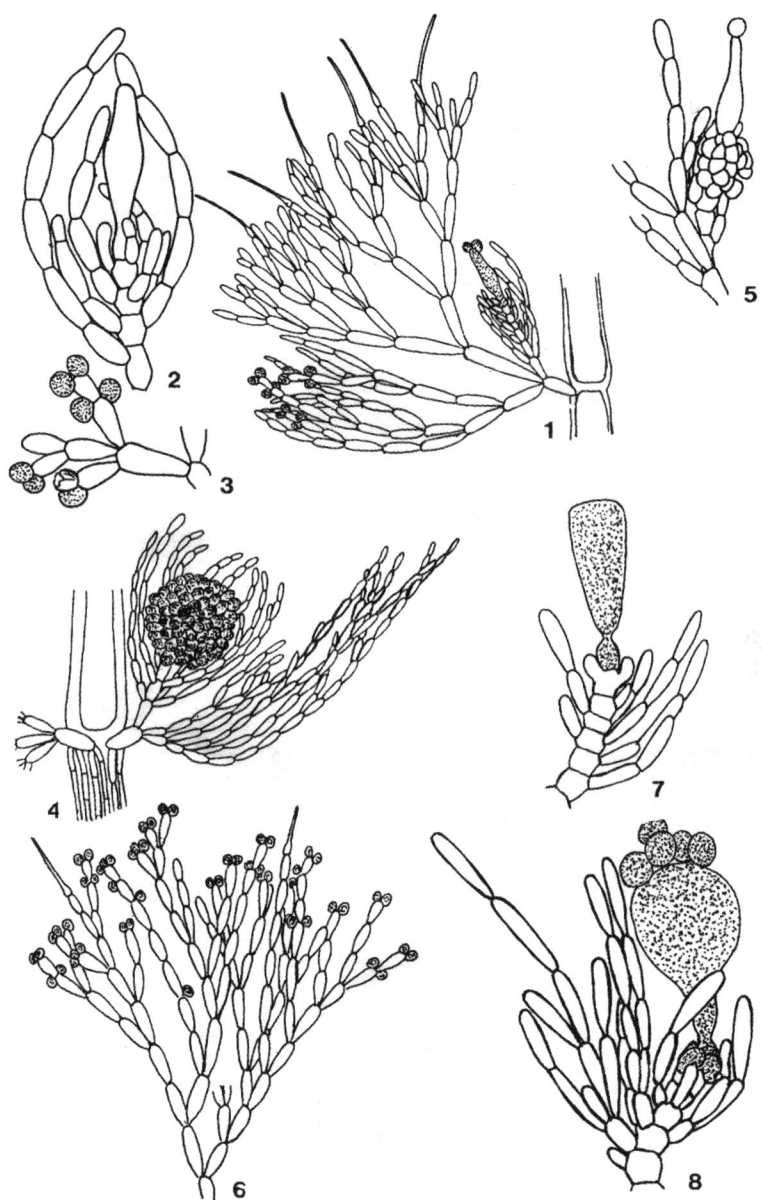

Plate 46
Batrachospermum gelatinosum as *B. moniliforme*, figs 1–4 (Sirodot 1884)
1. the portion of a whorl showing spermatangia terminal on primary fascicle, a differentiated carpogonium-bearing branch composed of barrel-shaped cells, 2. a carpogonium with a trichogyne, 3. spermatangia subterminal on primary fascicle, 4. a carposporophyte inserted within whorl.

Batrachospermum gelatinosum as *B. densum*, fig. 5 (original author, Sirodot 1884)
5. a fertilized carpogonium with an attached spermatium.

Batrachospermum sinense, figs 6–8 (original author, Jao 1941)
6. spermatangia terminal or subterminal on fascicle, 7. a carpogonium-bearing branch with a cuneate trichogyne, 8. a carpogonium-bearing branch with an inflated trichogyne and gonimoblast filaments arising from a fertilized carpogonium.

Batrachospermales

(1884:212, pl. I, figs 1–2); *B. moniliforme* var. *helminthoideum* Sirodot (1884:212, pl. IV, figs 1–2); *B. moniliforme* var. *scopula* Sirodot (1884:213, pl. IX, figs 1–5); *B. decaisneanum* Sirodot (1884:214, pl. I, fig. 4, pl. X, figs 1–10); *B. radians* Sirodot (1884:218, pl. I, fig. 5, pl. II, figs 4, 6–13); *B. reginense* Sirodot (1884:219, pl. XV, fig. 5, pl. XVI, figs 6–10); *B. corbula* Sirodot (1884:226, pl. V, figs 1–3, pl. VI, figs 9–11); *B. densum* Sirodot (1884:228, pl. XII, figs 1–2, pl. XIII, figs 1–11, pl. XIV, figs 1–8); *B. pygmaeum* Sirodot (1884:230, pl. XIX, figs 1–7); *B. pyramidale* Sirodot (1884:232, pl. XV, figs 1–4, pl. XVI, figs 1–5, pl. XVII, figs 1–6), nom. illeg.; *B. godronianum* Sirodot (1884:235, pl. XVIII, figs 1–9); *B. moniliforme* var. *isoeticola* Skuja (1928:205); *B. corbula* var. *alcoense* Reis (1954:70); *B. arcuatoideum* Reis (1973:139, tab. I, a–c); *B. japonicum* Mori (1975:470, pl. III, 1–12); *B. polycarpum* Mori, (1975:474, pl. I, 15, pl. II, 8, 11–12, pl. V, 1–11, pl. VI, 4); *B. moniliforme* var. *obtrullatum* Kumano et Watanabe (1983:89, figs 14–17, 18–27); *B. moniliforme* var. *trullatum* Shi *et al.* (1994:211, pl. 1, 1–4, pl. II, 1–6).

plate 46, figs 1–4 (original author, Sirodot 1884) as *B. moniliforme*,
fig. 5 (original author, Sirodot 1884) as *B. densum*
plate 47, figs 1–12 (Vis, Sheath & Cole 1996a) as *B. moniliforme*

Plants monoecious, whorls distant to confluent, sometimes pressed, spherical or barrel-shaped, 257–972 μm in diameter with 1–11 carposporophytes exserted or within the whorls at various distance from the axis; main axis with cortication, consisting of cylindrical cells only; carposporophytes spherical, pedicellate, 40–139 μm in diameter, gonimoblast filaments of 2–4 cylindrical cells; carpogonia 20–68 μm long with club-shaped or lanceolate trichogynes 5–17 μm in diameter; carpogonium-bearing branch undifferentiated, 3–10 cells long; carposporangia obovoidal, 8–16 μm long, 6–12 μm in diameter. (after Vis *et al.* 1995)

Type Locality: a locality in northern Europe

Lectotype: OXF Dillenius herbarium, specimen on which tab. 7, fig. 42 from Dillenius Historia Muscorum is based, designated by Compère (1991).

Distribution: Known from temperate regions such as France, Germany, Italy, Poland, Belgium, Sweden in Europe; India, Iraque in Asia; Korea, China, Japan in Eastern Asia; Australia; Rhode Island, Baffin Island, Texas, Louisiana, USA in North America; South America. In Arctic regions such as Yukon, Newfoundland, Canada and Alaska, USA in North America.

In China, on rocks in a stony pool, in which water slowly flowing, well shaded by trees, Nonyoh in Human. In Japan, Inokashira, Shakujii, Kokubunji in Tokyo, Northern region of Chiba, Kamo in Aichi, Ohno in Fukui, Okada, Imajyuku, Chounotsubo, Himeji, Yumesaki-gawa River, Tasuno, Shingu, Otama-no shimizu in Hyogo, Kuse in Okayama, Tamabuchi-sen, Minami-doi in Ehime.

15) ***Batrachospermum anatinum*** Sirodot (1884:249, pl. XXXII, figs 1–7, pl. XXXIII, figs 1–4) *emend.* Vis *et al.* (1995:54)

The synonymy proposed by Vis *et al.* (1995) as follows;
Synonym: *B. ectocarpum* Sirodot (1884:222, pl. VII, figs 1–5, pl. VIII, figs 1–7), nom. illeg.

plate 48, figs 1–5 (original author, Sirodot 1884) as *B. ectocarpum*
plate 49, figs 1–6 (Vis, Sheath & Cole 1996b)

Plants monoecious; main axis with inflated irregular cortication; whorls confluent or distant, spherical, 479–994 μm in diameter with 1–11 carposporophytes scattered within and exserted from the whorl; carposporophytes spherical, pedicellate, 69–140 μm in diameter, gonimoblast filaments of 2–4 cylindrical cells, carpogonia 17–41 μm long with inflated clavate or lanceolate

1. *Batrachospermum*

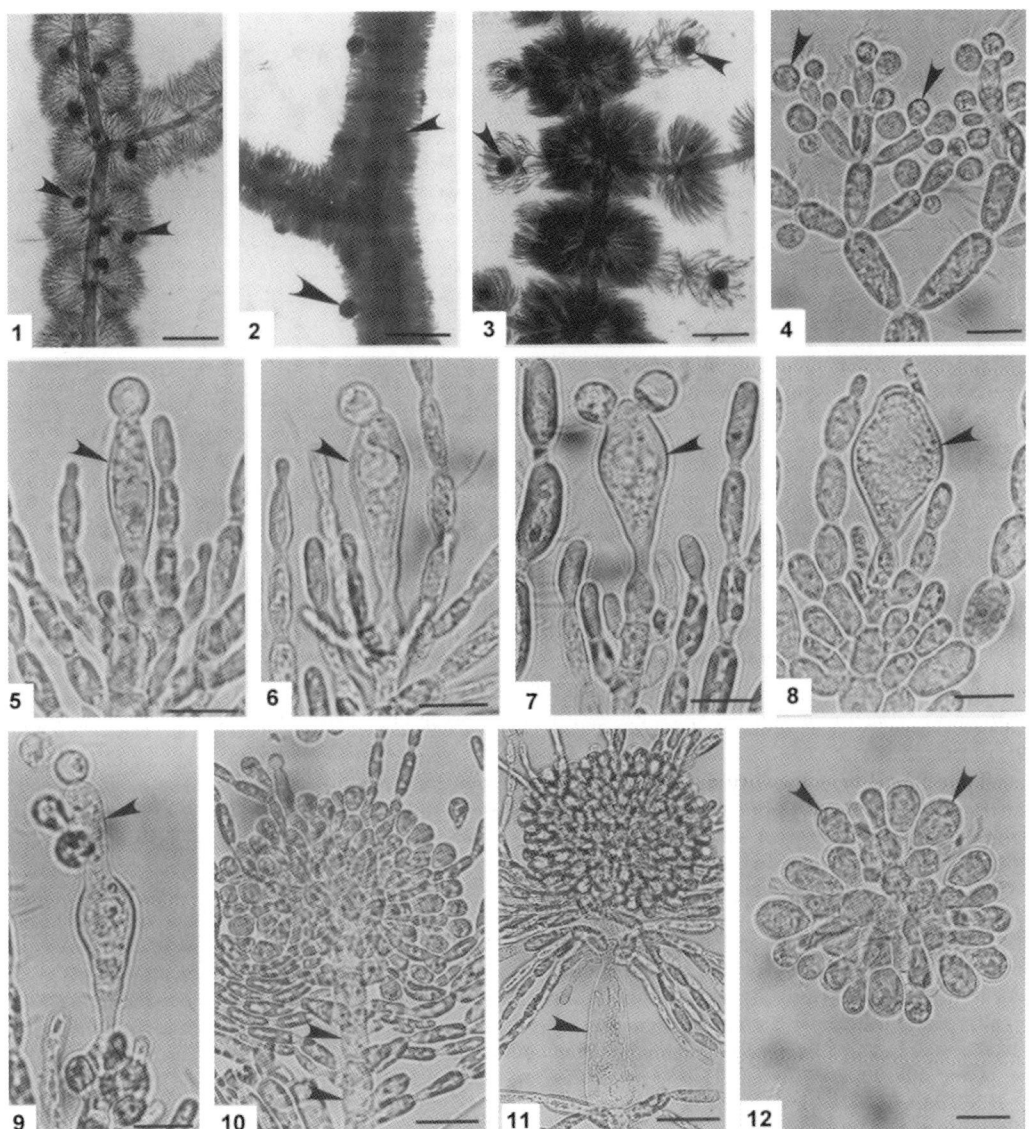

Plate 47
Batrachospermum gelatinosum as *B. moniliforme*, figs 1–12 (Vis, Sheath & Cole 1996a)
1. main axis with barrel-shaped whorls containing numerous carposporophytes (arrowheads), Connecticut, 2. main axis with brown cortication, carposporophytes exerted (large arrowhead) and within (arrowhead) indistinct whorls, Northwest Territories, 3. main axis with barrel-shaped whorls and exerted carposporophytes (arrowheads), Missouri, 4. fascicles with abundant spermatangial development (arrowheads), Northwest Territories, 5. fertilized carpogonium with fusiform trichogyne (arrowhead), Virginia, 6. fertilized carpogonium with fusiform trichogyne (arrowhead), Maine, 7. fertilized carpogonium with slightly inflated, club-shaped, lanceolate trichogyne (arrowhead), Nova Scotia, 8. fertilized carpogonium with inflated, ellipsoidal trichogyne (arrowhead), Northwest Territories, 9. fertilized carpogonium with elongate, lanceolate trichogyne (arrowhead), Alaska, 10. carposporophyte on branch with cylindrical cells (arrowheads), Missouri, carposporophyte on branch with cylindrical cells (arrowheads), Missouri, 11. carposporophyte on branch with inflated, elongate cells (arrowheads), Missouri, 12. compact carposporophyte with obovoidal carposporangia (arrowheads).
Scale bars =300 µm (figs 1–2); 250 µm (fig. 3); 25 µm (fig. 11); 20 µm (fig. 10); 10 µm for figs 4–9, 12).

Batrachospermales

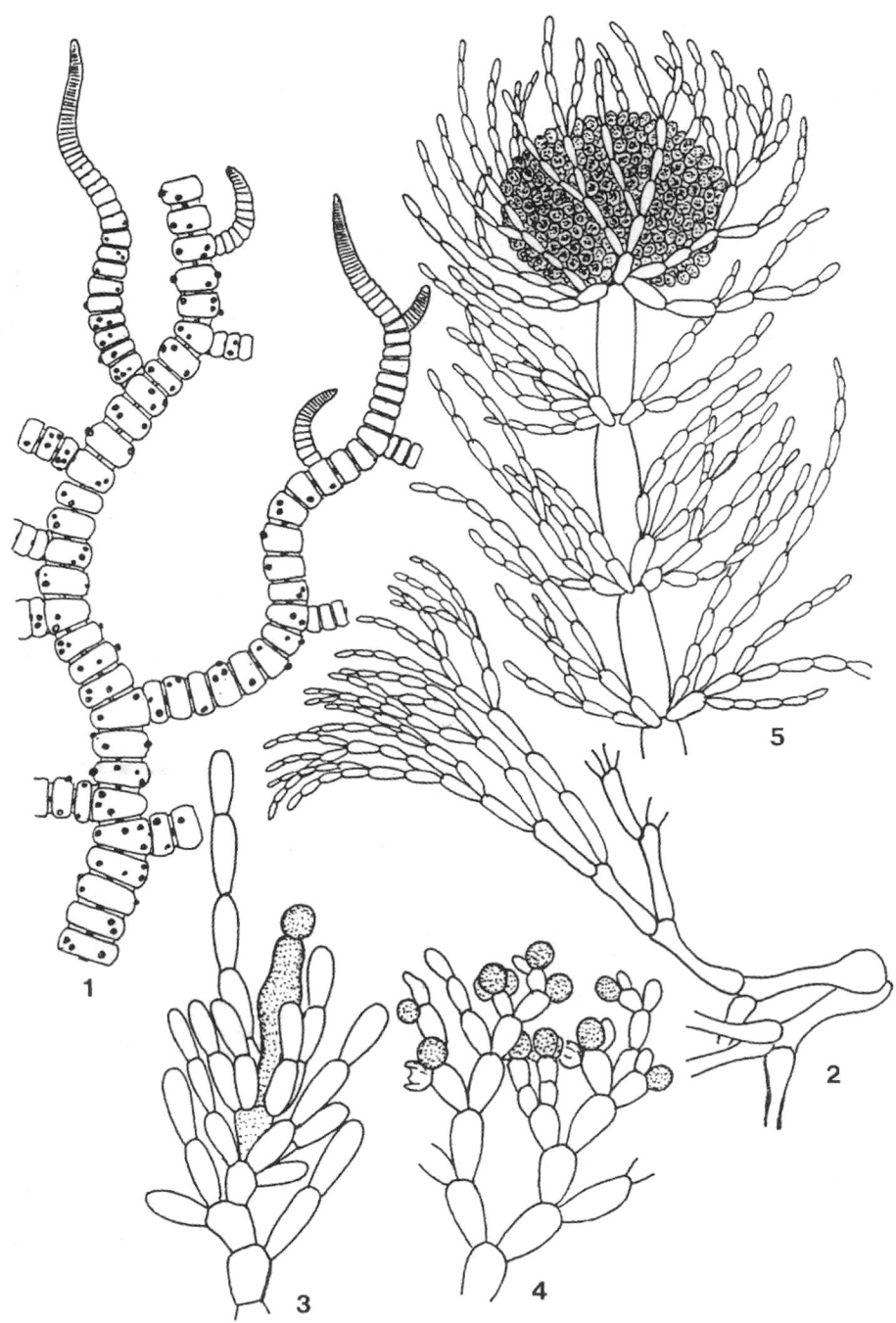

Plate 48
Batrachospermum anatinum as *B. ectocarpum,* figs 1–5 (original author, Sirodot 1884)
1. a main axis with barrel-shaped and compressed whorls with many carposporophytes within or exerted from whorls, 2. fascicles arising from a periaxial cell, 3. undifferentiated carpogonium-bearing branches composed of the similar shaped cells of fascicle and a carpogonium with a lanceolate trichogyne, 4. spermatangia terminal on fascicles, 5. a stalked carposporophyte with short gonimoblast filaments terminated with carposporangia.

1. *Batrachospermum*

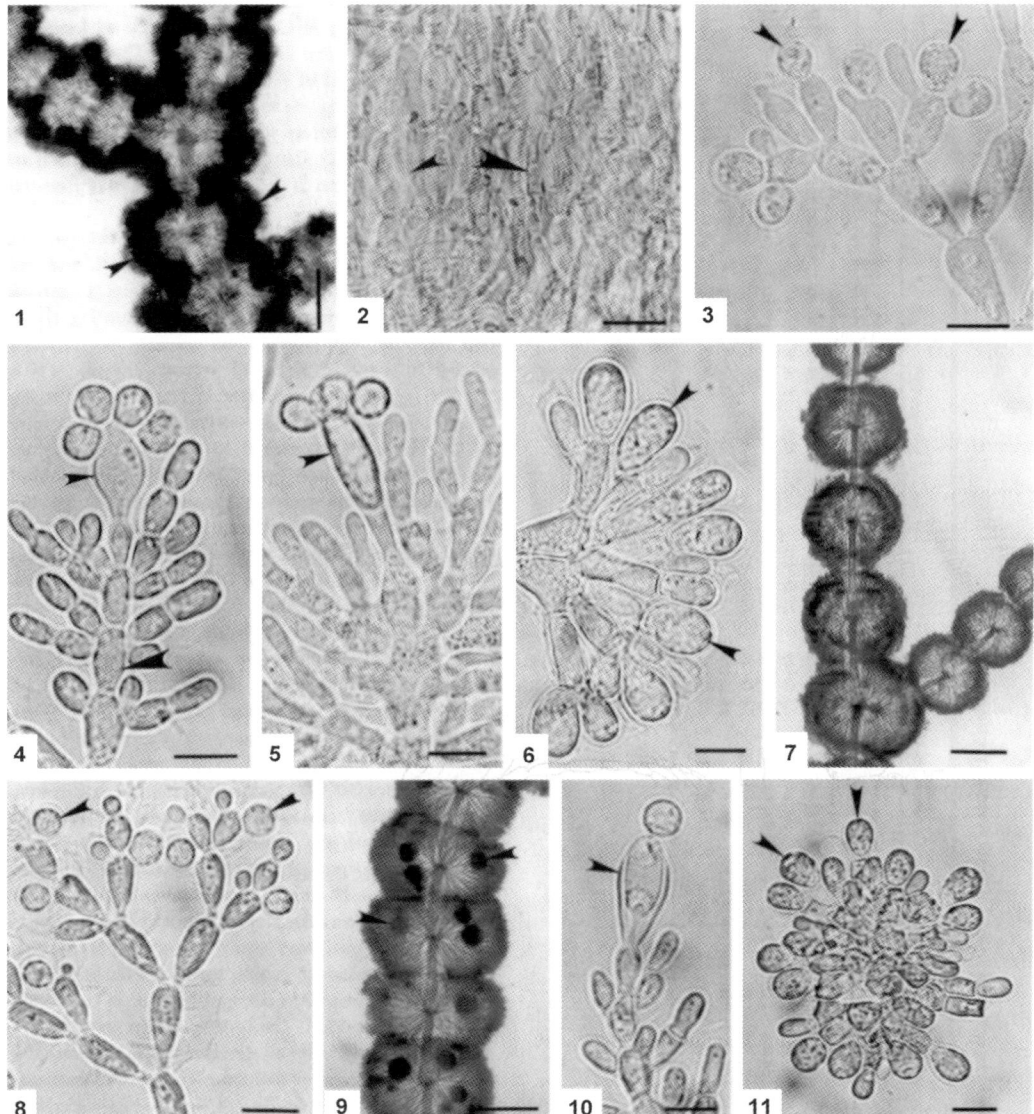

Plate 49
Batrachospermum anatinum, figs 1–6 (Vis, Sheath & Cole 1996b)
1. main axis with confluent, spherical whorls containing numerous spherical carposporophytes (arrowheads), 2. cortication of main axis with cylindrical (large arrowhead) to bulbous (small arrowhead) cell, 3. spermatangia (arrowheads) terminal on fascicle, 4. fertilized carpogonium with club-shaped, lanceolate trichogyne (small arrowhead) on undifferentiated carpogonium-bearing branch (large arrowhead), 5. fertilized carpogonium with lanceolate trichogyne (arrow head), 6. carposporophyte with 2 celled gonimoblast filaments having terminal carposporangia (arrowheads).

Batrachospermum arcuatum, figs 7–11 (Vis, Sheath & Cole 1996b)
7. male plant main axis with confluent, spherical whorls, 8. spermatangia (arrowheads) terminal on fascicle, 9. female plant main axis with confluent, barrel-shaped whorls containing numerous spherical carposporophytes (arrowheads), 10. fertilized carpogonium with club-shaped, lanceolate trichogyne (arrowhead), 11. carposporophyte with short dense, 2–3 celled gonimoblast filaments having terminal carposporangia (arrowheads). Scale bars: 300 μm (figs 1, 7, 9); 50 μm (fig. 2); 10 μm (figs 3–6, 8, 10–11).

trichogynes, 6–10 μm in diameter; carpogonial branch undifferentiated, 3–8 cells long; carposporangia obovoidal, 8–16 μm long, 7–14 μm in diameter.

Type Locality: Vau-de-Mau stream, near Monfort, France in Europe.
Lectotype: PC (designated by Vis *et al.* 1995: 36)
Distribution: Known from Portugal, France, Belgium, Sweden, Poland, Latvia in Europe; India in Asia; China, Japan in Eastern Asia; Australia; various localities such as Missouri, Arkansas, Virginia, USA and Canada in North America.

In China, on rocks in a mountain stream derived from underground springs, in front of Tsui-wong-ting, Lan-ya-shan in Chuchhou and beside Chin-lian Temple, Sin-tien-tze, Pa-hsin in Szechwan.

In Japan, known from spring-fed streams in Ohono in Fukui, Taniyatsu, Chiba, Yoshioka, Shikaido in Chiba, Sukain, Himeji in Hyogo, Kitakawa-cho in Miyazaki.

16) ***Batrachospermum fluitans*** Kerner (1882:362) *emend.* Vis *et al.* (1995:54)

Plants monoecious, main axis with inflated irregular cortication; whorls confluent, spherical, 466–664 μm in diameter, with 1–30 carposporophytes; carposporophytes spherical, pedicellate, 64–93 μm in diameter; carpogonia 52–65 μm long with lanceolate trichogynes 11–15 μm in diameter; carpogonium-bearing branch undifferentiated, 4–6 cells long.

Type Locality: near Mühlau, Innsbruck, Austria in Europe.
Syntype: BM.
Distribution: Known from Europe.

17) ***Batrachospermum nova-guineense*** Kumano et Johnstone (1983:66, figs 1–20)

plate 51, figs 1–5 (Kumano 1996ze)
figs 6–9 (original authors, Kumano & Johnstone 1983)

Plants dioecious, very mucilaginous, dull green, grayish blue, brown, 2–5 cm high, 300–550 μm in diameter, more or less irregularly branched; whorls ellipsoidal and separated, sometimes more or less compressed; primary fascicles abundantly branched, 8–18 cell-storeys, cells lanceolate-ellipsoidal, fusiform or ovoid, terminal hairs absent; cortical filaments well-developed; secondary fascicles rare; spermatangia spherical, 6–7 μm in diameter; carpogonium-bearing branches arising from periaxial cells and fascicles, 20–100 μm long, consisting of 2–7 barrel-shaped cells, slightly curved; involucral filaments numerous, elongated, unilaterally issued; carpogonia 3–5 μm in diameter at the base, 4–8 μm in diameter at the apex, 16–24 μm long, trichogyne ellipsoidal, distinctly stalked; carposporophytes single or rarely paired, inserted within whorls, spherical, 80–150 μm in diameter; carposporangia ellipsoidal or ovoid, 9–12 μm in diameter, 12–17 μm long.

Type Locality: Ove Ove Creek, a tributary of Veimauri River, about 50 km north-west of Port Moresby, in Papuan lowland, Papua New Guinea.
Holotype: UPNG 413a.
Distribution: Known from the type locality only.
Note: Johnstone *et al.* (1980) assigned this species to the section *Hybrida*, however, Kumano & Johnstone (1983) preferred to assign it to the section *Batrachospermum.*

18) ***Batrachospermum arcuatum*** Kylin (1912:22, fig. 7 a–e) *emend.* Vis *et al.* (1995:52)

plate 49, figs 7–11 (Vis, Sheath & Cole 1996b)
plate 50, figs 1–4 (original author, Kylin 1912)

Plants dioecious; whorls confluent, barrel-shaped, 428–727 μm in diameter with 1–3 peripheral carposporophytes; main axis with cortication consisting of cylindrical cells only; carposporophytes spherical, pedicellate, 92–129 μm in diameter, gonimoblast filaments of 2–4 cylindrical cells; carpogonia 28–39 μm long with clavate trichogynes 6–12 μm in diameter; carpogonium-bearing branch undifferentiated, 5–9 cells long; carposporangia obovoidal, 10–15 μm long, 7–12 μm in diameter.

Type Locality: Hör, Skane, Sweden in Europe.

Lectotype: PC; Wittrock & Nordstedt, Algae exsiccatae no. 1356b (see Entwisle & Foard 1997)

Distribution: Known from Portugal, Sweden, Belgium, Crimea and Poland in Europe; Japan in Eastern Asia; Australia; from Alaska, Woodman Hollow Creek, Iowa, USA to central Mexico in North America. In Japan, in Makomanai, Sapporo in Hokkaido, Seki in Tokyo, Ohwada, Tsukuba in Ibaragi, Ohshimizu, Ohno in Fukui, Mt. Iwawaki in Osaka, Kamigohri, Okada, Himeji, Nomura-yakujin, Yakogawa, Mondo-yakujin, Nishinomiya in Hyogo, Zentuji, Marugame in Kagawa, Tamabichi-sen in Ehime. Known from spring-fed streams.

Note: *B. arcuatum* and *B. anatinum*, in which the carpogonia are subtended by a relatively short filaments of unmodified cells, are included in *Batrachospermum arcuatum* group by Entwisle & Foard (1997).

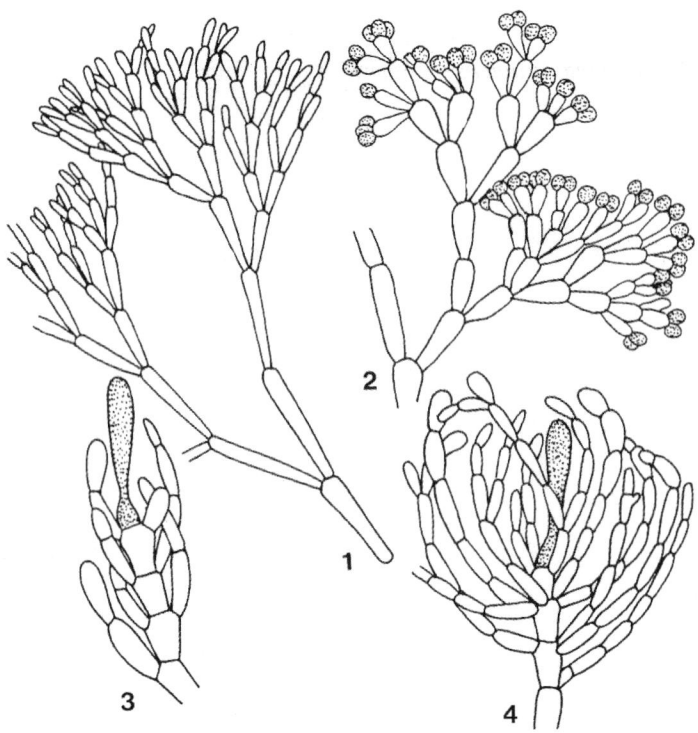

Plate 50
Batrachospermum arcuatum, figs 1–4 (original author, Kylin 1912)
1. a primary fascicle, 2. spermatangia terminal on fascicles, 3–4. carpogonium-bearing branches.

Batrachospermales

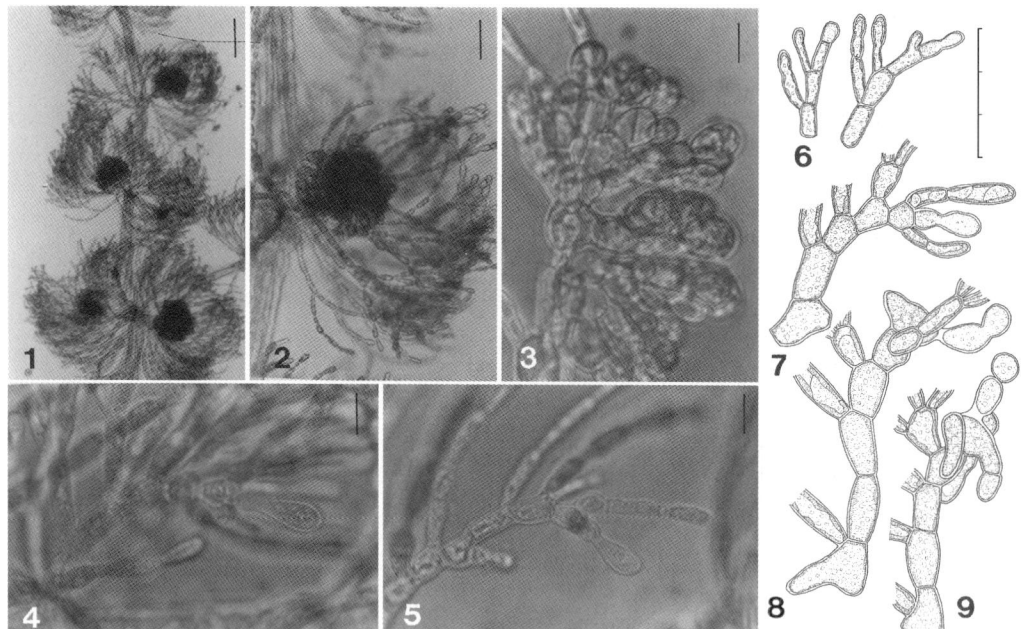

Plate 51

Batrachospermum nova-guineense, figs 1–5 (Kumano 1996ze), figs 6–9 (original authors, Kumano & Johnstone 1983)
1. the structure of whorls showing carposporophytes inserted within whorls, 2. a carposporophyte inserted within a whorl, 3. carposporangia terminal on radially branched determinate gonimoblast filaments, 4–5. a slightly curved carpogonium bearing branch with many elongated involucral filaments issued unilaterally, 6. the early stages in development of slightly curved carpogonium-bearing branches with many elongated involucral filaments issued unilaterally, 7. a mature carpogonium-bearing branch with an ellipsoidal or urn-shaped trichogyne, 8. a carpogonium bearing branch with mature and fertilized carpogonia, 9. the stage in development of a gonimoblast initial.
Scale bars = 100 µm (fig. 1); 40 µm (fig. 2); 30 µm (figs 6–9); 20 µm (figs 3–5).

19) ***Batrachospermum longipedicellatum*** Hua et Shi (1996:324, pl. I, 1–10, pl. II, 1–4, pl. III, 1–6)

plate 52, figs 1–10 (original authors, Hua & Shi 1996)
plate 53, figs 1–10 (original authors, Hua & Shi 1996)

Plants monoecious, 8–10 cm high, caespitose, densely and irregularly branched, with branches arising not only from periaxial cells but also from cells of cortical filaments. Whorls spherical or ellipsoidal, usually separated, 600–1000 µm in diameter. Primary fascicles consisting of 10–14 cell-storeys, di- rarely trichotomously branched, cells obovoidal or ellipsoidal, 7.5–15 µm in diameter, 10–25 µm long. Terminal hairs numerous, 20–150 µm long. Cortical filaments well-developed. Secondary fascicles numerous, sometimes sparse. Spermatangia spherical, 7–9 µm in diameter, terminal or lateral on primary and secondary fascicles. Carpogonium-bearing branch straight, arising from periaxial cells, proximal cells of fascicles and cells of cortical filaments. Trichogyne 4.5–6 µm in diameter, 20–30 µm long, ellipsoidal, elongate cylindrical, semilunate or sub-lunate when young, varied in shape at maturity. Carposporophytes one or more, involucral filaments numerous, usually exserted from whorl, diameter 1–2 times the radius of whorl. Carposporangia obovoidal, 8–9 µm in diameter, 10–11 µm long.

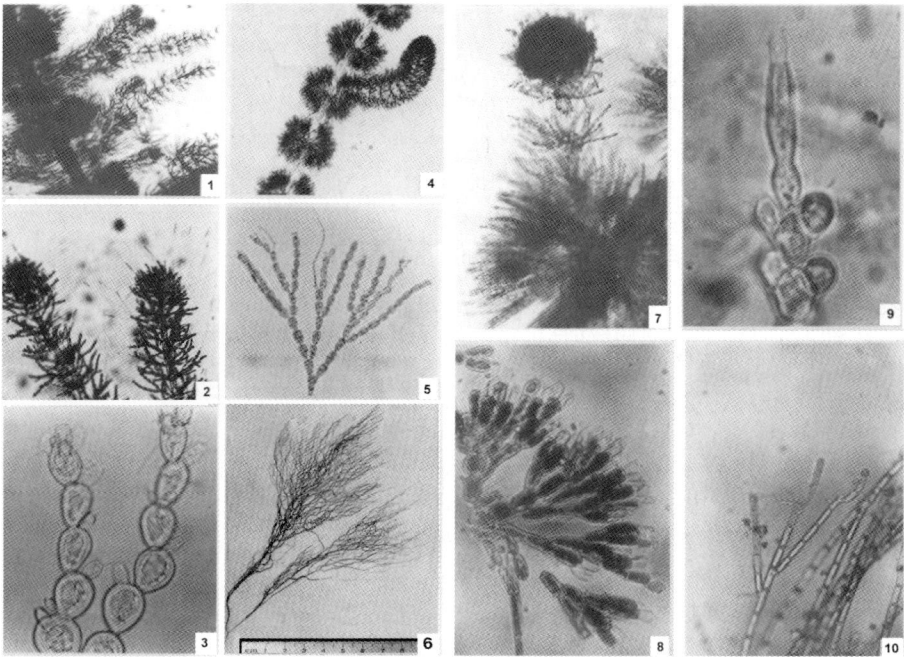

Plate 52
Batrachospermum longipedicellatum, figs 1–10 (original authors, Hua & Shi 1996)
1. secondary fascicles and a branch arising from the internode, 2. terminal hairs, 3. spermatangia terminal and lateral on fascicle, 4. spherical whorls, 5. fascicle dichotomously branched, 6. a plant showing the habit of ramification, 7. a carposporophyte with numerous involucral filaments and obviously extending out of a whorl, 8. carposporangia, 9. a carpogonium with a narrow cylindrical trichogyne, 10. chantransia phase.

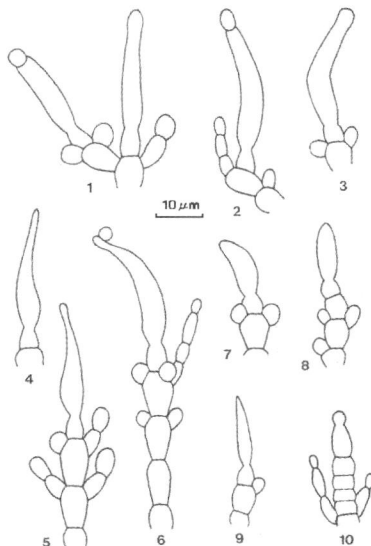

Plate 53
Batrachospermum longipedicellatum, figs 1–10 (original authors, Hua & Shi 1996)
1–3, 6. a carpogonium with a trichogyne narrow cylindrical and curved into luniform, 4–5. a lanceolate trichogyne, 7–9. a young trichogyne with lunate (7), ellipsoidal (8), subuliform (9), 10. the carpogonium with a papilla initial of trichogyne at the very early stage in development.

Batrachospermales

Type Locality: Xuzhou, Jiangsu, China in Eastern Asia. Growing in spring in calcareous mountains.
Holotype: XZTC 9102.
Distribution: Known from the type locality only.

20) ***Batrachospermum carpoinvolucrum*** Sheath et Vis in Vis & Sheath (1996:128, figs 16–24)

plate 54, figs 1–9 (original authors, Vis & Sheath 1996)

Plants dioecious, male and female plants with barrel-shaped, contiguous whorls. Main axis covered by heterocortication, cylindrical to bulbous cells. Spermatangia terminal on fascicles. Mature female whorls 489–853 µm. Carpogonium-bearing branch appearing to continue growth after degeneration of carpogonium. Carpogonia 6–8 µm in diameter, 15–26 µm long with sessile, clavate trichogynes. Carposporophytes spherical, 1–3 per whorl, protruding beyond whorl. Mature carposporophytes 128–232 µm in diameter. Carposporangia 7–11 µm in diameter, 10–14 µm long, terminal on 3–4 celled gonimoblast filaments.
Type Locality: Montezuma Well outflow canal, Arizona, USA in North America.
Holotype: UBC A81613.
Distribution: Known from the type locality only.

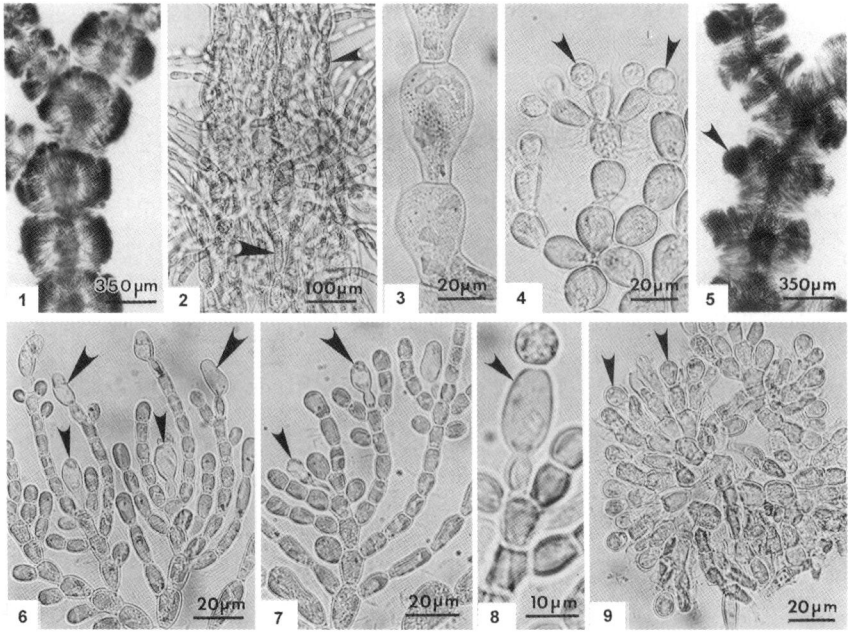

Plate 54
Batrachospermum carpoinvolucrum, figs 1–9 (original authors, Vis & Sheath 1996)
1. a male plant, main axis with confluent, barrel-shaped whorls, 2. the cortication of main axis with cylindrical (large arrow) to bulbous (small arrow) cells, 3. bulbous cortical cells, 4. abundant spermatangia (arrows) terminal on fascicle, 5. a female plant, a main axis with confluent, barrel-shaped whorls having large spherical exerted carposporophytes (arrow), 6. carpogonia (arrows) with long involucral filaments terminated with carpogonia (large arrows), 7. a degenerating carpogonium pushed aside (small arrow) by an elongating carpogonium-bearing branch and involucral filaments terminated with carpogonia (large arrows), 8. a fertilized carpogonium with an ovoidal trichogyne (arrow), 9. a carposporophyte with dense 2–4 celled gonimoblast filaments terminated with carposporangia (arrows).

1. *Batrachospermum*

21) ***Batrachospermum boryanum*** Sirodot (1874:136, 1884:246, pl. XXIX, figs 1–3, pl. XXX, figs 3–5, pl. XXXI, figs 1–6) *emend*. Vis *et al*. (1995:54)
Synonym: *B. ectocarpoideum* Skuja ex Flint (1949:552).

>plate 55, figs 1–6 (Kumano 1996e); figs 7–8 (original author, Sirodot 1884)
>plate 56, figs 1–7 (Vis, Sheath & Cole 1996b)
>plate 57, figs 1, 5 (original author, Sirodot 1884); figs 2–4 (Starmach 1976)

Plants dioecious; main axis with inflated, irregular cortication; whorls confluent, spherical, 337–1034 μm in diameter with 1–30 carposporophytes scattered within the whorl; carposporophytes spherical, pedicellate, 68–139 μm in diameter, gonimoblast filaments of 2–3 cylindrical cells; carpogonia 16–35 μm long with inflated clavate trichogyne 6–10 μm in diameter; carpogonium-bearing branch undifferentiated, 4–8 cells long; carposporangia obovoidal, 8–17 μm long, 7–14 μm in diameter.

Type Locality: near Bourg-des-Comptes, south of Rennes, France in Europe.
Lectotype: PC herb. Thuret.
Distribution: Known from Portugal, Belgium, Crimea, France, Poland, Romania and Sweden in Europe; Australia; from Newfoundland, British Columbia, Canada to Georgia, Michigan, USA, from tundra or tropical rain forests in North America; Brazil in South America.

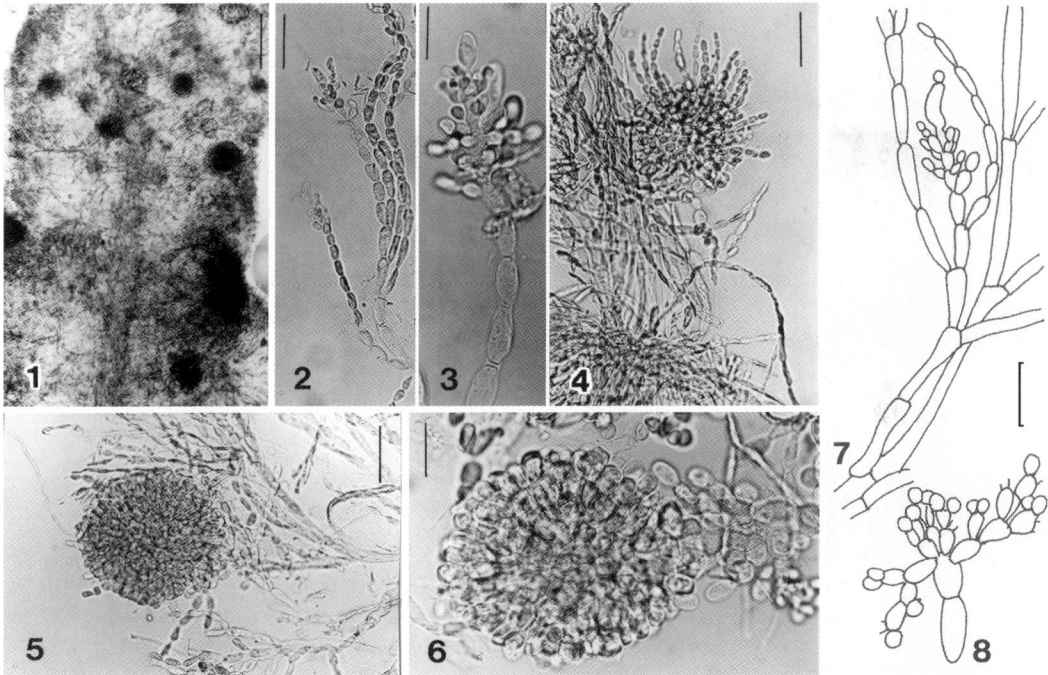

Plate 55
Batrachospermum boryanum, figs 1–6 (Kumano 1996e), figs 7–8 (original author, Sirodot 1884)
1, 4–6. a main axis with spherical whorls showing many small carposporophytes within whorls, 2, 8. spermatangia terminal on primary fascicles, 3, 7. a carpogonium and an undifferentiated carpogonium-bearing branch, young and matured carposporophytes with carposporangia terminal on gonimoblast filaments.

Batrachospermales

22) *Batrachospermum heterocorticum* Sheath et Cole (1990:566, figs 9–18)

plate 40, figs 1–6 (Vis, Sheath & Cole 1996b)
plate 58, figs 1–10 (original authors, Sheath & Cole 1990)

Plants dioecious, 2–6 cm high, 385–647 µm in diameter, olive-green. Whorls distinct at branch tips but poorly defined in old plants due to dense secondary fascicles. Fascicles uncurled, dichotomous or trichotomous, consisting of 4–7 cell-storeys. During initial stages of growth, cortical filaments cylindrical, 5.0–7.9 µm in diameter. In older plants, cortical cells enlarging

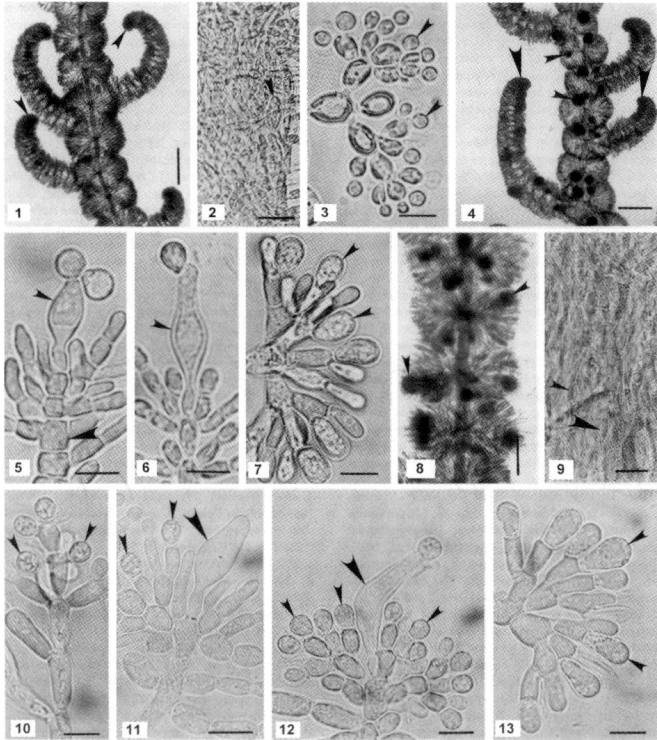

Plate 56
Batrachospermum boryanum, figs 1–7 (Vis, Sheath & Cole 1996b)
1. male plant main axis with confluent whorls and secondary branches curled at the tip (arrowheads), 2. main axis cortication containing bulbous cells (arrowhead), 3. spermatangia (arrowheads) terminal on fascicle, 4. female plant main axis with confluent whorls with carposporophytes (small arrow head) and secondary branches curled at the tips (arrowheads), 5. fertilized carpogonium with club-shaped, lanceolate trichogyne (small arrowhead) on undifferentiated carpogonium-bearing branch (large arrowhead), 6. fertilized carpogonium with inflated, lanceolate trichogyne (arrowhead), 7. carposporophyte with short, 2–3 celled gonimoblast filaments having terminal carposporangia (arrowheads).

Batrachospermum confusum, figs 8–13 (Vis, Sheath & Cole 1996b)
8. main axis with confluent whorls containing numerous spherical carposporophytes (arrowheads), 9. cortication of main axis with cylindrical (large arrowhead) to bulbous (small arrowhead) cells, 10. spermatangia (arrowheads) terminal on fascicle, 11. carpogonium with club-shaped, lanceolate trichogyne (large arrowhead), involucral filaments with terminal spermatangia (arrowheads), 12. fertilized carpogonium with bent, lanceolate trichogyne (large arrowhead), involucral filaments with abundant terminal spermatangia (small arrowheads), 13. carposporophyte with short, dense, 3 celled gonimoblast filaments having terminal carposporangia (arrowheads).
Scale bars = 400 µm (fig. 4); 350 µm (fig. 8); 300 µm (fig. 1); 50 µm (figs 2, 9); 10 µm (figs 3, 5–7, 10–13).

1. *Batrachospermum*

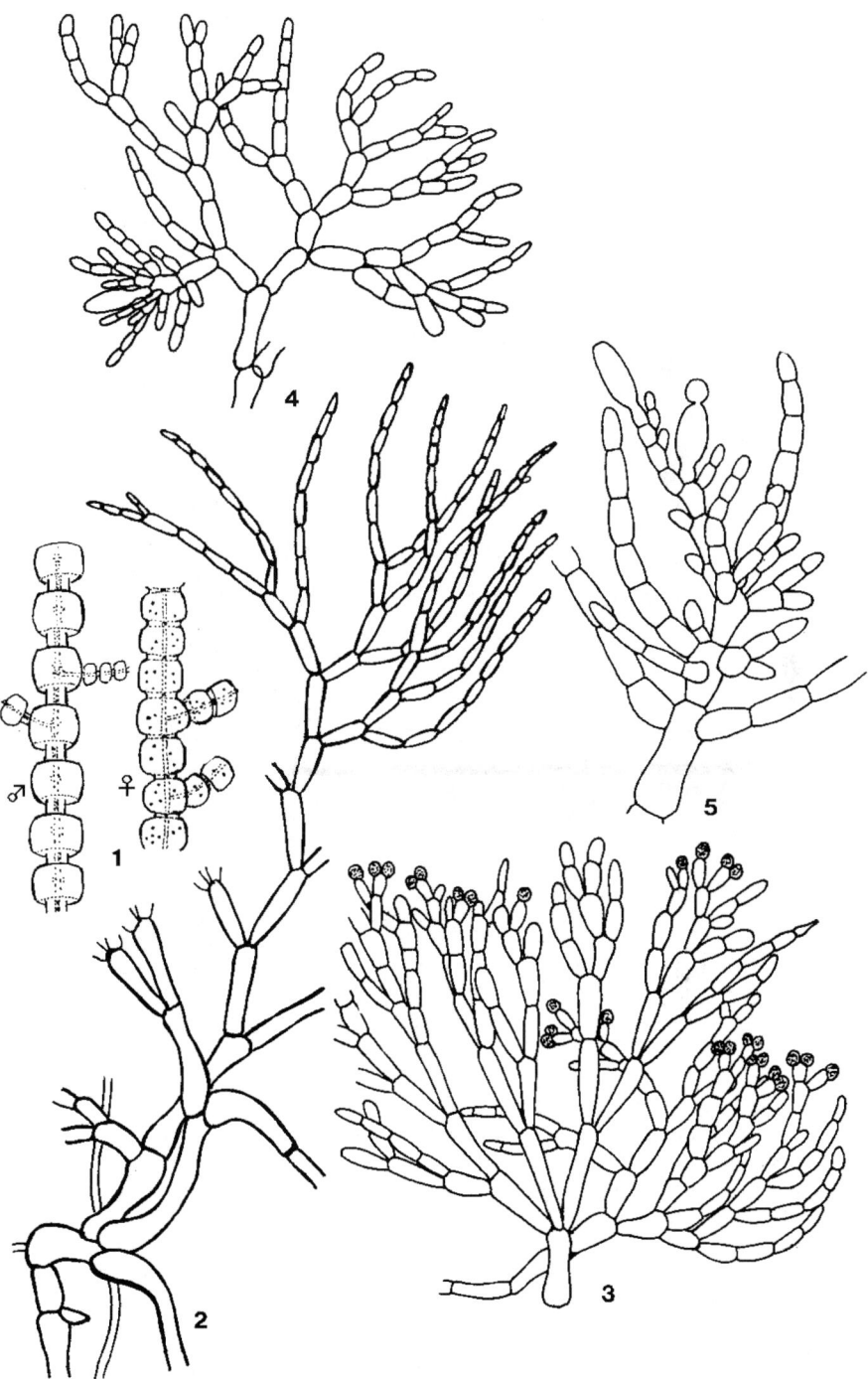

Plate 57
Batrachospermum boryanum, figs 1, 5 (original author, Sirodot 1884), figs 2–4 (Starmach 1976)
1. a main axis with whorls barrel-shaped and separated (male plant) or somewhat confluent (a female plant), 2. primary fascicle, 3. spermatangia terminal on primary fascicles, 4–5. carpogonia and undifferentiated carpogonium-bearing branches.

Batrachospermales

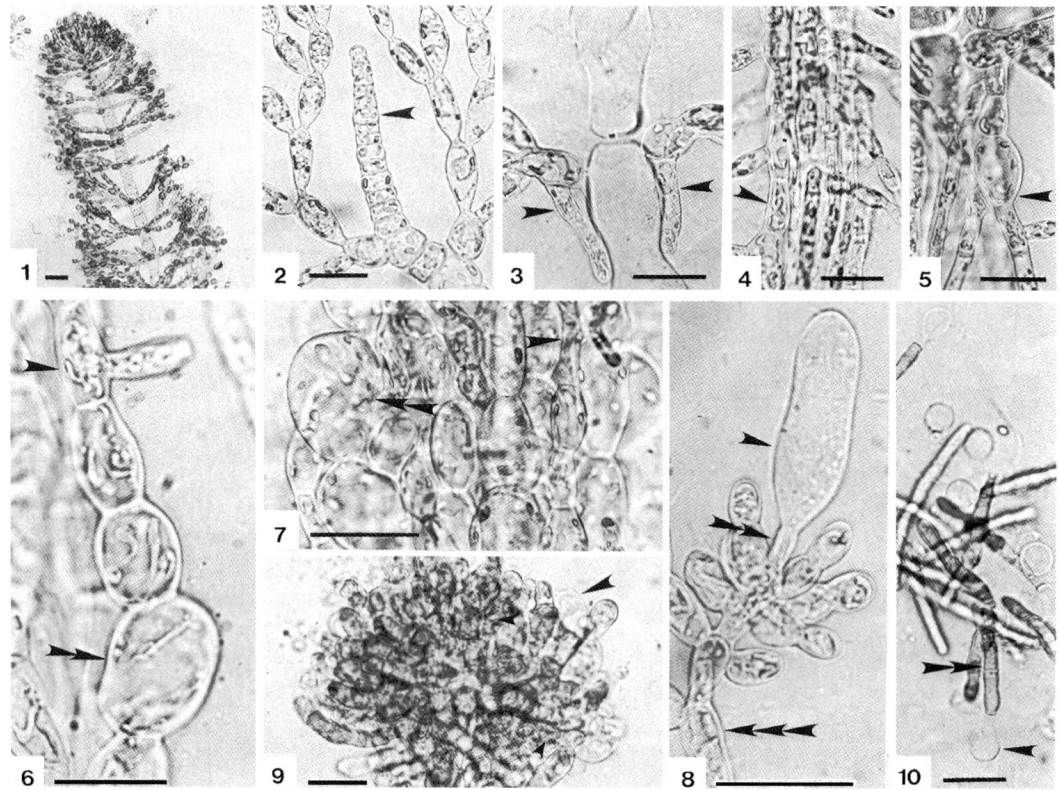

Plate 58
Batrachospermum heterocorticum, figs 1–10 (original authors, Sheath & Cole 1990)
1. an apical portion with no cortication and little densely branched fascicles, 2. fascicle initial (arrow) with no branches, 3. initials (arrows) of cortical cell, 4. 2. the early stage of the cortical cell development with cylindrical cells producing secondary fascicles (arrow), 5. the early stage of the cortical cell enlargement (arrow) and ramification, 6. cortical filaments showing cylindrical cells (arrow) and progressively enlarged cells (double arrow), 7. the mature cortication containing few cylindrical cells (arrow) intermixed with mostly enlarged cells (double arrow), 8. a carpogonium having an ellipsoidal trichogyne (arrow) with a distinctive stalk (double arrow) on a 4–celled carpogonium-bearing branch (triple arrow), 9. a mature carposporophyte with empty carposporangia (arrow), 10. germinating carpospores (arrow) forming *chantransia* filaments (double arrow).

considerably, ellipsoidal, 12.9–24.1 µm in diameter, consisting mostly of enlarged cells intermixed with a few cylindrical cells, with the heterogeneous nature of the cortication being the most distinctive feature. Spermatangia ovoid, formed at fascicle tips of male plants. Carpogonium-bearing branches arising from proximal cells of the fascicles, straight and composed of 1–4 cells. Carpogonia 31–46 µm long, with a cylindrical, lanceolate trichogyne 7.3–12.4 µm in diameter, a distinctive stalk and a small conical base. Carposporophytes spherical, 109–198 µm in diameter, in mid to outer portions of lateral whorls. Carposporangia terminal on short lateral branches, ovoid and 6.3–9.4 µm in diameter, 9.2–15.0 µm long.

Type Locality: Morman Branch at Route 19, 3 km north of state highway 40 in Marion, Florida, USA in North America.

Holotype: UBC A70042.

Distribution: Known from five localities in Florida, Arkansas and Arizona in southern USA in North America.

1. *Batrachospermum*

Note: Sheath & Cole (1990) assigned this species to the section *Batrachospermum*, because of straight carpogonium-bearing branches, large lateral whorls and presence of several carposporophytes in the middle to outer portion of whorls. On the other hand, the shape of trichogyne of this species is similar to that of *B. sirodotii* Skuja ex Reis (1974) of the section *Virescentia*.

Entwisle & Foard (1997) proposed the following groupings (intended to facilitate identification, not to indicate phylogenetic relationships) for species described from Australia and New Zealand.

Key to the groups of the genus ***Batrachospermum*** proposed by Entwisle & Foard (1997)

1. Fascicles up to 6 cells long, closely pressed to main axis; carposporophyte protruding from whorl; carpogonium 2-3 cells from periaxial ....*B. atrum* and related species (ref. p. 70)
1. Fascicles 3–22 cells long, not pressed to main axis; carposporophyte within or protruding from whorl; carpogonium 3–17 cells from periaxial cell...2
2. Carposporophyte hemispherical (rarely spherical) or diffuse, centre adjacent to thallus axis (rarely distant from axis and then carposporophyte always hemispherical); carpogonia subtended by 2–7 modified cells; eastern Australia..3
2. Carposporophyte spherical (rarely some hemispherical), centre not (or rarely within a specimen) adjacent to thallus axis; carpogonia subtended by 3–17 modified or unmodified cells; Australia or New Zealand...4
3. Thallus somewhat rigid; axial cells 9–23 μm in diameter, rhizoidal filaments inflated, 6–23 μm in diameter; distal cells spherical or conical-spherical...5) *Batrachospermum wattsii* group
3. Thallus flaccid; axial cells 18–80 μm in diameter; rhizoidal filaments cylindrical, 4–13 μm in diameter; distal cells not spherical...2) *Batrachospermum gelatinosum* group (in part)
4. Fascicle cells cylindrical with distal expansion; fascicle branching unilateral; secondary fascicle absent (rarely present and then sparse in overmature thalli); trichogyne club-shaped..3) *Batrachospermum antipodites* group (in part)
4. Fascicle cells and branching otherwise; secondary fascicles present or absent; trichogyne variously shaped..5
5. Fascicles in conical whorls (at least in some parts of thallus), branching 2–5 times...................6
5. Fascicles in globose, barrel-shaped or disc-shaped whorls, branching 3–11 times....................7
6. Carpogonia subtended by 7–22 cells; trichogyne pedicellate; rhizoidal filaments 5–6 μm in diameter; carposporophyte diameter less than half whorl radius..6) *Batrachospermum cayennense* group
6. Carpogonia subtended by 6–11 cells; trichogyne sessile; rhizoidal filaments 5–12 μm in diameter; carposporophyte similar in diameter to whorl radius, with some indeterminate gonimoblast filaments...4) *Batrachospermum prominens* group (in part)
7. Trichogyne bi-laterally asymmetrical (check a few trichogyne to ensure correct orientation).....................3) *Batrachospermum antipodites* group (in part)
7. Trichogyne symmetrical (occasionally distorted or lumpy)...8

8. Cells of carpogonium-bearing branch similar in shape
to other fascicle cells; New Zealand..............................1) *Batrachospermum arcuatum* group
8. Cells of carpogonium-bearing branch modified in
shape and size; Australia or New Zealand...9
9. Inner cells of fascicle quite different in size and shape from outer cells
(without graduation between the two layers); rhizoidal filaments
7–26 μm in diameter, consisting of cylindrical and inflated cells;
carposporangia 8–11 μm in diameter............3) *Batrachospermum antipodites* group (in part)
9. Fascicles without clearly defined inner and outer layers of cells;
rhizoidal filaments 2–10 μm in diameter, consisting of
cylindrical cells; carposporangia 6–26 μm in diameter..10
10. Whorls less than 300 μm in diameter; carposporophyte
2 or more times as broad as whorl radius;
trichogyne ellipsoidal......................................4. *Batrachospermum prominense* group (in part)
10. Whorls more than 300 μm in diameter; carposporophyte
up to 1.6 times as broad as whorl radius; trichogyne
rarely ellipsoidal...2) *Batrachospermum gelatinosum* group (in part)

(1) *Batrachospermum arcuatum* group

Carpogonia subtended by a relatively short filaments of unmodified cells.

Key to the species of the *Batrachospermum arcuatum* group

1. Cells of carpogonium-bearing branch are differentiated from
those of fascicles, from Australia and New Zealand.........................1) *B. discorum* (see p. 60)
1. Cells of carpogonium-bearing branch are not differentiated
from those of fascicles, from New Zealand...2
2. Cortical filament cells always cylindrical, 4–7 μm in diameter,
involucral filaments 2 or 3 cells long, not extending beyond
carpogonium; dioecious...2) *B. arcuatum*
2. Cortical filament cells inflated in mature plants, 6–32 μm
in diameter, involucral filaments 4–11 cells long, extending
beyond carpogonium; monoecious..3) *B. anatinum*

1) *Batrachospermum discorum* Entwisle et Foard (1997:341).............................(ref. p. 111, pl. 64)
2) *Batrachospermum arcuatum* Kylin (1912:22)..(ref. p. 93, pl. 50)
3) *Batrachospermum anatinum* Sirodot (1884:249)..(ref. p. 88, pl. 48)

(2) *Batrachospermum gelatinosum* group

Carpogonia subtended by a relatively short filament of modified cells.

Note: Entwisle & Foard (1997) stated that the carpogonia are subtended by a relatively short filament of modified cells, and the carposporophyte size and position in the whorl is extremely variable. This is a taxonomically difficult group in Australia and New *Zealand*, ranging from specimens resembling Northern Hemisphere *B. gelatinosum* to Northern Hemisphere *B. turfosa*, with an almost complete range of intermediates. The full resolution of this group must await intensive field collecting and further scoring and analysis of data (e.g. morphometrics).

(2) *Batrachospermum gelatinosum*

Key to the species of the *Batrachospermum gelatinosum* group

1. Carposporophyte hemispherical (rarely spherical) or diffuse, centre adjacent to thallus axis (rarely distant from axis and then carposporophyte always hemispherical); carpogonium-bearing branch consisting of 2–7 modified cells; eastern Australia..2
1. Carposporophyte spherical (rarely some semi-spherical), centre not (or rarely within a specimen) adjacent to thallus axis; carpogonium-bearing branch consisting of 3–17 modified or unmodified cells; Australia or New Zealand..3
2. Trichogyne club-shaped (sometimes narrowly so), 7–14 µm in diameter; carposporangia always terminal on gonimoblast filaments, 17–30 µm long, 10–18 µm in diameter..............................1) *B. theaquum* (in part)
2. Trichogyne linear, 3–5 µm in diameter; carposporangia arising in a lateral series from gonimoblast filaments; 10–16 µm long, 7–9 µm in diameter..2) *B. debilis*
3. Trichogyne club-shaped; base of carpogonium 5–11 µm in diameter; carposporophyte always adjacent to or overlapping the axis; fascicles branching 2–5 times; eastern Australia..............................1) *B. theaquum* (in part)
3. Trichogyne variously shaped (sometimes club-shaped); base of carpogonium 2–7 µm in diameter; carposporophyte variously positioned (sometimes adjacent to or overlapping the axis); fascicles branching 4–8 times; eastern Australia and New Zealand..4
4. Trichogyne 4–6 µm in diameter; carposporophyte less than 190 µm in diameter, 0.2–0.5 times the whorl radius, remaining within inner cortex; Australia..3) *B. gelatinosum*
4. Trichogyne 4–12 µm in diameter; carposporophyte less than 60–345 µm in diameter, 0.4–1.8 times the whorl radius, sometimes extending into outer cortex; New Zealand..............................4) *B. campyloclonum*

1) ***Batrachospermum theaquum*** Skuja ex Entwisle et Foard (1997:357, fig. 12, A–G, fig. 13, A–D)

plate 59, figs 1–7 (original authors, Entwisle & Foard 1997)

Plants monoecious (rarely dioecious), 2–5 cm high, grass-green (yellow-green when dry); apices acute or obtuse, apical cell protruding from, more or less flush with, or embedded within primary fascicles. Fascicle whorls usually conical or cylindrical, sometimes spherical or barrel-shaped, 102–759(–2000) µm in diameter; confluent or separated; internodes 125–540 µm long. Axial cells colourless (rarely brownish), 18–57 µm in diameter; cortical filaments cylindrical (swelling only where secondary fascicles arise), 6–14 µm in diameter, in several layers in older plants (sometimes free from axial cells and extending through whorl). Secondary fascicles rare to common, sometimes as long as primary fascicles. Primary fascicles 3 or 4 per periaxial cell, more or less straight distally, of 4–26 cells storeys; branching di- or trichotomous, 2–5 times; proximal cell cylindrical, ellipsoidal or obovoidal, 11–43 µm long, 3–15 µm in diameter; intermediate cells ellipsoidal, fusiform or obovoidal (rarely cylindrical), 7–44 µm long, 4–11 µm in diameter; distal cells ellipsoidal or obovoidal, 5–15 µm long, 4–9 µm in diameter; hairs 8–130 µm long. Monosporangia absent. Spermatangia on primary or secondary fascicles, sometimes on short, specialized filaments (aborted carpogonial branches), rarely on involucral filaments, clustered or scattered, spherical, 5–10 µm in diameter.

Batrachospermales

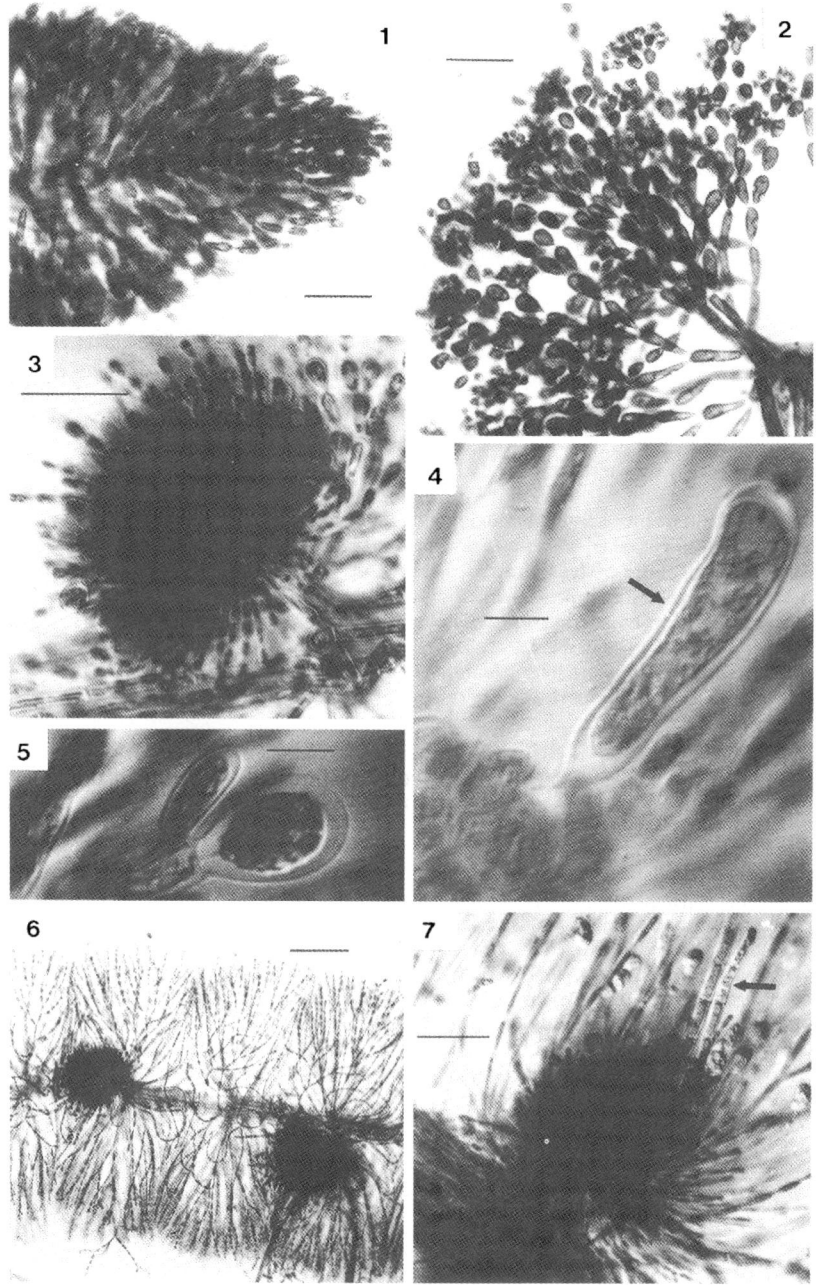

Plate 59
Batrachospermum theaquum variant 1, figs 1–2 (original authors, Entwisle & Foard 1997)
1. an apex of a plant, 2. spermatangia on a primary fascicle.

Batrachospermum theaquum variant 2, figs 3–5 (original authors, Entwisle & Foard 1997)
3. a carposporophyte, 4. a carpogonium (arrow to trichogyne), 5. a carposporangium.

Batrachospermum theaquum variant 4, figs 6–7 (original authors, Entwisle & Foard 1997)
6. a mature plant with carposporophytes, 7. ca carposporophyte with chantransia phase(arrow) arising from carpospores germinating *in situ*.
Scale bars = 500 μm (fig. 6); 200 μm (fig. 7); 100 μm (figs 1–3); 20 μm (figs 4–5).

Carpogonium-bearing branch arising from usually 3–7 cells from the periaxial cell, rarely from intercalary cells of fascicles, rarely slightly curved, consisting of 2–6 (rarely more) modified cells, these barrel-shaped, ellipsoidal, cylindrical or discoid, 5–14 μm long, 6–12 μm in diameter. Involucral filaments arising from all modified cells bearing carpogonium, 1–5 cells long, usually not extending beyond carpogonium. Carpogonium more or less straight, 30–78 μm long, (3–)5–11 μm in diameter at base, trichogyne sessile, radially symmetrical, usually club-shaped (occasionally lumpy), 6–15 μm in diameter at broadest part. Carposporophytes 1 or 2 per whorl, in inner and outer cortex and sometimes exserted from whorl, usually semi-spherical (sometimes spherical), compact or loose, 115–348 μm in diameter, 0.3–2.3 times whorl radius, centre 0–130 μm from node; gonimoblast filaments usually audouinelloid, 2–9 cells long; post-fertilization cells of carpogonium-bearing branch spherical, barrel-shaped or somewhat turbinate, often producing rhizoidal filaments soon after fertilization. Carposporangia obovoidal (rarely spherical or tadpole-shaped), 17–33 μm long, 10–18 μm in diameter.

Type Locality: Gordon, Middle Harbor, Sydney, New South Wales, Australia.
Holotype: NSW A008816.
Distribution: Known from Eastern Australia, usually in humid streams.

2) *Batrachospermum debilis* Entwisle et Foard (1997:362, fig. 14, A–E)

plate 60, figs 1–5 (original authors, Entwisle & Foard 1997)

Plants monoecious, 2–3 cm high, green to dark brown, flaccid or firm, predominantly dichotomous; apices obtuse, apical cell embedded within primary fascicles. Fascicle whorls barrel- to disc-shaped, 300–550 μm in diameter, confluent; internodes 110–180 μm long. Axial cells 24–30 μm in diameter, cortical filament cells cylindrical, 4–6 μm in diameter. Secondary fascicles common, some as long as primary fascicles. Primary fascicles 3 per periaxial cell, more or less straight distally, of 6–12 cell storeys; branching 3 or 4 times di- or trichotomous; proximal cell fusiform, 24–40 μm long, 4–6 μm in diameter; intermediate cells fusiform or obovoidal, 10–30 μm long, 3–6 μm in diameter; distal cells elongate-obovoidal, 8–12 μm long, 2–5 μm in diameter; hairs 5–80 μm long. Monosporangia absent. Spermatangia on primary or secondary fascicles, scattered, spherical, 5–6 μm in diameter. Carpogonia on carpogonium-bearing branch arising from periaxial cells, consisting of 5–7 modified cells, these discoidal, 4–7 μm long, 4–5 μm in diameter. Involucral filaments arising from all subtending cells, ca. 3 cells long, not extending beyond carpogonium. Carpogonium more or less straight, 32–60 μm long, 3–4 μm in diameter at base, trichogyne sessile (but base of trichogyne attenuate and narrow), radially symmetrical or curved, narrowly club-shaped to linear, 3–5 μm in diameter. Carposporophytes 1 per whorl, in inner or outer cortex, hemispherical, loosely aggregated, 160–200 μm in diameter, 0.6–1 times whorl radius, centre 60–80 μm from node; gonimoblast filaments 4–11 cells long, producing succession of carposporangia laterally, then one terminally; post-fertilization cells of carpogonium-bearing branch barrel-shaped, covered by cortical filaments. Carposporangia obovoidal, 10–16 μm long, 7–9 μm in diameter.

Type Locality: Creek flowing into Bathurst Narrows, on southern side of Mt. Rugby, Southwestern National Park, Tasmania, Australia.
Holotype: dry MEL 2033517, wet MEL 2033518.
Distribution: Known from the type locality only.
Note: According to Entwisle & Foard (1997), the carposporangia arising in a lateral series are unique to this species.

Batrachospermales

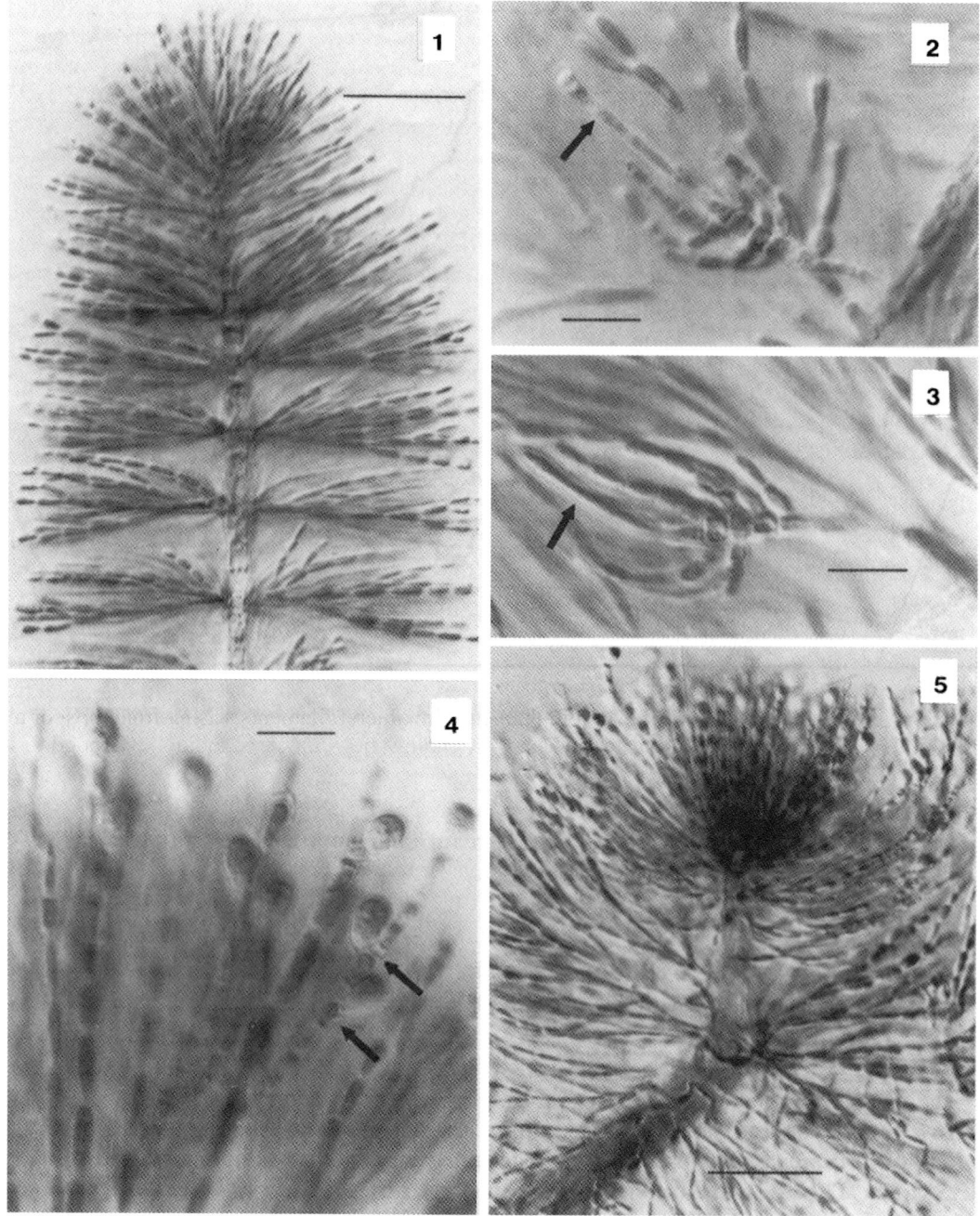

Plate 60
Batrachospermum debilis, figs 1–5 (original authors, Entwisle & Foard 1997)
1. an apex of a plant, 2–3. carpogonia (arrows to trichogynes) and carpogonium-bearing branches, 4. gonimoblast filaments giving rise to carposporangia laterally (arrows), 5. a whorl with a carposporophyte. Scale bars = 200 µm (figs 1, 5); 50 µm (figs 2–4).

3) ***Batrachospermum gelatinosum*** (Linnaeus) DeCandolle (1801:21) *sensu* Entwisle & Foard (1997)
Basionym: *Conferva gelatinosa* Linnaeus (1753: pl. 1166),
The synonymy proposed by Entwisle & Foard (1997) is as follows;
Synonym: *B. helminthoideum* (Sirodot) Mori (1975:474). (See also p. 86, the synonymy proposed by Vis *et al.* 1995)

4) ***Batrachospermum campyloclonum*** Skuja ex Entwisle et Foard (1997:364, fig. 17, A–E)

plate 61, figs 1–5 (original authors, Entwisle & Foard 1997)

Plants monoecious or rarely dioecious, red, brown or green; apices acute or obtuse, apical cell protruding through, more or less flush with, or embedded within primary fascicles. Fascicle whorls barrel- to disc-shaped or spherical, 170–670(–920) μm in diameter, confluent; internodes 58–440 μm long. Axial cells 7–54 μm in diameter, cortical filaments sometimes profuse (up to 11 layers), cells cylindrical, 2–9 μm in diameter. Secondary fascicles common or rare, shorter than or as long as primary fascicles. Primary fascicles 2–6 per periaxial cell, more or less straight or slightly curved distally, of 7–16(–25) cell storeys; branching 2–4-chotomous, 4–8 times; proximal cell fusiform, bone-shaped, cylindrical or ellipsoidal, 12–91 μm long, 3–11 μm in diameter; intermediate cells similar in shape, 6–27(–63) μm long, 3–12 μm in diameter, distal cells obovoidal, ellipsoidal or cylindrical, 6–16(–21) μm long, 2–7 μm in diameter; hairs sometimes present, 5–30 μm long. Monosporangia absent. Spermatangia on primary fascicles, scattered, spherical, 4–9 μm in diameter. Carpogonium-bearing branch arising from periaxial cells, rarely from intercalary cells of fascicles, consisting of 4–11 modified cells, these cylindrical, discoidal or barrel-shaped, 3–9 μm long, 4–11 μm in diameter. Involucral filaments arising from most or all subtending cells, 1–6 cells long, sometimes extending beyond carpogonium. Carpogonium usually straight, 21–60 μm long, 3–9 μm in diameter at base, trichogyne sessile, radially symmetrical or lumpy (possibly due to poor preservation), fusiform to cylindrical or club-shaped, 4–9(–12) μm in diameter at broadest part. Carposporophytes usually 1 or 2 per whorl, but sometimes up to 5, in inner and outer cortex, sometimes exserted from whorl, spherical or apparently semi-spherical, compact or loose, 60–280(–345) μm in diameter, 0.4–1.8 times whorl radius, centre 0–203 μm from node; gonimoblast filaments 2–12 cells long (possibly longer in very old plants); post-fertilization cells of carpogonial branch mostly spherical, cylindrical or ellipsoidal, sometimes sheathed by cortical filaments. Carposporangia obovoidal to ellipsoidal or oblong (rarely tadpole-shape), 7–32 μm long, 6–17 μm in diameter.
Type Locality: Waikoropupu Springs, Takaka, north-western of Nelson, South Island, New Zealand.
Holotype: MEL 2026362.
Distribution: Widespread throughout New Zealand.
Note: According to Entwisle & Foard (1997), the individual carposporophytes sometimes intermingle in older plants creating a single very large carposporophyte mass. This feature and the variability in carposporophyte size make this a difficult taxon to circumscribe. It is similar to (and somewhat intermediate between) *B. gelatinosum* and *B. theaquum* in Australia. Further collecting and study are needed to clarify relationships within the *B. gelatinosum-campyloclonum* complex. The extremes of both species are clearly distinct, but the diverse range of intermediates make delineation of taxa extremely difficult.

Batrachospermales

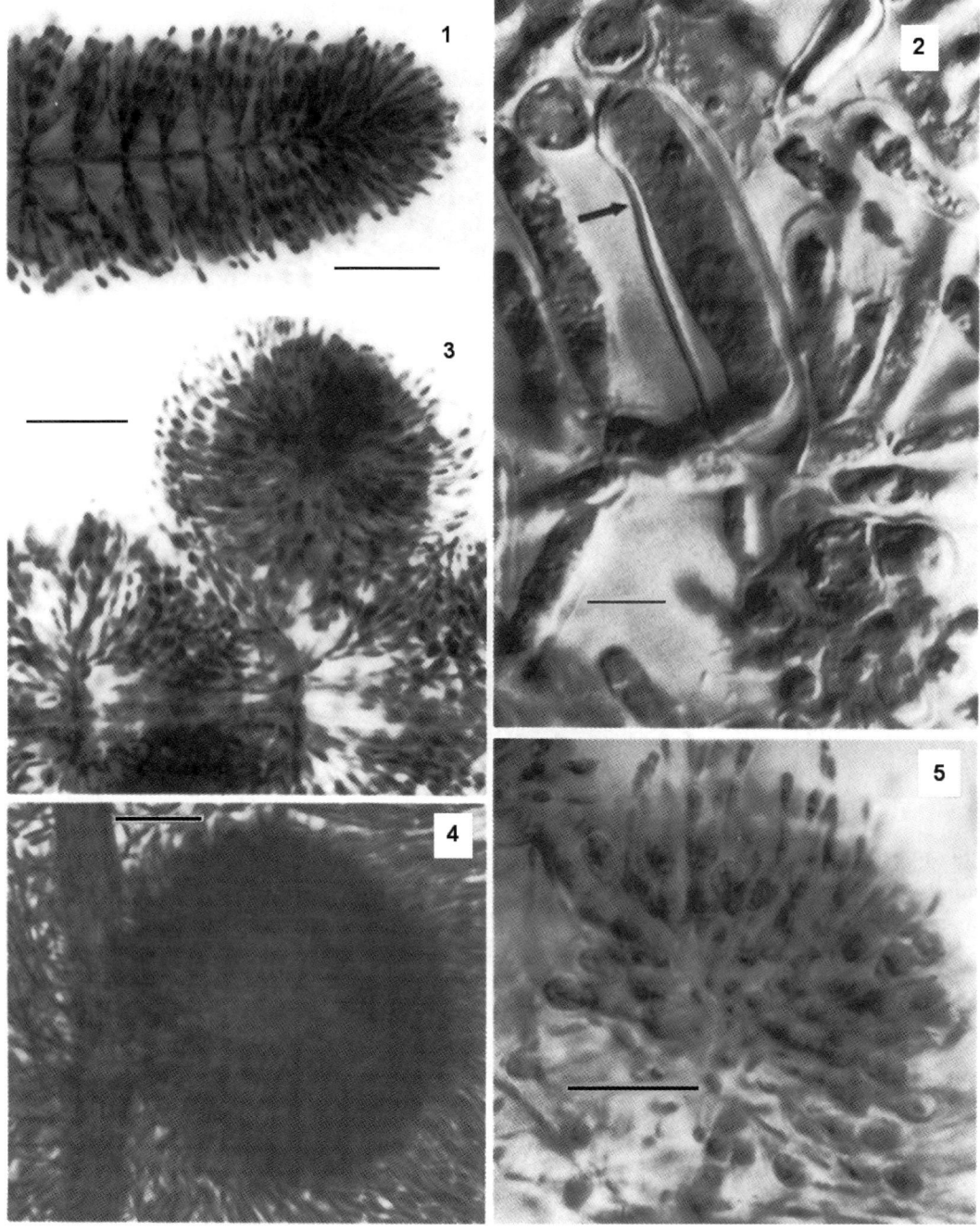

Plate 61
Batrachospermum campyloclonum, figs 1–5 (original authors, Entwisle & Foard 1997)
1. an apex of a plant, 2. a carpogonium (arrow to trichogyne) with attached spermatia, 3. a carposporophyte exerted from a whorl, 4–5. carposporophytes.
Scale bars = 200 μm (figs 1, 3); 100 μm (figs 4–5); 20 μm (fig. 2).

(3) *Batrachospermum antipodites* group

Carpogonia subtended by a relatively short filaments of modified cells.

Key to the species of the group *Antipodites*

1. Fascicle cells cylindrical with distal expansion; fascicle branching unilateral; secondary fascicle absent (rarely present and then sparse in overmature plants); trichogyne club-shaped......................1) *B. antipodites*
1. Fascicle cells and branching otherwise; secondary fascicles present or absent; trichogyne variously shaped..2
2. Trichogyne bi-laterally symmetrical..2) *B. discorum*
2. Trichogyne asymmetrical..3
3. Trichogyne oblique-ellipsoidal to scimitar-shaped; carposporangia < 30 μm long...3) *B. kraftii*
3. Trichogyne foot- or sausage-shaped; carposporangia > 30 μm long..4
4. Trichogyne curved sausage-shaped, 8–10 μm in diameter; whorl dense..4) *B. ranuliferum*
4. Trichogyne foot-shaped, 10–16 μm in diameter at broadest part; whorls relatively open..5) *B. antiquum*

1) ***Batrachospermum antipodites*** Entwisle (1995:291, fig. 1, a–e, fig. 2, a–d)
Misapplied names: *Batrachospermum ectocarpum* auct. non Sirodot; Entwisle & Kraft (1984:228); *B. boryanum* auct. non Sirodot, Entwisle (1989:42); see also Entwisle & Kraft (1984:254).

plate 62, figs 1–5 (original authors, Entwisle 1995)
plate 63, figs 1–4 (original authors, Entwisle & Foard 1997)

Plants monoecious, 1–5 cm high, red to dark grayish (purple when dry). Whorls spherical to barrel- or disc-shaped, confluent or separated, 330–800 μm in diameter, internodes 230–400 μm long; axial cells 34–120 μm in diameter; cortical filaments 5–8 μm in diameter, covering axial cell. Primary fascicles 2–3 per periaxial cell, somewhat audouinelloid, with more or less straight distal ends, without any clearly defined outer and inner cell layers, of 8–13 cell-storeys; branching lateral, dichotomous or (sometimes) trichotomous, 3–7 times. Periaxial cells ovoid to cylindrical; proximal cells cylindrical with apical dilation at lateral branch insertion, 16–30 μm long, 3–5 μm in diameter; intermediate cells similar in shape to proximal cells, 11–13 μm long, 2–5 μm in diameter; distal cells cylindrical to obovoidal (apex rounded), 6–9 μm long, 3–4 μm in diameter; terminal hairs absent. Secondary fascicles absent (or rarely present in overmature, mostly denuded axes). Spermatangia on primary or rarely (in over-mature plants) on secondary fascicles, spherical, 5–8 μm in diameter. Carpogonium-bearing branch more or less straight, arising from periaxial cells; consisting of 4–7 barrel-shaped to ellipsoid cells, 6–7 μm long, 4–7 μm in diameter. Involucral bracts 1–2 cells long, arising from all cells. Carpogonium more or less straight 19–39 μm long; base symmetric, 3–5 μm in diameter; trichogyne sessile, club-shaped, 4–8 μm in diameter at broadest part. Carposporophytes pedicellate, 1 or (rarely) 2 per whorl, in inner or outer cortex, centre 70–100 μm from node, spherical, 70–200 μm in diameter, 0.5–1 times the whorl radius; gonimoblast filaments determinate, 4–5 cells long; post-fertilization cells of carpogonial branch spherical to barrel-shaped, or slightly constricted in middle. Carposporangia obovoidal, 10–18 μm long, 6–12 μm in diameter.

Batrachospermales

Type Locality: Kondalilla Falls, Skene Creek, Kondalilla National Park, Queensland, Australia.
Holotype: MEL 2020014.
Distribution: Known from New South Wales, Queensland, Tasmania, and Victoria, Australia; New Zealand.

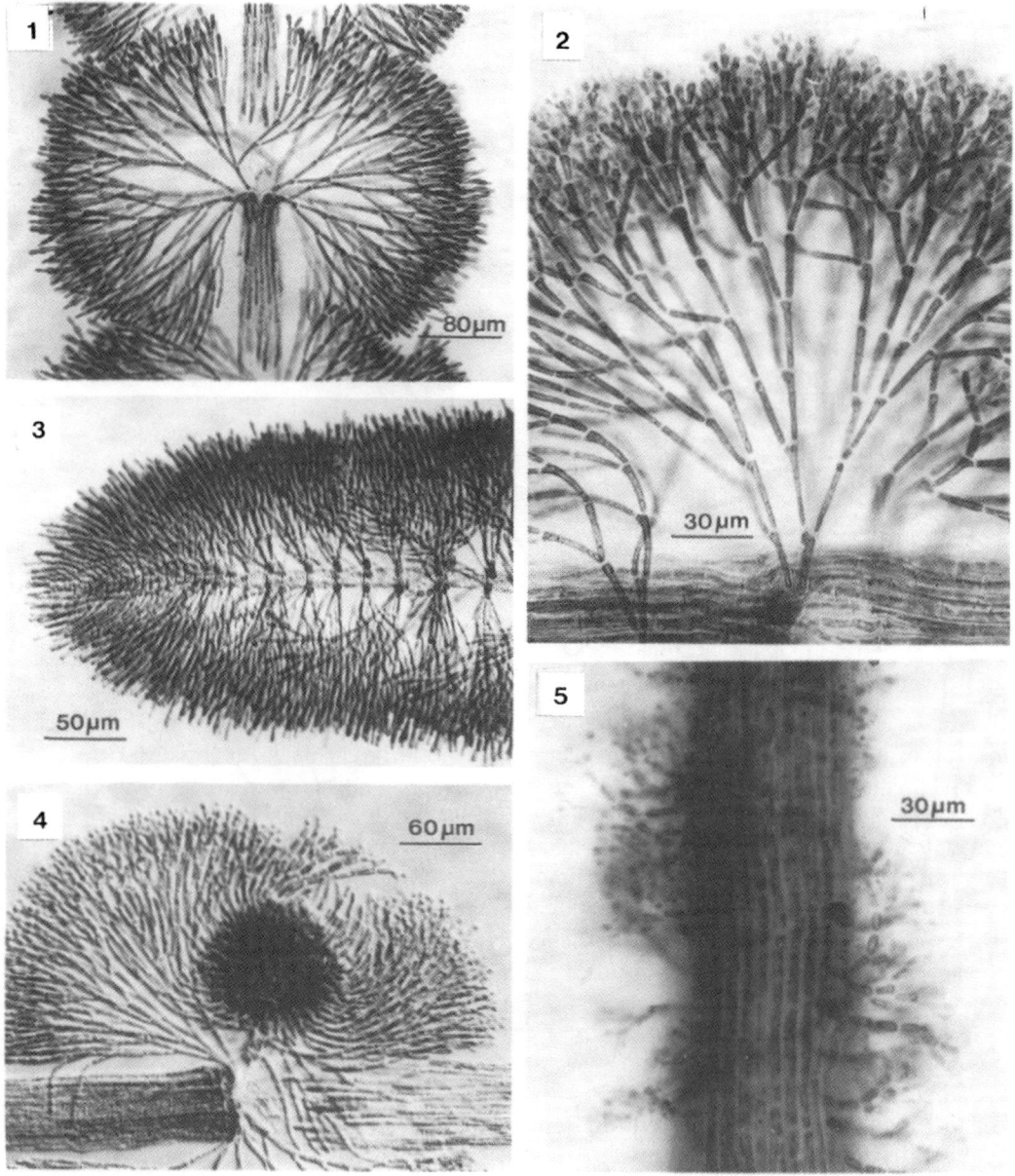

Plate 62
Batrachospermum antipodites, figs 1–5 (original authors, Entwisle & Foard 1995)
1. a whorl, 2. a fascicle terminated with spermatangia, 3. an apex of a plant, 4. a carposporophyte within a whorl, 5. secondary fascicles bearing spermatangia in an overmature plant.

2) ***Batrachospermum discorum*** Entwisle et Foard (1997:341, fig. 5, A–E)

plate 64, figs 1–5 (original authors, Entwisle & Foard 1997)

Plants dioecious, 2–4 cm high, yellow-green when dry, flaccid or firm, monopodial; apices obtuse; apical cell embedded within primary fascicles. Fascicle whorls barrel- to disc-shaped, 620–1080 μm in diameter, confluent; internodes 460–600 μm long. Axis colourless, axial cells 28–32 μm in diameter, cortical filaments 2 or 3 layers around axial cells, cortical cells cylindrical or inflated, 7–26 μm in diameter. Secondary fascicles common, some as long as primary fascicles. Primary fascicles 3 or 4 per periaxial cell, with inner cortical cells quite different in size

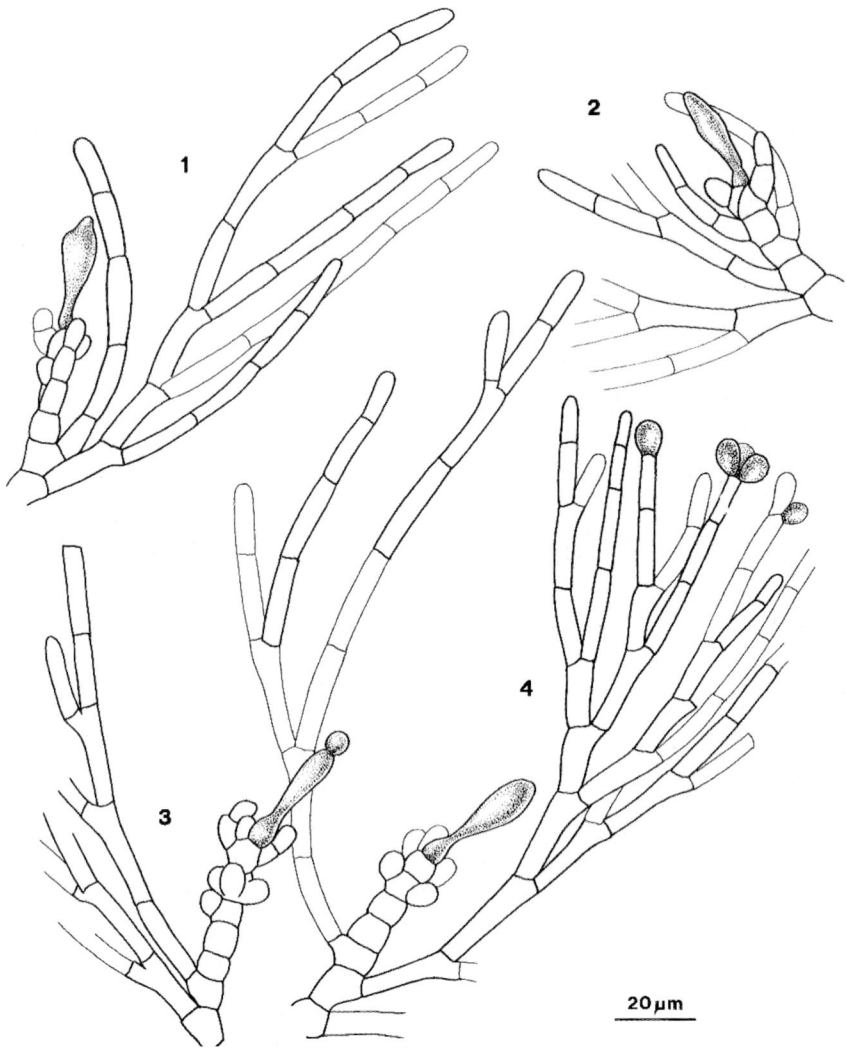

Plate 63
Batrachospermum antipodites, figs 1–4 (original authors, Entwisle & Foard 1997)
1. a fascicle bearing a carpogonium, 2. a carpogonium-bearing branch with a carpogonium, 3. a carpogonium bearing branch with a spermatium, attached to a carpogonium, 4. fascicles bearing terminal spermatangia and a carpogonium bearing branch with a carpogonium.

Batrachospermales

and shape to outer cortical cells and not graduating into one another, more or less straight distally, of 12–22 cell storeys; branching dichotomous or trichotomous; 6–11 times; proximal cell bone-shaped (but swollen mainly at proximal end only), 14–64 μm long, 6–10 μm in diameter; intermediate cells fusiform to obovoidal, 10–60 μm long, 4–10 μm in diameter; distal cells obovoidal to spherical, 6–15 μm long, 5–12 μm in diameter; terminal hairs 14–40 μm long or absent. Monosporangia absent. Spermatangia on primary and secondary fascicles, abundant, spherical, 7–9 μm in diameter. Carpogonia on primary and secondary fascicles, 5–14 cells from periaxial cells. Carpogonium-bearing branch curved, consisting of 1–14 modified cells, these, barrel-shaped to discoidal, 8–21 μm long, 10–23 μm in diameter. Involucral filaments arising

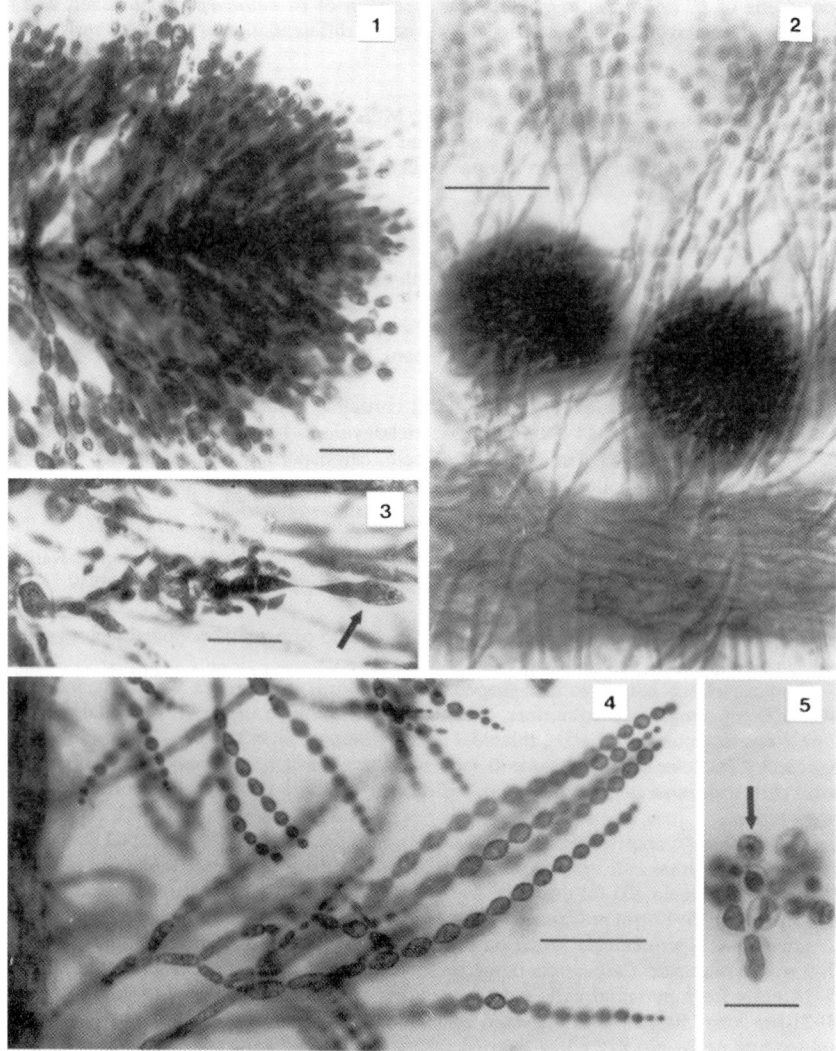

Plate 64
Batrachospermum discorum, figs 1–5 (original authors, Entwisle & Foard 1997)
1. an apex of a plant, 2. whorls with carposporophytes, 3. a carpogonium (arrow to trichogyne) and a carpogonium-bearing branch, 4. a fascicle showing spherical outer cells (right) and more elongate inner cells (left), 5. spermatangia (arrow to one) on a fragment of primary fascicle.
Scale bars = 100 μm (figs 1, 3, 5); 200 μm (figs 2, 4).

from all subtending cells, not extending beyond carpogonium, 1–5 cells long and branched, often bearing 1 or more additional carpogonia. Carpogonia more or less straight, (30–)44–60 μm long, 5–8 μm in diameter at base, trichogyne sessile, club-shaped, ellipsoidal or fusiform, 8–14 μm in diameter at broadest part. Carposporophytes 1 or 2 per whorl (with more internodally), in inner cortex, spherical, compact, 120–185 μm in diameter, 0.2–0.4 times whorl radius, centre 120–230 μm from node; gonimoblast filaments ca. 4 or 5 cells long; post-fertilization cells of carpogonial branch spherical, cylindrical or barrel-shaped. Carposporangia obovoidal to pear-shaped, 17–22 μm long, 8–11 μm in diameter.

Type Locality: Canungra Creek, Box Forest Circuit, Green Mountains, Lamington National Park, Queensland, Australia.

Holotype: MEL 2019946.

Distribution: Disjunct, far Known from central North Island, New Zealand, and near the New South Wales-Queensland border, Australia.

Note: Entwisle & Foard (1997) stated as follows: this species is allied to *B. arcuatum* in the system of Vis *et al.* (1995), but differs in having longer carpogonia, larger carposporangia and almost certainly in some of its vegetative features.... With its inflated rhizoidal cells, this species might be confused with *B. heterocorticum* Sheath & Cole but *B. discorum* has fewer gonimoblasts per whorl, modified carpogonial branch cells and sessile trichogynes. Compared with *B. gelatinosum [sensu* Entwisle et Foard (1997)], *B. discorum* is always dioecious and has generally longer internodes; broader and sometimes inflated cortical filaments; often spherical, generally broader, distal cells; larger spermatangia and broader trichogynes.

3) ***Batrachospermum kraftii*** Entwisle et Foard (1997:344, fig. 7, A–E)

plate 65, figs 1–6 (original authors, Entwisle & Foard 1997)

Plants dioecious or rarely some individuals with carpogonia and few spermatangia, 1–10 cm high, olive-green or brown (red or purple when dry), flaccid or firm, monopodial; apices obtuse; apical cell protruding through, flush or embedded within primary fascicles. Fascicle whorls barrel- to disc-shaped, 420–1300 μm in diameter, confluent; internodes 200–530 μm long. Axis colourless or brown, axial cells 23–52 μm in diameter, cortical filament cells cylindrical, 4–9 μm in diameter, sometimes arising from intercalary cells of fascicles. Secondary fascicles absent or few, always shorter than primary fascicles. Primary fascicles 3 or 4 per axial cell, more or less straight distally, of 11–20 cell storeys; branching di- or trichotomous, 4–8 times; proximal cells bone-shaped, 25–50 μm long, 2–6 μm in diameter; intermediate cells bone-shaped, cylindrical, fusiform or obovoidal, 6–90 μm long, 3–7 μm in diameter; distal cells ellipsoid or obovoidal, 5–12 μm long, 3–4 μm in diameter; terminal hairs 4–200 μm long. Monosporangia absent. Spermatangia on primary fascicles, scattered, spherical, 5–8 μm in diameter. Carpogonium-bearing branches on primary fascicles or replacing them, arising from periaxial cell, 5–14 cells long. Carpogonium-bearing branch sometimes slightly curved, 5–9 cells subtending carpogonium modified, usually ellipsoidal, sometimes barrel-shaped, cylindrical or obovoidal, 7–22 μm long, 6–13 μm in diameter. Involucral filaments profuse, arising from all modified cells, 2–5 cells long, upper filaments not extending beyond carpogonium. Carpogonium more or less straight, 20–40 μm long, 3–5 μm in diameter at base, trichogyne sessile or pedicellate, bent or almost straight, oblique-ellipsoidal (sometimes with apical pimple) to scimitar shaped, 6–11 μm in diameter at broadest part. Carposporophytes 1 or 2 (or 3) per whorl, in inner cortex (rarely extending into outer cortex when large), spherical, compact, 85–260 μm in diameter, 0.1–0.9 times whorl radius, centre 90–180 μm from node; gonimoblast filaments 2–4 cells long; post-fertilization cells of carpogonium-bearing branch cylindrical or barrel-shaped. Carposporangia obovoidal or tadpole-shaped, 18–26 μm long, 7–14 μm in diameter.

Batrachospermales

Type Locality: Arve River, Hartz Mountain area, Tasmania, Australia.
Holotype: MEL 2027351.
Distribution: Known from South-western Australia, Victorian central highlands, north-western and southern Tasmania in Australia, New Zealand

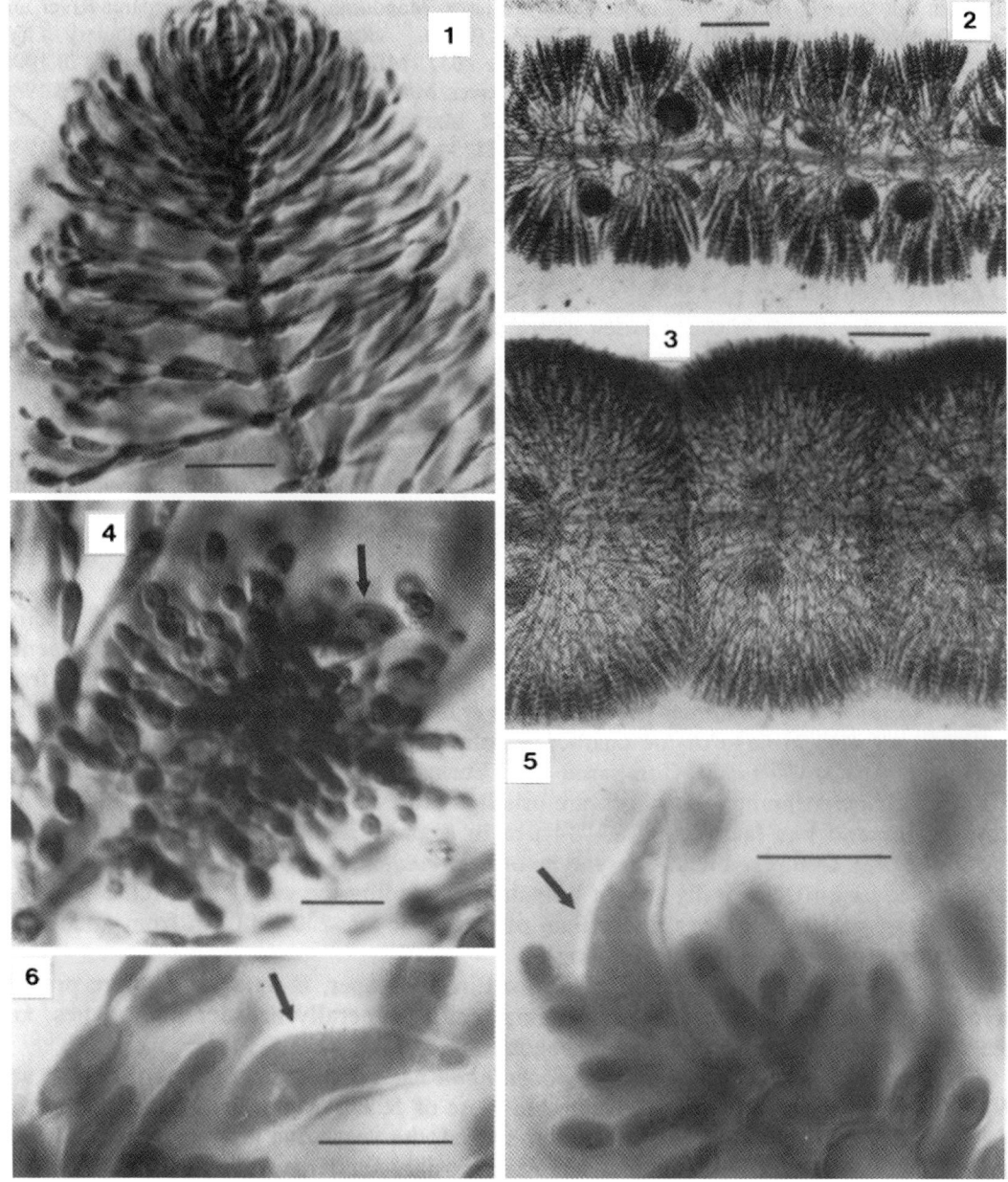

Plate 65
Batrachospermum kraftii, figs 1–6 (original authors, Entwisle & Foard 1997)
1. an apex of a plant, 2–3. mature plants with carposporophytes, 4. a carpogonium (arrow to trichogyne) with spermatia attached, and a bearing branch with profuse involucral filaments, 5–6. trichogynes (arrows). Scale bars = 500 µm (figs 2–3); 100 µm (fig. 1); 50 µm (figs 4–6).

(3) *Batrachospermum antipodites*

4) **Batrachospermum ranuliferum** Entwisle et Foard (1997:347, fig. 8, A–F)

plate 66, figs 1–6 (original authors, Entwisle & Foard 1997)

Plants monoecious, 2–3 cm high, olive to blue-green when dry, flaccid, monopodial; apices obtuse, apical cell protruding through primary fascicles. Fascicle whorls spherical, barrel- or disc-shaped, 270–330 μm in diameter, confluent; internodes 180–230 μm long. Axis colourless, axial cells 12–22 μm in diameter, cortical filament cells cylindrical, 7–10 μm in diameter. Secondary

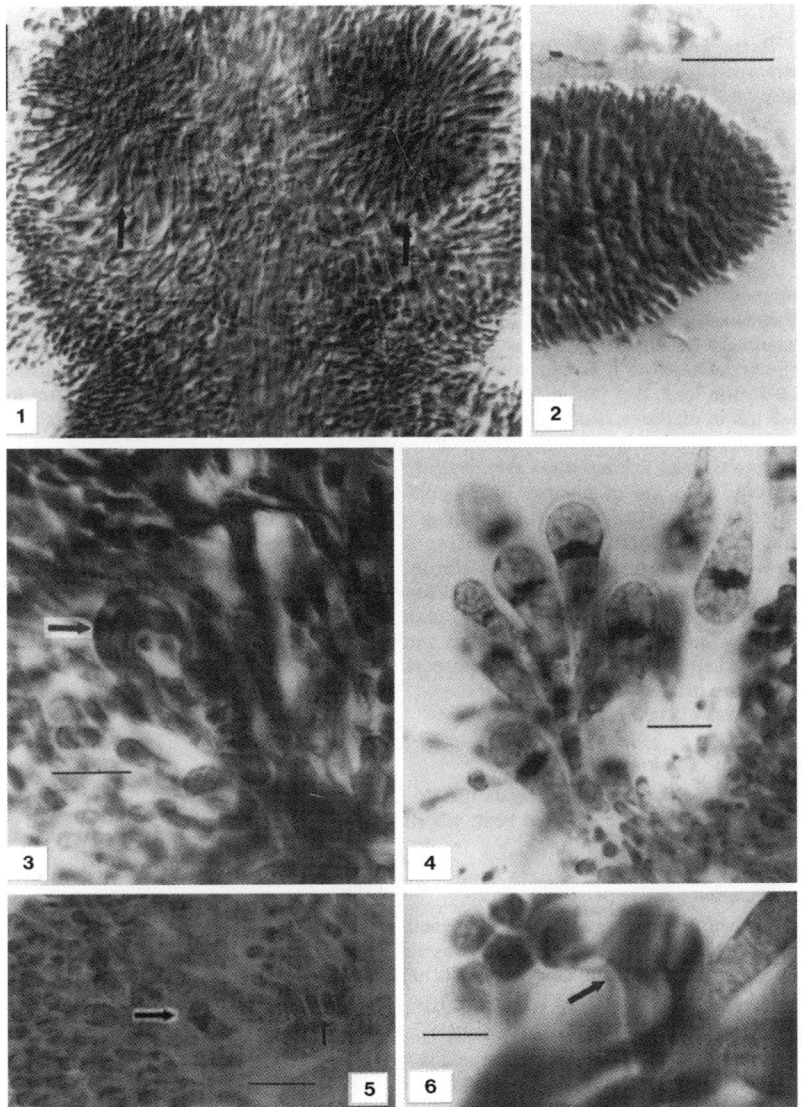

Plate 66
Batrachospermum ranuliferum, figs 1–6 (original authors, Entwisle & Foard 1997)
1. a dense plant with carposporophytes (arrows), 2. an apex of a plant, 3. a carpogonium (arrow to trichogyne) and a carpogonium-bearing branch, 4. tadpole-shaped carposporangia, 5. a carpogonium (arrow to trichogyne) and a bearing branch (smaller arrow), 6. a trichogyne (arrow to bend in trichogyne) with spermatia attached.
Scale bars = 200 μm (fig. 1); 100 μm (fig. 2); 50 μm (figs 3–5); 20 μm (fig. 6).

fascicles common, some as long as primary fascicles. Primary fascicles 3 per periaxial cell, of 7–9 cell storeys; branching di- or trichotomous, 5 or 6 times; proximal cell cylindrical (but often slightly broader distally), 27–35 µm long, 8–9 µm in diameter; intermediate cells elongate obovoidal, 8–28 µm long, 2–8 µm in diameter; distal cells obovoidal, 4–11 µm long, 2–4 µm in diameter; terminal hairs sparse, ca. 15 µm long. Monosporangia absent. Spermatangia apparently on primary fascicles, spherical, 3–5 µm in diameter. Carpogonium-bearing branch arising from periaxial cell, consisting of modified 4–5 cells 2–5 µm long, 5–8 µm in diameter. Involucral filaments arising from all subtending cells, 3–4 cells long; carpogonium 25–46 µm long, 3–7 µm in diameter at base, trichogyne sessile, curved, sausage-shaped, 8–10 µm in diameter. Carposporophytes 1–4 per whorl, in inner and outer cortex, spherical, compact, 130–160 µm in diameter, 0.6–1 times whorl radius, centre 75–90 µm from node; gonimoblast filaments 2 or 3 cells long; post-fertilization cells of carpogonial branch cylindrical. Carposporangia tadpole-shaped, clavate, 36–46 µm long, 12–18 µm in diameter.

   Type Locality: Eve River, tributary of Adam River and Gordon River, Tasmania, Australia.
   Holotype: NSW A008809.
   Distribution: Known from a single collection in southwestern Tasmania.

5) *Batrachospermum antiquum* Entwisle et Foard (1997:349, figs A–F)

   plate 67, figs 1–6 (original authors, Entwisle & Foard 1997)

   Plants monoecious, 2–3 cm high, grass-green, flaccid, monopodial; apices obtuse, apical cell slightly embedded within fascicles. Fascicle whorls barrel- to disc-shaped, 400–500 µm in diameter, confluent; internodes 200–300 µm long. Axis colourless, axial cells 15–18 µm in diameter, cortical filament cells cylindrical, 4–9 µm in diameter. Secondary fascicles common, some as long as primary fascicles. Primary fascicles 3 per periaxial cell, more or less straight distally, of 7–12 cell storeys; branching 4–6 times dichotomous; proximal cell ellipsoidal or fusiform, 30–60 µm long, 6–9 µm in diameter; intermediate cells ellipsoidal, fusiform or obovoidal, 12–45 µm long, 6–9 µm in diameter, distal cells obovoidal, 9–12 µm long, 3–6 µm in diameter; terminal hairs sparse, 24–30 µm long. Monosporangia absent. Spermatangia on primary fascicles and involucral filaments, scattered, spherical, 5–6 µm in diameter. Carpogonium-bearing branch straight, arising from periaxial cell, consisting of 4–6 modified cells, discoidal, 3–10 µm long, 6–10 µm in diameter. Involucral filaments arising from all subtending cells, 2–4 cells long, not extending beyond carpogonium before fertilization (afterwards extending far beyond the carposporophytes and resembling vegetative fascicles). Carpogonium curved, 44–53 µm long, 4–6 µm in diameter at base, trichogyne pedicellate, bent (at attachment point), foot-shaped (or in some views with a conical protuberance at base), 10–16 µm in diameter at broadest part. Carposporophyte 1, sometimes 2, per whorl, in inner cortex, spherical, compact, 120–175 µm in diameter, 0.4–0.5 times whorl radius, centre 90–125 µm from node; gonimoblast filaments 1–2 cells long; post-fertilization cells of carpogonium-bearing branch spherical or cylindrical, covered in cortical filaments. Carposporangia tadpole-shaped (or elongate obovoidal), 30–45 µm long, 18–20 µm in diameter.

   Type Locality: Old River, flowing into Bathurst Harbor, Southwest National Park, Tasmania, Australia.
   Holotype: MEL 2033387.
   Distribution: Known from a single river in south-western Tasmania.

(3) *Batrachospermum antipodites*

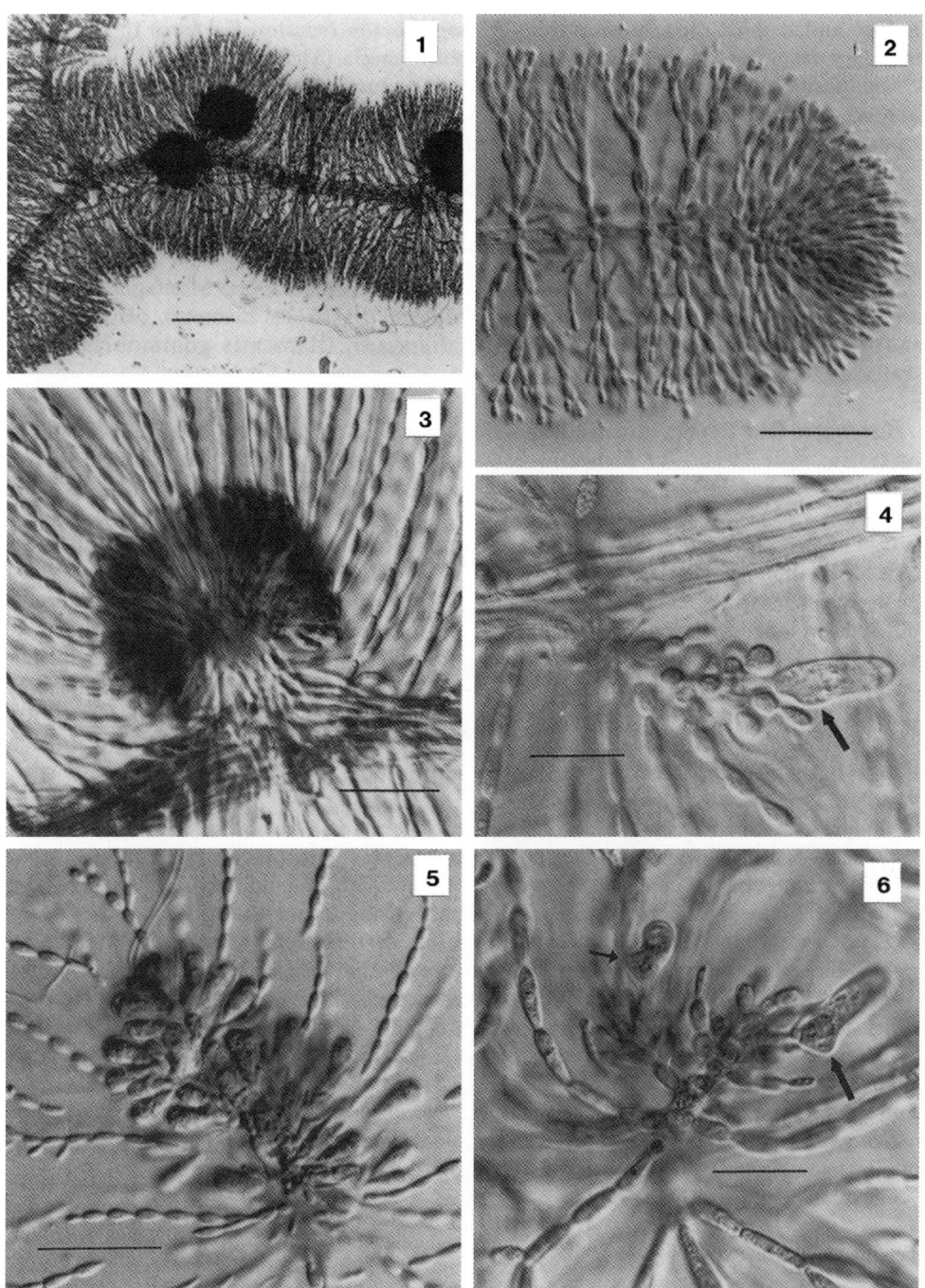

Plate 67
Batrachospermum antiquum, figs 1–6 (original authors, Entwisle & Foard 1997)
1. a mature plant with carposporophytes, 2. an apex of a plant, 3. a whorl with a carposporophyte, 4. a carpogonium (arrow to obscured 'heel' of trichogyne) and a carpogonium-bearing branch, 5. a squashed carposporophyte showing elongate carposporangia, 6. a developing carpogonium (smaller arrow to 'heel' of trichogyne) and a mature carpogonium (arrow to 'heel' of trichogyne) and a carpogonium-bearing branch.
Scale bars = 500 μm (fig. 1); 200 μm (figs 2–3, 5); 100 μm (figs 4, 6)

Batrachospermales

(4) *Batrachospermum prominens* group

1) ***Batrachospermum prominens*** Entwisle et Foard (1997:368, fig. 18, A–G)

        plate 68, figs 1–7 (original authors, Entwisle & Foard 1997)

Plants 1–2 cm high, grass-green, olive to blue-green when dry, flaccid, monopodial; apices obtuse, apical cell more or less flush with primary fascicles. Fascicle whorls barrel- to disc-shaped, 200–250 μm in diameter, confluent to separated; internodes 180–350 μm long. Axis colourless to light tan, axial cells 22–30 μm in diameter, cortical filament cells cylindrical, ca. 7 μm in diameter. Secondary fascicles common, some as long as primary fascicles. Primary fascicles 4 per periaxial cell, more or less straight distally, of 4–6 cell storeys; branching 3–5 times di- or trichotomous; proximal cell ellipsoidal or fusiform, 8–12 μm in diameter, 19–44 μm long; intermediate cells ellipsoidal or obovoidal, 6–12 μm in diameter, 9–28 μm long; distal cells ellipsoidal, obovoidal or spherical, 7–10 μm in diameter, 3–13 μm long; hairs 14–18 μm long. Monosporangia absent. Dioecious. Spermatangia borne on primary fascicles, scattered, spherical, 7–8 μm in diameter. Carpogonium-bearing branch curved or straight, arising from periaxial cell, consisting of 3–5 modified cells, discoid to cubed, 8–11 μm in diameter, 8–15 μm long. Involucral filaments arising from all subtending cells, 3 or 4 cells long, not extending beyond carpogonium. Carpogonium more or less straight, 35–40 μm long, ca. 7 μm in diameter at base, trichogyne sessile, radially symmetrical, ellipsoidal, 8–13 μm in diameter at broadest part. Carposporophyte 1 per whorl, extending through entire cortex and exserted from whorl, spherical, loosely aggregated, 160–300 μm in diameter, 2–2.5 times whorl radius, centre 72–135 μm from node; gonimoblast filaments 2–5 cells long; post-fertilization cells of carpogonial branch cylindrical. Carposporangia obovoidal or ellipsoidal, 19–22 μm in diameter, 22–34 μm long, sometimes germinating *in situ*.

    Type Locality: Deep River, Fernhook Falls, Western Australia, Australia.
    Holotype: MEL 2020289.
    Distribution: Known from a single river in south-western Australia.
    Note: Entwisle & Foard (1997) stated as follows: With its large single carposporophytes it has affinities with section *Virescentia* and *Turfosa*. It differs from species in the former section in having carpogonia with sessile, ellipsoidal trichogynes, carposporophyte-centre distant from whorl and in being dioecious. With only 4–6 cell storeys and a carposporophyte approximately as broad as the whorl diameter, it resembles some species in the *B. atrum* complex and according to Sheath *et al.* (1993) and Sheath and Vis (1995) it would belong in 'section *Setacea*' (included within section *Virescentia* by Necchi & Entwisle 1990).

2) ***Batrachospermum terawhiticum*** Entwisle et Foard (1997:371, fig. 19, A–F)

        plate 69, figs 1–6 (original authors, Entwisle & Foard 1997)

Plants monoecious, up to 4 cm high, red or purple when dry, monopodial; apices obtuse or acute, apical cell more or less flush with primary fascicles. Fascicle whorls conical (or in part somewhat spherical), 232–437 μm in diameter, separated or confluent; internodes 130–460 μm long. Axis colourless to brown, axial cells 25–58 μm in diameter, cortical filament cells cylindrical, 5–12 μm in diameter. Secondary fascicles common, always shorter than primary fascicles. Primary fascicles 3 per periaxial cell, more or less straight or slightly curved distally, of 6–10 cell storeys; branching 3–5 times di or trichotomous; proximal cell bone-shaped, obovoidal, cylindrical or fusiform, 4–10 μm in diameter, 12–30 μm long; intermediate cells bone-shaped to obovoidal, 10–26 μm long, 4–10 μm in diameter; distal cells obovoidal, fusiform or cylindrical,

(4) *Batrachospermum prominens*

Plate 68
Batrachospermum prominens, figs 1–7 (original authors, Entwisle & Foard 1997)
1. an apex of a plant, 2. a whorl with a carpogonium protruding, 3. a whorl, 4. a carpogonium and a carpogonium-bearing branch, 5. an inflated trichogyne in focal plane, 6. a carposporophyte above an axis and a whorl below, 7. whorls with spermatangia.
Scale bars = 200 μm (figs 6–7); 150 μm (fig. 2); 100 μm (figs 1, 3–5).

Batrachospermales

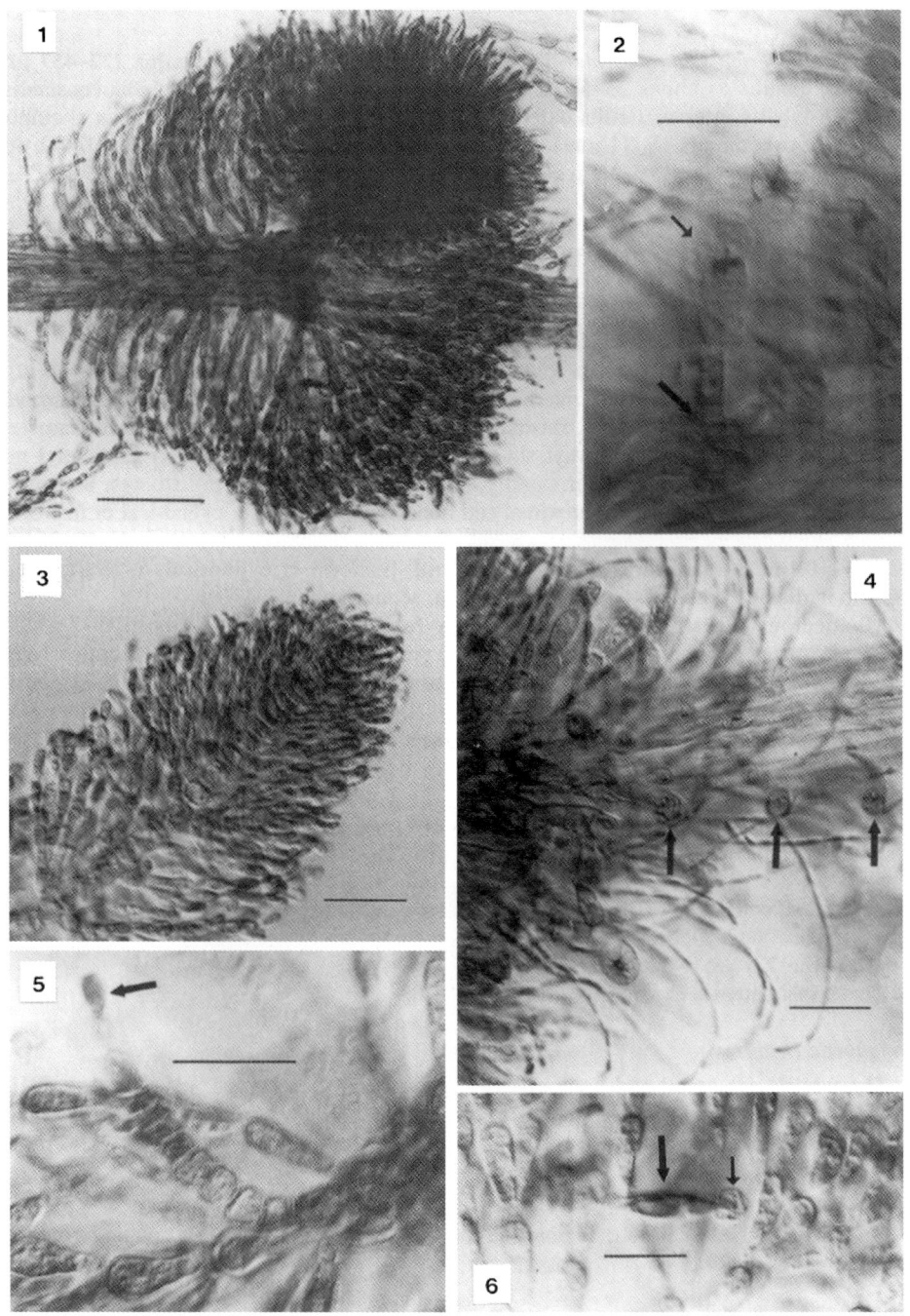

Plate 69
Batrachospermum terawhiticum, figs 1–6 (original authors, Entwisle & Foard 1997)
1. a whorl, 2. carposporangia (arrow to one) on short filaments arising laterally (small arrow) from an indeterminate gonimoblast filament, 3. an apex of a plant, 4. indeterminate gonimoblast filaments with carposporangial short filaments in an optical section (arrows), 5. an immature carpogonium (arrow to a developing trichogyne) and a carpogonium-bearing branch, 6. a trichogyne (arrow) with attached spermatia (small arrow). Scale bars = 200 µm (fig. 1); 100 µm (figs 2–4, 6); 50 µm (fig. 5).

6–13 µm long; hairs 5–6 µm long or absent. Monosporangia absent. Spermatia borne on primary 3–6 µm in diameter, fascicles, scattered, spherical, 5–7 µm in diameter. Carpogonium-bearing branch sometimes curved, arising from periaxial cell, consisting of 6–11 modified cells, these barrel-shaped to discoidal, 6–11 µm in diameter, 4–11 µm long. Involucral filaments arising from all subtending cells, 1–4 cells long, not extending beyond carpogonium. Carpogonium usually straight, 23–42 µm long, 4–7 µm in diameter at base, trichogyne sessile, club-shaped or cylindrical, 5–8 µm in diameter at broadest part. Carposporophyte 1 or 2 per whorl, in inner and outer cortex, sometimes exserted from whorl, spherical (rarely hemispherical), compact, 90–232 µm in diameter, 0.7–1.3 times whorl radius, centre (0–)75–120 µm from node (i.e. only rarely almost 'sessile' -usually when supporting filament curves toward axis); gonimoblast filaments 2–6 cells long or indeterminate and creeping along axis; post-fertilization cells of carpogonial branch spherical to cylindrical. Carposporangia obovoidal to spherical, 11–18(–22) µm in diameter, 18–31 µm long.

Type Locality: Te Rawhiti Creek, Bay of Islands, North Island, New Zealand.
Holotype: AKU 101795.
Distribution: Known from North Island, New Zealand: common in the Bay of Islands area with a couple of outliners further south.

Note: *Batrachospermum terawhiticum* has a clearly defined carposporophyte mass produced by determinate gonimoblast filaments (each terminated by a carposporangium) as well as some indeterminate (but relatively short) gonimoblast filaments which extend beyond the primary mass (fig. 19B, D). This feature distinguishes *B. terawhiticum* from all other Australian and New Zealand taxa, and most taxa elsewhere. *B. wattsii* from Australia has a loose, diffuse carposporophyte of more or less indeterminate filaments, but it is unlike *B. terawhiticum* in vegetative and reproductive morphology. Necchi (1990) lists the following taxa as producing both determinate and indeterminate gonimoblast filaments; *B. orthosticum* Skuja, *B. periplocum* (Skuja) Necchi, *B. vagum sensu* Kumano *et al.* (1970), *B. tapirense* Kumano et Phang, *B. keratophytum sensu* Necchi (1990). Sheath *et al.* (1993, 1994c), however, found no indeterminate filaments in type material of *B. orthosticum, B. keratophytum* or *B. vagum*. The characterization and taxonomic importance of indeterminate gonimoblasts needs further study, but *B. terawhiticum* should be compared with the taxa listed by Necchi (1990). *B. orthosticum sensu* Necchi (1990) is superficially most similar, but has generally less-developed whorls (cf. 3–6 cell storeys, branching 1–3 times), generally broader trichogynes (cf. 6–12 µm in diameter); semi-spherical rather than spherical carposporophytes, 'centered'; and smaller carposporangia (cf. 10–14 µm long, 7–11 µm in diameter).

(5) *Batrachospermum wattsii* group

Gonimoblast development obscure but unique; fascicle morphology characteristic.

1) ***Batrachospermum wattsii*** Entwisle et Foard (1997:374, fig. 21, A–F)
Misapplied name: *Batrachospermum keratophytum sensu* Entwisle et Kraft (1984:234); Entwisle (1989:45), non Bory (1808:328).

plate 70, figs 1–6 (original authors, Entwisle & Foard 1997)

Plants dioecious, dark brown, firm to rigid; apices obtuse, apical cell protruding through primary fascicles. Fascicle whorls ovoidal, to pyriform (occasionally barrel-shaped or cylindrical in older plants), 150–450 µm in diameter, confluent; internodes 200–500 µm long. Axis colourless to brown, axial cells 9–23 µm in diameter, cortical filaments 6–34 µm in diameter, consisting of mostly inflated, ellipsoidal cells.

Batrachospermales

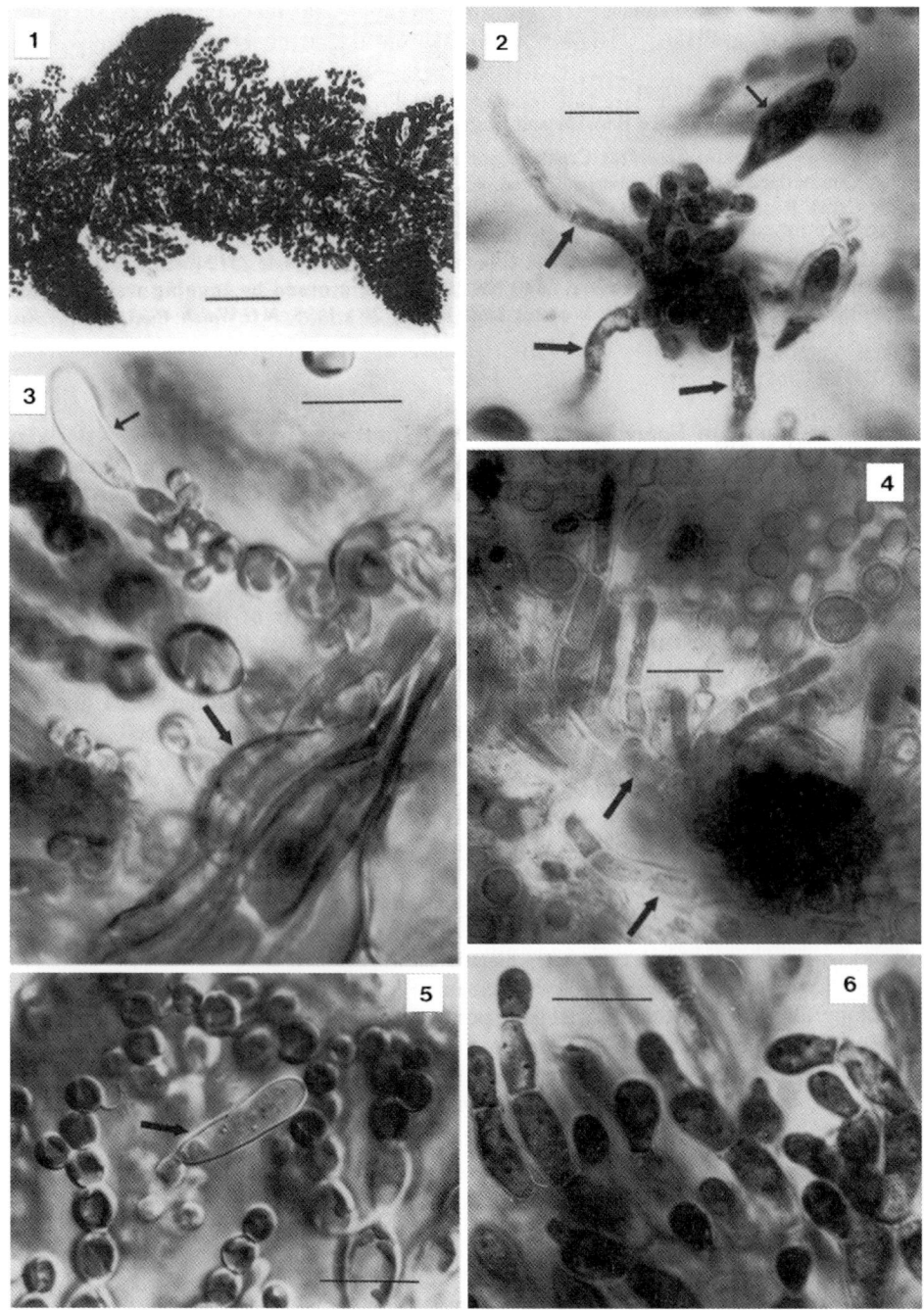

Plate 70
Batrachospermum wattsii, figs 1–6 (original authors, Entwisle & Foard 1997)
1. a plant with young short branches, 2. gonimoblast filaments (arrows) arising from a fertilized carpogonium (small arrow to trichogyne), 3. a carpogonium (smaller arrow to trichogyne), carpogonium bearing branches and an inflated cortical cell (arrow), 4. gonimoblast filaments (arrows), 5. a carpogonium (arrow to base of trichogyne) and distal cells of fascicles, 6. relatively compact array of gonimoblast filaments with swollen apical cells. Scale bars = 100 μm (figs 1, 4); 20 μm (figs 2–3, 5–6).

Secondary fascicles common, not as long as primary fascicles. Primary fascicles 2–6 per periaxial cell (2 or 3 fascicles borne at apex of cell, 2–4 produced laterally), more or less straight distally, of 3–6 cell storeys; branching 3–7 times 2–4-chotomous (rarely more); proximal cells irregularly cylindrical, ellipsoidal or obovoidal, 4–16 µm in diameter, 12–65 µm long; intermediate cells cylindrical, or ellipsoidal to obovoidal-spherical, 5–13 µm in diameter, 8–60 µm long; distal cells conical-spherical to spherical, 3–14 µm in diameter, 3–15 µm long. Hairs short or absent. Spermatangia abundant, borne on primary fascicles and secondary fascicles, scattered, spherical, 5–6 µm in diameter. Carpogonium-bearing branch sometimes slightly curved, arising from periaxial cell, consisting of 3–7 cells, all (or sometimes not proximal 1 or 2) cells modified, discoid to cubed (or somewhat turbinate), ca. 6 µm in diameter, ca. 6 µm long. Involucral filaments common, 1–3 cells long, not extending beyond carpogonium. Carpogonium usually more or less straight, base symmetrical, 20–56 µm long, 3–6 µm in diameter at base, trichogyne sessile, radially symmetrical, club-shaped to elongate-ellipsoidal, 5–8 µm in diameter at broader part. Carposporophyte diffuse, difficult to observe, consisting of apparently determinate filaments 5–12 cells long. Carposporangia rare, apparently obovoidal or ellipsoidal, 5–9 µm in diameter, 8–15 µm long. *Chantransia phase* always abundant. Epiphytic *Audouinella* (cells 3 µm in diameter, ca. 25 µm long) observed in some specimens.

Type Locality: Slip Creek, Road 1, Maroondah Reservoir catchment, Victoria in Australia.
Holotype: MEL 2026652.
Distribution: Known from central and southern Victoria in Australia.
Note: According to Entwisle & Foard (1997), despite intensive study of all available material, details of the post-fertilization development remain obscure and carposporangium production has yet to be unequivocally observed. The carpogonial branch and carpogonium are similar to those of species in the section *Turfosa*, but the loose, diffuse (all somewhat indeterminate) gonimoblast filaments are more like those of the genus *Nothocladus* (but never as well developed).

(6) *Batrachospermum cayennense* group (ref. p. 163)

Carpogonia subtended by a relatively long filament of modified cells.

Note: According to Entwisle & Foard (1997), the Australian population resembles *B. cayennense* sensu Sheath, Vis & Cole (1994b) but the whorls are generally narrower. The type of *B. cayennense* has slightly longer (19–22 µm) carpogonia but is otherwise a good match. All contemporary authors (e.g. Necchi & Entwisle 1990, Kumano 1993, Sheath *et al.* 1994b) include *B. cayennense* in the section *Aristata*. Due to the variability in the number and modification of cells subtending the carpogonium in species with carposporophytes borne away from the axis, the section *Aristata* is not used by Entwisle & Foard (1997).

2. Section *Setacea* De Toni (1897:57)

Type: *Batrachospermum dillenii* Bory

Synonym: section *Moniliformia* subsection *Capillacea* Sirodot (1873);

section *Moniliformia* subsection *Setacea* Sirodot (1875);

section '*Setaces*' Sirodot (1884:253).

Fascicles reduced and very short. Carpogonium-bearing branches very short, arising from the periaxial cells, well-differentiated and reduced to a few cells. Carpogonia with club- or urn-shaped trichogyne. Carposporophytes appearing as wart-like protuberances on central axis.

Batrachospermales

Note: Section *Setacea* is included in the section *Virescentia* by Necchi & Entwisle (1990), however, Vis *et al.* (1998) supported to keep the sections separated. A re-evaluation of *B. atrum* and *B. puiggarianum* including the description of *B. diatyches* from Tasmania, Australia was made by Entwisle (1992). From an analysis of six North American populations of the section *Setacea* and eight type specimens, Sheath, Vis & Cole (1993c) recognized 4 species, *B. androinvolucrum*, *B. diatyches*, *B. atrum* and *B. puiggarianum*. Entwisle & Foard (1998) established *B. latericium* sp. nov. from Tasmania, Australia, with new observations on *B atrum* and a discussion of their relationships. The author's circumscription of *B. atrum* is somewhat different.

Key to the species of the section *Setacea*

1. Spermatangia restricted to involucral filaments of the
   carpogonium-bearing branch..1) *B. androinvolucrum*
1. Spermatangia terminal on only primary or secondary fascicle..2
2. Axial cells < 2.5 times the diameter of the cortical filaments..3
2. Axial cells > 2.5 times the diameter of the cortical filaments..4
3. Primary fascicle consist of 2 (–4) cells..2) *B. latericium*
3. Primary fascicle consist of 4–6 cells..3) *B. diatyches*
4. Whorls non adherent to main axis..4) *B. atrum*
4. Whorls adherent to main axis..5) *B. puiggarianum*

1) ***Batrachospermum androinvolucrum*** Sheath, Vis et Cole (1993c:722, figs 3–7)

plate 71, figs 1–5 (original authors, Sheath Vis et Cole 1993c)

Plants monoecious with spermatangia formed on the involucral filaments of the carpogonium-bearing branch. Whorls non adherent, 70–164 μm in diameter and composed of 3–6 fascicle cells. Carpogonia 3.2–6.4 μm in diameter, 13–8–32.6 μm long. Trichogynes club-shaped, obovoidal and sessile. Carpogonium-bearing branch 1–3 cells long. Carposporophytes 124–142 μm in diameter, 33–103 μm high, with 1–3 gonimoblast cells. Carposporangia 5.3–8.4 μm in diameter, 7.0–13.0 μm long.

Type Locality: Last Shoe Creek, at Route 4, 8 km north-east of Ucluelet, Vancouver Island, British Columbia, Canada in North America.

Holotype: UBC A80771.

Distribution: Known from the type locality and Staghorn Creek, Grice Bay Road, 17 km north of Ucluelet, Vancouver Island, British Columbia, Canada, Skagit River, Washington and Swift Creek, Alabama, USA in North America.

2) ***Batrachospermum latericium*** Entwisle et Foard (1998:28, fig. 1, a–i, fig. 2, a)
Synonym: *B. nothogeae* Skuja *nom. nud., pro parte*.

plate 73, figs 1–9 (original author, Entwisle & Foard 1998)

Plants monoecious, 1–5 cm high, 99–180 μm in diameter, olive-green, apical cell 11–15 μm in diameter, 6–12 μm long, protruding through primary fascicles. Whorls separated, conical, very small and often barely arising above cortical filaments, internodes 120–170 μm long. Primary fascicles 2 or 3 per periaxial cell, of 2(–4) cell storeys, non- or once dichotomously branched; proximal cell obovoidal to spherical, about 8 μm in diameter, about 8 μm long; distal cells spherical to hemispherical or dome-shaped, about 6 μm in diameter, 6–10 μm long. Terminal

2. *Setacea*

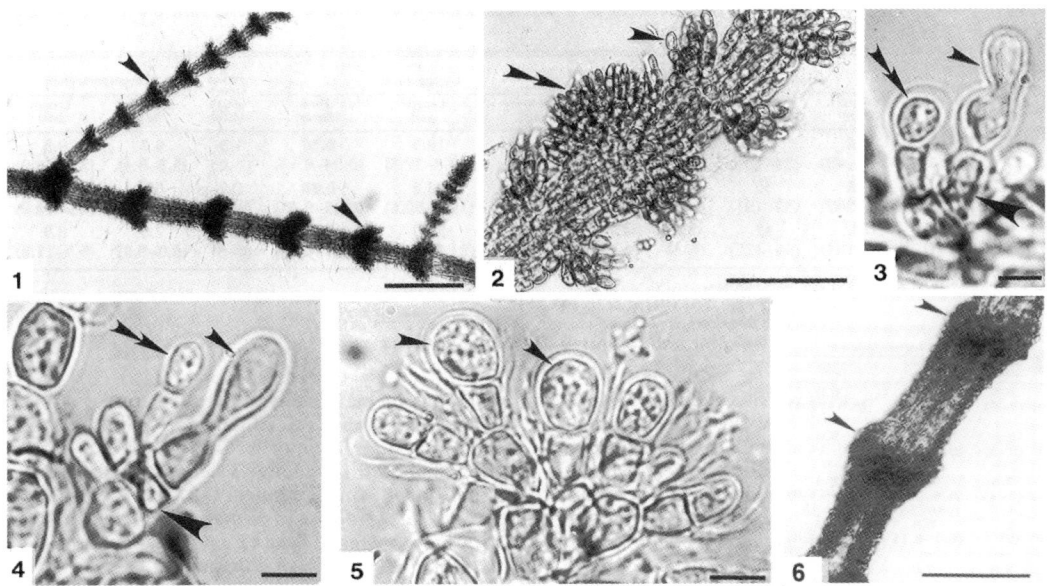

Plate 71
Batrachospermum androinvolucrum, figs 1–6 (original authors, Sheath, Vis & Cole 1993c).
1. a main axis with branches showing non-adherent whorls (arrows), 2. whorls with three to four fascicle cells (arrow) and a carposporophyte protruding beyond a whorl (double arrow), 3. an immature carpogonium (arrow) on a two celled carpogonium-bearing branch (large arrow) with a spermatium terminal on an involucral filament (double arrow), 4. a mature carpogonium (arrow) on a two celled carpogonium bearing branch (large arrow) with a spermatangium terminal on involucral filaments (double arrow), 5. a carposporophyte with carposporangia (arrow) and short gonimoblast filaments, 6. *B. anglense* with adherent whorls (arrows).
Scale bars = 200 μm (figs 1, 6); 100 μm (fig. 2); 5 μm (figs 3–5).

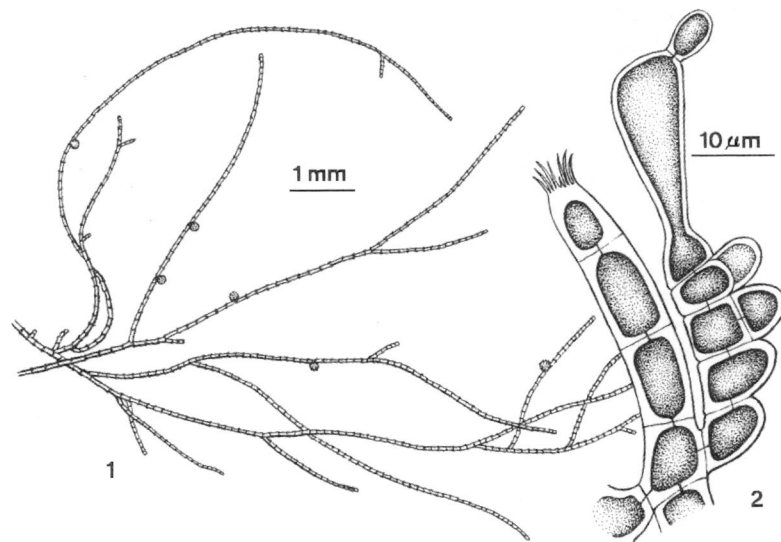

Plate 72
Batrachospermum diatyches, figs 1–2 (original author, Entwisle 1992)
1. the habit of a plant showing lateral, spherical carposporophytes, 2. a carpogonium-bearing branch (right) arising from a proximal cell of a sterile lateral and a carpogonium with an attached spermatium.

Batrachospermales

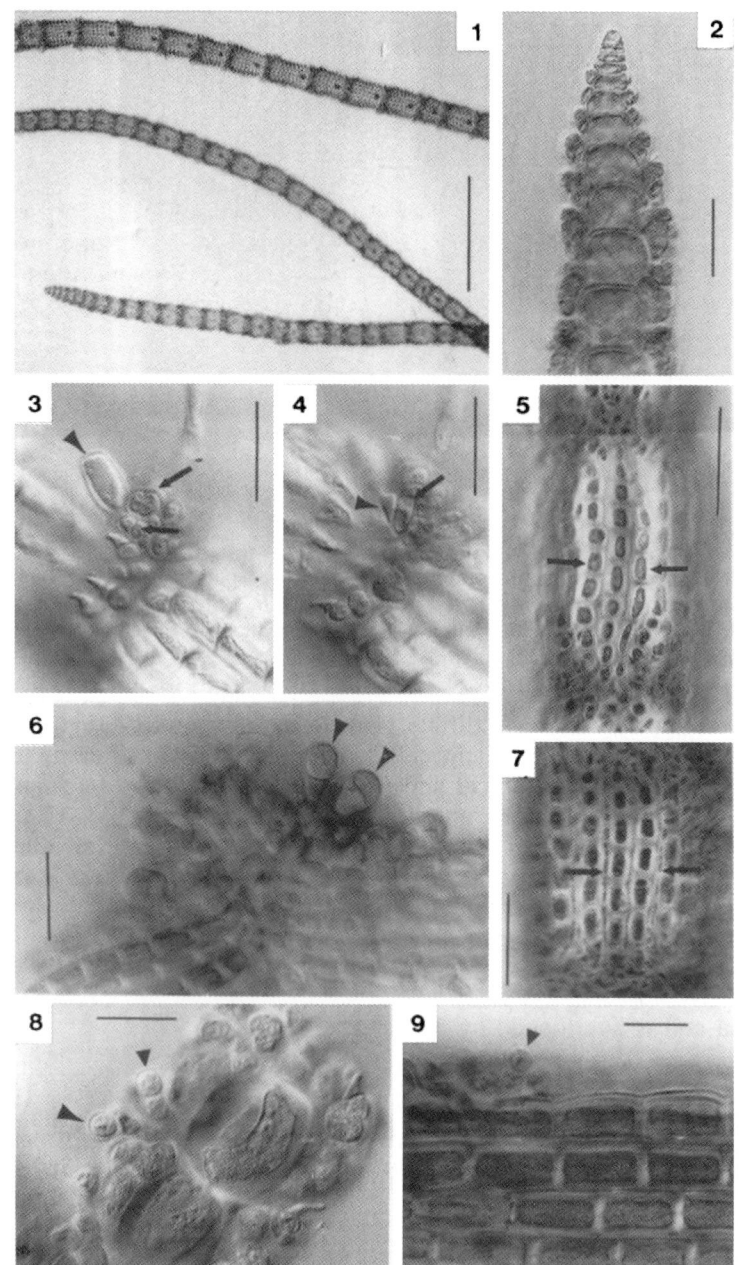

Plate 73
Batrachospermum latericiuum Entwisle, figs 1–9 (original author, Entwisle 1998)
1. thallus showing reduced whorls and regular cortical filaments cells, 2. apex showing fascicle initiation and structure, 3. nodal portion of thallus showing trichogyne (arrowhead), involucral filaments (arrow) and reduced fascicles, 4. same as 3 but focusing on the base of the carpogonium (arrowhead) and carpogonium-bearing branch (arrow), 5. brick-like cortical filament cells, 6. gonimoblast filaments with carposporangia (arrowheads), 7. brick-like cortical filament cells and narrow secondary cortical filaments (arrows) in older portion of thallus, 8. spermatangia (arrowheads) on primary fascicle in young portion of thallus, 9. spermatangium (arrowhead) on secondary fascicle (older portion than 8) arising from brick-like cortical cells. Scale bars = 1400 μm (fig. 1); 200 μm (fig. 7); 150 μm (figs 2–6); 100 μm (fig. 8); 50 μm (fig. 9).

hairs common in young portion of thallus, up to 12 about 8 µm in diameter, about 8 µm long. Axes colourless, axial cells 40–80 µm in diameter, cortical filament cells brick-like, 8–15 µm in diameter, 7–12 µm long, narrow secondary cortical filaments sometimes present between primary cortical filament in older thalli. Secondary fascicles either a single cell cut-off from primary cortical filament or a short filament arising from secondary cortical filament. Monosporangia absent. Spermatangia clusters on primary and secondary fascicles, spherical, about 8 µm in diameter. Carpogonium-bearing branch consisting of scarcely modified 1–2 cells, 6–8 µm in diameter, 4–5 µm long, arising from periaxial cell; involucral filaments 1–2 cells long, arising from all subtending cells, not extending beyond carpogonium. Carpogonium more or less straight, about 22–24 µm long, symmetrical or slightly oblique, about 6 µm in diameter at base, trichogyne not stalked, obovoidal to fusiform, 7–10 µm in diameter at broadest portion. Carposporophytes single, exserted from whorl, hemispherical, compact, about 160 µm in diameter, centered on node, gonimoblast filaments 2–3 cells long; post-fertilization cells of carpogonium-bearing branch obscure; carposporangia obovoidal or spherical, 16–18 µm in diameter, 20–26 µm long. Chantransia phase rarely observed, sparsely branched; cells cylindrical, 8–10 µm in diameter, 38–56 µm long.

Type Locality: first major riffle above Bathurst Harbour, Old River, Southwest National Park, Tasmania in Australia.

Holotype: MEL, Entwisle 2507, 4/III, 1996.

Isotype: HO.

Distribution: Restricted to stream in the far south-west corner of Tasmania, sometimes occurring with *B. atrum*. The streams are all humid-rich (black water) and flow through a submerged vasicular plant, *Gymnoschoenus sphaerocephalus,* heath and cool temperate rainforest in Australia.

Note: This species is similar to *B. diatyches* in morphology but primary fascicles of former species consist of 1(–4) cells, latter species 4–6 cells.

3) ***Batrachospermum diatyches*** Entwisle (1992:426, fig. 1, a–f, fig. 2, a–b)
Synonym: *B. nothogeae* Skuja *nom. nud., pro parte.*

> plate 72, figs 1–2 (original author, Entwisle 1992)
> plate 74, figs 1–6 (original author, Entwisle 1992)

Plants monoecious, entangled, firm, wiry, up to 6 cm high, 70–110 µm in diameter, dark brown, irregularly, sparsely, acutely to perpendicularly branched. Fascicles 2 per periaxial cell, whorls inconspicuous, separated, covered by regular layers of cortical filaments. Primary fascicles finger-like, non-branched, 4–6 cell-storeys, proximal cells cylindrical, intercalary cells more or less cylindrical, terminal hairs long. Secondary fascicles rare, usually single celled. Spermatangia ellipsoidal, 8–10 µm in diameter, terminal on fascicles. Carpogonium-bearing branches rare, arising from proximal cells of primary fascicles, consisting of 2–3 cells, straight or curved, with compact, few-celled involucral filaments. Carpogonia 44–46 µm long, trichogyne elongate to swollen, fusiform to obovoidal, 5–10 µm in diameter, 32–38 µm long. Carposporophytes usually sparse, more or less spherical, protruding wart-like from whorls, 80–220 µm in diameter. Carposporangia obovoidal to spherical, 6–18 µm in diameter, 15–19 µm long.

Type Locality: Lake Meston, north-central Tasmania, Australia. Epilithic in littoral zone of northern portion of lake.

Holotype: MEL 1587821.

Distribution: Known at elevations above 300 m in western Tasmania in Australia.

Batrachospermales

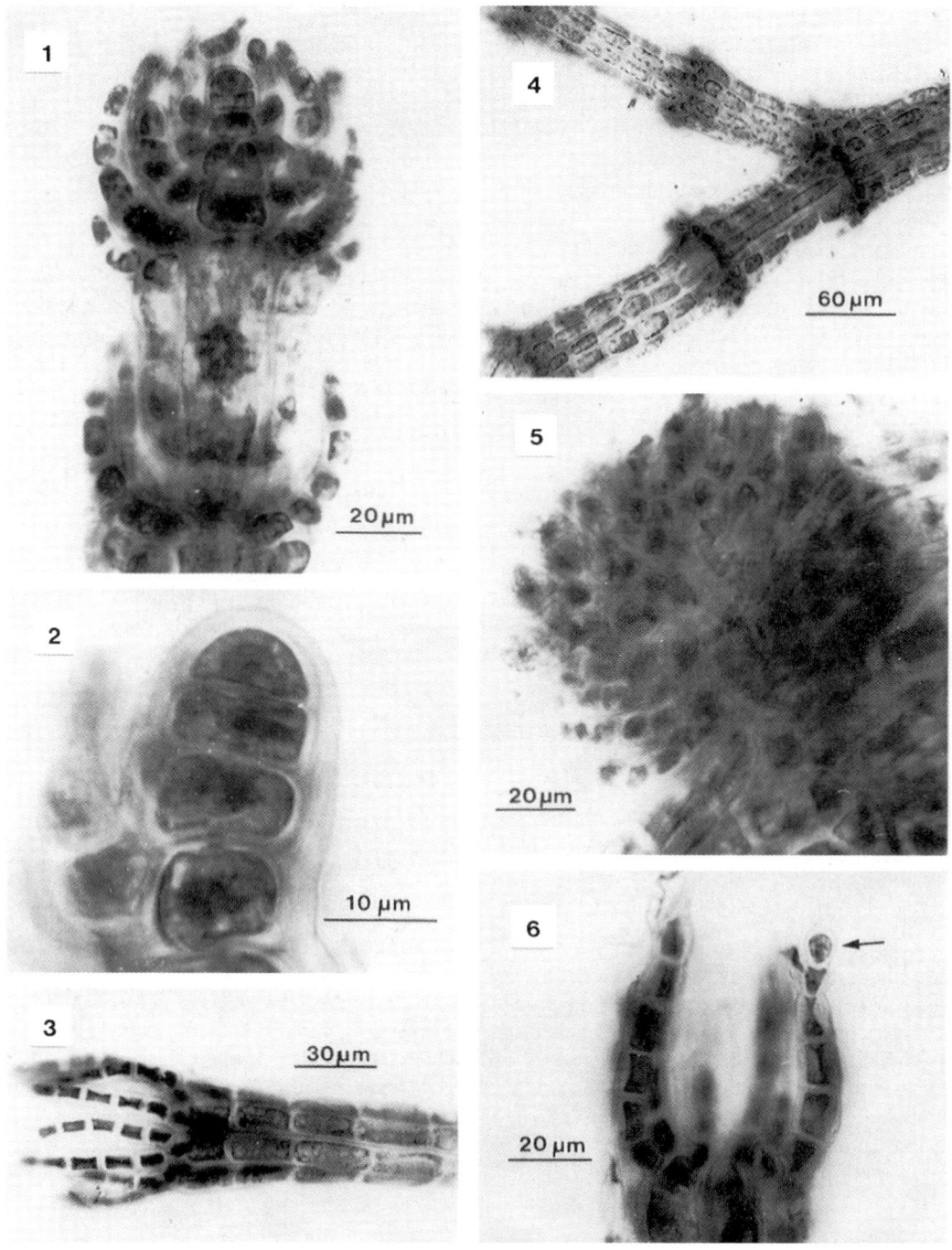

Plate 74
Batrachospermum diatyches, figs 1–6 (original author, Entwisle 1992)
1. a plant apex with an apical cell overtopped by young fascicles, 2. an apical cell and first axial cells of a plant, 3. fascicles and cortical filaments in a disassociated fragment, 4. a portion of a mature plant with axial cells in focus towards right and cortical filaments towards left, 5. a carposporophyte attached to a plant at bottom right, 6. a spermatangium (arrow) terminal on a lateral fascicle.

2. Setacea

Note: This species is similar to *B. atrum* in habit and morphology but has blunt apices with an embedded, large apical cell more than 10 μm in diameter; apical cell less than 2.5 times the diameter of the broad cortical filaments, more than 9 μm in diameter; and fascicle filaments gradually tapering and not constricted at crosswalls.

4) ***Batrachospermum atrum*** (Hudson) Harvey (1841:120)
Basionym: *Conferva atra* Hudson (1898:597).
The synonymy proposed by Kumano (1980) is mainly as follows;
Synonym: *B. tenuissimum* Bory (1823:227); *B. dillenii* Sirodot (1884:254, pl. 20, figs 1–2, pl. 21, figs 1–12, pl. 22, figs 8–13); *B. gallaei* Sirodot (1884:256, pl. 22, figs 1–7); *B. angolense* W. West et G. S. West (1897:2); *Sirodotia angolensis* (W. West et G. W. West) Skuja in Reis (1960:53).

plate 75, figs 1–9 (Kumano 1997b)
plate 76, figs 1–8, (Entwisle 1992)
plate 77, figs 1–3 (Entwisle 1992)
figs 4–7 (original author, Sirodot 1884), as *B. gallaei*

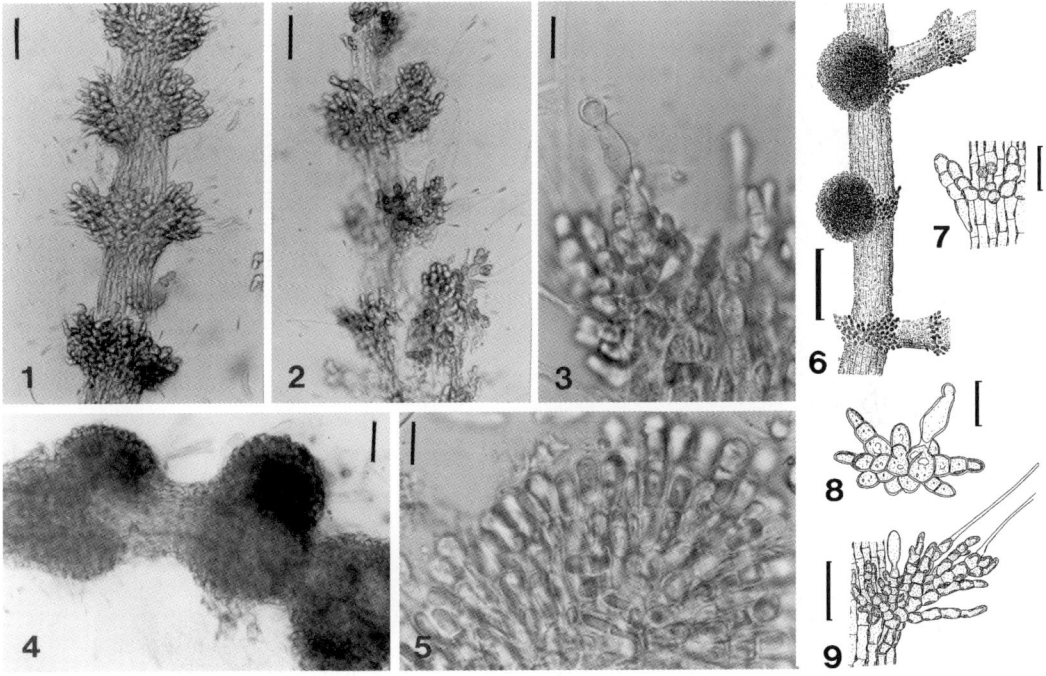

Plate 75
Batrachospermum atrum, figs 1–9 (Kumano 1997b)
1, 2, 9. a main axis with whorls showing primary fascicles with terminal hairs and a differentiated short carpogonium bearing branch, 3, 8. a fertilized carpogonium and a short carpogonium-bearing branch, 4–6. wart-like carposporophytes showing carposporangia terminal on gonimoblast filaments, 7. spermatangia terminal on very short fascicles.
Scale bars = 200 μm (fig. 6); 100 μm (fig. 4); 40 μm (figs 1–2); 20 μm (figs 7–9); 10 μm (figs 3, 5).

Batrachospermales

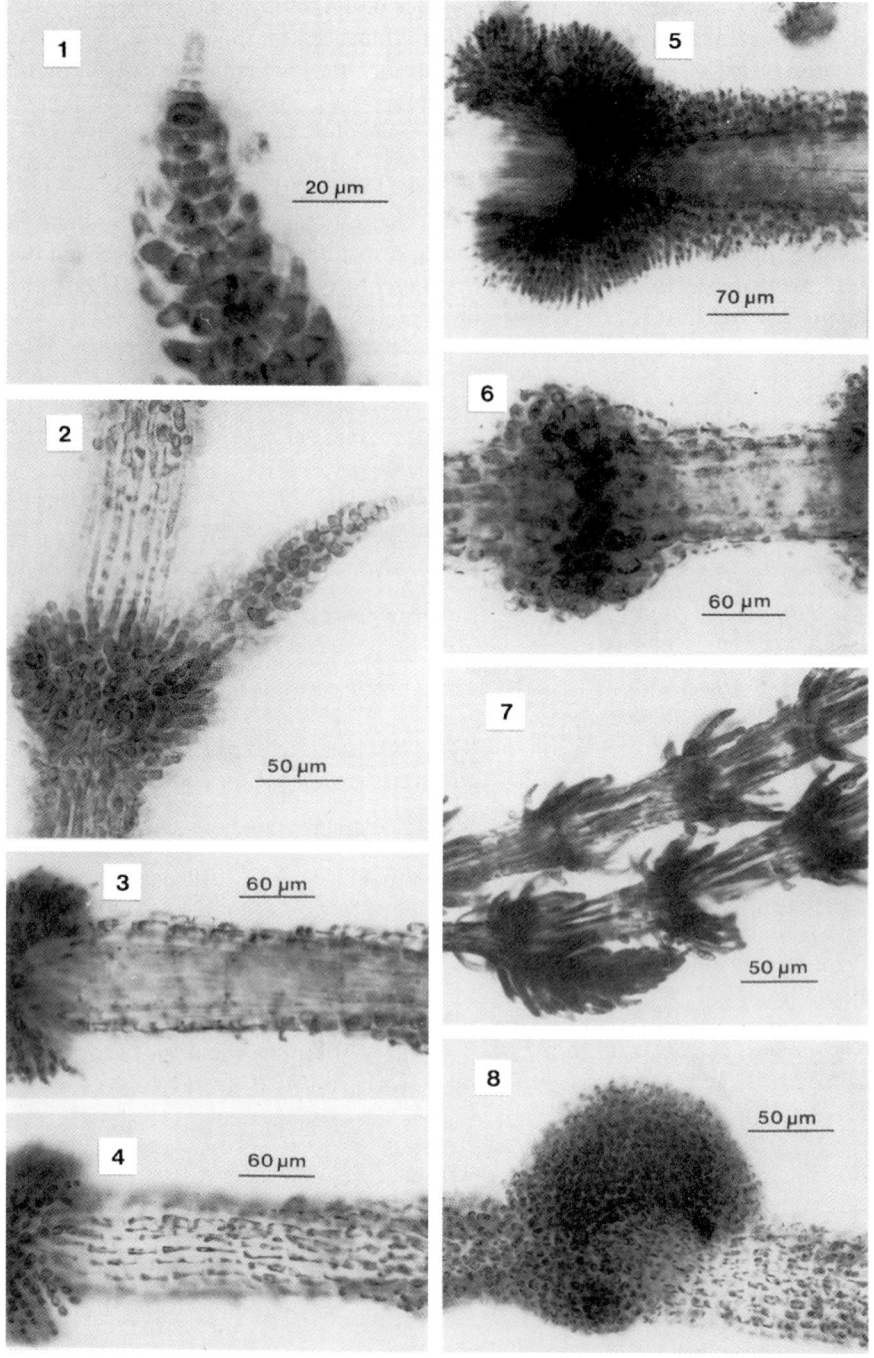

Plate 76
Batrachospermum atrum, figs 1–8 (Entwisle 1992)
1. an apex of a plant with a protruding apical cell and an immediate derivative, 2. a whorl and a young indeterminate branch, 3–4. an internode of a plant at two different foci showing axial cell (a) and cortical filaments (c), 5. a relatively open whorl, 6. a relatively compact whorl, 7. an open, loose whorl with long, cylindrical fascicle cells, 8. a compact whorl and a protruding carposporophyte.

Plants monoecious or dioecious, green to reddish-brown, almost black, up to 4 cm high, 120–240 µm in diameter, irregularly, sparsely to frequently branched; fascicles 2–3, whorls ovoid to conical and separated except in aged plants, compressed; primary fascicles abundantly branched, 3–6 cell-storeys, proximal cells spherical to dome-shaped, terminal hairs absent, sparse or abundant; intertwining cortical filaments well-developed and covering all central axes; secondary fascicles frequent; spermatangia spherical to obovoidal, 4–6 µm in diameter, terminal on fascicles and involucral filaments; carpogonium-bearing branches arising from periaxial cells and fascicles, consisting of 3–5 cells, slightly curved, with few-celled involucral filaments; carpogonia 4–7 µm in diameter at the apex, 15–27 µm long, trichogyne inflated-ovoid, urn-shaped or ellipsoidal; carposporophytes single or paired, hemispherical, protruding wart-like from whorls, 100–140 µm in diameter, 50–70 µm high; carposporangia obovoidal, 6–7 µm in diameter, 7–11 µm long.

Type Locality: a little well, in the plain called Gors Bach, between Llanfaethly and Trefadog, Wales, UK in Europe.

Lectotype: BM herb. Sloane, designated by Sheath & Cole 1993.

Distribution: Known from Angola in Africa, UK, France, Belgium, Germany, Poland, Portugal, Sweden in Europe; China, Korea, Japan in Eastern Asia; Australia; New Zealand; Brazil in South America. In Japan, Morioka in Iwate, Inokashira, Koishikawa in Tokyo, Anjyo, Toyohashi in Aichi, Okada, Himeji, Shingu, Hirotani-gawa River, Suzuran-dai, Kobe in Hyogo, Matsue in Shimane. Known from spring-fed streams.

In China, Nan-chiao-shan, Chienkiang in Kiangsu. Occurs on rocks in a small pool in connecting with under ground springs.

Note: Sheath *et al.* (1993) considered *B. orthostichum* to be another synonym of *B. atrum* based on the multivariate association, lack of significant difference in whorl size, and absence of indeterminate gonimoblast filaments in the type specimen of *B. orthostichum*. According to Entwisle (1992), two species of the section *Turfosa*, *B. orthostichum* and *B. keratophytum* sensu Necchi (1990), can have reduced whorls, however, are quite different from *B. atrum*. *B. orthosticum*, although producing *B. atrum*-like fascicles, has relatively well-developed, confluent whorls. It also has club-shaped apices and both determinate and indeterminate gonimoblast filaments (Necchi 1990), as figured and described by Skuja (1931), so that this species may be assigned to the section *Turfosa*.

5) ***Batrachospermum puiggarianum*** Grunow in Wittrock et Nordstedt (1889:1, no. 5–1)
Synonym: *B. schwacheanum* Möbius (1892:20, pl. 1, figs 1–8); *B. nigrescens* W. West et G. S. West (1897:2); *Sirodotia nigrescens* (W. West et G. S. West) Skuja in Reis (1960:54); *B. atrum* var. *puiggarianum* (Grunow) Necchi (1989:25).

plate 77, figs 8–9 (Starmach 1977)
figs 10–11 (Entwisle 1992)
plate 78, figs 1–6 (Entwisle 1992)

Plants monoecious or dioecious, up to 6 cm high, 80–300 µm in diameter, irregularly, abundantly, acutely to perpendicularly branched. Fascicles 2–3, whorls obconical or pear-shaped and separated, contiguous and compressed. Primary fascicles abundantly branched, 2–5 cell-storeys, proximal and intercalary cells cylindrical or barrel-shaped, terminal hairs long, absent to abundant. Secondary fascicles frequent. Spermatangia spherical to obovoidal, 4–7 µm in diameter, sub-terminal on primary and secondary fascicles. Carpogonium-bearing branches arising from periaxial cells and proximal cells of fascicles, consisting of 1–3 cells, straight or curved, with few-celled involucral filaments. Carpogonia 5–9 µm in diameter at the apex, 8–32 µm long, trichogyne club-shaped or ellipsoidal. Carposporophytes 1–2(3), hemispherical,

Batrachospermales

protruding wart-like from whorls, 90–220 μm in diameter, 50–130 μm high. Carposporangia obovoidal to ellipsoidal, 6–11 μm in diameter, 8–13 μm long.

Type Locality: Apiaí, São Paulo, Brasil in South America.

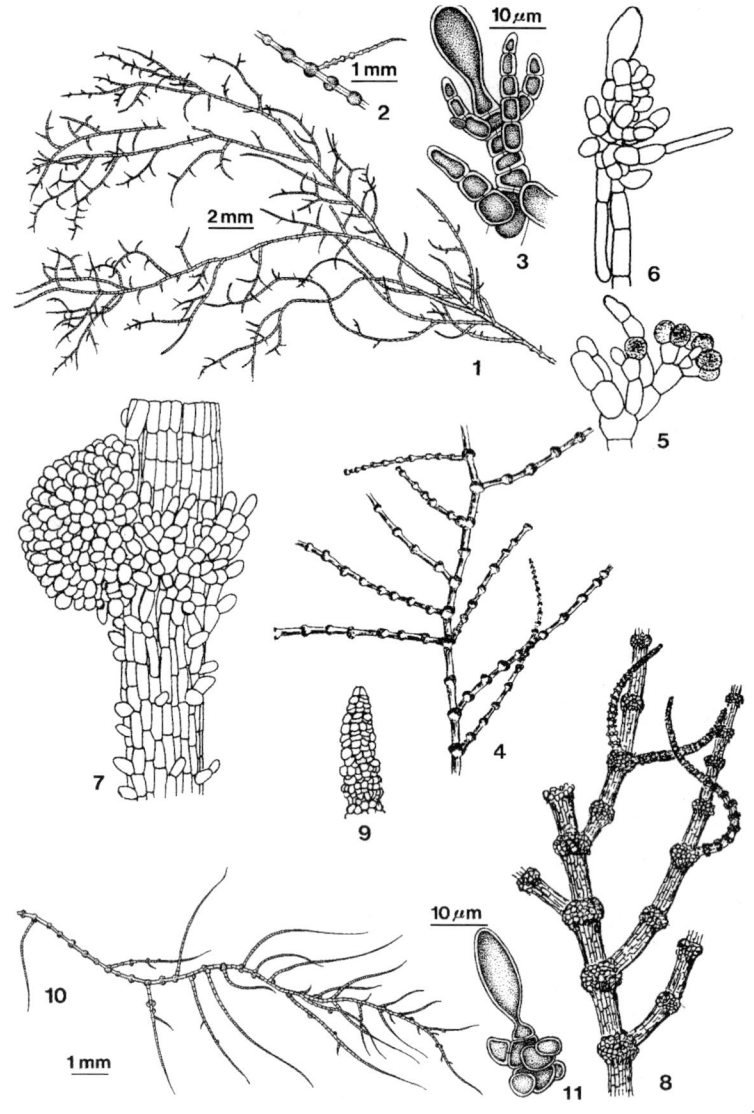

Plate 77
Batrachospermum atrum, figs 1–3 (Entwisle 1992)
1. the habit of plants, 2. wart-like carposporophytes on a main branch, 3. a carpogonium-bearing branch with 3–5 celled, occasionally branched involucral filaments.

Batrachospermum atrum as *B. gallaei*, figs 4–7 (original author, Sirodot 1884)
4. the habit of a plant showing reduced whorls, 5. spermatangia subterminal on fascicles, 6. a differentiated short carpogonium-bearing branch, 7. wart-like hemispherical carposporophytes.

Batrachospermum puiggarianum, figs 8–9 (Starmach 1977), figs 10–11 (Entwisle 1992)
8. the habit of a plant showing reduced whorls, 9. an apex of a plant, 10. the habit of a plant showing reduced whorls, 11. a carpogonium-bearing branch.

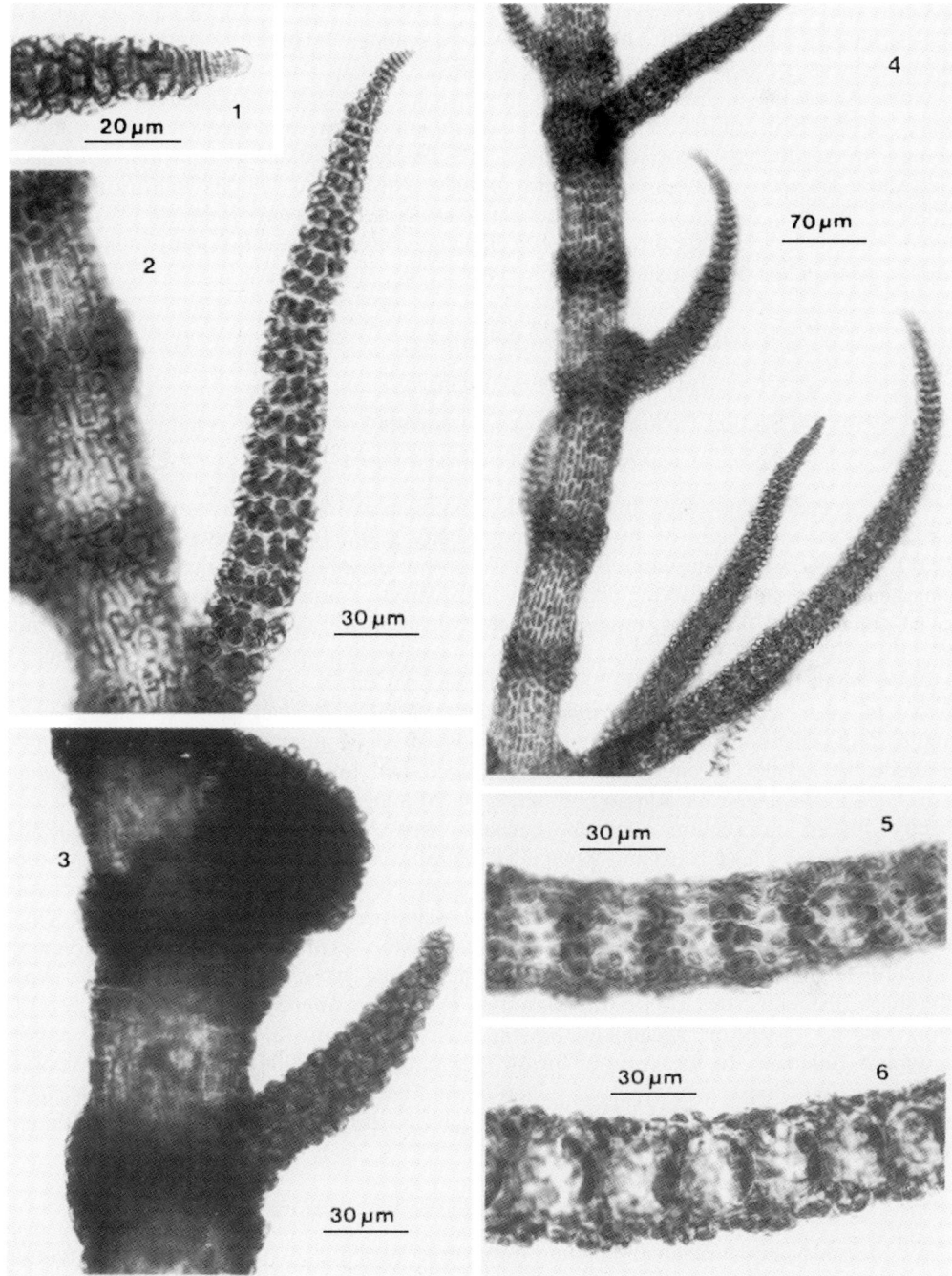

Plate 78
Batrachospermum puiggarianum, figs 1–6 (Entwisle 1992)
1. an apex of a plant with a protruding apical cell and immediate derivatives, 2. the whorl structure in a young indeterminate fascicle and a mature axis, 3. a carposporophyte protruding out of a whorl and an indeterminate branch arising from a whorl base, 4. the plant habit showing an appearance, 5–6. an internode of a plant at two different foci showing cortical filaments and young cortical filaments and compact fascicles.

Lectotype: S coll. Y. I. Puiggari.

Distribution: Known from Angola in Africa; Argentina, Brazil, Uruguay in South America. Attached to rocks in moderate to fast flowing streams.

Note: According to Entwisle (1992), this species has been distinguished from *B. atrum* primary on the basis of its very compact whorls, in which individual fascicles cannot be discerned. In addition to whorl compactness, the illustrations of *B. puiggarianum* by Necchi (1990) show an apex with globular lateral initials, no 2-celled laterals until some 13 cells proximal to the apex, and an indiscernible fascicle structure. Although Starmach (1977) used the absence of secondary laterals to characterize *B. puiggarianum*, Necchi (1990) found secondary laterals to be abundant in the materials he observed.

3. Section *Turfosa* Sirodot *emend.* Necchi (1990a:124).

Type: *Batrachospermum turfosum* Bory

Synonym: Section *Turfosa* Sirodot (1873), section *Turficola* De Toni (1897:58);

section '*Turficoles*' Sirodot (1884:259).

Plants pseudo-dichotomously branched. Carpogonium-bearing branches straight, short, arising from periaxial cells of fascicles. Carpogonia sessile or indistinctly stalked, elongate conical trichogyne with distal large diameter. Carposporophytes big, spherical or semi-spherical, single, sometimes paired, inserted centrally within whorls. Carposporophytes consisting of two types of gonimoblast filaments, radially branched determinate ones and prostrate indeterminate ones.

Note: Some species originally classified in section *Turfosa* (e.g. Sirodot 1873) have been removed to other sections because they have characters other than those already given. For example, *B. guyanense* (Montagne) Kumano, *B. nodiflorum* Montagne and *B. torridum* Montagne have helically twisted carpogonium-bearing branch and are classified in section *Contorta*. *B. cayennense* Montagne ex Kützing, *B. excesum* Montagne and *B. oxycladum* Montagne are presently placed in section *Aristata* because they have relatively long, differentiated carpogonium-bearing branches and pedicellate carposporophytes (Kumano 1990). Moreover, according to the definition of the section *Turfosa sensu* Necchi (1990a), taxa originally classified in the section *Turfosa*, such as *B. vogesiacum* T. G. Schultz ex Skuja and *B. gombakense* Kumano et Ratnasabapathy must be transferred to the section *Virescentia*, as redefined in Kumano & Phang (1987) and Necchi (1990a), because these taxa have only a determinate type of gonimoblast filaments. Necchi (1990) reclassified *B. orthostichum* Skuja from section *Setacea* to section *Turfosa* based on specimens having both indeterminate and determinate types of gonimoblast filaments. In the section *Turfosa sensu* Necchi (1990a), the following taxa were recognized as possessing two types of gonimoblast filaments (determinate and indeterminate): *B. orthostichum* Skuja, *B. periplocum* (Skuja) Necchi, *B. turfosum* Bory, *B. tapirense* Kumano et Phang.

Key to the species of the section *Turfosa*

1. Carpogonium-bearing branch descending from periaxial cell..................................1) *B. tapirense*
1. Carpogonium-bearing branch ascending from periaxial cell..2
2. Whorls reduced; primary fascicles (2–)3–7(–8) cell storeys..3
2. Whorls well developed; primary fascicles 8–15(–17) cell storeys..4
3. Fascicles chantransia-like...2) *B. orthostichum*
3. Fascicles not chantransia-like...3) *B. keratophytum*
4. Cortical filaments creeping along peripheral portion of frond..........................4) *B. periplocum*
4. Cortical filaments creeping along axial cell...5) *B. turfosum*

3. *Turfosa*

1) ***Batrachospermum tapirense*** Kumano et Phang (1987:259, figs 1–17)

plate 79, figs 1–4 (Kumano 1996zh)
figs 5–9 (original authors, Kumano & Phang 1987)

Plants monoecious, moderate mucilaginous, green bluish, rigid in consistency ca. 6 cm high, 80–170 μm in diameter, more or less trichotomously branched; whorls obconical, contiguous, more or less compressed in mature plants; primary fascicles di- or trichotomously branched, 4–5 cell-storeys, cells fusiform or ellipsoidal, terminal hairs present; cortical filaments well-developed; secondary fascicles 2–4 cell-storeys, covering all the internodes; spermatangia spherical, 4–6 μm in diameter, terminal on primary and secondary fascicles; carpogonium-bearing branches consisting of 4–6 disc- or barrel-shaped cells, arising from periaxial cells. Involucral filaments very short, more or less laterally issued; carpogonia 4–5 μm in diameter at the base, 5–6 μm in diameter at the apex, 30–40 μm long, trichogyne club-shaped, more or less indistinctly stalked, often bent at the base; carposporophytes indefinite in form, indistinguishable from and the same length as whorls; gonimoblast filaments of two kinds: radially branched determinate filaments, and diffuse, indeterminate filaments creeping along cortical filaments; carposporangia spherical or ellipsoidal, 5–8 μm in diameter, 8–12 μm long.

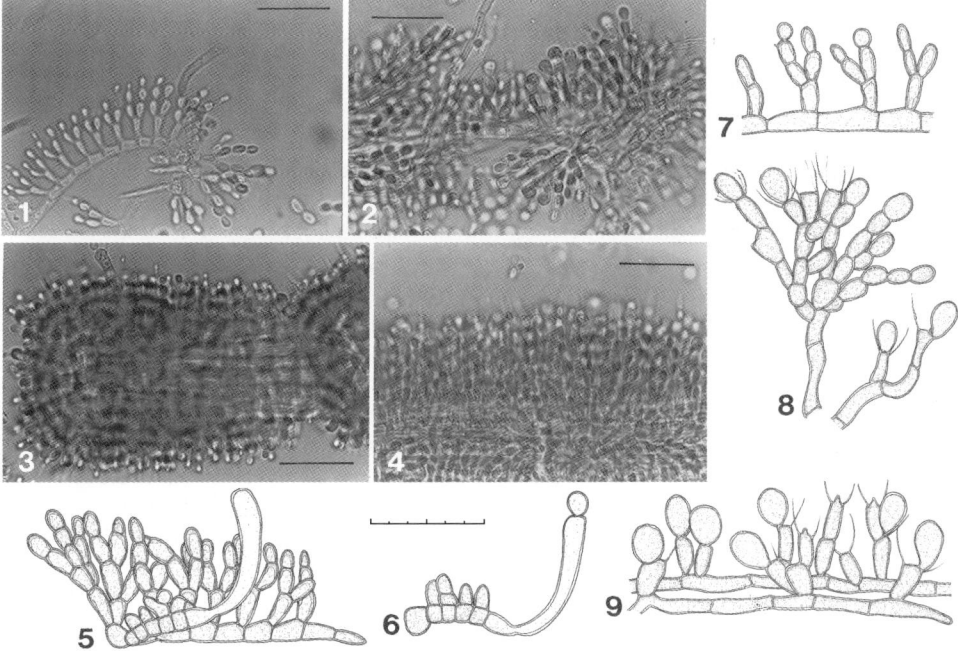

Plate 79
Batrachospermum tapirense, figs 1–4 (Kumano 1996zh), figs 5–9 (original authors, Kumano & Phang 1987)
1. a portion of a plant showing primary, secondary fascicles, a carpogonium bearing branch and cortical filaments, 2. an indeterminate gonimoblast filament and a determinate gonimoblast filament, 3. the structure of a plant showing the extruding terminal portion trichogyne, 4. a carposporophyte indistinguishable from the whorl fascicles, 5. a portion of a plant showing primary, secondary fascicles, cortical filaments and a carpogonium-bearing branch, which grows toward the same direction as cortical filaments elongates, 6. a carpogonium-bearing branch with a fertilized trichogyne, 7. spermatangia terminal on secondary fascicles, 8. Carposporangia terminal on determinate, radially branched gonimoblast filaments, 9. carposporangia terminal on laterals of indeterminate, defused gonimoblast filaments.
Scale bars = 40 μm (figs 1–4); 30 μm (figs 5–9).

Batrachospermales

Type Locality: Sungai Tapir, a tributary of Sungai Endau, Johor, Malaysia in Southeast Asia.
Holotype: Kobe University, Phang Siew Moi, no. 216, 30/IX 1985.
Distribution: Known from the type locality only.
Note: The initial of the carpogonium-bearing branch of this species is produced from the rear side of the periaxial cell, which is on the same side where the initial of the cortical filaments is formed, and grows downward in the same way as the cortical filaments elongate.

2) *Batrachospermum orthostichum* Skuja (1931:84, pl. 2, figs 1–15)

plate 80, figs 1–5 (original author, Skuja 1931)

Plants monoecious, weakly mucilaginous, rigid in consistency, 1–7 cm high, 80–280 μm in diameter, pseudo-dichotomously and sparsely branched. Fascicles 2–3, whorls reduced, dense, obconical, pear-shaped, contiguous, compressed. Primary fascicles chantransia-like unilaterally branched, curved, (2–)3–6(–7) cell-storeys, proximal cells barrel-shaped, obovoid or ellipsoidal, distal cells obovoid, subspherical to spherical, terminal hairs numerous, short to long. Cortical filaments well-developed. Secondary fascicles numerous, reaching the length of primary fascicles. Spermatangia spherical, 6–10 μm in diameter, terminal or subterminal on secondary, rarely primary, fascicles. Carpogonium-bearing branches straight, short, 12–27 μm long, consisting of 2–8 disc- or barrel-shaped cells, arising from periaxial cells, or rarely from proximal cells of primary fascicles or cortical filaments. Carpogonia 4–5 μm in diameter at the base, 8–12 μm in diameter at the apex, 30–53 μm long, trichogyne club-shaped, sessile. Carposporophytes sessile, single, rarely paired, dense, hemispherical, larger than radius of whorl, 60–120 μm in diameter, 90–230 μm high. Gonimoblast filaments dimorphic, radially branched determinate with 3–5 ellipsoidal or barrel-shaped cells and diffuse, indeterminate and creeping with cylindrical cells. Carposporangia obovoid, 7–10 μm in diameter, 10–14 μm long.

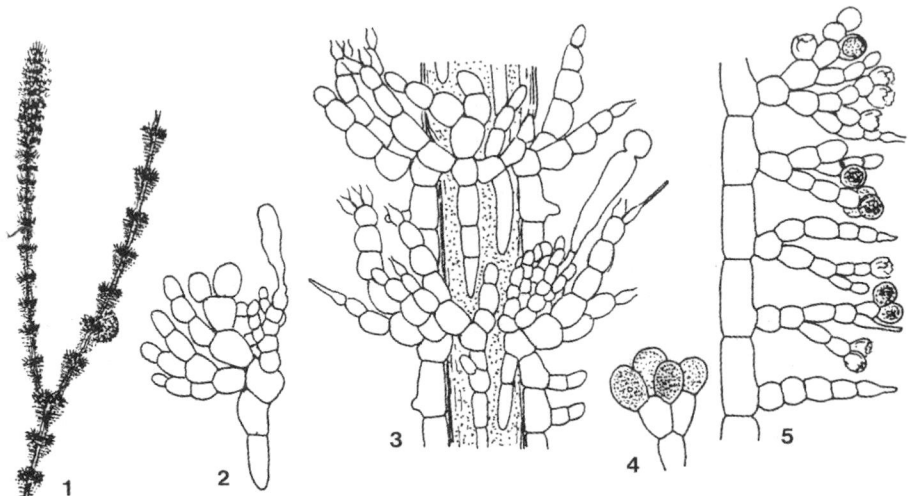

Plate 80
Batrachospermum orthostichum, figs 1–5 (original author, Skuja 1931)
1. a portion of a plant showing obconical whorls and a wart-like hemispherical carposporophyte, 2–3. primary fascicles and a carpogonium-bearing branch with a fertilized carpogonium, 4. carposporangia, 5. spermatangia terminal on secondary fascicles.

Type Locality: Santa Teresa, Espírito Santo, Brazil in South America.
Holotype: UPS, O. Conde, X. 1928.
Distribution: Known from the type locality and in Espírito Santo, Rio de Janeiro, São Paulo, and Sergipe, Brazil in South America.
Note: Although Skuja (1931a), Necchi & Kumano (1984) and Necchi (1988) originally assigned this species with *B. atrum*-like fascicles to the section *Setacea*, Sheath *et al.* (1993) considered *B. orthostichum* to be another synonym of *B. atrum* based on the multivariate association, lack of significant difference in whorl size, and absence of indeterminate gonimoblast filaments in the type specimen of *B. orthostichum*. It seems better, however, to assign it to the section *Turfosa* because of the shape of trichogyne and the thin, long carpogonium-bearing branch, and especially because of the fact that some indeterminate gonimoblast filaments extend out from the globular carposporophyte (Skuja 1931a, Necchi 1990).

3) ***Batrachospermum keratophytum*** Bory (1808:328, pl. 31, fig. 2) *emend.* Sheath, Vis et Cole (1994c:879, figs 13–14)
Synonym: *B. vagum* (Roth) Agardh var. *keratophytum* (Bory) Sirodot (1884:264, pl. XXXIV, pl. XXXV, fig. 3, pl. XXXVII, fig. 1, 5); *B. suevorum* Kützing (1849:536).

plate 81, figs 1–7 (Sirodot 1884) as *B. vagum* var. *keratophytum*
plate 82, figs 1–3, 5–7 (Necchi 1988)
figs 4, 8–12 (Necchi 1990) as *B. keratophytum*

Plants monoecious, weakly or moderately mucilaginous, rigid in consistency, 3–7.5 cm high, 160–500 μm in diameter, branching pseudo-dichotomous, sparse, main branch indistinct; apices straight. Whorls reduced, dense, obconical, pear-shaped or compressed and indistinct, contiguous. Internodes 200–450 μm long. Periaxial cell spherical or ovoidal, 2–3 primary fascicles. Cortical filaments well developed. Primary fascicles straight, (3–)4–7(–8) cell storeys; proximal cells cylindrical or ellipsoidal, 6–12 μm in diameter, 13–30 μm long; distal cells obovoidal, subspherical or spherical, 5–12 μm in diameter 6–17 μm long; branching di- rarely trichotomous, 1–3(–4). Terminal hairs numerous, short or long, base inflated, 1–2 in each terminal cell. Secondary fascicles numerous, on all internodes, reaching the length of primary fascicles. Spermatangia spherical, terminal or subterminal on secondary, rarely on primary fascicles, 6–9 μm in diameter. Carpogonium-bearing branch straight, on pericentral, rarely on proximal cells, 3–7-celled, short, 17–30 μm long; cells barrel- or disc-shaped, 4–6 μm in diameter 3–5 μm long. Involucral filaments numerous, short, 2–5 barrel-shaped cells. Carpogonium symmetric; 36–47 μm long, base 4.5–6 μm in diameter, apex 6.5–10 μm in diameter. Trichogyne club-shaped, sessile. Carposporophytes sessile, 1 per whorl, semi-spherical, larger than the radius of the whorl, 100–250 μm high, 55–130 μm wide. Gonimoblast filaments of 2 types: radially branched determinate filaments 4–6(–7) composed of ellipsoidal or barrel-shaped cells, 5–12 μm in diameter, 9–18 μm long; diffuse indeterminate ones composed of cylindrical cells, 4–7 μm in diameter, 15–25 μm long. Carposporangia subspherical or spherical, 9–12 μm in diameter, 10–13 μm long.
Type Locality: Marensin, Landes, France in Europe.
Lectotype: PC herb. Thuret, for *B. keratophytum*, designated by Sheath *et al.* 1994.
Distribution: Known from Europe, Asia, North America and South America. In East Asia, in wetland along Numasiri-gawa River in Shumshir Island, northern Kuriles.

Batrachospermales

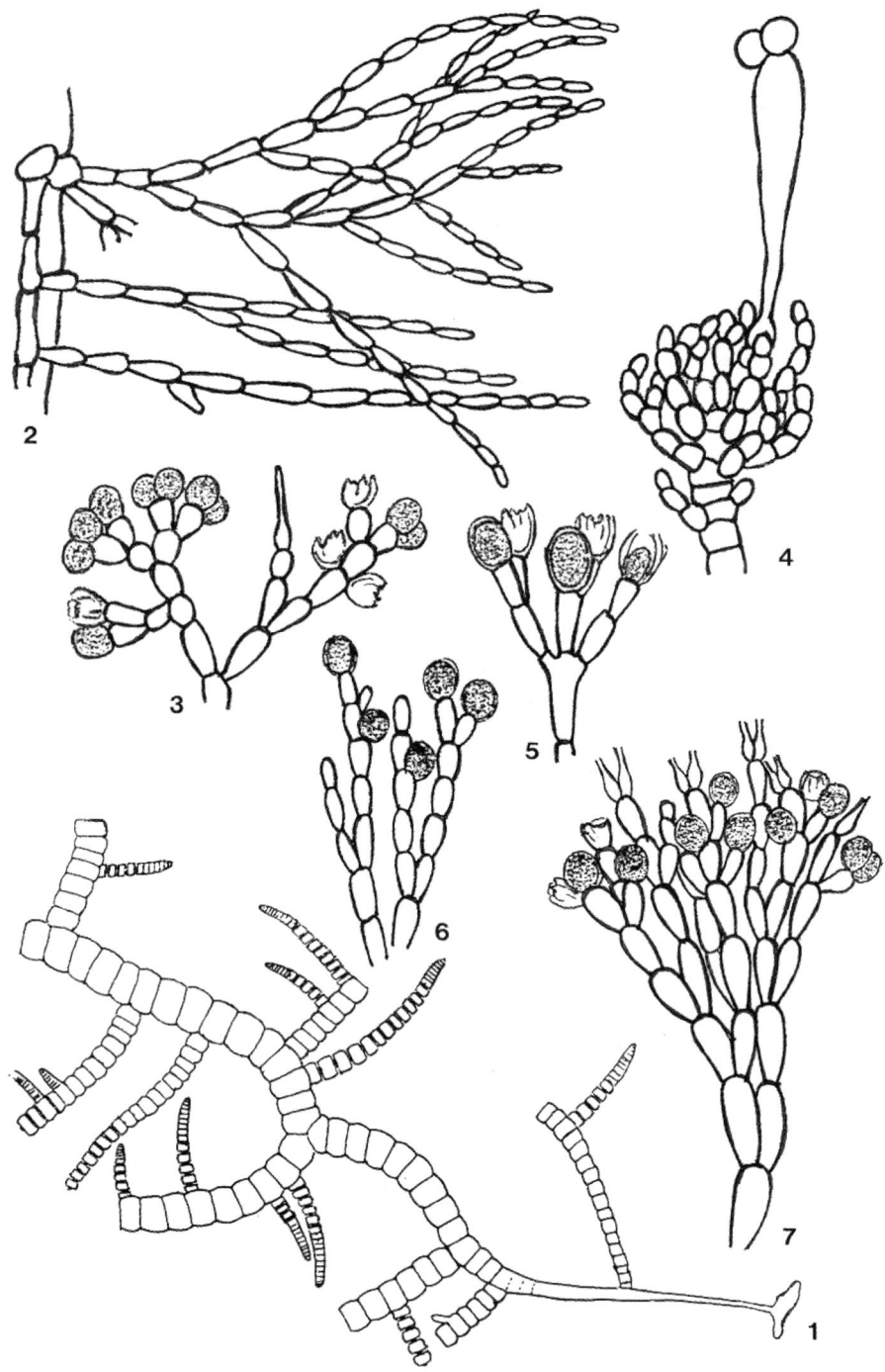

Plate 81
Batrachospermum keratophytum as *B. vagum* var. *keratophytum*, figs 1–7 (Sirodot 1884)
1. the habit of a plant showing compressed and barrel-shaped whorls, 2. a primary fascicle, cortical filaments and secondary fascicles, 3. spermatangia terminal on fascicles, 4. a carpogonium-bearing branch with a fertilized trichogyne and short involucral filaments, 5. carposporangia, 6–7. monosporangia terminal on fascicles.

3. *Turfosa*

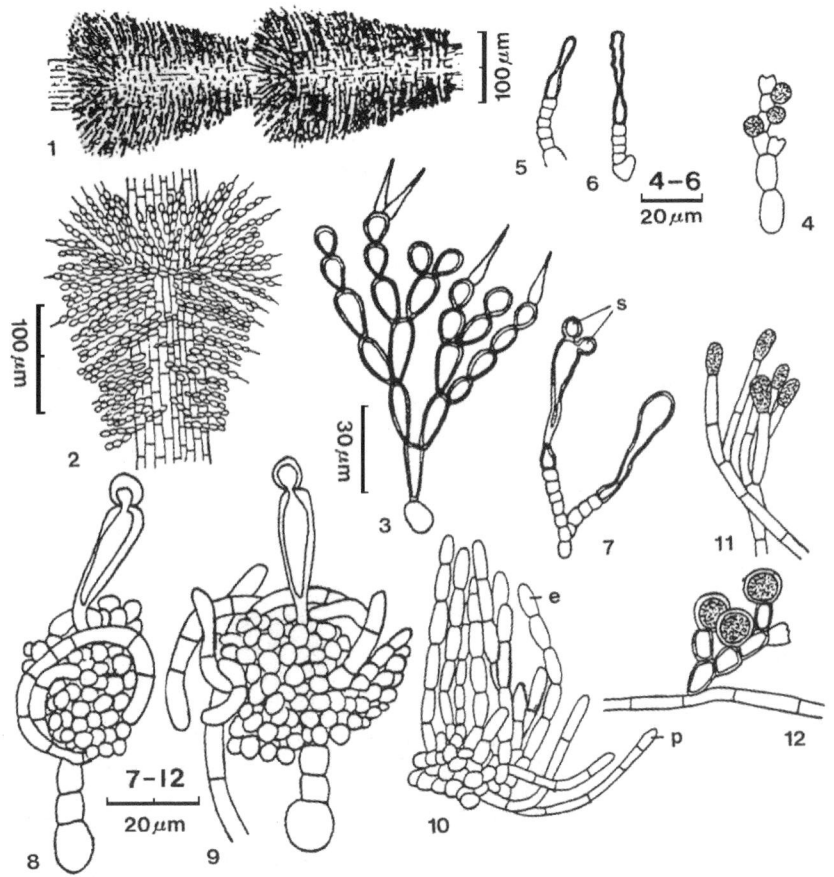

Plate 82
Batrachospermum keratophytum, figs 1–3, 5–7 (Necchi 1988), figs 4, 8–12 (Necchi 1990)
1. general view of whorls, 2. structure of a whorl showing cortical filaments, primary and secondary fascicles, 3. detail of a primary fascicle, 4. distal cells of a primary fascicle with spermatangia, 5–6. young carpogonia, 7. mature and fertilized carpogonia with attached spermatia (s), 8–10. early stage in the development of carposporophyte showing prostrate (p) and erect (e) gonimoblast filaments, later stage in the development of carposporophyte showing prostrate (p) and erect (e) gonimoblast filaments, 11. erect gonimoblast filaments with immature carposporangia, 12. prostrate gonimoblast filaments with carposporangia.

4) ***Batrachospermum periplocum*** (Skuja) Necchi (1990:139, figs 218–227)
Basionym: *B. vagum* var. *periplocum* Skuja (1969:62, figs 1–15).

plate 83, figs 1–3, 5–6, 9–10 (original author, Skuja 1969)
figs 4, 7–8 (Necchi 1990)
plate 84, figs 1–15 (original author, Skuja 1969) as *B. vagum* var. *periplocum*

Plants monoecious, moderately mucilaginous, rigid in consistency, 2.5–9.5 cm high, 400–1300 µm in diameter, pseudodichotomously and sparsely branched. Fascicles 3–4, whorls well-developed, dense, conical, barrel-shaped, contiguous, compressed. Primary fascicles dichotomously branched, straight, 8–15 cell-storeys, proximal cells cylindrical or ellipsoidal, distal cells ellipsoidal, obovoid or pear-shaped, terminal hairs numerous, short. Cortical filaments

Batrachospermales

well-developed. Secondary fascicles numerous, reaching the length of primary fascicles. Monosporangia spherical or hemispherical, 8–10.5 μm in diameter, 8–11 μm long, terminal or subterminal on fascicles. Spermatangia ovoid or spherical, 5–9 μm in diameter, terminal or subterminal on secondary fascicles, rarely primary ones. Carpogonium-bearing branches straight, short, 20–75 μm long, consisting of 5–11 disc- or barrel-shaped cells, arising from periaxial cells, rarely proximal cells of primary fascicles or cortical filaments. Involucral filaments numerous, short, consisting of 2–3 barrel-shaped cells. Carpogonia 4.5–7 μm in diameter at the base, 7–12 μm in diameter at the apex, 40–60 μm long, trichogyne club-shaped, sessile. Carposporophytes single, lax, hemispherical, larger than radius of whorl, 200–260 μm high, 400–550 μm wide. Gonimoblast filaments consisting of two types: radially branched determinate filaments with 3–5 cell-storeys and cylindrical or ellipsoidal cells and creeping indeterminate filaments with cylindrical cells. Carposporangia ellipsoidal or obovoid, 9–13 μm in diameter, 12–16 μm long.

Type Locality: Rio Negro, Amazonas, Brazil in South America.

Lectotype: Figs 1–15, p. 62 for *B. vagum* var. *periplocum* in Skuja, designated by Necchi 1990.

Distribution: Known from the type locality in Manaus and Forest Reserve Adolfo Ducke, Amazonas, Itanbe do Mato Dentro and Santa Barbara, Minas Gerais, Brazil in South America.

Note: Skuja (1969) and Necchi (1990) observed some diffuse, indeterminate gonimoblast filaments extending out from globular carposporophytes of this species.

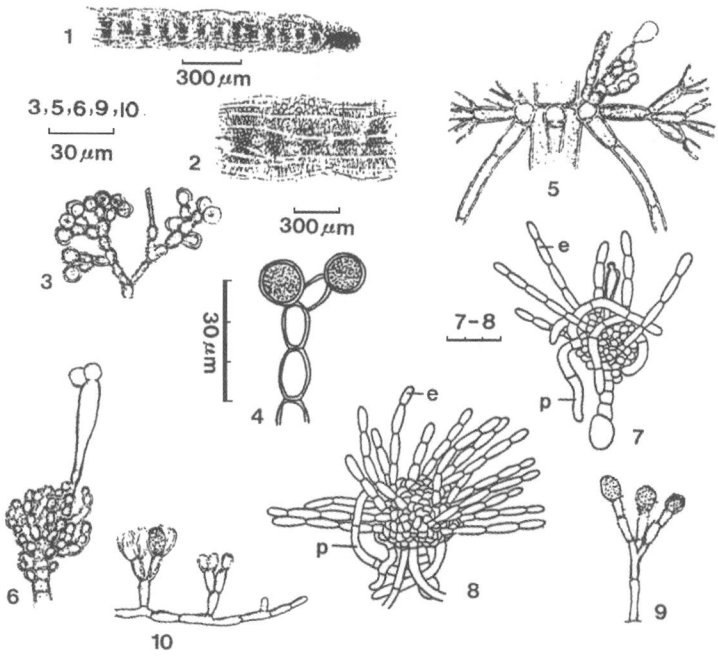

Plate 83

Batrachospermum periplocum, figs 1–3, 5–6, 9–10 (original author, Skuja 1969), figs 4, 7–8 (Necchi 1990) 1–2. the general views of whorls showing peripheral cortical filaments, 3. distal cells of a primary fascicle with spermatangia, 4. distal cells of a primary fascicle with monosporangia, 5. a young whorl showing a carpogonium and cortical filaments free from the axial cell, 6. a fertile carpogonium with attached spermatia showing involucral filaments and a carpogonium-bearing branch, 7–8. the early stages in development of a carposporophyte showing erect (e) determinate and prostrate (p) indeterminate gonimoblast filaments, 9. an erect determinate gonimoblast filament with carposporangia, 10. prostrate indeterminate gonimoblast filaments with carposporangia.

3. Turfosa

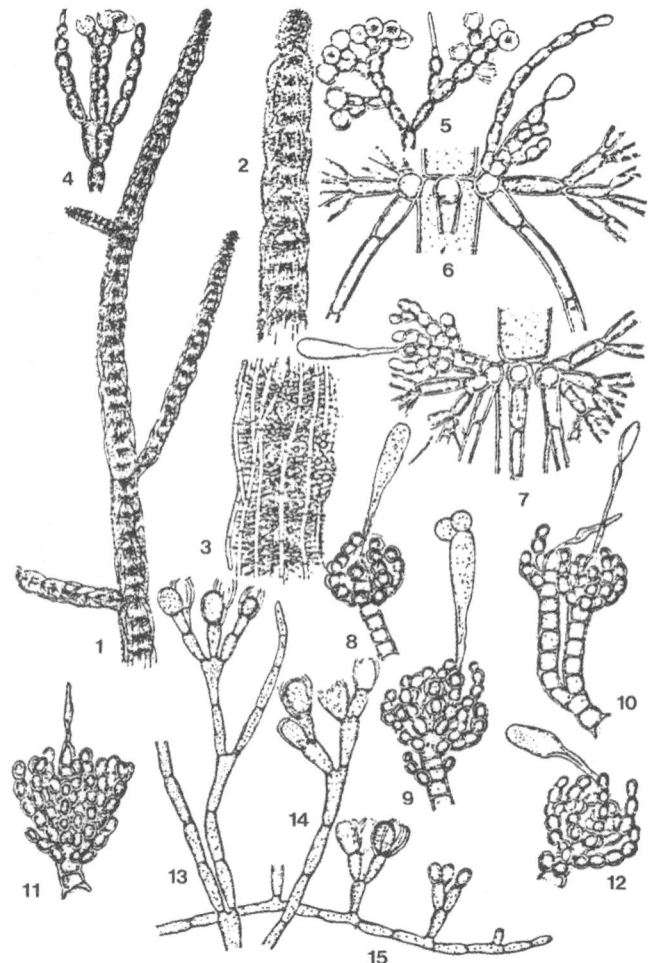

Plate 84
Batrachospermum periplocum as *B. vagum* var. *periplocum,* figs 1–15 (original author, Skuja 1969)
1–3. the general view of a whorl showing peripheral cortical filaments, 4–5. distal cells of a primary fascicle with spermatangia, 6–7. a young whorl showing a carpogonium and cortical filaments free from the axial cell, 8, 10–12. a carpogonium showing involucral filaments and a carpogonium-bearing branch, 9. a fertile carpogonium with attached spermatia showing involucral filaments and a carpogonium-bearing branch, 13–14. an erect determinate gonimoblast filament with carposporangia, 15. prostrate indeterminate gonimoblast filaments with carposporangia.

5) ***Batrachospermum turfosum*** Bory (1808:327) *emend.* Sheath, Vis et Cole (1994c:882, figs 18–19)
Synonym: *Chara gelatinosa* var. *vaga* Roth (1797); *B. vagum* (Roth) C. Agardh (1812: 41); *B. moniliforme* var. *vagum* Roth (1800:482); *B. turfosum* var. *undulato-pedicellatum* Kumano et Watanabe (1983:87, figs 1–13) .

plate 85, figs 1–9 (Kumano 1997g) as *B. vagum*

Plants monoecious, moderately mucilaginous, consistency rigid, 2.5–5 cm high, 300–1000 µm in diameter, branching pseudo-dichotomous, sparse, main branch indistinct, apices straight, base denuded. Whorls well developed, dense, obconical, barrel-shaped or compressed and indistinct, contiguous. Internodes 250–400 μm long. Periaxial cells spherical or ovoidal, 3–4 primary fascicles. Cortical filaments well developed. Primary fascicles straight, 8–13 cell storeys, proximal cells ellipsoidal, 7.5–15 μm in diameter, 20–45 μm long; distal cells ellipsoidal, obovoidal or semi-spherical, 4–10 μm in diameter 10–20 μm long; branching 3–7 times di- rarely trichotomous. Hairs numerous, short, with inflated base. Secondary fascicles numerous, along the length of the internode, reaching the length of the primary fascicles. Spermatangia spherical, terminal or subterminal, on primary or secondary fascicles, 6–9 μm in diameter. Carpogonium-bearing branch straight, on periaxial cell, 3–7 celled, short, 18–40 μm long, cells disc- or barrel-shaped, 6–8 μm in diameter, 6–9 μm long. Involucral filaments numerous, short, 1–4 barrel-shaped or subspherical cells. Carpogonium symmetrical, 35–60 μm long, base 5–6 μm in diameter, apex 7–10 μm in diameter. Trichogyne club-shaped, sessile. Carposporophytes abortive. Gonimoblast filaments and carposporangia not observed (Necchi 1990).

Type Locality: near Dax, France in Europe.

Lectotype: PC herb. Thuret, for *B. turfosum*, designated by Compère 1991b.

Distribution: Known from France, Ireland, Portugal, Belgium, Sweden in Europe; India in Asia; Japan in Eastern Asia; highland in Papua New Gunea; Canada, USA in North America and Brazil in South America. In Japan, Hachimantai in Iwate, Mt. Naeba in Niigata, Tanayama Highland in Aichi, Mida-ga-hara, Mt. Tateyama in Toyama, Sukain, Hirotani-gawa River, Ohara, Kondoh-ike, Himeji, Hirota, Jyurinji, Nishinomiya, Oosoma-ike, Kobe in Hyogo. Known from humid ponds and streams.

Note: Kumano *et al.* (1970) and Kumano (1978, 1979) described the occurrence of determinate and indeterminate gonimoblast filaments in this species.

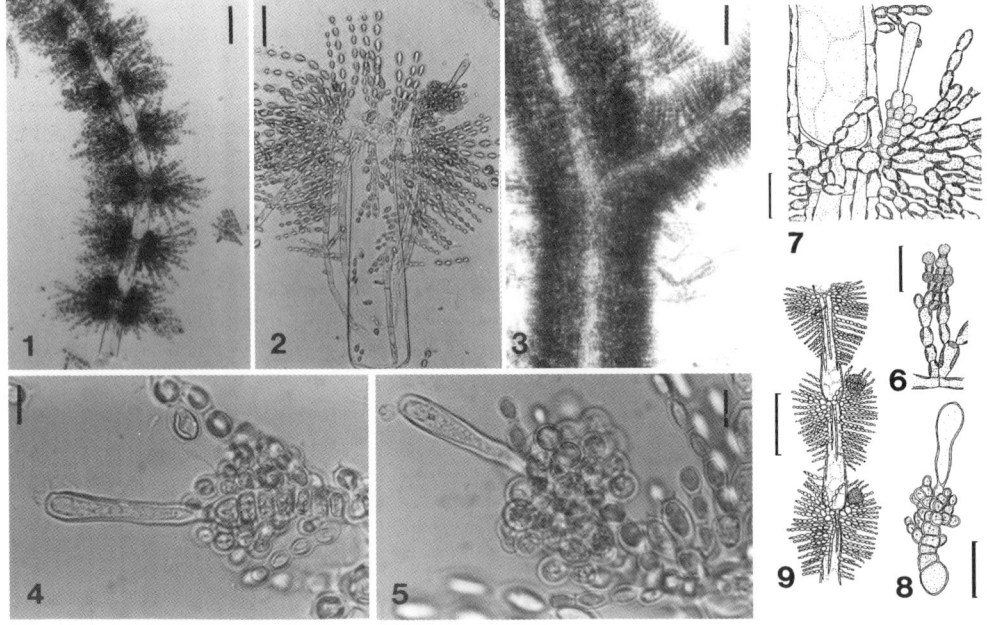

Plate 85
Batrachospermum turfosum as *B. vagum*, figs 1–9 (Kumano 1997g)
1, 9. a main axis with immature whorls, 2, 7. whorls showing axial cells, primary fascicles, cortical filaments, a carpogonium with an obconical trichogyne, 3. cylindrical and confluent mature whorls, 4, 5, 8. a carpogonium with an obconical trichogyne and a carpogonium-bearing branch with involucral filaments consisting of spherical cells, 6. spermatangia terminal on secondary fascicles.
Scale bars = 200 μm (fig. 9); 100 μm (figs 1, 3); 40 μm (fig. 2); 20 μm (figs 6–8); 10 μm (figs 4, 5).

4. Section *Virescentia* Sirodot (1873)

Type: not designated in 1873, but only one species cited *Batrachospermum coerulescens* Sirodot by Sirodot

Synonym: section *Viridia* De Toni (1897:60); section '*Verts*' Sirodot (1884:259); secto *Claviformis* Reis (1973).

Plants intense green. Carpogonium-bearing branches differentiated, short, arising from periaxial cells. Carpogonia with distinctly pedunculate, cylindrical trichogynes. Carposporophyte big, spherical, single, rarely paired, inserted centrally within a whorl.

Note: From the analysis on 39 North American populations of the section *Virescentia* Sheath, Vis & Cole (1994a) recognized only two species, *B. helminthosum* Bory and *B. elegans* Sirodot. The existing taxa of this section including the above mentioned two species are shown here.

Key to the species of the section ***Virescentia***

1. Fascicles curled..1) *B. crispatum*
1. Fascicles not curled..2
2. Carpogonium-bearing branch often reduced to one cell.................................2) *B. gombakense*
2. Carpogonium-bearing branch not reduced to one cell...3
3. Carpogonium < 40 µm long..4
3. Carpogonium > 40 µm long..8
4. Carpogonium 20–45 µm long..3) *B. vogesiacum*
4. Carpogonium 20–30 µm long..5
5. Carposporophyte up to 300 µm wide..4) *B. gulbenkianum*
5. Carposporophyte up to 200 µm wide...6
6. Carpogonium-bearing branch short, up to 5 celled..5) *B. bakarense*
6. Carpogonium-bearing branch long, up to 12 celled...7
7. Monoecious..6) *B. azeredoi*
7. Polyoecious...7) *B. ferreri*
8. Carpogonium-bearing branch arising from periaxial cell and cell of fascicle.........8) *B. elegans*
8. Carpogonium-bearing branch arising from periaxial cell..9
9. Carpogonium-bearing branch curved..9) *B. transtaganum*
9. Carpogonium-bearing branch not curved..10
10. Carpogonium-bearing branches composed of 1–5 short cells,
    carposporangia < 30 µm long ....10) *B. helminthosum*
10. Carpogonium-bearing branches composed of 5–10 cells,
    carposporangia > 30 µm long ....11) *B. desikacharyi*

1) ***Batrachospermum crispatum*** Kumano et Ratnasabapathy in Ratnasabapathy & Kumano (1982a:18, fig. 1, A–I)

plate 86, figs 1–5 (Kumano 1996j)
figs 6–9 (original authors, Ratnasabapathy & Kumano 1982a)

Plants trioecious, strongly mucilaginous, 2–13 cm high, 20–350 µm in diameter, more or less dichotomously branched, deep green with a mixture of blue; whorls pear-shaped, very often touching each other; primary fascicles curled, unilaterally branched, 5–13 cell-storeys, cells cylindrical or fusiform, terminal hairs absent; secondary fascicles numerous, curled, covering all internodes; spermatangia spherical or ovoid, 4–6 µm in diameter, 6–8 µm long, lateral on

Batrachospermales

fascicles; carpogonium-bearing branches short, consisting of 3–4 barrel-shaped cells, arising from periaxial cells; involucral filaments numerous, very short; carpogonia 5–9 µm in diameter at the base, 5–7 µm at the apex, 54–75 µm long, trichogyne cylindrical, indistinctly stalked; carposporophytes single or in pairs, spherical or hemispherical, 140–190 µm in diameter, centrally inserted; carposporangia club-shaped or obovoidal, 9–10 µm in diameter, 17–30 µm long.

Type Locality: Sungai Ayer Besar, Pulau Tioman, Malaysia in Southeast Asia. Attached to submerged rocks and stones in upper reaches.

Holotype: Kobe University, Ratnasabapathy no. 21, 24/V 1974.

Distribution: Known from the type locality only.

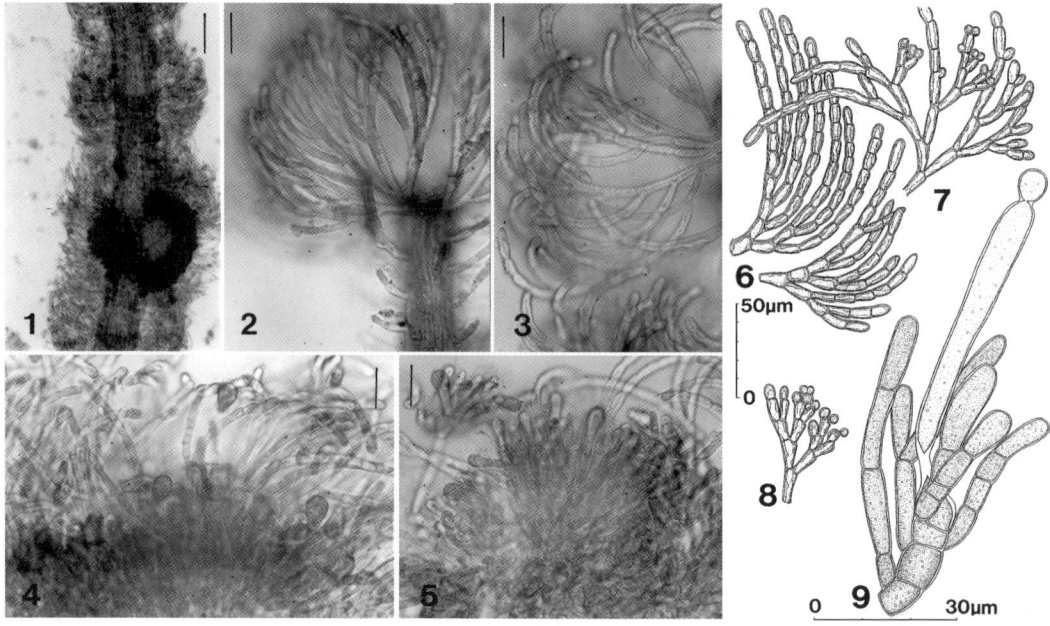

Plate 86
Batrachospermum crispatum, figs 1–5, (Kumano 1996j), figs 6–9 (original authors, Ratnasabapathy & Kumano 1982a)
1. the structure of whorls showing curled or hooked primary and secondary fascicles and carposporophytes inserted centrally, 2–3, 6. primary and secondary fascicles curled, hooked and unilaterally branched, 7–8. spermatangia lateral on primary and secondary fascicles unilaterally branched, 4–5. clavate carposporangia terminal on gonimoblast filaments, 9. a fertilized carpogonium with cylindrical trichogyne with an attached spermatium and showing gonimoblast initials.
Scale bars = 100 µm (fig. 1); 50 µm (figs 7–8); 30 µm (fig. 9); 20 µm (figs 2–5).

2) ***Batrachospermum gombakense*** Kumano et Ratnasabapathy in Ratnasabapathy & Kumano (1982b:119, fig. 2, A–E)

plate 87, figs 1–5 (Kumano 1996p)
figs 6–7 (original authors, Ratnasabapathy & Kumano 1982b)

Plants dioecious, not very mucilaginous, 1–2 cm high, 200–400 µm in diameter, more or less dichotomously branched, green; whorls touching each other, more or less compressed; primary fascicles unilaterally branched, 9–15 cell-storeys, cells cylindrical or barrel-shaped, 5–7 µm in diameter, 5–10 µm long, terminal hairs absent; secondary fascicles numerous, covering all internodes; spermatangia spherical, 4–8 µm in diameter, terminal on lateral branchlets or in clusters on shortened branchlets; carpogonium-bearing branches very short, consisting of 1–2 hexagonal cells, arising from periaxial cells; carpogonia 3–5 μm in diameter at the base, 10–15 μm at the apex, 40–50 μm long, trichogyne inverted-conical or club-shaped, indistinctly stalked; carposporophytes single, ellipsoidal, big, 140–210 μm in diameter, 185–330 μm long, centrally inserted; carposporangia obovoidal, 8–11 μm in diameter, 20–25 μm long.

Type Locality: Field Study Centre, Sungai Gombak, Selangor, Malaysia in Southeast Asia.

Holotype: Kobe University, Ratnasabapathy no. 1220, 31/V 1976.

Distribution: Known from type locality and also in an upper tributary of Sungai Tabin, Sabah, East Malaysia in Southeast Asia (Anton *et al.* 1999, Sato *et al.* 1999).

Note: This species is assigned to the section *Virescentia* because the carpogonium is indistinctly stalked with an inverted-conical trichogyne and the single large carposporophyte is inserted centrally in a whorl.

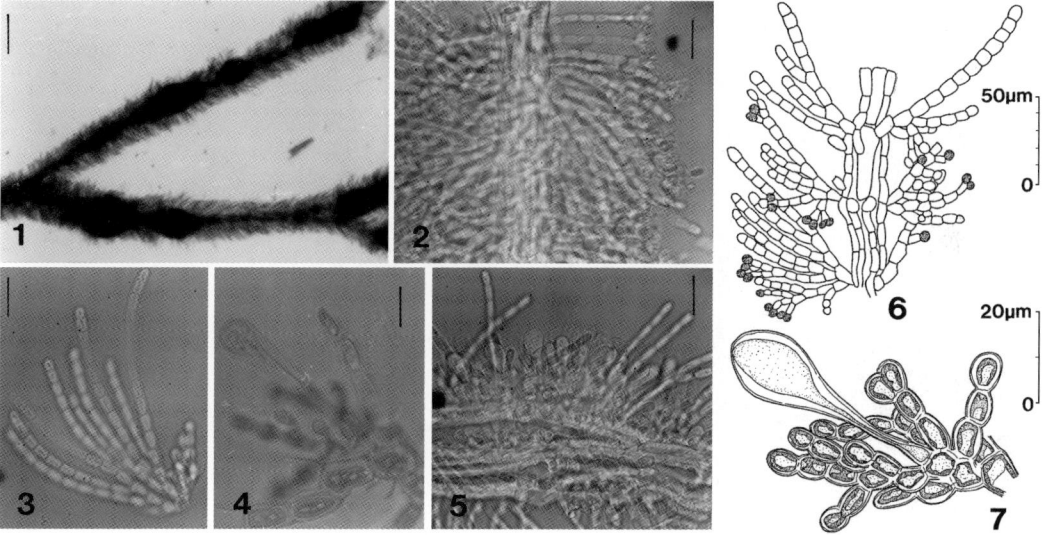

Plate 87
Batrachospermum gombakense, figs 1–5, (Kumano 1996p), figs 6–7 (original authors, Ratnasabapathy & Kumano 1982b)
1. a main axis with confluent whorls and axial carposporophytes, 2, 6. the structure of a whorl showing spermatangia terminal or lateral on primary and secondary fascicles, 3. primary fascicles unilaterally branched, 4, 7. a carpogonium with an obconical trichogyne and a carpogonium-bearing branch consisting of two cells, 5. carposporangia.
Scale bars = 200 μm (fig. 1); 50 μm (fig. 6); 30 μm (fig. 9); 20 μm (figs 2–3, 5); 10 μm (fig. 4).

3) ***Batrachospermum vogesiacum*** Schultz ex Skuja (1938:623)
Synonym: *B. vagum* var. *flagelliforme* Sirodot (1884:263, pl. XXXV, fig. 1, pl. XXXVI, figs 1–6, pl. XXXVII, figs 10–14, pl. XXXIX, figs 1–20); *B. flagelliforme* (Sirodot) Necchi (1988:11), *nom. illeg.*.

plate 88, figs 1–2 (original author, Sirodot 1884) as *B. vagum* var. *flagelliforme*

Plants monoecious, very mucilaginous, 2–10 cm high, 340–700 μm in diameter, richly branched, grayish or brownish olive or pale green. Whorls ellipsoidal or barrel-shaped, more or less separated or confluent. Fascicles 3, primary fascicles richly branched, 6–11 cell-storeys, cells cylindrical, fusiform, elongate-ellipsoidal, terminal hairs frequent. Secondary fascicles numerous,

Batrachospermales

finally reaching the length of primary ones. Spermatangia spherical, 5–6 μm in diameter, 6–8.5 μm long, terminal on fascicles at different level. Carpogonium-bearing branches straight, consisting of 4–8 cells, arising from periaxial cells. Involucral filaments numerous, short. Carpogonia 4–7 μm in diameter at the base, 20–45 μm long, trichogyne elongate-conical, finally almost cylindrical, indistinctly or shortly stalked. Carposporophytes single, sometimes in pairs, globose, 140–280(–330) μm m in diameter, centrally inserted. Carposporangia ellipsoidal, 8.5–13(–17) μm in diameter, 13–19(–25) μm long.

Type Locality: Logerie-Haute, France in Europe.
Lectotype: PC herb Thuret
Distribution: Known from various localities in France and Belgium in Europe.

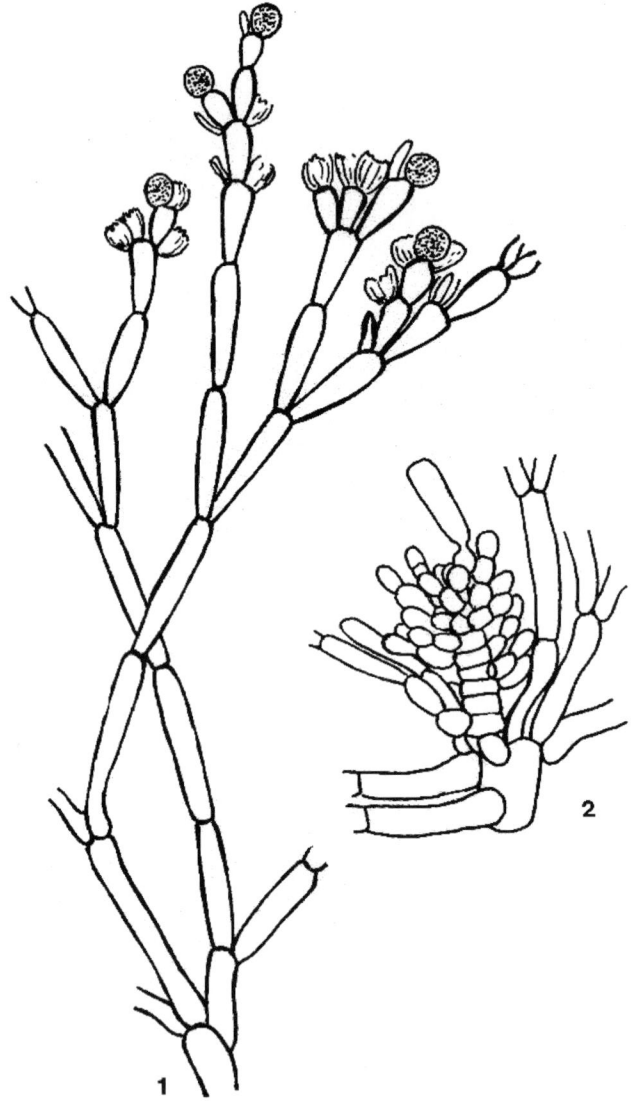

Plate 88
Batrachospermum vogesiacum as *B. vagum* var. *flagelliforme*, figs 1–2 (original author, Sirodot 1884)
1. spermatangia terminal on fascicles, 2. a carpogonium-bearing branch.

4) ***Batrachospermum gulbenkianum*** Reis (1965b:31, tab. I, a–c, tab. II, a–b, tab. III, a–e, tab. IV, a–d, tab. V, a–b)

Plants 4–5 cm high, isolated or in tufts, intense green in low light, green-yellow in high light, mucilaginous, naked proximal, attached to substrate by rhizoidal pad. Ramification paniculiform, irregular, branches generally inserted at right angles, very slender. Primary branches rarely numerous (as a result of the development of adventitious branches), either simple, long and flagelliform, equalling and sometimes surpassing the main axis or branched. Secondary branches usually simple, flagelliform, less often branched and then with a branch distribution identical to that of primary branches or the main axis. Final order branches basely attenuate, cylindrical or conical. Whorls not only contiguous, distinct, spherical to elongate-ellipsoidal, slightly crowded, and radiating in all directions, sometimes with truncate internodal filaments, but also, in thin fronds, distant or rarely discoid and not always complete. Interverticillary filaments generally sparse. Cortical filaments cylindrical and very coherent. Periaxial cells 10, 5 large cells alternating with 5 small cells producing 1–3 (4?) secondary fascicles (excluding carpogonium-bearing branches) pseudo-, di-, trichotomously branched internally, but simple, flagelliform peripherally, these branches formed of cylindrical cells basely, obovoidal, large cells in the mid-portions, dilating toward the terminal portion, and fusiform or obovoidal, small cells at the peripheral part. Apices of axes piliferous; hairs short or longer, slightly inflated at base, variable in abundance. Monoecious. Spermatangia distributed at outer part of whorl, and very abundant at involucral filaments. Carpogonium-bearing branches upright resembling small branchlets, enlarging little by little, inserted on periaxial cells, and very rarely on cells of secondary fascicles, formed by 6–12 cells broader than long, producing ultimate short (4–7 cells), dense, involucral filaments embracing carpogonium. Carpogonium truncate and 5–6.6 µm long. Trichogyne with very short stalk, truncate, very rarely cylindrical, generally club-shaped, 20–33 µm long. Carposporophytes 1–3 on inner part of each whorl, more or less central, spherical or hemispherical, variable in size, 150–300 µm in diameter, sometimes equaling the radius of whorl. Gonimoblast filaments consisting of large, spherical or ovoidal cells in the central part, with small, truncate, fusiform or cylindrical cells distributed irregularly towards the outer part. Carposporangia generally obovoidal, very rarely spherical, 6.5–10 µm in diameter, 10–16.5 µm long.

Type Locality: Rio Alfusqueiro, tributary of river Àgueda, at Comfulcos, near Vouzela, Portugal in Europe.

Holotype: COL 456.

Distribution: Known from the type locality only.

Note: Reis (1965) assigned this species to the section *Turfosa*. However, the author assigns it to section *Virescentia,* because it has radially branched and determinate gonimoblast filaments.

5) ***Batrachospermum bakarense*** Kumano et Ratnasabapathy (1984:20, figs 1–14)

plate 89, figs 1–6 (Kumano 1996d)
figs 7–9 (original authors, Kumano & Ratnasabapathy 1984)

Plants monoecious, slightly mucilaginous, about 1.5 cm high, 70–300 µm in diameter, more or less irregularly branched, blackish green; whorls barrel-shaped, touching each other in mature plants; primary fascicles abundantly branched, 7–9 cell-storeys, cells fusiform, 10–12 µm long, terminal hairs absent; cortical filaments well-developed; secondary fascicles numerous, 5–7 cell-storeys, covering all internodes; spermatangia spherical, ca. 4 µm in diameter, terminal on fascicles; carpogonium-bearing branches short, slightly curved, 12–30 µm long, consisting of 2–5 barrel-shaped cells, arising from periaxial cells and fascicles; involucral filaments numerous,

Batrachospermales

elongated, unilaterally issued; carpogonia 4–6 μm in diameter at the base, 4–6 μm at the apex, 23–36 μm long, trichogyne club-shaped, indistinctly stalked; carposporophytes single, hemispherical, ca. 60 μm high, ca. 90 μm in diameter, centrally inserted; carposporangia club-shaped or obovoidal, 9–10 μm in diameter, 13–18 μm long.

Type Locality: Sungai Bakar, Kelantan, Malaysia in Southeast Asia.
Holotype: Kobe University, Ratnasabapathy no. 13, 3/VI 1982.
Distribution: Known from the type locality only.

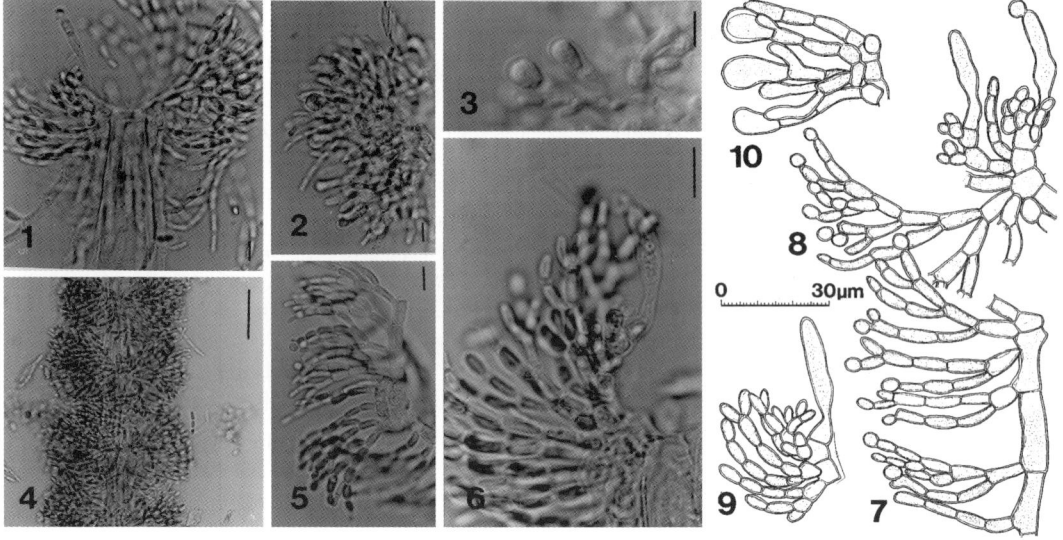

Plate 89
Batrachospermum bakarense, figs 1–6 (Kumano 1996d), figs 7–10 (original authors, Kumano & Ratnasabapathy 1984)
1, 6. a portion of a plant showing an axial cell, primary fascicles and a carpogonium-bearing branch with a fertilized trichogyne, 2–3, 10. carposporangia terminal on gonimoblast filaments, 4. a main axis with barrel-shaped whorls, 5, 7. spermatangia terminal on primary and secondary fascicles, 8. a portion of a plant showing spermatangia terminal on primary fascicles and two carpogonium-bearing branches, one of which issued as an involucral filament of the other, 9. a slightly curved carpogonium-bearing branch showing involucral filaments unilaterally issued and a club-shaped trichogyne indistinctly stalked, 10. carposporangia terminal on gonimoblast filaments. Scale bars = 100 μm (fig. 4); 30 μm (figs 7–10); 10 μm (figs 1–3, 5, 6).

6) ***Batrachospermum azeredoi*** Reis (1967:168, tab. I, a–c, tab. II, a–b, tab. III, a, tab. IV, a–b)

Plants monoecious, 5–6 cm high, blue-gray, drying blue-violet, moderately mucilaginous, caespitose, enlarged at the base by cortical filaments, truncate, naked eventually, and attached to the substrate by a rhizoidal pad. Female and bisexual plants similar in appearance, male plants less branched than female ones. Whorls generally separated, lax, very rarely disc-shaped, elongate-ellipsoidal, sometimes spherical. Periaxial cells ovoidal or cylindrical, producing 3–5 primary fascicles, apparently di- trichotomously branched, consisting of elongate conical or pear-shaped at inner 2/3 of whorls, and generally obovoidal or fusiform cells at outer 1/3. Terminal hairs rare, short and slender, inflated at the base. Secondary fascicles absent or rare. Spermatangia occurring on outer parts of whorls. Female plants abundantly and densely branched in lower half, rarely in upper part of plants. Whorls contiguous, very rarely disc-shaped, elongate-ellipsoidal, with the exception of ultimate branchlets and thin axes with discoid and spherical whorls respectively. Secondary fascicles generally sparse. Cortical filaments consisting of large cylindrical cells. Periaxial cells cylindrical, rarely ovoidal, producing 2–5 primary fascicles, whorls frequently pseudo-, di-, trichotomously branched, consisting of pear-shaped or elongate conical, rarely cylindrical in inner 2/3, and obovoidal, ellipsoidal or spherical in outer 1/3. Carpogonium-bearing branches inserted on periaxial cells and beginning of primary fascicles, composed of 3–5 cells, with the 2 uppermost cells producing short involucral filaments embracing the carpogonium, and consisting of truncate-cupula-shaped cells. Trichogyne sessile or with short stalk, club-shaped or cylindrical, 23.3–33.3 µm long. Carposporophytes rarely numerous, centrally inserted, 1–2 at each whorl, spherical or hemispherical, variable in size, 120–180 µm in diameter, equal to half of the radius of whorl. Gonimoblast filaments consisting of cylindrical cells distally, truncate cells proximal. Carposporangia obovoidal, 8.3–10 µm in diameter, 13–16.6 µm long.

    Type Locality: on rocks in Caster R., near Vila da Feira, Portugal in Europe.
    Holotype: COL 482.
    Distribution: Known from the type locality and in various localities, Portugal in Europe.
    Note: For *B. azeredoi* Reis and *B. ferreri* Reis having a carpogonium with a sessile and club-shaped trichogyne and central carposporophytes, the section *Claviformia* was established by Reis (1973). According to these circumscription, I agree with Necchi's opinion (1990a) that the section *Claviformia* can be merged in the section *Virescentia*, whose members have a stalked or sessile, cylindrical or club-shaped trichogyne and sessile or central carposporophytes.

7) ***Batrachospermum ferreri*** Reis (1967:174, tab. V, a–c, tab. VI, a–d, tab. VII, Tab, VIII, a–d)

    Plants polyoecious, 4–7 cm high, intense green, blue-violet when dry, very mucilaginous, solitary or in small tufts. Whorls truncate at base, eventually naked, terminating in disc. Branching irregular, patent initially pyramidal, becoming corymb-pyramidal. Branches inserted frequently at right angles and directed in all directions. Secondary branches, simple, flagelliform or sparingly branched, tertiary branches simple, or rarely with one or more branchlets. Whorls of male plants contiguous and compressed, elongate-ellipsoidal, but separated and spherical or hemispherical on slender branches. Periaxial cells cylindrical, dilated at base, producing 2–5 fascicles, pseudo-trichotomously branched at the peripheral fascicles, consisting of cells diverse in shape: cylindrical, fusiform, elongate-conical, pear-shaped in the lower 1/2, but fusiform or obovoidal in the upper 1/3. Terminal hairs not numerous, short and long, slightly inflated at base. Cortical filaments cylindrical, abundant in the lower part of plants. Spermatangia occurring at periphery of branchlets of whorl. Female plants, whorls generally distinct, spherical or elongate-ellipsoidal, discoidal at terminal parts of plants and truncate at lower part of primary axis. Periaxial cells cylindrical or ovoidal, producing 2–5 fascicles, frequently much curved at periphery, pseudo-trichotomously branched and consisting of pear-shaped or fusiform cells in the interior 2/3 ellipsoidal or obovoidal cells in the outer 1/3. Terminal hairs rare and very short. Secondary fascicles none or rare. Cortical filaments cylindrical, numerous at lower parts of plants. Carpogonium-bearing branches arising from periaxial cells, rarely on principally primary fascicles, consisting of 3–12 cells, short, 10–16 µm in diameter, producing from upper 1/2–1/3 short involucral filaments embracing the carpogonium, and from lower part unilateral filaments. Carpogonium truncate-cupul-shaped, 4–6.6 µm in diameter. Trichogyne shortly stalked, club-shaped or truncate, rarely cylindrical, 20.3–33.3 µm long. Carposporophytes rare, 1–2 each whorl, occupying more or less central part, irregular hemispherical, variable in size, 130–180 µm in diameter, 1/3–1/2 of the radius of whorl. Carposporangia obovoidal, 6.6–10 µm in diameter, 10–13.3 µm long. Bisexual plants resembling male plants.

Batrachospermales

Type Locality: in a stream called da Mina do Pintor, near Val de Cambra, Portugal in Europe.
Holotype: COL 430 et 485.
Distribution: Known from various localities in Portugal in Europe.
Note: For *B. azeredoi* Reis and *B. ferreri* Reis having a carpogonium with a sessile and club-shaped trichogyne and central carposporophytes, the section *Claviformia* was established by Reis (1973). According to these circumscription, I agree with Necchi's opinion (1990a) that the section *Claviformia* can be merged in the section *Virescentia*, whose members have a stalked or sessile, cylindrical or club-shaped trichogyne and sessile or central carposporophytes.

8) ***Batrachospermum elegans*** Sirodot (1884:273, pl. XLIV, figs 1–5) *emend.* Sheath, Vis et Cole (1994a:115, figs 11–18)

The synonymy proposed by Sheath, Vis & Cole (1994a) is as follows;
Synonym: *B. coerulescens* Sirodot (1884:270, pl. XL, figs 1–2, pl. XLI, figs 1–5) *nom. illeg.*

plate 90, figs 1–8 (Sheath, Vis & Cole 1994a)
plate 91, figs 1–4 (original author, Sirodot 1884) as *B. coerulescens* Sirodot

Plants monoecious, dioecious with brownish main axis; barrel-shaped, confluent whorls, 316–1002 μm in diameter with 8–17 primary fascicle cells but few secondary fascicles. Carpogonium-bearing branches consisting of 1–3 short cells, arising from either periaxial cells or proximal cells of the fascicle branch. Carpogonia 35.5–65.4 μm long with pedicellate, cylindrical, club-shaped trichogynes having 1–3 basal knobs or branches and being 4.5–9.7 μm in diameter along the main axis. Carposporophytes axial, 135–317 μm in diameter and 106–341 μm high and with 2–6 cylindrical gonimoblast cells. Carposporangia obovoidal, 6.6–14.6 μm in diameter and 9.9–23.5 μm long.
Type Locality: near Compénéac, Morbihan, France in Europe.
Lectotype: PC, for *B. elegans,* designated by Sheath *et al.* 1994.
Distribution: Known from France in Europe; Japan in Eastern Asia. In Japan, in Tokyo, Katori in Chiba, Hamada-gawa River, Toyokawa in Aichi, Kagawa-cho in Kagawa, Okuchi-sen, Matsuyama in Ehime, Shimonoseki in Yamaguchi, Kurume in Fukuoka, Matsubase, Udo, Kagami-cho, Chuoh-mura in Kumamoto. Known from spring-fed streams.

9) ***Batrachospermum transtaganum*** Reis (1970:23, tab. I, a–d, tab. II, a–d, tab. III, a–c, tab. IV, a–c)

plate 97 (a), figs 1–4 (original author, Reis 1970)

Plants polyoecious, 2–3 cm high, light green when exposed to sunlight and retaining its colour upon drying, dark brown in dark places and drying dark violet, very mucilaginous, separated or caespitose, attenuate and sometimes naked at the base; proliferous. Branches inserted here and there at near right angles and at all directions, very undulate and slightly attenuate. Whorls of male plants, separated and spherical, but discoid toward the tips. Secondary fascicles none or rare. Periaxial cell of primary fascicles ovoidal, dilated at the base, very rarely cylindrical, producing 2–4 primary fascicles, pseudo-tetrachotomously branched, with the branches consisting of cylindrical cells in the inner 1/3, and ovoidal or ellipsoidal cells in the outer 2/3 of the whorls. Terminal hairs numerous, long and slender, inflated at the base. Spermatangia rare, produced in the interior and at the periphery of fascicles. Female plants: whorls separate or contiguous, compressed (in winter), conical, disc-shaped toward the tips; interverticillary filaments numerous, occupying the upper part of the internode, becoming shorter and shorter, reaching the lower whorl

4. *Virescentia*

and often becoming at the base equal to the radius of the whorl. Periaxial cells, ovoidal, swelling at base, producing 2–4 primary fascicles, with these pseudo-tetrachotomously branched, consisting of cylindrical cells in the inner 1/3, and ellipsoidal or ovoidal in the outer 2/3 of whorls.

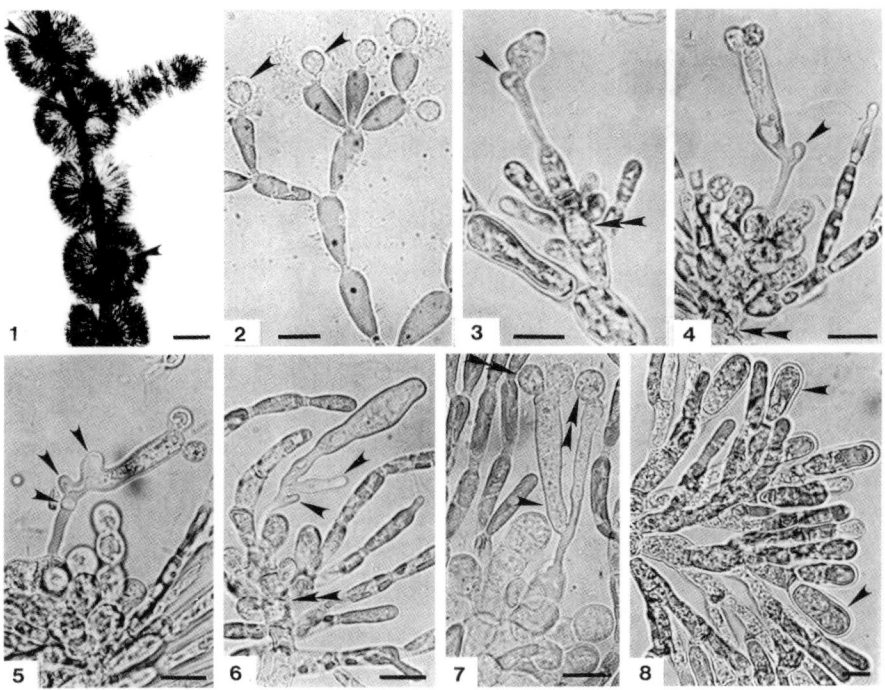

Plate 90
Batrachospermum elegans, figs 1–8 (Sheath, Vis & Cole 1994a)
1. a main axis and a branch showing barrel-shaped, confluent whorls and young carposporophytes (arrow), 2. a fascicle apex with colourless spermatangia (arrows), 3. an immature carpogonium showing knob (arrow) already formed at base of a trichogyne and a three celled carpogonium-bearing branch (double arrow) arising from a proximal cell of fascicle, 4. a fertilized carpogonium with a single knob (arrow) at base of a trichogyne and a three celled carpogonium-bearing branch arising from a periaxial cell (double arrow), 5. a fertilized carpogonium with three knobs (arrows) at base of a trichogyne, 6. a mature carpogonium with two branches (arrows) from base of a trichogyne and a three celled carpogonium-bearing branch (double arrows) from a periaxial cell, 7. a fertilized carpogonium with one large branch (arrow) at base of a trichogyne, both the tip of a trichogyne have attached spermatia (double arrows), 8. a portion of a carposporophyte with cylindrical gonimoblast cells and obovoidal carposporangia (arrows). Scale bars = 200 µm (fig. 1); 10 µm (figs 2–8).

Terminal hairs rare and short. Cortical filaments cylindrical and yellowish. Carpogonium-bearing branches generally curved, arising from periaxial cells of whorls, consisting of 3–5 short cells, producing from the upper 2 or 3 short involucral filaments, which envelope the carpogonium and from the lower cells one or more long filaments. Carpogonia with ovoidal or elongate ellipsoidal trichogyne, pedicellate, cylindrical, 46–60 µm long. Carposporophytes numerous, green-coloured, irregularly hemispherical, 165–220 µm in diameter, 2/3 (rarely equal) the radius of whorls. Carposporangia obovoidal, 10–11 µm in diameter, 16.6–18 µm long, ellipsoidal, about 10 µm in diameter, 20–23 µm long. Secondary fascicles of monoecious plants and cortical filaments of the same shapes as those of female plants. Spermatangia rare, occurring peripherally on primary and secondary fascicles.

Batrachospermales

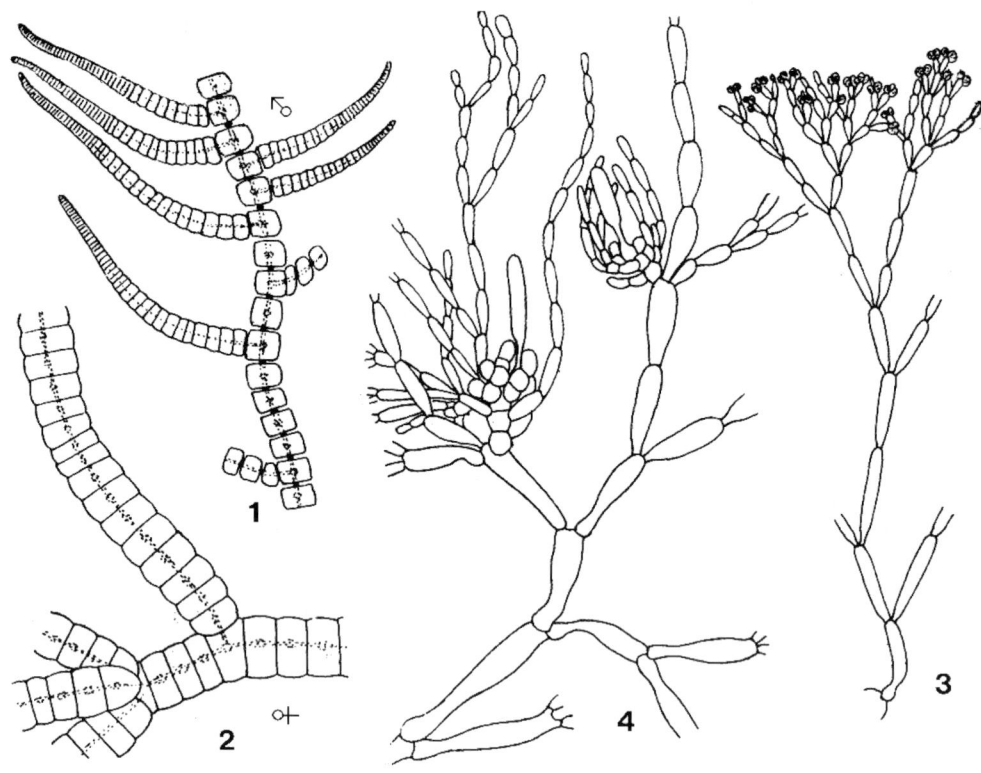

Plate 91
Batrachospermum elegans as *B. coerulescens,* figs 1–4 (original author, Sirodot 1884)
1. a male plant with barrel-shaped, compressed whorls, 2. a female plant with disc-shaped, more compressed whorls than male one, 3. spermatangia terminal on fascicles, 4. a carpogonium-bearing branches arising from intercalary cells of fascicles.

Carpogonium-bearing branches curved, arising from periaxial cells of whorls, having the same structure as those of female plants. Carposporophytes reduced in numbers, irregularly spherical, 1/2 as wide (rarely the same width) as the radius of the whorls.
    Type Locality: on rocks in River Torgal, a tributary of River Mira, near Odemira, Portugal in Europe.
    Holotype: COL 542A.
    Distribution: Known from the type locality and in Ribeiro do Torgal, another tributary of River Mira, Portugal in Europe.

10) ***Batrachospermum helminthosum*** Bory (1808:316) *emend.* Sheath, Vis et Cole (1994a:115, figs 3–10) (non *B. helminthosum* Sirodot 1884:240)

The synonymy proposed by Sheath, Vis & Cole (1994a) is as follows;
Synonym: *B. graibussoniense* Sirodot (1884: 278, pl. XLVII, figs 1–10, pl. XLVIII, figs 1–6); *B. bruziense* Sirodot (1884:281, pl. XLV, figs 1–7, pl. XLVI, figs 1–8); *B. testale* Sirodot (1884:284, pl. XLII, figs 1–5, pl. XLIII, figs 1–8); *B. virgatum* Sirodot (1884: 286, pl. XLIV, figs 1–6, pl. L, figs 1–5) nom. illeg.; *B. sirodotii* Skuja ex Flint (1950:775, figs 13–20).

    plate 92, figs 1–8 (Sheath, Vis & Cole 1994a)
    plate 93, figs 1–9 (Kumano 1997f) as *B. sirodotii*

4. *Virescentia*

Plants monoecious, dioecious with brownish main axis; barrel-shaped, confluent whorls, 306–794 μm in diameter with 7–21 primary fascicles. Carpogonium-bearing branches composed of 1–5 short cells arising from periaxial cell or proximal cell of fascicle branch. Carpogonium 40.2–79.0 μm long with pedicellate, cylindrical to slightly club-shaped trichogynes, 4.8–13.7 μm in diameter. Carposporophytes axial, 120–415 μm in diameter and 102–420 μm high and with 1–8 cylindrical gonimoblast cells. Carposporangia obovoidal, 5.3–16.9 μm in diameter and 9.8–27.6 μm long.

Type Locality: Riviere de Coesne, à Fougères, France in Europe.
Lectotype: PC herb. Thuret, designated by Compère 1991.
Distribution: Known from France, Belgium, Germany, Italy, Poland, Sweden in Europe; China, Korea, Japan in Eastern Asia; Canada, USA in North America; Brazil in South America.

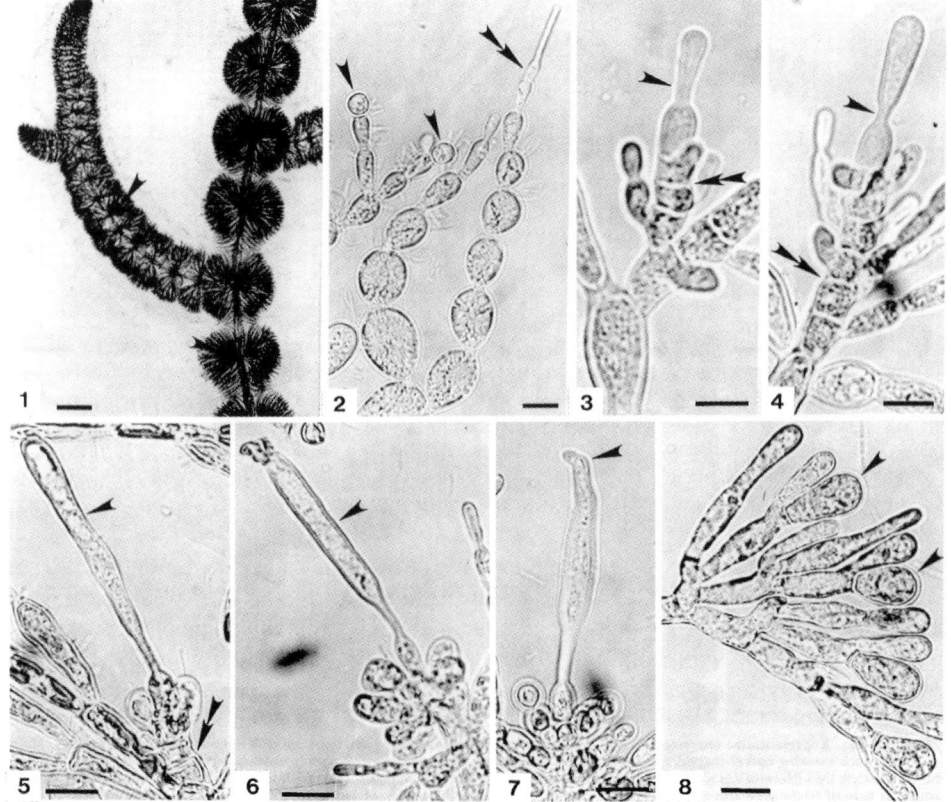

Plate 92
Batrachospermum helminthosum, figs 1–8 (Sheath, Vis & Cole 1994a)
1. a main axis and a branch showing barrel-shaped, confluent whorls and axial carposporophytes (arrow), 2. a fascicle tip with colourless spermatangia (arrows) and a single terminal hair (double arrows), 3. the early stage of carpogonium development with a small trichogyne (arrow) and a three celled carpogonium-bearing branch (double arrow) arising from a proximal cell of a fascicle, 4. the intermediate stage of carpogonium develop-ment with a stalk formation (arrow) between a trichogyne and base, a five celled carpogonium-bearing branch arising from a proximal cell of a fascicle (double arrow), 5. a mature carpogonium with single trichogyne constriction (arrows), and a three celled carpogonium-bearing branch (double arrow) arising from a periaxial cell of a fascicle, 6. a mature carpogonium with a cylindrical stalked trichogyne (arrows), 7. a carpogonium with a stalked trichogyne having an apical hook (arrow), 8. a portion of a carposporophyte with cylindrical gonimoblast cells and obovoidal carposporangia (arrows).
Scale bars = 200 μm (fig. 1); 10 μm (figs 2–8).

Batrachospermales

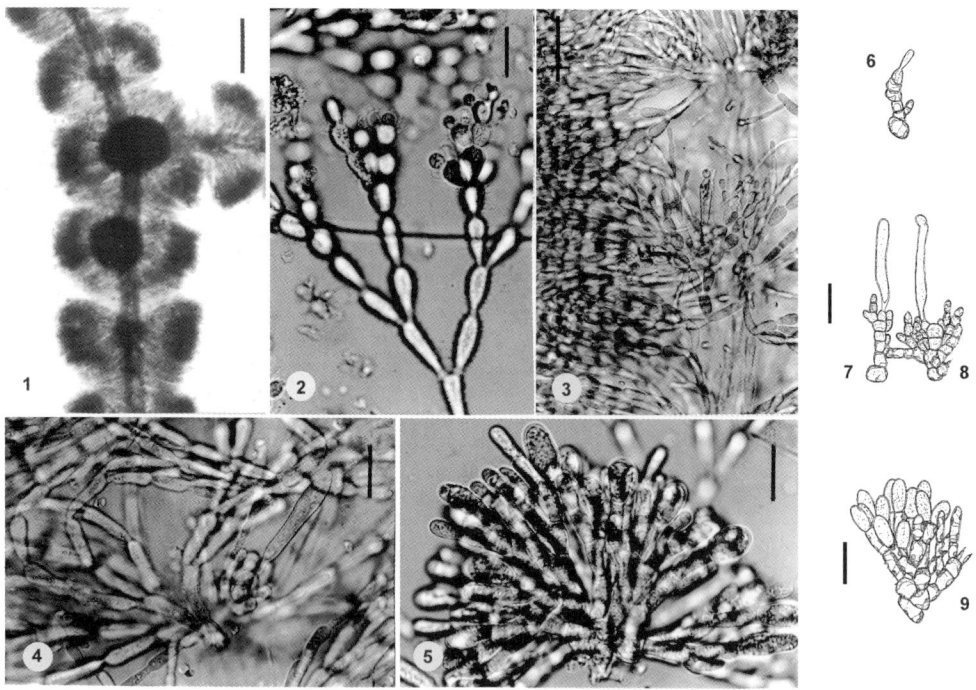

Plate 93
Batrachospermum helminthosum as *B. sirodotii*, figs 1–9 (Kumano 1997f)
1. a main axis with barrel-shaped whorls with hemispherical axial carposporophytes, 2. spermatangia terminal on fascicles, 3, 4. a whorl showing primary fascicles and a carpogonium with a cylindrical trichogyne, 5, 9. carposporangia terminal on gonimoblast filaments, 6–8. the development of carpogonium with a cylindrical, stalked trichogyne and a differentiated carpogonium-bearing branch consisting of barrel-shaped cells. Scale bars = 100 μm (fig. 1); 50 μm (fig. 3); 20 μm (figs 2, 4–9).

In China, Hua-kai-shan, Kiangtsin in Szechwan. Occurs on rocks in a mountain stream, well shaded by overhanging shrubs and grasses. In Japan, Asahikawa in Hokkaido, Shakujii in Tokyo, Owada, Tsukuba in Ibaragi, Ohwada, Tsukuba in Ibaragi, Shikaido, Kongoh-ike, Ichihara in Chiba, Anjyo, Toyokawa in Aichi, Okada, Sukain, Hayashida-gawa River, Yumekasi-gawa River, Himeji, Mond-yakujin, Nishinomiya in Hyogo, Kohnan-cho in Kagawa, Sairinji in Ehime, Udo, Yatyssiro in Kumamoto, Kobayashi in Miyazaki. Known from spring-fed streams.

11) ***Batrachospermum desikacharyi*** Sankaran (1984:169: Figs 1–4)

plate 94, figs 1–4 (Sankaran 1984)

Plants monoecious, 15–20 cm high, pink to violet, profusely branched. Whorls 700 μm in diameter. Axial cells up to 700 μm long. Spermatangia solitary or paired, 6.0–8.5 μm in diameter. Carpogonium-bearing branches straight, single or rarely paired arising from periaxial cell, consisting of 5–10 cells; cells with paired involucral filaments, becoming enveloped by cortical filaments after fertilization. Carpogonium 6.0–8.5 μm in diameter, 8.5–12.0 μm long. Trichogyne club-shaped, stalked, 11.5–14 μm in diameter, 40–46 μm long. Carposprophyte 120 μm in diameter at the top, with corticated stalk. Carposporangia 27 μm in diameter, 34–47.6 μm long.

Type Locality: a hill stream near a water fall in Valparai (Tamil Nadu) at an altitude of about 1000 m, India in Asia.

Holotype: no. S 1/1, coll. V. Sankaran, deposited in the Herbarium of Centre for Advanced Study in Botany, University of Madras.

Distribution: Known from the type locality only.

Note: This species is nearer to section *Virescentia* based on the shaped of trichogyne (Sankaran 1984). However, the carpogonium-bearing branch of this species is comparatively longer than those of section *Virescentia*, and similar to that of section *Batrachospermum*.

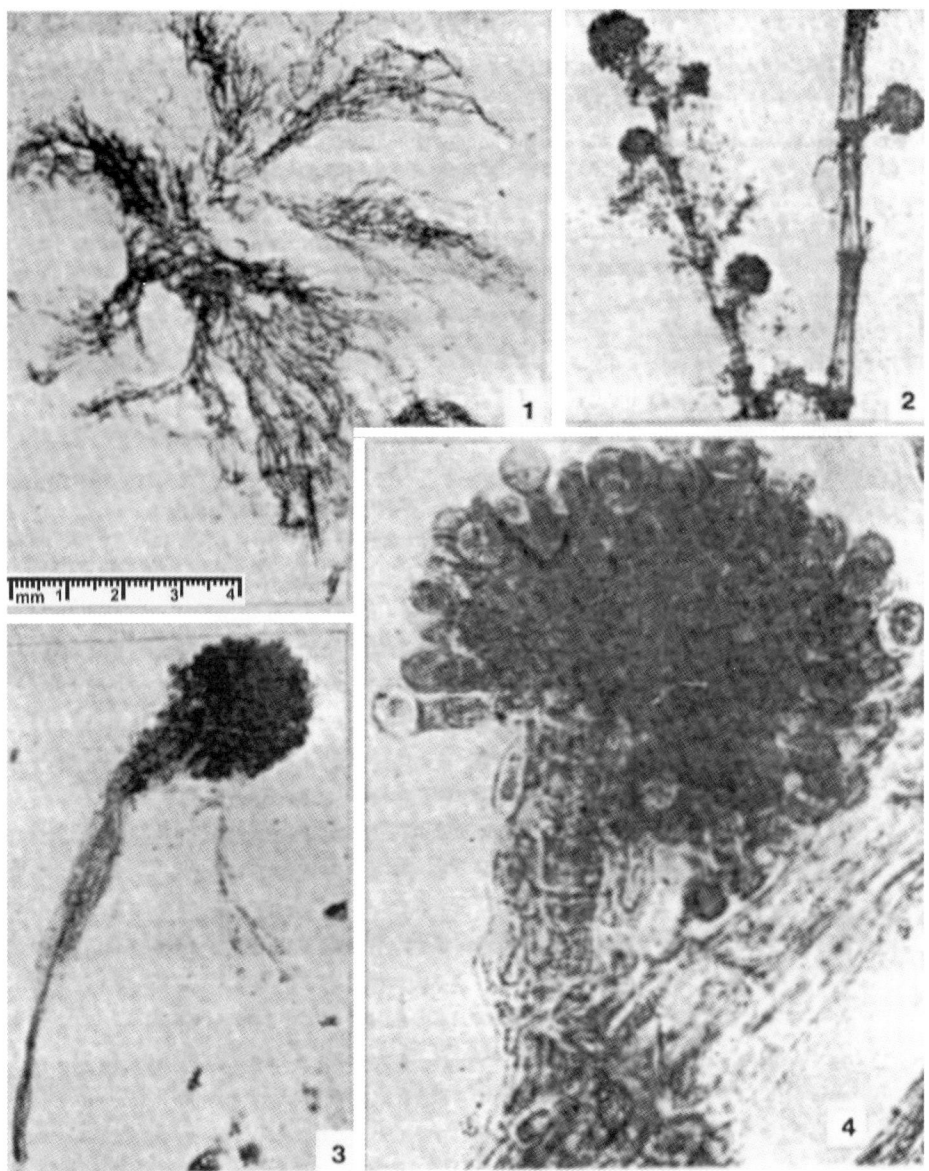

Plate 94
Batrachospermum desikacharyi, figs 1–4, (original author, Sankaran 1984)
1. thallus spread out, 2. portion of the thallus showing main branches with distinct nodes and internodes and stalked carposporophytes clusters, 3. a well-developed carposporophytes cluster with stalk inserted by cortical filaments, 4. a mature carposporophyte cluster arising from a periaxial cell with the stalk completely invested by many rows of cortical filaments.

Batrachospermales

5. Section *Gonimopropagulum* Sheath et Whittick (1995:38)

Type: *Batrachospermum breutelii* Rabenhorst

Sheath & Whittick (1995) established this section, which is named to denote the zonately divided propagules formed by the gonimoblast.

1) ***Batrachospermum breutelii*** Rabenhorst (1855: 282)
Synonym: *B. dimorphum* Kützing (1857:36, pl.91, fig.2)

plate 95, figs 1–6 (Kumano 1997c); figs 7–9 (Skuja 1933)
plate 96, figs 1–12 (Sheath & Whittick 1995)

Plants monoecious, weakly mucilaginous, 10–15 cm high, 400–550 μm in diameter, loose, separated from each other or coalescent in lower part of plants, bluish or olive green; fascicles 4–6, primary fascicles di- or trichotomously branched, 4–5 cell-storeys, proximal cells cylindrical, 5.5–8 μm in diameter, apical cells pear-shaped, 9–12 μm in diameter; cortical filaments developed; secondary fascicles not numerous; Spermatangia spherical or pear-shaped, 7–9 μm in diameter, terminal on fascicles; carpogonium-bearing branches consisting of 3–6 barrel-shaped cells, arising from periaxial cells; carpogonia 15–20 μm in diameter at the apex, 70–95 μm long, trichogyne obconical or club-shaped, shortly stalked; carposporophytes consisting of gonimoblast filaments and clusters of five or more radiating gonimoblast propagules, obovoidal or fusiform, 65–73 μm in diameter, 150–200 μm long, with 4–6 chambers.

Type Locality: Gnadenthal, South Africa in Africa.
Holotype: G herb. De Candolle.
Distribution: Known from the type locality and also in Steenbok River in Bainskloof, Cape of Good Hope, South Africa in Africa.

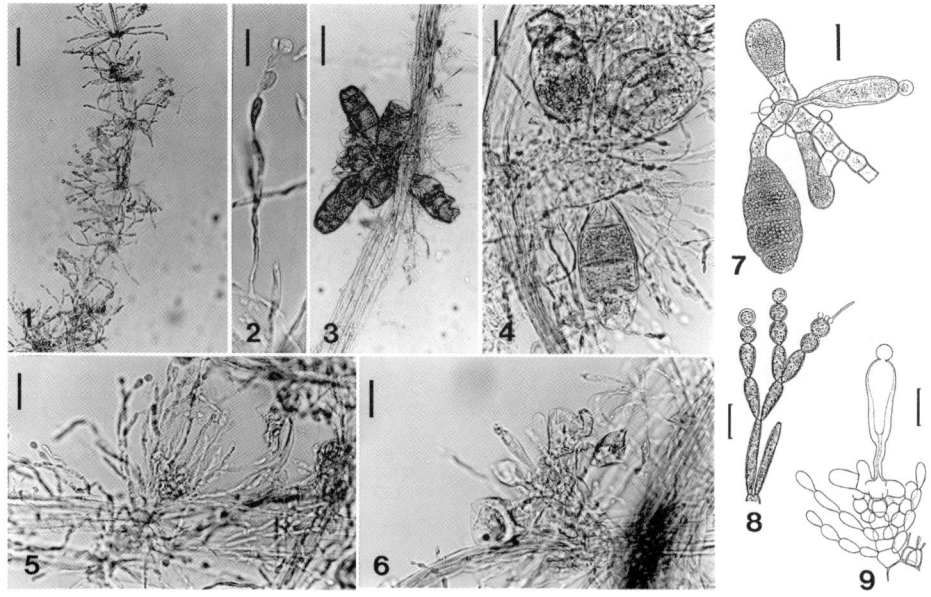

Plate 95
Batrachospermum breutelii, figs 1–6 (Kumano 1997c), figs 7–9 (Skuja 1933)
1. a main axis with whorls, 2, 8. spermatangia terminal on fascicles, 3–4. a carposporophyte with four chambered propagules, 5–6, 9. a carpogonium with an obconical trichogyne, 7. a fertilized carpogonium producing two one chambered and one four chambered propagules.
Scale bars = 200 μm (fig. 1); 100 μm (fig. 3); 40 μm (figs 4–9); 20 μm (fig. 2).

5. *Gonimopropagulum*

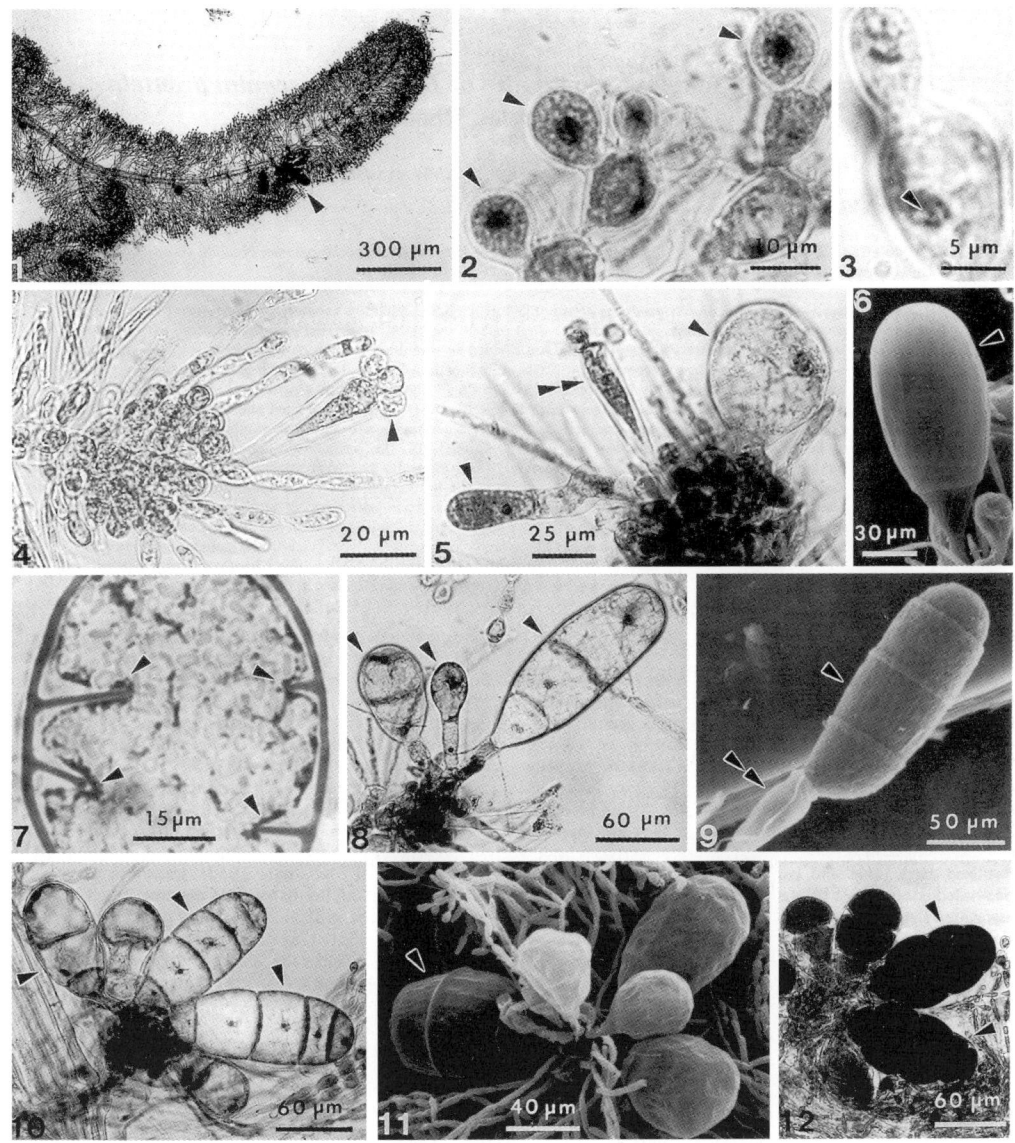

Plate 96
Batrachospermum breutelii, figs 1–12 (Sheath & Whittick 1995)
1. an apex of a plant with dense whorls and mature carposporophytes having propagules (arrow), 2. haematoxylin-stained spermatangia (arrows) terminal on fascicle, 3. a haematoxylin-stained spermatangial initial with approximately five chromosomes (arrows), 4. a fertilized carpogonium with attached spermatia (arrow), 5. a haematoxylin-stained fertilized carpogonium (double arrow) with adjacent uninucleate propagule initials (arrows), 6. a scanning electron micrograph of a single-chambered propagule initial (arrow), 7. a thick-sectioned, toluidine blue-stained propagule showing a simultaneous infurrowing of two septations (arrows), 8. a haematoxylin-stained immature gonimoblast with one- two- and three chambered propagules (arrows), each chamber with a single nucleus, the propagules are on a one-celled gonimoblast filament, 9. a scanning electron micrograph of four-chambered propagule (arrow) on a one-celled gonimoblast filament (double arrow), 10. a haematoxylin-stained mature gonimoblast with multiple-chambered propagules (arrows), 11. a scanning electron micrograph of a mature gonimoblast with one multi-chambered propagule (arrow), 12. an iodine-stained mature gonimoblast showing heavy staining of multi-chambered propagules (arrows).

Batrachospermales

6. Section *Hybrida* De Toni (1897:63)

Type: *Batrachospermum virgato-decaisneanum* Sirodot

Synonym: section '*Hybride*' Sirodot (1884:290)

Plants intense green. Carpogonium-bearing branches short, arising from periaxial cells. Carpogonia with trichogyne somewhat asymmetrical, sessile or indistinctly stalked, ellipsoidal. Carposporophytes spherical, big, single or paired, central within whorls.

1) *Batrachospermum virgato-decaisneanum* Sirodot (1884:290, pl. XXIII, figs 1–10)
Synonym: *B. mikrogyne* Flint et Skuja in Flint (1953:10, figs 1–10)

plate 97 (b), figs 1–10 (original author, Flint 1953) as *B. mikrogyne*
plate 98, figs 1–5 (Sheath & Vis 1995)
figs 6–9 (Sheath & Vis 1995) as *B. virgato-decaisneanum* var. *cochleophilum*,
figs 10–13 (Sheath & Vis 1995) as *B. mikrogyne*,
plate 99, figs 1–9 (Kumano 1996zn)
plate 100, figs 1–5 (original author, Sirodot 1884)

Plants monoecious, very mucilaginous, 2–8 cm high, 300–350 μm in diameter, richly or sparsely branched, bluish green. Whorls almost spherical, ellipsoidal or barrel-shaped, separated or confluent. Fascicles 3, primary fascicles 6–8 cell-storeys, cells rather uniform, shortly cylindrical, ellipsoidal or obovoidal, proximal cells 7–12 μm in diameter, 13–30 μm long, terminal hairs frequent or sparse. Secondary fascicles more or less frequent, finally covering all internodes, reaching the length of primary ones. Spermatangia spherical or ovoid, ca. 6 μm in diameter, terminal on fascicles at different levels. Carpogonium-bearing branches straight or slightly curved, consisting of 3–6 disc-shaped cells, arising from periaxial cells. Involucral filaments numerous, consisting of ellipsoidal cells, of varying length. Carpogonia 4–5 μm in diameter at the base, 15–26 μm long, trichogyne short, ellipsoidal, obovoidal, lanceolate, cylindrical, indistinctly stalked. Carposporophytes single or rarely in pairs, spherical or hemispherical, 100–120 μm in diameter, of densely aggregating gonimoblast filaments, centrally inserted. Carposporangia ellipsoidal, obovoidal or pear-shaped, 8–10 μm in diameter, 10.5–14 μm long.

Type Locality: not specified by Sirodot (1884), material collected from near Montfort and Rennes, France in Europe.
Lectotype: PC
Distribution: Known from France, Sweden in Europe; Oh-ike, Kobe in Hyogo, Japan in Eastern Asia; Australia; Brazil in South America.
[*B. mikrogyne* Flint et Skuja in Flint (1953:10, figs 1–10), Type Locality: small stream near Pollock, Grant Parish, Louisiana, USA in North America. Lectotype: NY. 22/III, 1950, coll. Flint, designated by Sheath & Vis (1995). Distribution: In streams on both sides of the Mississippi River, Louisiana, USA in North America]

2) *Batrachospermum abilii* Reis (1965:138, tab. I, a–d, tab. II, a–d, tab. III, a–b, tab. IV, a–b, tab. V, a–d)

Plants monoecious, 2–3 cm high, initially pyramidal, becoming caespitulose, extremely mucilaginous, moss-green. Ramification abundant, dense, irregular, frequently unilateral, alternate, opposite or, bi-, tri- chotomous in the same axis. Main axis disappearing among the upper branches, toward the base thickened by cortical filaments, solid-terete, finally naked, terminating in a disc.

6. Hybrida

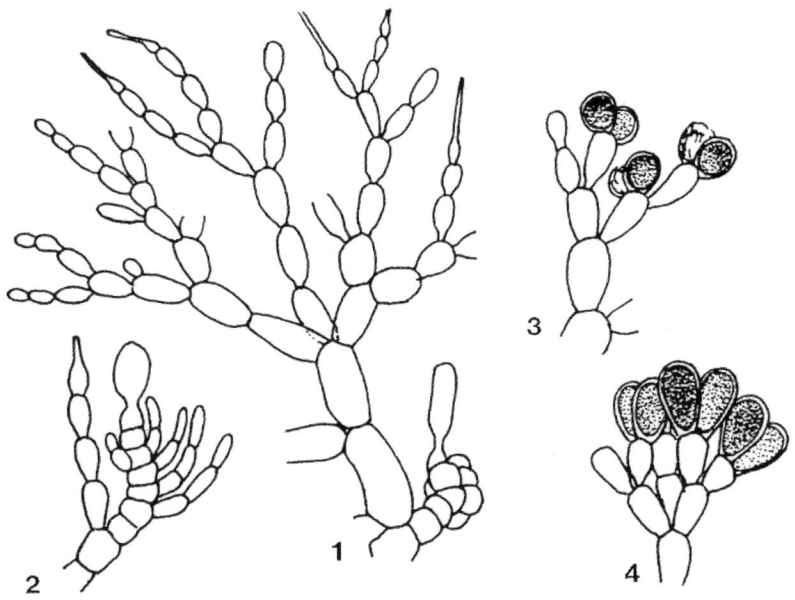

Plate 97(a)
Batrachospermum transtaganum, figs 1–4 (original author, Reis 1970)
1. a primary fascicle and a carpogonium-bearing branch arising from a periaxial cell, 2. a carpogonium-bearing branch, 3. spermatangia terminal on fascicle, 4. carposporangia.

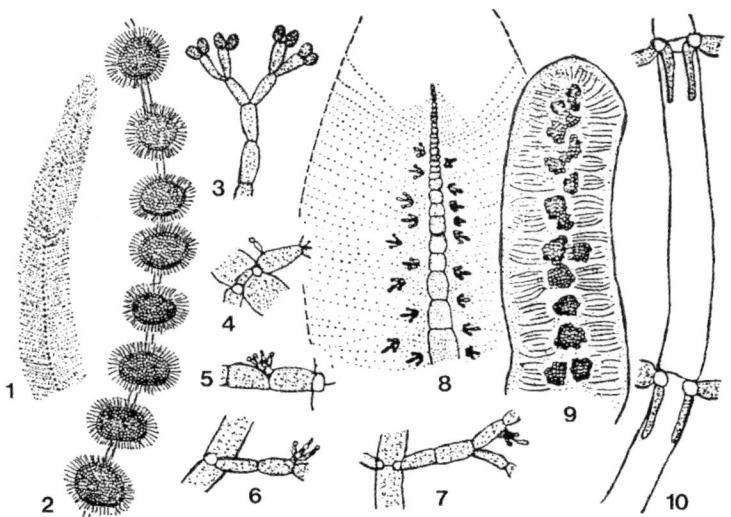

Plate 97(b)
Batrachospermum virgato-decaisneanum as *B. mikrogyne*, figs 1–10 (original author, Flint 1953)
1. an apical portion of a young plant, 2. a plant showing spherical whorls and carposporophytes inserted centrally within whorls, 3. carposporangia terminal on gonimoblast filaments, 4. a fertilized carpogonium with a trichogyne attached a spermatium, 5–7. spermatangia on the involucral filaments of a carpogonium-bearing branch arising from the 1/3 rd proximal cell of a fascicles, 8. a terminal portion of a mature plant showing location of sexual organs, 9. a terminal portion of mature plants, 10. a young plant showing cortical filaments in the early stage in development.

Batrachospermales

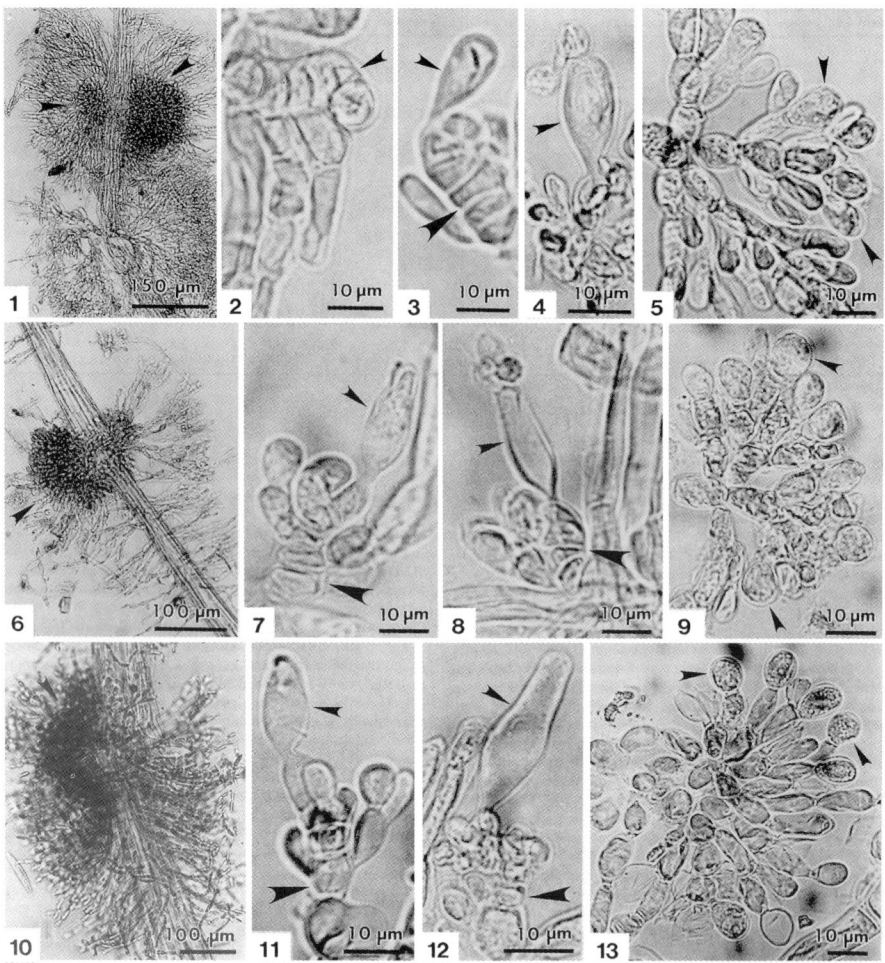

Plate 98
Batrachospermum virgato-decaisneanum, figs 1–5 (Sheath & Vis 1995)
1. obovoidal whorls with paired, axial carposporophytes (arrowheads), 2. curved immature carpogonium-bearing branch (arrowhead) arising from periaxial cell, 3. immature carpogonium with sessile trichogyne (small arrowheads), carpogonium-bearing branch 4-celled and straight (large arrowhead), 4. fertilized carpogonium with sessile clavate trichogyne (arrowhead), 5. carposporophytes with 3–4 celled gonimoblast filaments bearing terminal, obovoidal carposporangia (arrowheads).

Batrachospermum virgato-decaisneanum as *B. virgato-decaisneanum* var. *cochleophilum,* figs 6–9 (Sheath & Vis 1995)
6. obovoidal whorls with axial carposporophytes (arrowhead), 7. mature carpogonium with sessile, ellipsoidal trichogyne (small arrowhead), carpogonium-bearing branch mostly straight (large arrowhead), 8. fertilized carpogonium (small arrowhead) on a partially curved carpogonium-bearing branch (large arrowhead), 9. carposporophyte with 3–4 celled gonimoblast filaments bearing terminal, obovoidal carposporangia (arrowheads).

Batrachospermum virgato-decaisneanum as *B. mikrogyne,* figs 10–13 (Sheath & Vis 1995)
10. obovoidal whorls with axial carposporophyte (arrowhead), 11. immature carpogonium (small arrowhead) borne on a straight, 4-celled carpogonium-bearing branch (large arrowhead), 12. mature carpogonium with a sessile, lanceolate trichogyne (small arrowhead). carpogonium-bearing branch slightly twisted (large arrowhead), 13. carposporophyte with 3-4 celled gonimoblast filaments bearing terminal, obovoidal carposporangia (arrowheads).

6. Hybrida

Whorls contiguous, discoid, distinct before reproduction, often becoming indistinct afterwards. Secondary fascicles from the beginning numerous, immediately under whorls, gradually covering all internodes. Cortical filaments agglomerated especially at the base of plants, cylindrical and fairly coherent. Periaxial cells cylindrical, barrel-shaped, producing 2–4 secondary fascicles, with these apparently dichotomously branched, terminating in arcuate branchlets, curved or bent, consisting of cells that are either large, oblong-clavate, cylindrical, pear-shaped or pyramidal at the inner half of whorl, or small, obovoidal or ellipsoidal at the outer half of secondary fascicles. Terminal hairs rare, short, flexuose, arcuate or sharply bent, slightly inflated at the base. Spermatangia occurring on transformed peripheral branchlets of whorls and involucral filament, at maturity obovoidal, 2.5–4 μm in diameter, 3.5–6 μm long. Carpogonium-bearing branches arising from primary, rarely secondary fascicles, consisting of 2–8 cells, producing distally short, enveloping, involucral filaments; carpogonium cylindrical and 4.5–6 μm long, trichogyne obovoidal. Carposporophytes numerous, spherical, 150–300 μm in diameter, over 1/3 of the radius of whorls. Carpospores obovoidal, 8–10 μm in diameter, 13–16 μm long.

Type Locality: Vermoim, between Oliveira de Azeméis and Val de Cambra, Aveiro, Portugal in Europe.

Holotype: COL P. Reis & A. Nauwerck 135.

Distribution: Known from the type locality only.

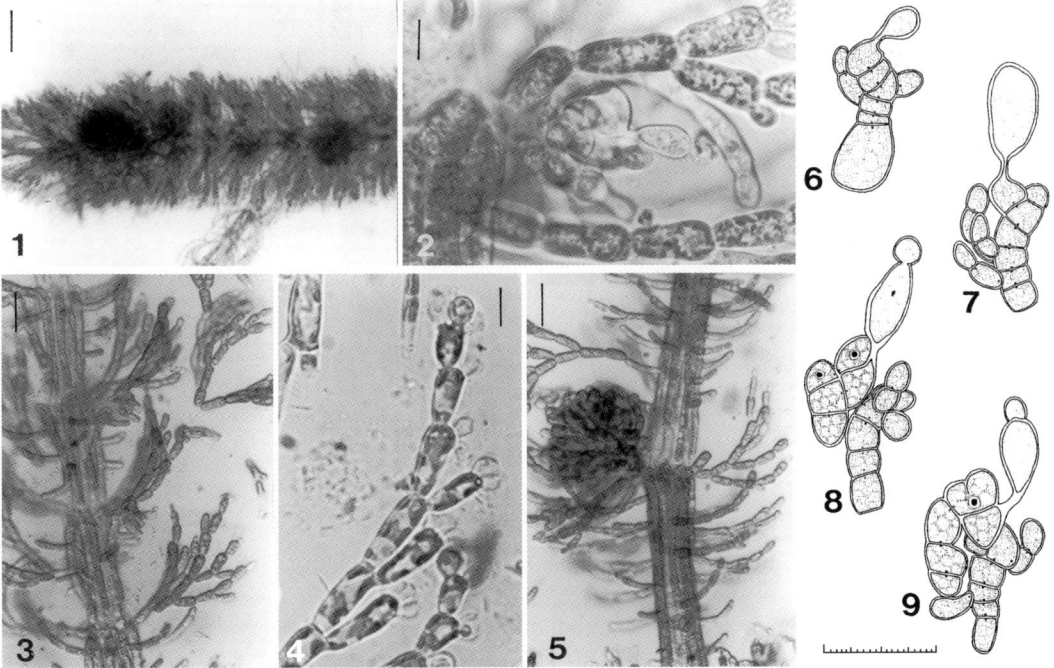

Plate 99

Batrachospermum virgato-decaisneanum, figs 1–9 (Kumano 1996zn)
1. a main axis with barrel-shaped or more or less confluent whorls, 2. a whorl showing a carpogonium with an obovoidal trichogyne, 5. axial carposporophytes, 3. whorls showing axial cells and primary fascicles, 4. spermatangia terminal or subterminal on primary fascicles, 6–7. the early stages in development of a slightly curved carpogonium-bearing branch, 8–9. a fertilized carpogonium with an attached spermatium and a gonimoblast filament.
Scale bars = 100 μm (fig. 1); 40 μm (figs 3, 5); 20 μm (figs 6–9); 10 μm (figs 2, 4).

Batrachospermales

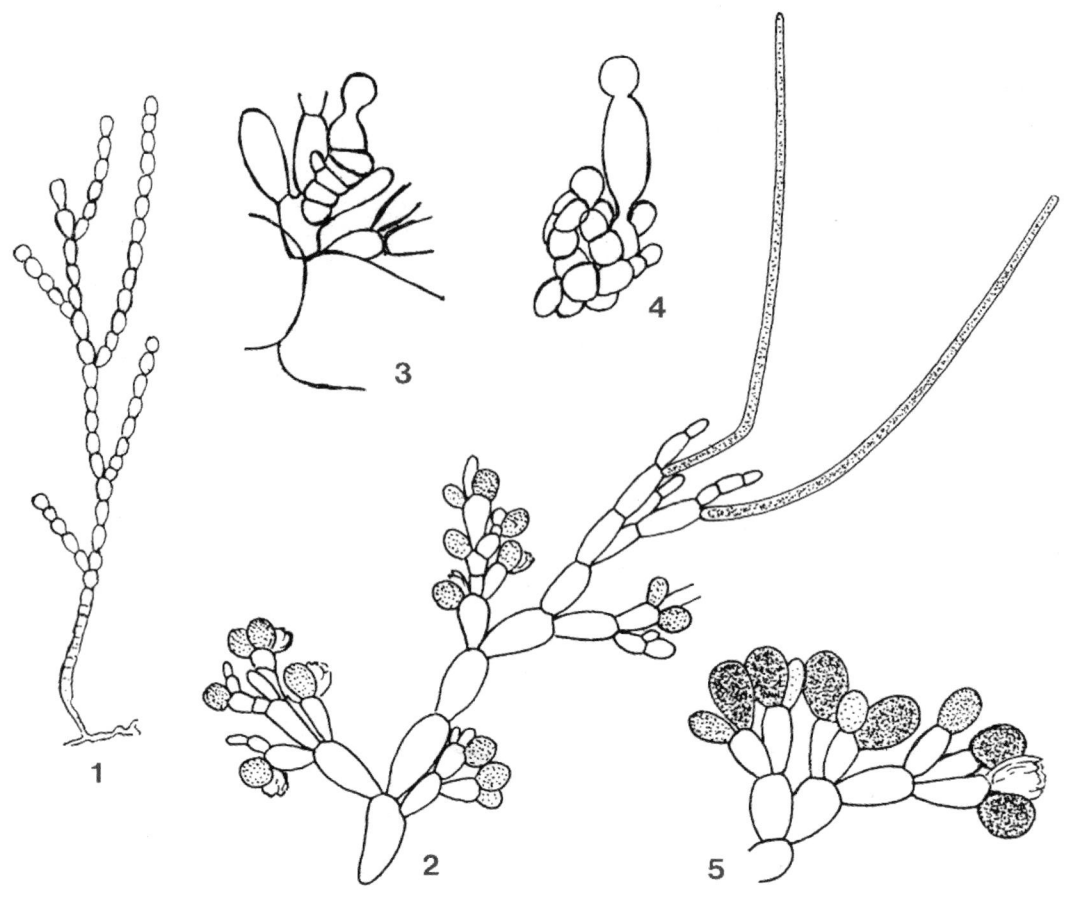

Plate 100
Batrachospermum virgato-decaisneanum, figs 1–5 (original author, Sirodot 1884)
1. a main axis and a branch showing barrel-shaped whorls, 2. spermatangia subterminal or lateral on fascicle, 3–4. somewhat curved carpogonium-bearing branches, 5. carposporangia.

7. Section *Aristata* Skuja (1933:365)

Lectotype: *Batrachospermum cayennense* Montagne

Plants irregularly branched. Carpogonium-bearing branches straight, long, differentiated from fascicles. Carpogonia symmetrical. Carposporophytes pedunculate, spherical.
Note: Sheath, Vis & Cole (1994b) examined nine North American populations and 8 type specimens, and recognized 6 species in section *Aristata* worldwide: *B. beraense* Kumano, *B. cayennense* Montagne ex Kützing (Synonym: *B. aristatum* Skuja), *B. hypogynum* Ratnasabapathy et Kumano, *B. longiarticulatum* Necchi, *B. macrosporum* Montagne (Synonym: *B. australe* Collins, *B. excelsum* Montagne, *B. oxycladum* Montagne) and *B. breutelii* Rabenhorst. *B. breutelii* is assigned the new section *Gonimopropagulum* established by Sheath et Whittick (1995) as mentioned the next section. The following tentative key to the species and subsections of the section Aristata is based on Kumano (1993).

Key to the subsections of the section *Aristata*

1. Hypogynous cells not in rosettes..7–1, subsection *Aristata*
1. Hypogynous cells in rosettes...7–2, subsection *Macrospora*

7–1. Subsection *Aristata* Kumano (1993:262)

Lectotype: *Batrachospermum cayennense* Montagne

Plants irregularly branched. Carpogonium-bearing branches straight, long, without rosette-like involucral filaments and differentiated from fascicles. Carpogonia symmetrical. Carposporophytes pedicellate, spherical.

Key to the species of the subsection *Aristata*

1. Whorls barrel-shaped and confluent with sparse secondary fascicles..2
1. Whorls obovoidal with well-developed secondary fascicles,
   carposporangia fewer than 20 µm long..1) *B. cayennense*
2. Fascicles 19–32 cell-storeys...2) *B. longiarticulatum*
2. Fascicles 10–15 cell-storeys...3
3. Carpogonium-bearing branch consisting of 5–8 cells,
   carpogonium 30–50 µm long..3) *B. turgidum*
3. Carpogonium-bearing branch consisting of 8–11 cells,
   carpogonium 20–27 µm long..4) *B. beraense*

1) ***Batrachospermum cayennense*** Montagne in Kützing (1849:537)

plate 101 (a), figs 1–3 (Bourrelly 1970)
plate 101 (b), figs 1–7 (Kumano 1996g)
plate 102, figs 1–4 (Sheath, Vis et Cole 1994b) as *B. cayennense*,
figs 5–8 (Sheath, Vis & Cole 1994b) as *B. aristatum* Skuja

Plants polyoecious or monoecious, very or moderately mucilaginous, 7–16.5 cm high, 800–2300 µm in diameter, irregularly branched. Whorls dense, obconical, pear-shaped, compressed, contiguous. Fascicles 2–3, primary fascicles straight, 10–17(–19) cell-storeys, proximal cells cylindrical or ellipsoidal, 5–10 µm in diameter, 25–60 µm long, distal cells ellipsoidal, obovoidal, hemispherical, 4–10 µm in diameter, 6–20 µm long, terminal hairs numerous, short. Cortical filaments well-developed. Secondary fascicles numerous, reaching the length of primary ones. Spermatangia spherical or ovoid, 5–6 µm in diameter, terminal or subterminal on fascicles. Carpogonium-bearing branches straight, long, differentiated from fascicles, 80–200 µm long, consisting of 8–30 cells, arising from periaxial cells, rarely from proximal cells. Involucral filaments numerous. Carpogonia 3–6 µm in diameter at the base, 7–9 µm in diameter at the apex, 27–40 µm long, trichogyne club-shaped, obovoidal, lanceolate, sessile. Carposporophytes pedunculate, single or in pairs, spherical, 80–220 µm in diameter, in middle or outer third of whorl. Carposporangia obovoidal or club-shaped, 8–10.5 µm in diameter, 12–18 µm long.

Type Locality: Cayenne, French Guiana in South America.
Holotype: PC herb Montagne, Le Prieur 348.
Distribution: Known from French Guiana, Brazil in South America; Malaysia in Southeast Asia and Australia.

Batrachospermales

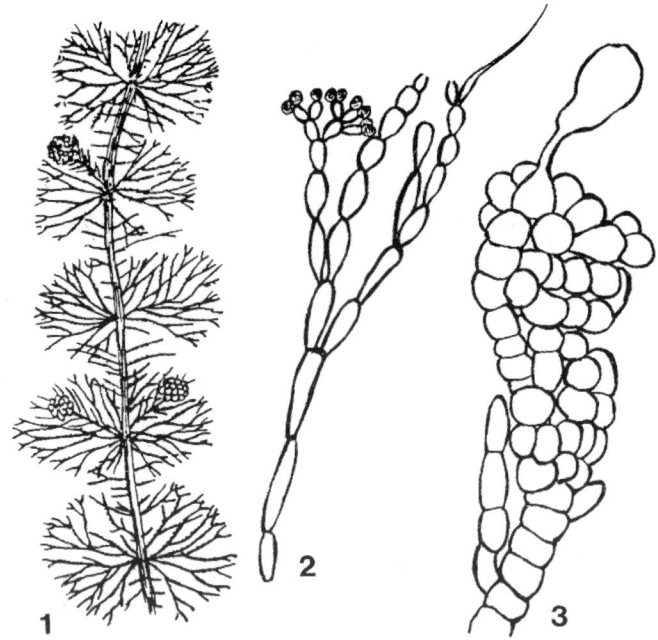

Plate 101 (a)
Batrachospermum cayennense, figs 1–3 (Bourrelly 1970)
1. a main axis with spherical whorls showing obconical carposporophytes with long stalks, 2. spermatangia terminal on fascicle, 3. a differentiated long carpogonium-bearing branch composed of barrel-shaped cells.

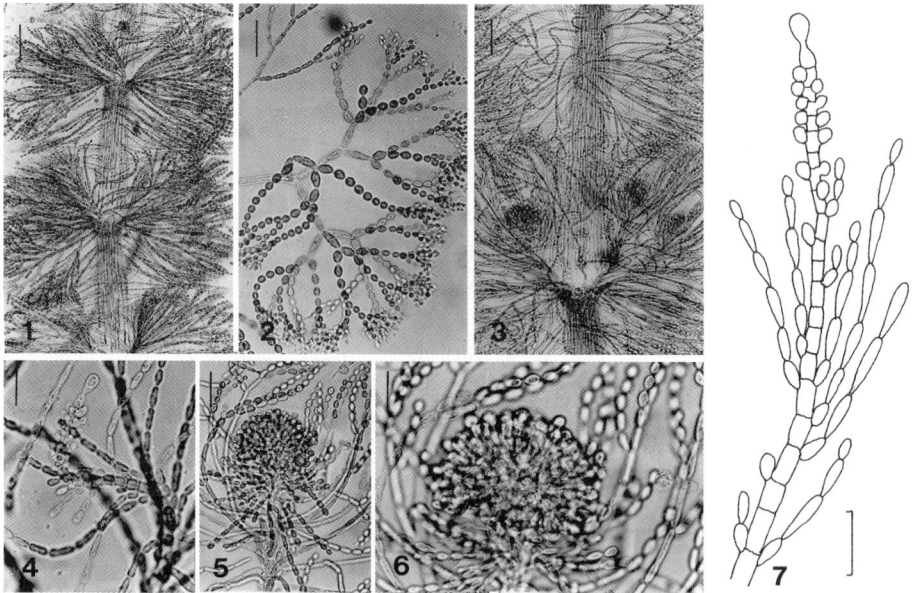

Plate 101 (b)
Batrachospermum cayennense, figs 1–7 (Kumano 1996g)
1. a main axis with spherical whorls, 2. spermatangia terminal on primary fascicles, 3. spherical whorls showing obconical carposporophytes with long stalks, 4, 7. a differentiated long carpogonium-bearing branch composed of barrel-shaped cells, 5–6. obconical carposporophytes with long stalks.
Scale bars = 200 µm (fig. 1); 100 µm (fig. 3); 40 µm (fig. 5); 20 µm (figs 2, 4, 6).

7–1. *Aristata*

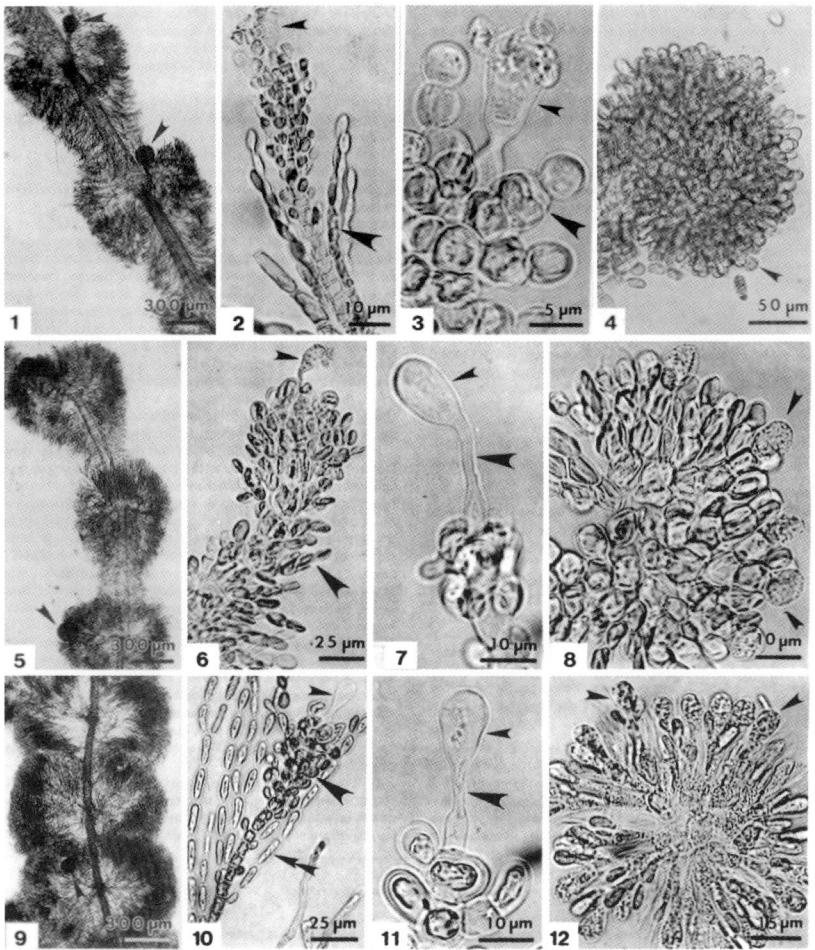

Plate 102
Batrachospermum cayennense type specimen, figs 1–4 (Sheath, Vis & Cole 1994b)
1. obovoidal whorls with prominently stalked carposporophytes (arrowheads), 2. fertilized carpogonium with ovoidal trichogyne (small arrowhead) on elongate, short-celled carpogonium-bearing branch having both short and long involucral filaments (large arrowhead), 3. fertilized carpogonium with pyriform, slightly stalked trichogyne (small arrowhead) on carpogonium-bearing branch having dense involucral filaments (large arrowhead), 4. dense carposporophyte with small, obovoidal carposporangia (arrowhead).

Batrachospermum cayennense as *B. aristatum*, figs 5–8 (Sheath, Vis & Cole 1994b)
5. obovoidal whorls with prominently stalked carposporophyte (arrowhead), 6. fertilized carpogonium with stalked, ovoidal trichogyne (small arrowhead) on elongate, short-celled carpogonium-bearing branch with dense involucral filaments (large arrowhead), 7. unfertilized, mature carpogonium with ovoidal trichogyne (small arrowhead) and elongate stalk (large arrowhead), 8. dense carposporophyte showing few-celled gonimoblast filaments with apical, ovoidal carposporangia (arrowheads).

Batrachospermum longiarticulatum, figs 9–12 (Sheath, Vis & Cole 1994b)
9. barrel-shaped, confluent whorls with prominently stalked carposporophytes (arrowhead), 10. unfertilized carpogonium with lanceolate stalked trichogyne (small arrowhead) on elongate, short-celled carpogonium-bearing branch having short, dense involucral filaments at the apex (large arrowhead) and elongate involucral filaments near the base (double arrowhead), 11. unfertilized, mature carpogonium with lanceolate trichogyne (small arrowhead) and mid-lengthened stalk (large arrowhead), 12. dense carposporophyte showing few-celled gonimoblast filaments and apical, obovoidal carposporangia (arrowheads).

Batrachospermales

2) ***Batrachospermum longiarticulatum*** Necchi (1990:31, figs 29, 39–42)

        plate 102, figs 9–12 (Sheath, Vis & Cole 1994b)
        plate 103 (a), figs 1–4 (original author, Necchi 1990)

Plants monoecious, very mucilaginous, 9–12 cm high, 1200–2500 μm in diameter, irregularly and abundantly branched. Whorls well-developed, barrel-shaped, contiguous. Fascicles 2–3, primary fascicles straight, distal end curved, 20–32 cell-storeys, proximal cells cylindrical or ellipsoidal, 6–18 μm in diameter, 30–120 μm long, distal cells ellipsoidal, obovoidal, subspherical, 5–8 μm in diameter, 8–15 μm long, terminal hairs numerous, short and long. Secondary fascicles few, sparse. Spermatangia spherical or ovoidal, 5–7 μm in diameter, terminal or subterminal on primary fascicles. Carpogonium-bearing branches straight, 120–250 μm long, consisting of 12–22 cells, arising from periaxial cells. Involucral filaments numerous. Carpogonia 3.5–5.5 μm in diameter at the base, 7 μm in diameter at the apex, 22–32 μm long, trichogyne club-shaped, lanceolate sessile. Carposporophytes pedunculate, single or in pairs, dense, spherical, 80–160 μm in diameter, in inner or middle third of whorl. Carposporangia obovoidal or club-shaped, 6.5–10 μm in diameter, 13–18 μm long.

    Type Locality: Manaus-Caracarai Road (Route BR–174) 115 km, Presidente Figueiredo, Amazonas, Brazil in South America.
    Holotype: SP 187156.
    Distribution: Known from the type locality and Humaita, Manaus in Amazonas, Brazil in South America.

3) ***Batrachospermum turgidum*** Kumano (1982b:291, fig. 2, A–L, fig. 3)

        plate 103 (b), figs 1–5 (Kumano 1996zm)
        figs 6–8 (original author, Kumano 1982b)

Plants trioecious, not very mucilaginous, 3–5 cm high, 400–470 μm in diameter, irregularly and abundantly branched, olive green; whorls ellipsoidal and separated, touching each other, more or less compressed; primary fascicles abundantly branched, 8–15 cell-storeys, cells lanceolate-ellipsoidal, fusiform or obovoidal, terminal hairs rare; secondary fascicles sparse; spermatangia spherical, 5–7 μm in diameter, terminal on fascicles; carpogonium-bearing branches consisting of 5–8 barrel-shaped cells, arising from periaxial cells; involucral filaments numerous, very short; carpogonia 4–5 μm in diameter at the base, 3–6 μm in diameter at the apex, 30–50 μm long, trichogyne irregularly cylindrical or twisted, swollen after fertilization and divided endogenously, indistinctly stalked; carposporophytes single or in pairs, spherical, 120–210 μm in diameter, inserted at periphery of whorl; carposporangia ovoidal, 9–13 μm in diameter, 16–20 μm long.

    Type Locality: Kita-dani, Mt. Gozaisyo-dake, Mie, Japan in Eastern Asia.
    Holotype: Kobe University, Seto, 11/V 1958.
    Distribution: Known from the type locality and Kiso Akasawa in Gifu, Japan in Eastern Asia.

7–1. *Aristata*

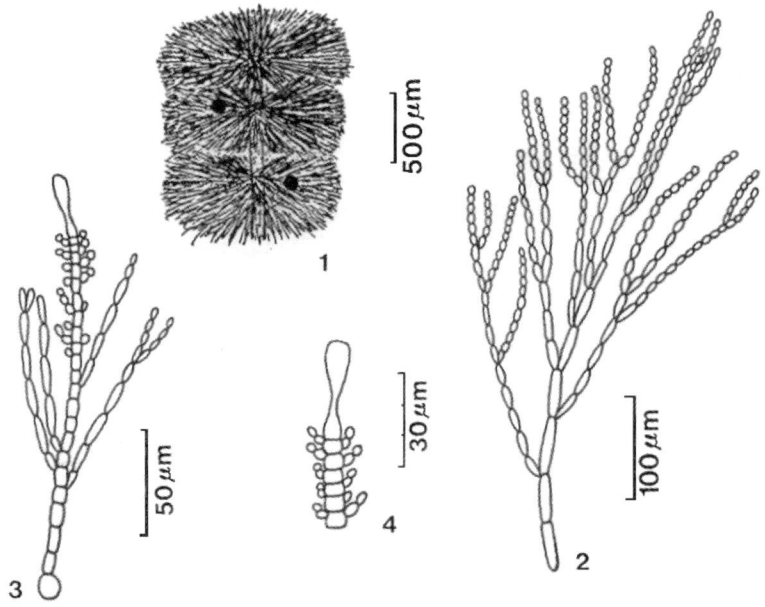

Plate 103 (a)
Batrachospermum longiarticulatum, figs 1–4 (original author, Necchi 1990)
1. whorls showing carposporophytes, 2. details of a primary fascicle, 3. a mature carpogonium showing a carpogonium-bearing branch, a trichogyne and involucral filaments, 4. a mature carpogonium showing a trichogyne and upper involucral filaments.

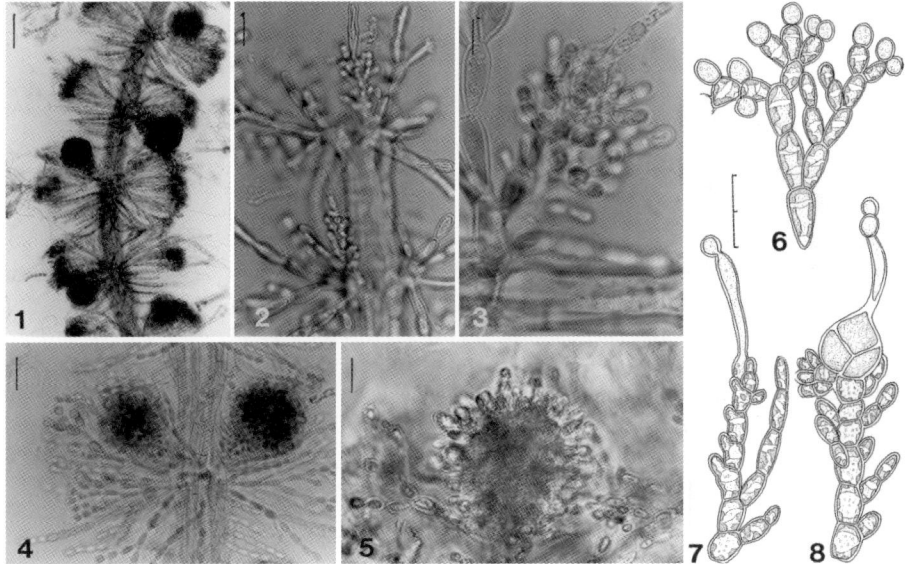

Plate 103 (b)
Batrachospermum turgidum, figs 1–5 (Kumano 1996zm), figs 6–8 (original author, Kumano 1982b)
1. a main axis with barrel-shaped whorls showing a spherical carposporophyte located peripheral portion within whorls, 2, 3, 7. a carpogonium with an elongate cylindrical trichogyne and differentiated carpogonium-bearing branches having short involucral filaments, 6. spermatangia terminal on primary fascicles, 8. a fertilized carpogonium, 4, 5. carposporophytes with carposporangia terminal on gonimoblast filaments. Scale bars = 100 μm (fig. 1); 40 μm (fig. 4); 20 μm (figs 2, 5–8); 10 μm (fig. 3).

Batrachospermales

4) ***Batrachospermum beraense*** Kumano (1978:98, fig. 2. A–C)

plate 104 (a), figs 1–5 (Kumano 1996d)
figs 6–8 (original author, Kumano 1978)

Plants monoecious, very mucilaginous, 5–17 cm high, 600–1000 µm in diameter, abundantly and irregularly branched, olive green; whorls ellipsoidal and separated, contiguous and more or less compressed; primary fascicles abundantly branched, 10–13 cell-storeys, proximal cells cylindrical, 8–10 µm in diameter, 50–90 µm long, distal cells fusiform or ellipsoidal, 6–8 µm in diameter, 8–25 µm long, terminal hairs rare; secondary fascicles sparse; spermatangia spherical, 4–5 µm in diameter, terminal on fascicles; carpogonium-bearing branches 60–90 µm long, consisting of 8–11 cells, arising from periaxial cells; involucral filaments numerous, elongated, embracing carposporophytes; carpogonia 5–6 µm in diameter at the base, 6–8 µm in diameter at the apex, 20–27 µm long, trichogyne urn-shaped, stalked; carposporophytes 2–3, spherical, 90–150 µm in diameter, inserted at periphery of whorl; carposporangia ovoidal, 10–12 µm in diameter, 19–22 µm long.

Type Locality: Fort Iskander, Tasek Bera, Pahang, East Malaysia in Southeast Asia.
Holotype: Kobe University, Kumano, 16/IV 1971.
Distribution: Known from the type locality and Saba, Malaysia in Southeast Asia.

7–2. Subsection *Macrospora* Kumano (1993:263)

Type: *Batrachospermum macrosporum* Montagne

Plants irregularly branched. Carpogonium-bearing branches differentiated from fascicles, straight, long, with rosette-like involucral filaments. Carpogonia symmetrical. Carposporophytes pedunculate, spherical, with large carposporangia.

Key to the species of the subsection *Macrospora*

1. Carpogonium-bearing branch consisting of up to 16 cells,
    carpogonium 30–50 µm, carposporangia 30–60 µm long..........................1) *B. macrosporum*
1. Carpogonium-bearing branch consisting of fewer than 9 cells..2
2. Carpogonium 25–35 µm long...2) *B. equisetifolium*
2. Carpogonium 40–45 µm long, carposporangia 50–65 µm long.........................3) *B. hypogynum*

1) ***Batrachospermum macrosporum*** Montagne (1850:293)
Synonym: *B. excelsum* Montagne (1850:291); *B. macrosporum* var. *excelsum* (Montagne) Sirodot (1884:268); *B. oxycladum* Montagne (1850:293): *B. macrosporum* var. *oxycladum* (Montagne) Sirodot (1884:269); *B. australe* Collins (1906:110).

plate 104 (b), figs 1–7 (Kumano 1996z)
plate 105, figs 1–3 (Sheath, Vis & Cole 1994b) as *B. macrosporum*,
figs 4–5 (Sheath, Vis & Cole 1994b) as *B. excelsum*,
figs 6–8 (Sheath, Vis & Cole 1994b) as *B. australe*,
figs 9–12 (Sheath, Vis & Cole 1994b) as *B. macrosporum*,
figs 13 (Sheath, Vis & Cole 1994b) as *B. oxycladum*

7–2. *Macrospora*

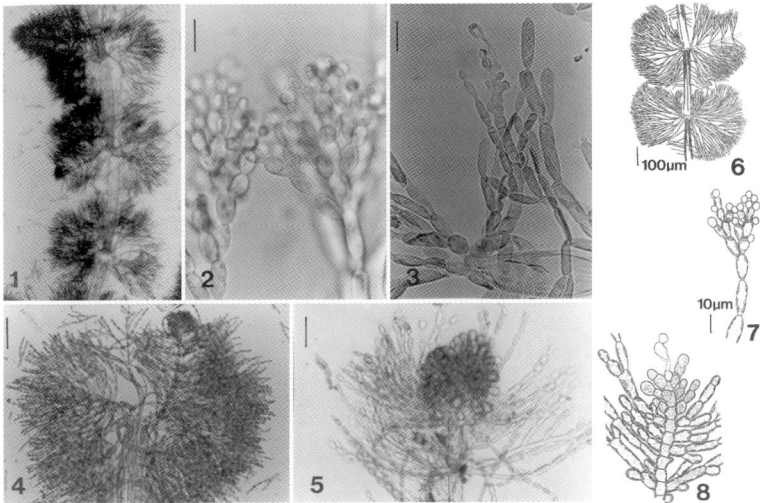

Plate 104 (a)
Batrachospermum beraense, figs 1–5 (Kumano, 1996d), figs 6–8 (original author, Kumano 1978)
1, 6. a main axis with spherical or barrel-shaped whorls showing primary fascicles and a carpogonium-bearing branch, 2, 7. spermatangia terminal on primary fascicles, 3, 8. a relatively long carpogonium-bearing branch with a fertilized carpogonium and numerous long involucral filaments, 4–5. a carposporophyte with a long stalk.
Scale bars = 200 μm (fig. 1); 100 μm (figs 4, 6); 40 μm (fig. 5); 20 μm (fig. 3); 10 μm (figs 2, 7–8).

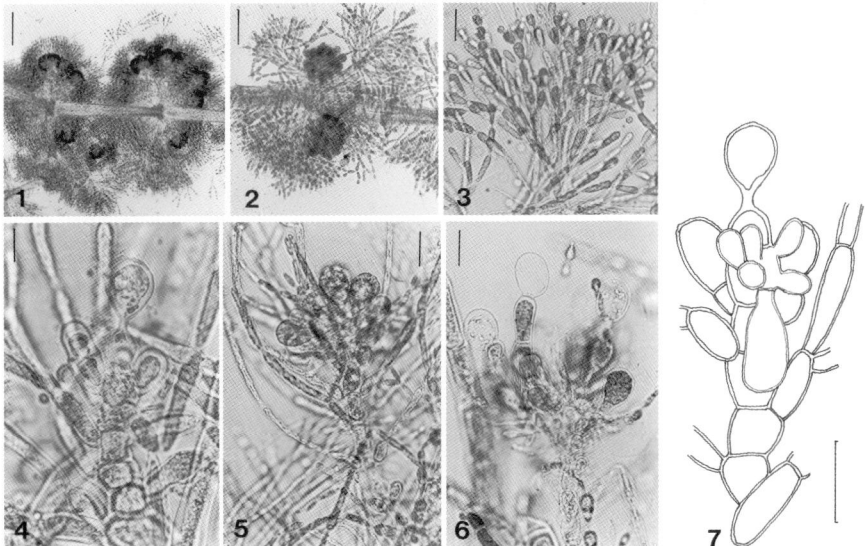

Plate 104 (b)
Batrachospermum macrosporum, figs 1–7 (Kumano 1996z)
1. a main axis with spherical whorls showing several carposporophytes within whorls, 2. two carposporophytes within a whorl having large carposporangia terminal on gonimoblast filaments, 3. spermatangia terminal on primary fascicles, 4. a carpogonium with an obovoidal trichogyne and a differentiated carpogonium-bearing branch having rosette like uppermost involucral filaments, 7. a fertilized carpogonium producing gonimoblast filaments, 5. large obovoidal carposporangia terminal on short gonimoblast filaments, 6. carpospores germinating *in situ*.
Scale bars = 200 μm (fig. 1); 100 μm (fig. 2); 40 μm (figs 3, 5–6); 20 μm (fig. 7); 10 μm (fig. 4).

Batrachospermales

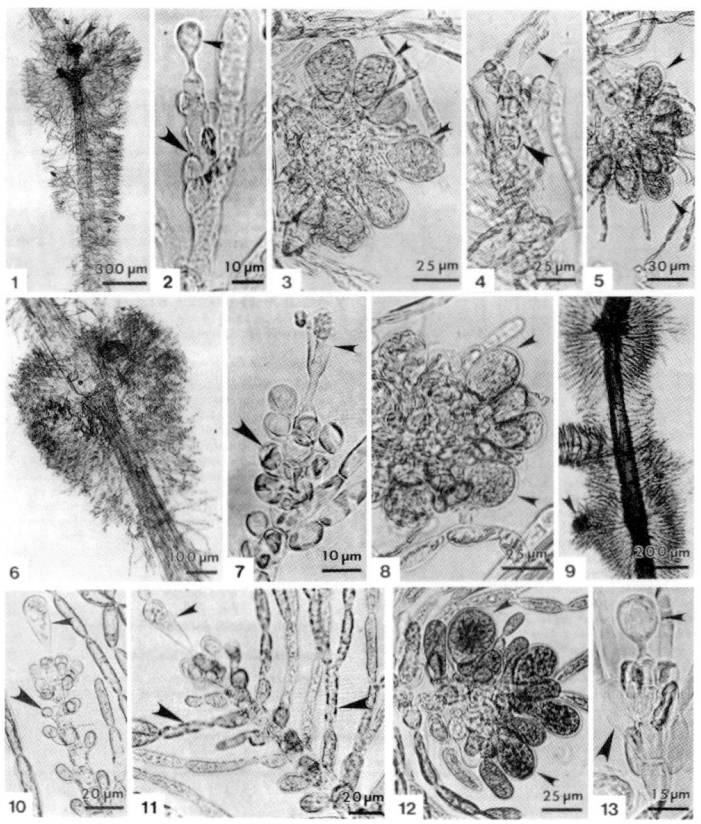

Plate 105
Batrachospermum macrosporum, type specimen, figs 1–3 (Sheath, Vis & Cole 1994b)
1. obovoidal whorl with prominently stalked carposporophyte (arrowhead), 2. developing carpogonium with stalked lanceolate trichogyne (small arrowheads) on about 5-celled carpogonium-bearing branch having short involucral filaments (large arrowhead), 3. carposporophyte with large, obovoidal carposporangia (arrowheads).

Batrachospermum macrosporum as *B. excelsum,* type specimen, figs 4–5 (Sheath, Vis & Cole 1994b)
4. fertilized carpogonium with stalked, lanceolate trichogyne (small arrowhead) on short-celled carpogonium-bearing branch (large arrowhead), 5. carposporophyte with large, obovoidal carposporangia (arrowheads).

Batrachospermum macrosporum as *B. australe,* type specimen, figs 6–8 (Sheath, Vis & Cole 1994b)
6. obovoidal whorl with prominently stalked carposporophyte (arrowhead), 7. fertilized carpogonium with stalked, lanceolate trichogyne (small arrowhead) on short-celled carpogonium-bearing branch having short involucral filaments (large arrowhead), 8. carposporophyte with large, obovoidal carposporangia (arrowheads).

Batrachospermum macrosporum, figs 9–12 (Sheath, Vis & Cole 1994b)
9. obovoidal whorls with prominent carposporophyte (arrowhead), 10. unfertilized carpogonium with clavate trichogyne (small arrowhead) on short-celled carpogonium-bearing branch having short involucral filaments (large arrowhead), 11. fertilized carpogonium with club-shaped trichogyne (small arrowhead) on short-celled carpogonium-bearing branch having mostly elongate involucral filaments (large arrowheads), 12. carposporophyte with large, obovoidal carposporangia (arrowheads).

Batrachospermum macrosporum as *B. oxycladum,* type specimen, figs 13 (Sheath, Vis & Cole 1994b)
13. carpogonium with stalked, orbicular trichogyne (small arrowhead) on carpogonium-bearing branch having elongate involucral filaments (large arrowhead).

Plants dioecious, rarely polyoecious, very (rarely moderately) mucilaginous, 3.5–15 cm high, 600–2500 μm in diameter, irregularly and abundantly branched; whorls well-developed, lax, rarely dense, obconical, pear-shaped or spherical, contiguous or separated; fascicles 2–4, primary fascicles straight, di- or trichotomously branched, 5–10(–12) cell-storeys, proximal cells ellipsoidal or cylindrical, 6–25 μm in diameter, 50–180 μm long, distal cells obovoidal, hemispherical or ellipsoidal, 7–30 μm in diameter, 10–50 μm long, terminal hairs short and long; secondary fascicles numerous, reaching or not reaching the length of primary ones; spermatangia spherical or obovoidal, 6–9 μm in diameter, terminal or subterminal on fascicles; carpogonium-bearing branches straight, long, 40–200 μm long, consisting of 3–16 barrel-shaped or cylindrical cells, arising from periaxial cells, rarely from proximal cells or cortical filament cells; carpogonia 6.5–13.5 μm in diameter at the base, 12–18 μm in diameter at the apex, 30–50 μm long, trichogyne club-shaped or ovoidal, lanceolate, sessile or indistinctly stalked; upper involucral filaments in rosettes, lower involucral filaments numerous; carposporophytes 1–2(–3), pedunculate, dense, spherical, 90–220 μm in diameter, extending from periphery of whorl; carposporangia obovoidal, pear- or club-shaped, very big, 20–45 μm in diameter, 32.5–60 μm long; carpospores germinating in situ.

Type Locality: River Orapu and Comté, Cayenne, French Guiana in South America.

Holotype: PC herb Montagne, Le Prieur 1105.

Distribution: Widespread from north Igarapé Tarumanzinho near Manaus in Amazonas, to south Port Alegre, Rio Grande do Sul in Brazil in South America; North Carolina, USA in North America.

2) ***Batrachospermum equisetifolium*** Montagne (1850:295)

plate 106 (b), figs 1–3 (Kumano 1990)

Plants very small, hair-like, arising from a disc, violet, slightly branched at the base, distally nearly unbranched; internodes 4 times longer than diameter; branchlets of whorls all curved. Spermatangia not observed. Carpogonium-bearing branch arising from the periaxial cell, consisting of 4–7 disc- or barrel-shaped cells, carpogonium 7–10 μm diameter at the base, 9–11 μm diameter at the apex, 25–35 μm long, trichogyne ellipsoidal or club-shaped, spatula-shaped, more or less indistinctly stalked. Hypogynous cells producing rosette-like involucral filaments. No carposporophytes or carposporangia observed.

Type Locality: On rocks in Gravier Creek in Kaw Mountains, French Guiana in South America.

Holotype: PC herb Montagne, Le Prieur 1109.

Distribution: Known from the type locality only.

3) ***Batrachospermum hypogynum*** Kumano et Ratnasabapathy in Ratnasabapathy & Kumano (1982b:122, figs 4, A–F)

plate 106 (a), figs 1–6 (Kumano 1996v)
figs 7–8 (original authors, Ratnasabapathy & Kumano 1982b)

Plants monoecious, strongly mucilaginous, 3–7 cm high, 300–570 μm in diameter, more or less irregularly branched, deep brown or wine red; whorls ellipsoidal and distant, touching each other or more or less compressed; primary fascicles abundantly and more or less unilaterally branched, 8–11 cell-storeys, proximal cells arcuate-club-shaped, 5–7 μm in diameter, 17–23 μm long, distal cells fusiform or ellipsoidal, 3–5 μm in diameter, 6–10 μm long, terminal hairs

Batrachospermales

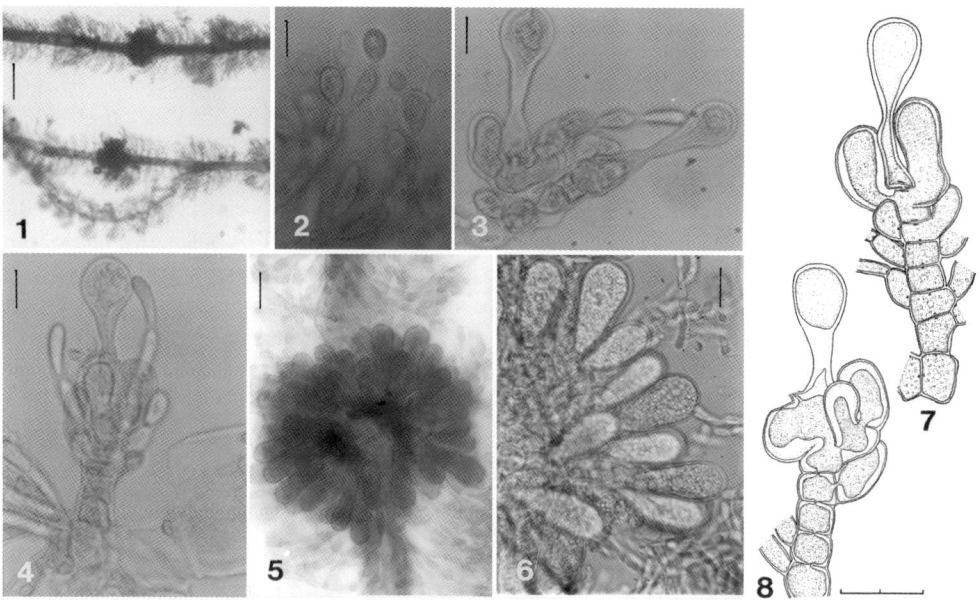

Plate 106 (a)
Batrachospermum hypogynum, figs 1–6 (Kumano 1996v), figs 7–8 (original authors, Ratnasabapathy & Kumano 1982b)
1. a main axis with subconical whorls and axial carposporophytes, 2. spermatangia terminal on fascicles, 3–4, 7. a mature carpogonium with an urn-shaped trichogyne and a differentiated carpogonium-bearing branch with rosette-like hypogynous cells, 5. an axial carposporophyte, 6. large obovoidal or club-shaped carposporangia terminal on gonimoblast filaments 8. a fertilized carpogonium with gonimoblast initials.
Scale bars = 200 µm (fig. 1); 40 µm (fig. 5); 20 µm (figs 6–8); 10 µm (figs 2–4).

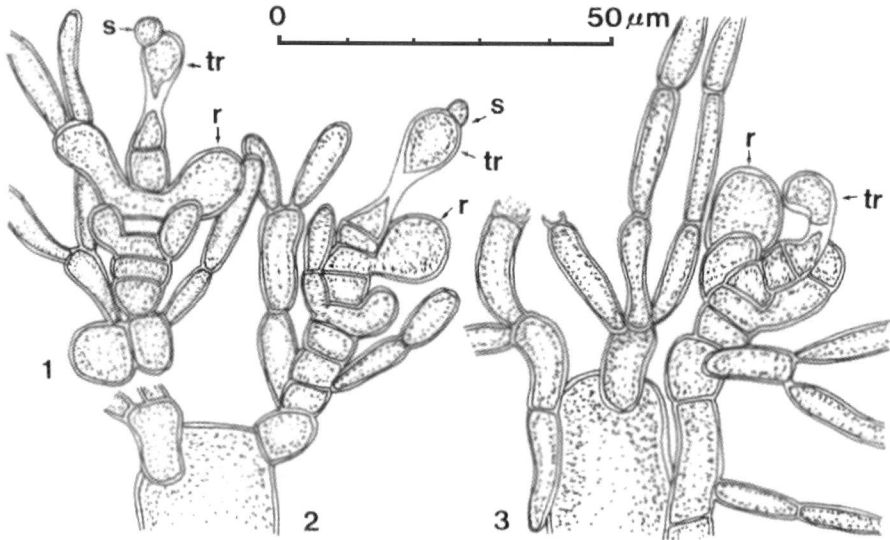

Plate 106 (b)
Batrachospermum equisetifolium, figs 1–3 (Kumano 1990)
1–3. a differentiated carpogonium-bearing branch with rosette-like involucral filaments and a carpogonium with a spatula-shaped trichogyne. (Scale bar = 50 µm for figs 1–3).

absent; secondary fascicles rare; spermatangia ovoidal or spherical, 4–6 μm in diameter, terminal or lateral on primary fascicles; carpogonium-bearing branches 40–70 μm long, consisting of 5–9 barrel-shaped cells, arising from periaxial cells; carpogonia 8–10 μm in diameter at the base, 10–13 μm in diameter at the apex, 40–4 μm long, trichogyne urn-shaped, indistinctly stalked; upper hypogynous cells forming rosettes of involucral filaments, lower involucral filaments numerous, laterally elongated, large; carposporophytes single, spherical, 100–200 μm in diameter, extending from periphery of whorl; carposporangia ovoidal or club-shaped, large, 18–35 μm in diameter, 50–65 μm long.

Type Locality: Sungai Pusu, near Kampon Sungai Pusu, Selangor, Malaysia in Southeast Asia. Attached to rocks in a clean stream, 30–40 cm wide, 15 cm depth.

Holotype: Kobe University, Ratnasabapathy no. 1201b, 2/VI 1979.

Distribution: Known from the type locality and Sungai Air Terjun, Pulau Lankawi, Malaysia in Southeast Asia.

8. Section *Contorta* Skuja (1931a:81)

Type: *Batrachospermum procarpum* Skuja

Plants irregularly or pseudodichotomously branched. Carpogonium-bearing branches curved, strongly undulated, or spirally coiled, differentiated from fascicles. Carpogonium asymmetrical. Carposporophytes sessile, hemispherical. Gonimoblast filaments of determinate radially branched type.

Key to the subsections of the section *Contorta*

1. Monosporangia present (part) ....8-1, subsection *Intorta*
1. Monosporangia absent..2
2. Carpogonium-bearing branches curved...8-2, subsection *Torrida*
2. Carpogonium-bearing branches spirally coiled, fascicles
    consisting of chantransia-like cells..8-3, subsection *Procarpa*
2. Carpogonium-bearing branches strongly undulated...3
3. Carposporophytes loosely agglomerated..8-4, subsection *Kushiroense*
3. Carposporophytes compactly agglomerated...8-5, subsection *Ambigua*

8-1. Subsection *Intorta* Kumano (1993:264)

Type: *Batrachospermum intortum* Jao

Plants irregularly branched. Monosporangia terminal on involucral filaments of fascicles. Carpogonium-bearing branches spirally undulated or coiled, differentiated from fascicles. Carpogonia asymmetrical. Carposporophytes sessile, hemispherical. Gonimoblast filaments of radially branched type.

Key to the species of the subsection *Intorta*

1. Monosporangia terminal on involucral filaments, sometimes on
    primary and secondary fascicles...2
1. Monosporangia terminal on primary and secondary fascicles..3
2. Monosporangia 11-15 μm long...1) *B. intortum*

Batrachospermales

2. Monosporangia 13-23 μm long..2) *B. pseudocarpum*
3. Carpogonium-bearing branch consisting of 4–7 cells..........................3) *B. woitapense*
3. Carpogonium-bearing branch consisting of 6–14 cells........................4) *B. lusitanicum*

1) ***Batrachospermum intortum*** Jao (1941:259, tab. VI, figs 39–45)

plate 107 (a), figs 1–10 (Kumano 1996w)
plate 107 (b), figs 1–4 (original author, Jao 1941)

Plants monoecious, moderately mucilaginous, up to 7 cm high, 300–500 μm in diameter, alternately and irregularly branched, obconical, confluent, deep olive-green; primary fascicles dicho-tomously branched, short, 5–12 cell-storeys, cells fusiform, ellipsoidal; cortical filaments very much developed; secondary fascicles numerous, unbranched or slightly branched, 2–8 cell-storeys, cells cylindrical or ovoidal, terminal hairs rare; spermatangia spherical or hemispherical, (5–)7–9 μm in diameter, terminal or subterminal on fascicles; carpogonium-bearing branches constantly spiral, long, consisting of (6–)8–11 cells, arising from periaxial cells, or rarely terminal on fascicles; carpogonia with club-shaped, shortly stalked trichogyne; carposporophytes hemispherical, 150–190 μm in diameter, 87–100 μm high; carposporangia obovoidal or club-shaped, 12–14 μm in diameter, 18–23 μm long; monosporangia numerous, obovoidal, (9–)10–12 μm in diameter, 11–15 μm long, usually terminal on fascicles, or on involucral filaments.

Type Locality: near Lung-Chu Sze, Pa-hsin, Szechwan, China in Eastern Asia. Growing on the submerged roots of willow in a spring-fed pond.

Holotype: SC 1114.

Distribution: Known from the type locality and several localities in the mountainous eastern part of Cuba, Oriente (Rieth 1979) in Central America.

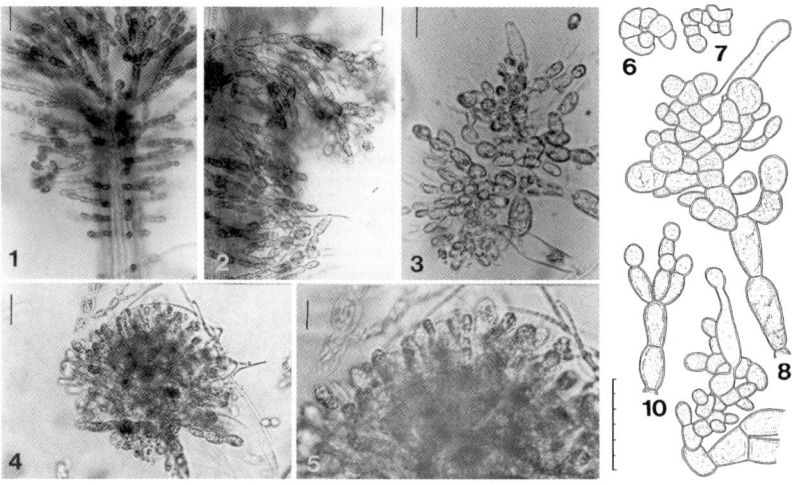

Plate 107 (a)
Batrachospermum intortum, figs 1–10 (Kumano 1996w)
1. a main axis with a subconical whorl, 2. spermatangia terminal or subterminal on fascicles, 4–5. carposporangia terminal on gonimoblast filaments, 6–7, 8. a carpogonium-bearing branch arising from a terminal cell of fascicle, with monosporangia terminal on involucral filaments, 3, 9. a coiled carpogonium-bearing branch and a fertilized carpogonium from which a few gonimoblast filaments are produced, 10. spermatangia terminal on primary, secondary fascicles.
Scale bars = 30 μm (figs 6–10); 20 μm (figs 1–2, 4); 10 μm (figs 3, 5).

8–1. *Intorta*

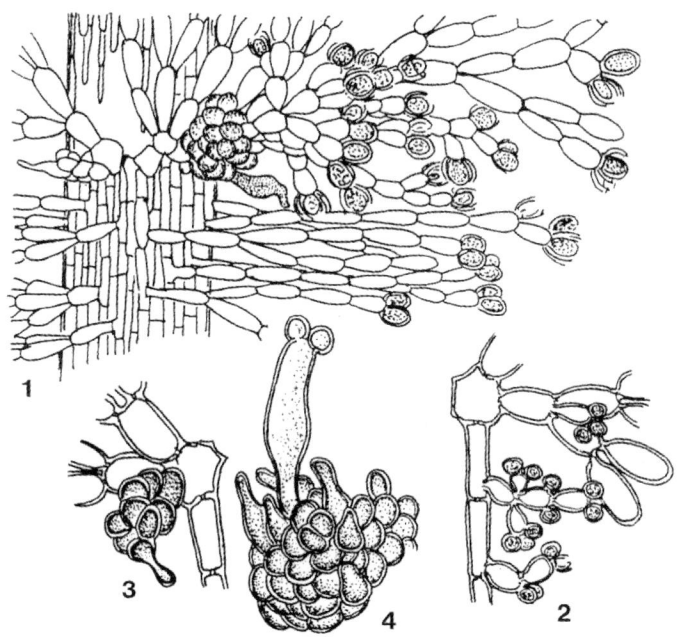

Plate 107 (b)
Batrachospermum intortum, figs 1–4 (original author, Jao 1941)
1. a portion of a plant showing monosporangia terminal on primary and secondary fascicles, a young carpogonium-bearing branch and an immature carposporophyte, 2. spermatangia terminal on primary, secondary fascicles, 3. coiled carpogonium-bearing branches at the very early stages in development, 4. an immature carposporophyte.

2) ***Batrachospermum pseudocarpum*** Reis (1973:146, tab. IV, a–e, tab. V, a–b)

Plants monoecious. 4–5 cm high, at first pyramidal, afterwards in tufts, moss-green, blue-gray when dry, very mucilaginous. Main axis densely branched, with branches patent, inserted here and there at right angles. Lower primary branches abundantly branched and the rest less branched towards the apex. Secondary branches generally simple, long and flagelliform. Adventitious branchlets distributed along principal axis and primary branches. Whorls contiguous and obtruncate. Secondary fascicles numerous immediately under the whorls. Cortical filaments cylindrical. Periaxial cells short, rarely ovoidal, dilated at base, producing 2–4, generally 3 primary fascicles, apparently di- tetrachotomously branched, consisting of cylindrical cells at inner half of whorl, elongate obovoidal or asymmetrical obovoidal or ellipsoidal cells at outer half. Piliferous apices with terminal hairs rare, very short or long, inflated at base. Spermatangia very rare, large and spherical, 10 µm in diameter, borne on peripheral branchlets of whorl. Carpogonium-bearing branches spiral, inserted on periaxial cells. Involucral filaments: shorter, incurved into a ball shape and longer, slightly curved, overtopping the former. Carpogonium asymmetrically truncate. Trichogyne with short stalk, cylindrical. True carposporophytes not observed. Monosporangium-bearing fascicles numerous, pseudo-carposporophytes apparently fixed on axis of whorl, semispherical or spherical, 200–300 µm in diameter, not over half of the radius of whorl, and also at periphery of whorl. Monosporangia obovoidal, 6–10 µm in diameter, 13–20 µm long, distributed on whorls.

Type Locality: Ribeira da Louçainha, Espinhal Mountains, near Penela, Portugal in Europe.
Holotype: COL.
Distribution: Known from the type locality only.

Batrachospermales

3) *Batrachospermum woitapense* Kumano (1983:76, figs 1–18)

plate 108, figs 1–6 (Kumano 1996zo)
figs 7–11 (original author, Kumano 1983)

Plants monoecious, mucilaginous, 3–6 cm high, 300–700 μm in diameter, more or less irregularly branched, olive-green; whorls pear-shaped, frequently touching each other; primary fascicles abundantly branched, 6–12 cell-storeys, proximal cells arcuate club-shaped, 3–5 μm in diameter, 20–40 μm long, distal cells fusiform or ellipsoidal, 3–5 μm in diameter, 10–20 μm long; secondary fascicles 5–7 cell-storeys, terminal hairs rare; spermatangia spherical, 5–7 μm in diameter, terminal or lateral on fascicles; carpogonium-bearing branches undulated, consisting of 4–7 disc- or barrel-shaped cells, arising from periaxial cells; carpogonia 5–8 μm in diameter at the base, 7–10 μm in diameter at the apex, 40–90 μm long, trichogyne club-shaped, cylindrical, indistinctly stalked; carposporophytes single, spherical or hemispherical, 250–700 μm in diameter, 150–700 μm high, inserted centrally; carposporangia ovoidal, 8–10 μm in diameter, 12–20 μm long; monosporangia spherical or ovoidal, 8–10 μm in diameter, 10–15 μm long, terminal on fascicles.

Type Locality: Woitape situated at about 100 km north of Port Moresby and about 1,500 m above sea level, Central District, Papua New Guinea in Oceania. Known from small stream in *Sphagnum* swamp.

Holotype: TNS AL–52622a.

Distribution: Known from the type locality only.

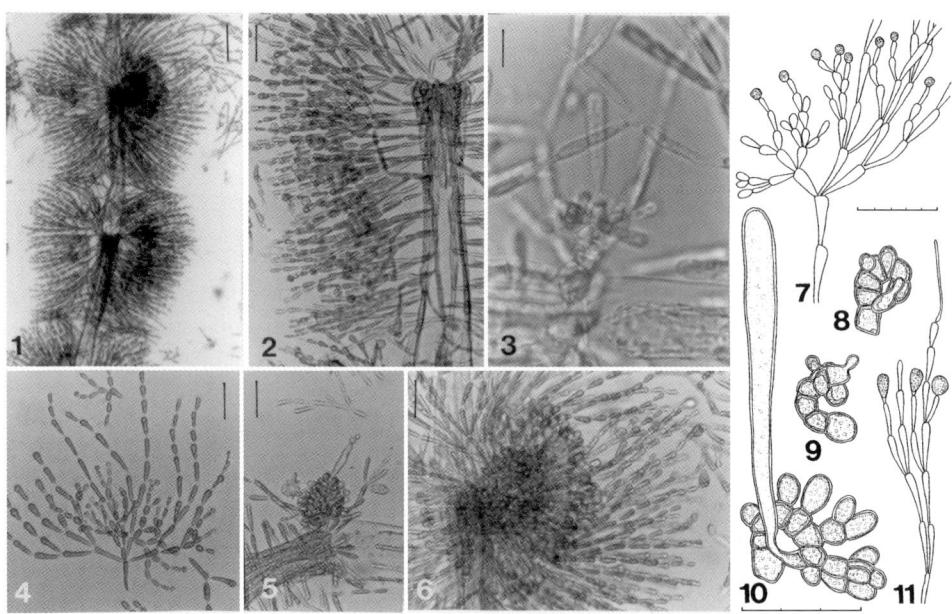

Plate 108
Batrachospermum woitapense, figs 1–6 (Kumano 1996zo), figs 7–11 (original author, Kumano 1983)
1, 6. the structure of whorls showing axial carposporophytes within a subconical whorl, 2. a portion of a whorl showing an axial cell, primary and secondary fascicles and cortical filaments, 3, 10. a strongly curved carpogonium-bearing branch with a mature trichogyne, 4, 7. spermatangia terminal on the laterals of primary fascicles, 5. a young carposporophyte with a fertilized trichogyne, 8–9. the early stages in developing of twisted carpogonium-bearing branches and a carpogonium with trichogyne initials, 10. a carpogonium-bearing branch with a mature trichogyne, 11. monosporangia terminal on primary fascicle.
Scale bars = 100 μm (fig. 1); 50 μm (figs 7, 11); 40 μm (figs 2, 4–6); 30 μm (figs 8–10); 20 μm (fig. 3).

4) ***Batrachospermum lusitanicum*** Reis (1965:141, tab. VI, a–d, tab. V, a–d, tab. VI, a–d, tab. VII, a–d, tab. VIII, a–d, tab. IX, a–d)

Plants dioecious, 3–4 cm high, moss-green, very mucilaginous, caespitulose (fixed to stones by means of rhizoidal callus consisting of descending cortical filaments), individuals initially separate and later confluent, forming tufts. Ramification generally corymbiform, rarely pyramidal, irregular: on same axis whorls alternately, unilaterally, oppositely, cross-oppositely and bi- tetrachotomously branched). Branches inserted here and there at right angles, sometimes unbranched and flagelliform: ultimate branchlets spine-like. Male plants less branched in the main axis than female plants. Whorls indistinct at base, generally distinct and barrel-shaped, rarely contiguous and transversely ellipsoid or spherical elsewhere. Periaxial cells ovoidal, producing 3 or 4 fascicles with apparent di-, tetrachotomously branching and consisting of cells generally cylindrical, sometimes pear-shaped but asymmetrically obovoidal in curved peripheral branchlets. Piliferous apices with terminal hairs short and inflated at base. Secondary fascicles numerous, covering upper part of the internode, and becoming ever shorter, reaching the lower contiguous whorl, at the base of plants equaling the radius of the whorl in length. Spermatangium-bearing branchlets inserted on base of whorls along secondary fascicles. Spermatangia spherical, 6.6–8.3 µm in diameter. Female plants extremely branched in the main axis. Whorls generally contiguous and elongate-discoidal, rarely indistinct at base. Periaxial cells ovoidal, producing 2–4 secondary fascicles, frequently curved and unilaterally branched at periphery. Piliferous apices with terminal hairs short and inflated at base. Cortical filaments abundant at lower part, cylindrical. Carpogonium-bearing branches developed spirally, frequently inserted on periaxial cells, rarely on secondary fascicles and cortical filaments, 6–14 celled. Involucral filaments: shorter, incurved into a ball shape and longer, slightly curved, overtopping the former. Carpogonium cylindrical or asymmetrically conical. Trichogyne with short stalk or sessile, cylindrical or elongate-ovoidal. Carposporophytes 1 or 2 at each whorl, apparently fixed along the axis of plants, spherical, 150–300 µm in diameter. Gonimoblast filaments consisting of cylindrical cells. Carpospores obovoidal, 10–20 µm in diameter, 10–13 µm in diameter, 20–23 µm long. Monosporangia formed on whorls at the base of plants, obovoidal, 13–16 µm in diameter, 20–26.6 µm long.

Type Locality: On rocks in river Alfusqueiro, tributary of river Àgueda, near village Vouzela, Portugal in Europe.

Holotype: COL, P. Reis & A. Santos 333.

Distribution: Known from the type locality only.

8-2. Subsection ***Torrida*** Kumano (1993:264)

Type: *Batrachospermum torridum* Montagne

Plants more or less irregularly branched. Carpogonium-bearing branches more or less curved, arising from periaxial cells. Trichogyne club-shaped, indistinctly stalked. Carpogonia asymmetrical. Carposporophytes hemispherical, big, inserted centrally.

Key to the species of the subsection ***Torrida***

1. Carposporangia up to 46 µm long...1) *B. henriquesianum*
1. Carposporangia up to 20 µm long...2
2. Carpogonium-bearing branch consisting of 2–4 cells...3
2. Carpogonium-bearing branch consisting of 5–11 cells...4
3. Carposporophytes 50–60 µm in diameter..2) *B. tortuosum* v. *turtuosum*

Batrachospermales

3. Carposporophytes 220–300 μm in diameter..3) *B. tortuosum* v. *majus*
4. Trichogyne bent at the base, monosporangia present.......................................4) *B. torridum*
4. Trichogyne not bent at the base...5
5. Carpogonium 30–50 μm long..5) *B. faroense*
5. Carpogonium 25–32 μm long..6) *B. curvatum*

1) ***Batrachospermum henriquesianum*** Reis (1972:181, tab. I, a–d, tab. II, a–c, tab. III, a–d)

Plants monoecious, 1.5–2.5 cm high, in direct light light-green, in shade violet on young plants, dark-violet on old plants, very mucilaginous, solitary, perennial. Ramification irregular. Main axis attenuate towards the base, naked in the lower part, generally disappearing among primary branches and the primary branches disappearing among upper secondary branches. Apices of branches obtuse. Whorls indistinct, inclined forwards along spine-shaped branchlets, becoming confluent at the apices of these. Secondary fascicles equal to the radius of whorl. Cortical filaments very numerous, cylindrical, forming successive layers; inner ones strongly adhering to central axis; outer ones appeared free and undulated. Periaxial cells ovoidal or spherical, producing 1–4 secondary fascicles, pseudodichotomously branched and consisting internally of obtruncate, cylindrical, pear-shaped cells, externally of fusiform, cylindrical, rarely obovoidal cells. Piliferous apices with terminal hairs rare, short, inflated at base. Spermatangia occurring at periphery of primary and secondary fascicles, spherical, 6.6–8.6 μm in diameter. Carpogonium-bearing branch inserted on periaxial cells, curved, 9–15 μm in diameter. Involucral filaments incurved into the form of a ball. Carpogonium asymmetric. Trichogyne with short stalk. Carposporophytes very rare, 1 at each whorl, semi-spherical, fixed on main axis of plants, variable in size, 120–300 μm in diameter. Gonimoblast filaments consisting of obtruncate, cylindrical cells. Carpospores obovoidal or ellipsoidal, 26.7–30 μm in diameter, 40–46 μm long.

Type Locality: In spring Fonte velha in the village Eirol near Aveiro, Portugal in Europe.
Holotype: COL, Reis & Ferreira 566.
Distribution: Known from the type locality only.

2) ***Batrachospermum tortuosum*** Kumano var. ***tortuosum*** (1978:101, fig. 4, A–D)

plate 109, figs 1–4 (Kumano 1996zl)
figs 5–6 (original author, Kumano 1978)

Plants dioecious, slightly mucilaginous, 3–7 cm high, 200–350 μm in diameter, abundantly and more or less irregularly branched, olive-green; whorls spherical or ellipsoidal, more or less separated; primary fascicles abundantly branched, 5–8 cell-storeys, cells lanceolate, ellipsoidal, fusiform, or ovoidal, terminal hairs rare; secondary fascicles numerous, varying in length; spermatangia spherical, 4–5 μm in diameter, terminal on fascicles; carpogonium-bearing branches strongly curved, 10–20 μm long, consisting of 3–4 barrel-shaped cells, arising from periaxial cells; involucral filaments numerous and very short; carpogonia 5 μm in diameter at the base, 4–5 μm in diameter at the apex, 30–35 μm long, trichogyne cylindrical, indistinctly stalked; carposporophytes single, rarely in pairs, spherical or hemispherical, 50–200 μm in diameter, 50–150 μm high, inserted centrally in whorl; carposporangia ellipsoidal or ovoidal, 7 μm in diameter, 8–9 μm long.

Type Locality: Tasek Bera, Fort Iskander, Pahang, Malaysia in Southeast Asia. Occurs on submerged leaves of *Pandanus hericopus* and *Cryptocoryne griffithii* in narrow rivulets covered by swamp forests.
Holotype: Kobe University, Kumano, 12/VII 1971.
Distribution: Known from the type locality only.

8–2. *Torrida*

3) ***Batrachospermum tortuosum*** Kumano var. ***majus*** Kumano (1982a:184, fig. 2, A–L, fig. 4, A–D)

plate 110, figs 1–5 (Kumano, 1996zk)
figs 6–9 (original author, Kumano 1982a)

Plants monoecious, slightly mucilaginous, 4–7 cm high, 330–600 μm in diameter, abundantly and irregularly branched, olive-green; whorls ellipsoidal and distant or touching each other, more or less compressed; primary fascicles abundantly branched, 10–12 cell-storeys, cells pear-shaped or obovoidal, terminal hairs absent; secondary fascicles long, numerous, soon covering all internodes; spermatangia spherical, 5–7 μm in diameter, terminal on fascicles; carpogonium-bearing branches more or less curved, short, consisting of 2–4 disc- or barrel-shaped cells, arising from periaxial cells; involucral filaments numerous and short; carpogonia 8–9 μm in diameter at the base, 6–9 μm in diameter at the apex, 33–60 μm long, trichogyne club-shaped, indistinctly stalked, often bent at the base; carposporophytes single, big, 220–300 μm in diameter, 170–280 μm high, inserted centrally in whorl; carposporangia spherical or ovoidal, 10–16 μm in diameter, 14–19 μm long.

Type Locality: Miyara-gawa River, Ishigaki Island in Okinawa, Japan in Eastern Asia. On gravel in a mountain stream.

Holotype: Kobe University, Kumano, 5/IV 1977.

Distribution: Known from the type locality and in Urauchi-gawa River, Iriomote Island in Okinawa, Japan in Eastern Asia.

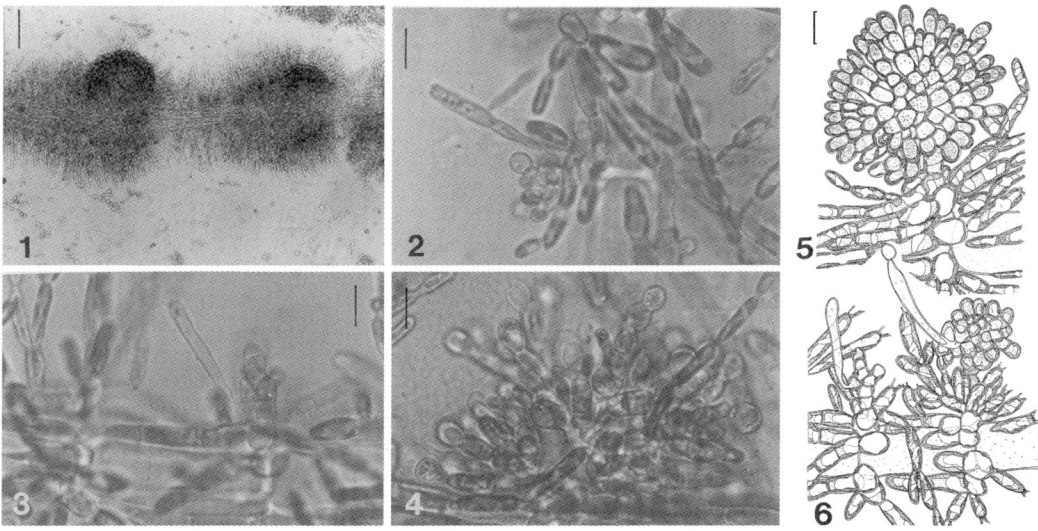

Plate 109
Batrachospermum tortuosum var. *tortuosum*, figs 1–4 (Kumano 1996zl), figs 5–6 (original author, Kumano 1978)
1. a main axis with ellipsoidal whorls showing spherical or hemispherical axial carposporophytes, 2, 3. a carpogonium with an elongate cylindrical trichogyne sometimes bent at the base and a curved carpogonium-bearing branch, 4, 5. a carposporophyte with carposporangia terminal on gonimoblast filaments, 6. a portion of a plant showing a young carpogonium, a fertilized carpogonium with an attached spermatium on a trichogyne and a young carposporophyte.
Scale bars = 100 μm (fig. 1); 10 μm (figs 2–6).

Batrachospermales

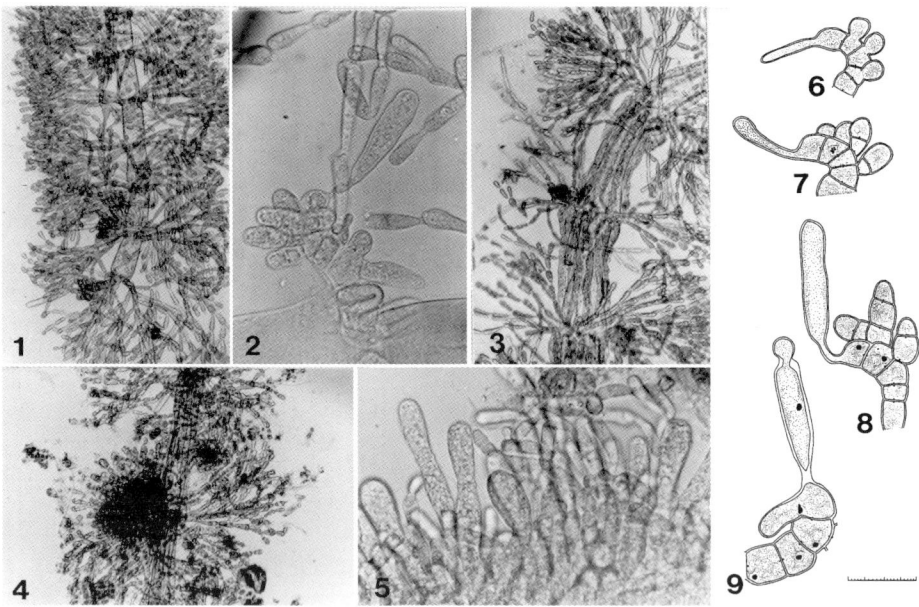

Plate 110
Batrachospermum tortuosum var. *majus*, figs 1–5 (Kumano 1996zk), figs 6–9 (original author, Kumano 1982a)
1, 3. a portion of a plant showing axial cells, primary fascicles, cortical filaments and carpogonia with well-developed club-shaped trichogynes, 2, 6–9. the various stages in development of a curved carpogonium-bearing branch with a long club-shaped trichogyne indistinctly stalked, 4. an axial carposporophyte, 5. elongate clavate carposporangia terminal on gonimoblast filaments.
Scale bars = 100 μm (fig. 4); 40 μm (figs 1, 3); 20 μm (figs 6–9); 10 μm (figs 2, 5).

4) ***Batrachospermum torridum*** Montagne (1850:292)
Synonym: *B. vagum* var. *torridum* (Montagne) Sirodot (1884:266)

plate 111 (a), figs 1–5 (Kumano 1996zj), figs 6–7 (Kumano 1990)

Plants dioecious?, slightly mucilaginous, 4–8 cm high, 400–700 μm in diameter, abundantly and irregularly branched; whorls ellipsoidal and touching each other, confluent; primary fascicles abundantly branched, 6–8 cell-storeys, cells cylindrical, fusiform, terminal hairs absent. Secondary fascicles long, numerous, soon covering all internodes; no spermatangia observed; carpogonium-bearing branches arising from periaxial cells, consisting of 6–9 disc- or barrel-shaped cells, slightly curved; carpogonia about 4 μm in diameter at the base, 9–12 μm in diameter at the apex, 35–40 μm long; trichogyne club-shaped, bent at the base, more or less indistinctly stalked; involucral filaments short; carposporophytes spherical or hemispherical, 170–350 μm high, 300–450 μm in diameter; gonimoblast filaments long, 5–10 cell-storeys radially branched, more or less loosely agglomerated, sometimes terminating in long terminal hair cells; carposporangia inverted pear-shaped or ellipsoidal, 7–10 μm in diameter, 10–13 μm long; monosporangia ellipsoidal, 6–9 μm in diameter, 7–11 μm long, borne as lateral of primary fascicles.
Type Locality: near Tigres Mountains in Cayenne, French Guiana in South America. On rocks in a quiet rivulet.
Holotype: PC herb. Montagne, Le Prieur coll. n° 833.
Distribution: Known from the type locality only.

5) ***Batrachospermum faroense*** Kumano et Bowden-Kerby (1986:123, figs 66–81)
Synonym: *B. doboense* Kumano et Bowden-Kerby (1986:112, figs 13–22)

> plate 112 (a), figs 1–8 (original authors, Kumano & Bowden-Kerby 1986)
> Kumano 1996n as *B. faroense*
> plate 112 (b), figs 1–8 (original authors, Kumano & Bowden-Kerby 1986)
> Kumano 1996l as *B. doboense*

Plants monoecious, mucilaginous, ca. 3.5 cm high, 300–500 μm in diameter, abundantly and irregularly branched, deep green; whorls barrel-shaped, touching each other in old plants; primary fascicles dichotomously branched, 7–10 cell-storeys, proximal cells lanceolate-club-shaped, distal cells pear-shaped or obovoidal, terminal hairs short; cortical filaments well-developed; secondary fascicles dichotomously branched, 5–10 cell-storeys, covering all internodes; spermatangia spherical, 4–5 μm in diameter, terminal or lateral on fascicles; carpogonium-bearing branches undulated, consisting of 5–10 disc- or barrel-shaped cells, arising from periaxial cells; involucral filaments more or less short; carpogonia 4–6 μm in diameter at the base, 5–9 μm in diameter at the apex, 30–40 μm long, trichogyne club-shaped, ellipsoidal, indistinctly stalked, often bent at the base; carposporophytes single, hemispherical, 200–250 μm in diameter, 150–200 μm high, inserted centrally in whorl, distal part of gonimoblast filaments more or less loosely agglomerated; carposporangia ovoidal, 7–11 μm in diameter, 12–15 μm long.

Type Locality: Faro Village, Tol Island, Truk, Caroline Islands, Micronesia in Pacific Ocean. Known from small rocks on the muddy bed of the slowing flowing rivulet from a taro swamp.

Holotype: Kobe University, Bowden-Kerby, 11/V 1982.

Distribution: Known from the type locality and in Dobo Spring, Guam Island in Mariana Islands (as *B. doboense*), Micronesia in Pacific Ocean.

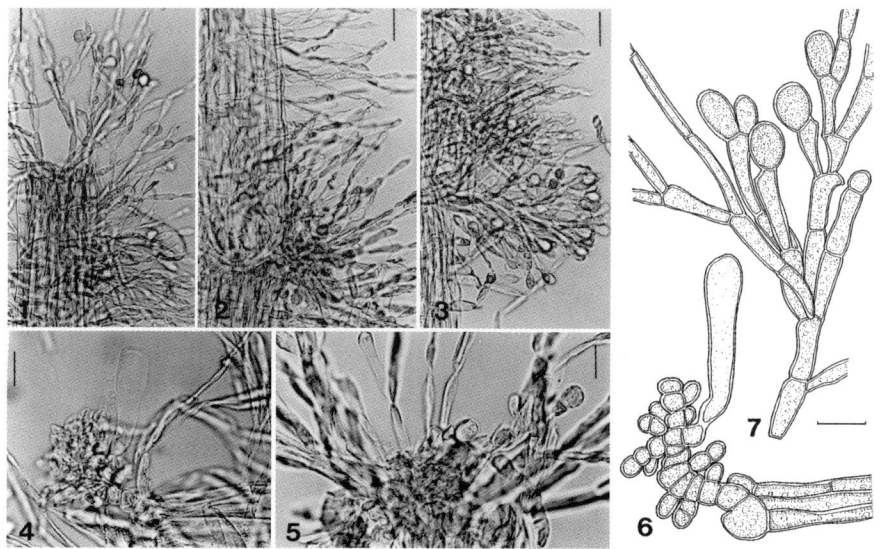

Plate 111 (a)
Batrachospermum torridum, figs 1–5 (Kumano 1996zj), figs 6–7 (Kumano 1990)
1. monosporangia terminal on primary fascicles, 2. a portion of a plant showing an axial cell, primary, secondary fascicles, cortical filaments, 3, 5, 7. carposporangia terminal on loosely agglomerated gonimoblast filaments, sometimes with terminal hairs and indistinguishable from primary fascicles, 4, 6. a carpogonium with a club-shaped trichogyne and a slightly curved carpogonium-bearing branch having many short involucral filaments.   Scale bars = 20 μm (figs 1–3); 10 μm (figs 4–7).

Batrachospermales

6) *Batrachospermum curvatum* Shi (1994:274, fig. 1, 1–4, fig. 2, 1–6)

plate 113, figs 1–10 (original author, Shi 1994)

Plants monoecious, caespitose, 2–4 cm high, mucilaginous, violet-green, densely and irregularly branched. Whorls conspicuous, pear-shaped or obconical, more or less ellipsoidal, generally compactly contiguous, 250–400 μm in diameter. Axial cells cylindrical, slightly constricted in middle, 50–130 μm in diameter, 260–550 μm long. Cortical filaments well-developed. Primary fascicles dichotomously branched, 6–10 cell storey; proximal cells large, obovoidal, 8–13 μm in diameter, 18–25 μm long, distal cells small, ellipsoidal or rounded, 6–10 μm in diameter, 10–15 μm long. secondary fascicles numerous, generally covering all internodes, unbranched or dichotomously branched, 5 or 6 cell-storeys, cells 4–7.5 μm in diameter, 9–20 μm long. Terminal hairs numerous 20–175 μm long. Spermatangia spherical, 4.5–6.5 μm in diameter, terminal on primary and secondary fascicles. Carpogonium-bearing branch conspicuously curved, 20–40 μm long, consisting of 5–7 disc- or barrel-shaped cells, arising from periaxial cells. Carpogonium 6.2–7.5 μm in diameter at base, trichogyne shortly stalked, narrowly ellipsoidal or narrowly ovoidal, with its anterior end usually attenuate and neck-shaped, 6.5–9 μm in diameter, 25–32 μm long. Carposporophytes single, sometimes two, hemispherical, inserted centrally within whorls, 63–175 μm wide, 60–160 μm high. Carposporangia obovoidal or ellipsoidal, 10–14 μm in diameter, 13–18 μm long.

Type Locality: Yuquansi in Dangyang Xian, Hubei, China in Eastern Asia. Known from a spring (Zhenzhuquan).

Holotype: HP 7138.

Distribution: Known from the type locality only.

Note: This species is similar to *B. turtuosum* Kumano in its curved carpogonium-bearing branches, but differs by having whorls pear-shaped or obconical (vs. spherical or ellipsoidal in the latter), terminal hairs numerous and longer (vs. rare in the latter), trichogyne narrowly ellipsoidal or narrowly ovoidal, with a short stalk and with its anterior end usually attenuate and neck-shaped (vs. cylindrical and curved at the base and without stalk in the latter).

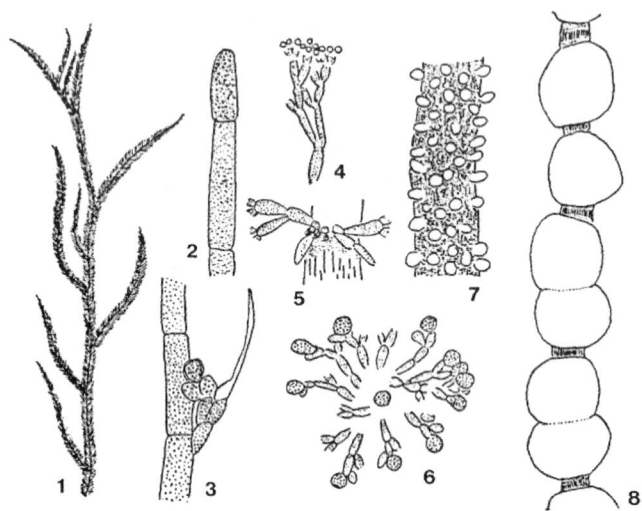

Plate 111 (b)

Batrachospermum louisianae, figs 1–8 (original author, Flint 1949)

1. the growth habit, a single plant, 2–3. a portion of *chantransia* phase, 4. spermatangia terminal on fascicles, 5. a carpogonium, 6. the carposporophyte development, 7. a portion of an old plant with the vegetative reproductive structure, 8. a portion of a plant with carposporophytes.

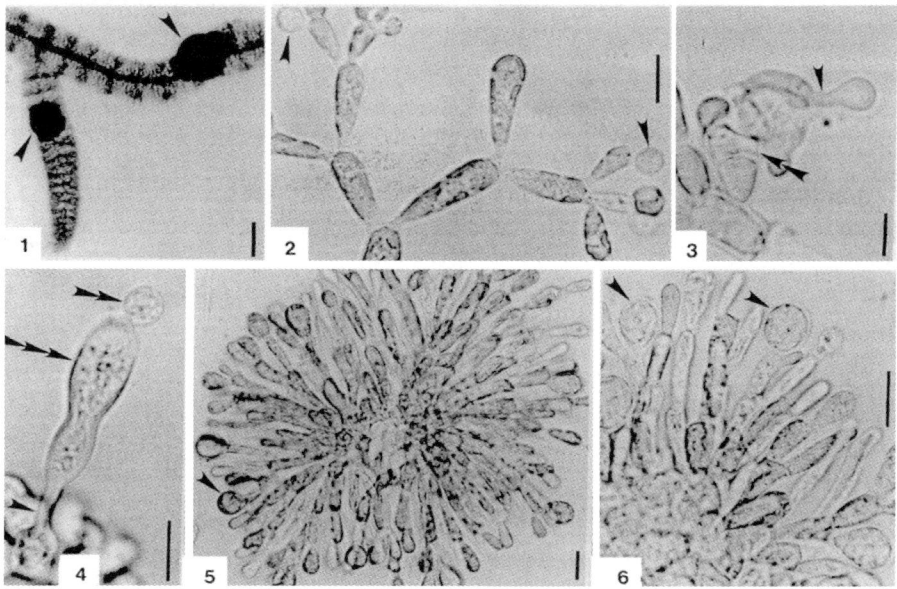

Plate 111 (c)
Batrachospermum louisianae, figs 1–6 (Sheath, Vis & Cole. 1992)
1. a main axis and a branch with compressed whorls and large carposporophytes extending beyond whorls (arrows), 2. a fascicle terminated with colourless spermatangia (arrows), 3. an immature carpogonium with a distinct stalk and a curved carpogonium-bearing branch, a portion of which out of a focus plane (double arrow), 4. a fertilized carpogonium (triple arrow) with an attached spermatium (double arrow) and a distinct stalk (arrow), 5–6. a mature carposporophyte with carposporangia (arrows).
Scale bars = 250 μm (fig. 1); 10 μm (figs 2–6).

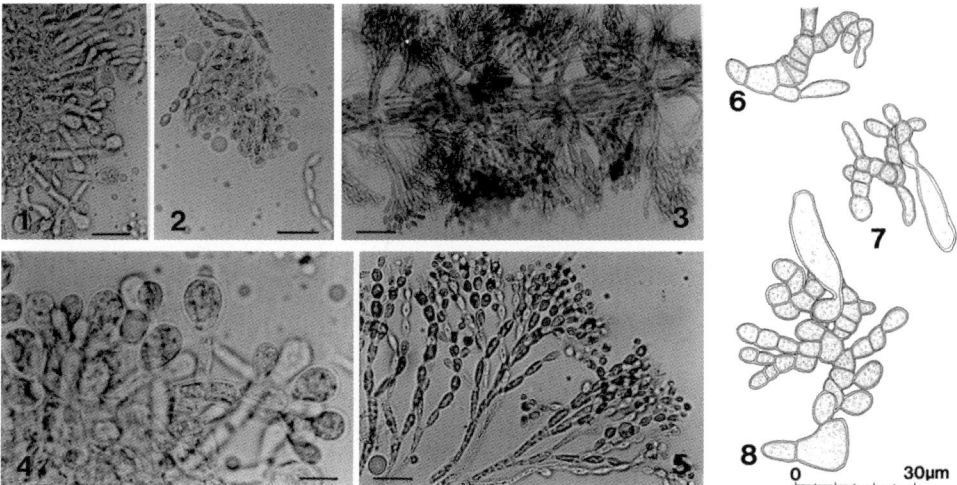

Plate 112 (a)
Batrachospermum faroense, figs 1–8 (original authors, Kumano & Bowden-Kerby 1986, Kumano 1996n)
1, 4. carposporangia terminal on more or less loosely agglomerated gonimoblast filaments, 2. a coiled carpogonium-bearing branch, 3. a portion of a plant showing barrel-shaped whorls, 5. spermatangia terminal and lateral on primary fascicles, 6. the early stage in development of a coiled carpogonium-bearing branch, 7–8. coiled carpogonium-bearing branches with mature trichogynes.
Scale bars = 40 μm (fig. 3); 20 μm (figs 1–2, 5); 10 μm (fig. 4).

Batrachospermales

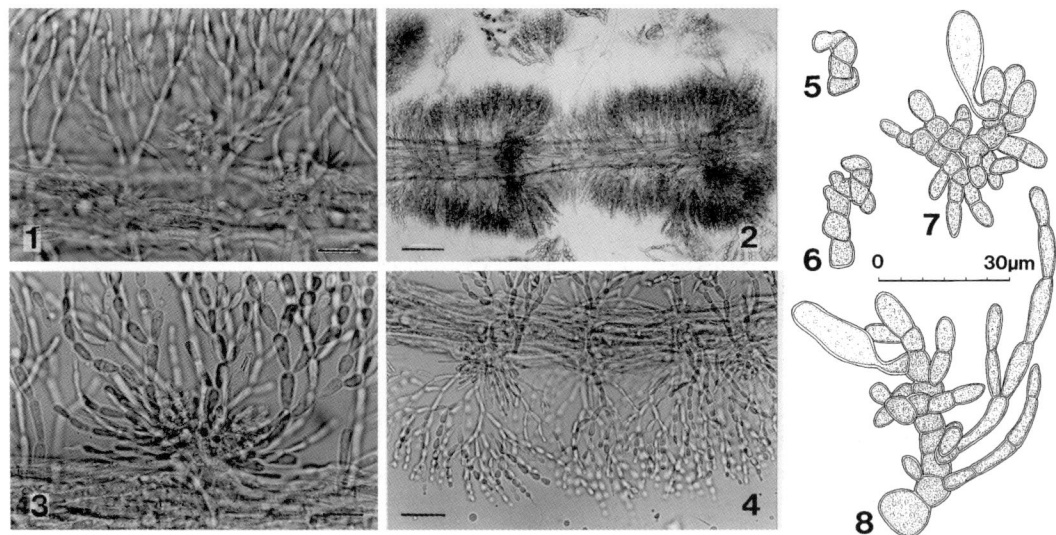

Plate 112 (b)
Batrachospermum faroense as *B. doboense,* figs 1–8 (original authors, Kumano & Bowden-Kerby 1986, Kumano 1996l)
1. a portion of a plant showing cortical filaments, primary fascicles and carpogonia, 2. a portion of a plant showing well-developed cortical filaments and pear-shaped whorls, 3. a carpogonium-bearing branch with a mature carpogonium, 4. a part of a young plant showing well-developed cortical filaments, primary fascicles and two carpogonia, 5–6. curved carpogonium-bearing branches at the very young stages in development, 7–8. curved carpogonium-bearing branches.
Scale bars = 100 µm (fig. 2); 40 µm (fig. 4); 20 µm (figs 1, 3).

8-3. Subsection *Procarpa* Kumano (1993:265)

Type: *Batrachospermum procarpum* Skuja

Plants more or less irregularly branched. Primary fascicles chantransia-like, alternately or unilaterally branched. Carpogonium-bearing branches undulated or spirally coiled, arising from periaxial cells. Trichogyne indistinctly stalked, ellipsoidal or club-shaped. Carposporophytes hemispherical, big, centrally inserted. Gonimoblast filaments long, more or less loosely agglomerated.

Key to the species of the subsection *Procarpa*

1. Carposporophytes 100–300 µm in diameter..2
1. Carposporophytes 300–900 µm in diameter..1) *B. equisetoideum*
2. Carpogonium > 35 µm long...2) *B. procarpum* v. *procarpum*
2. Carpogonium < 35 µm long...3) *B. procarpum* v. *americanum*

8–3. *Procarpa*

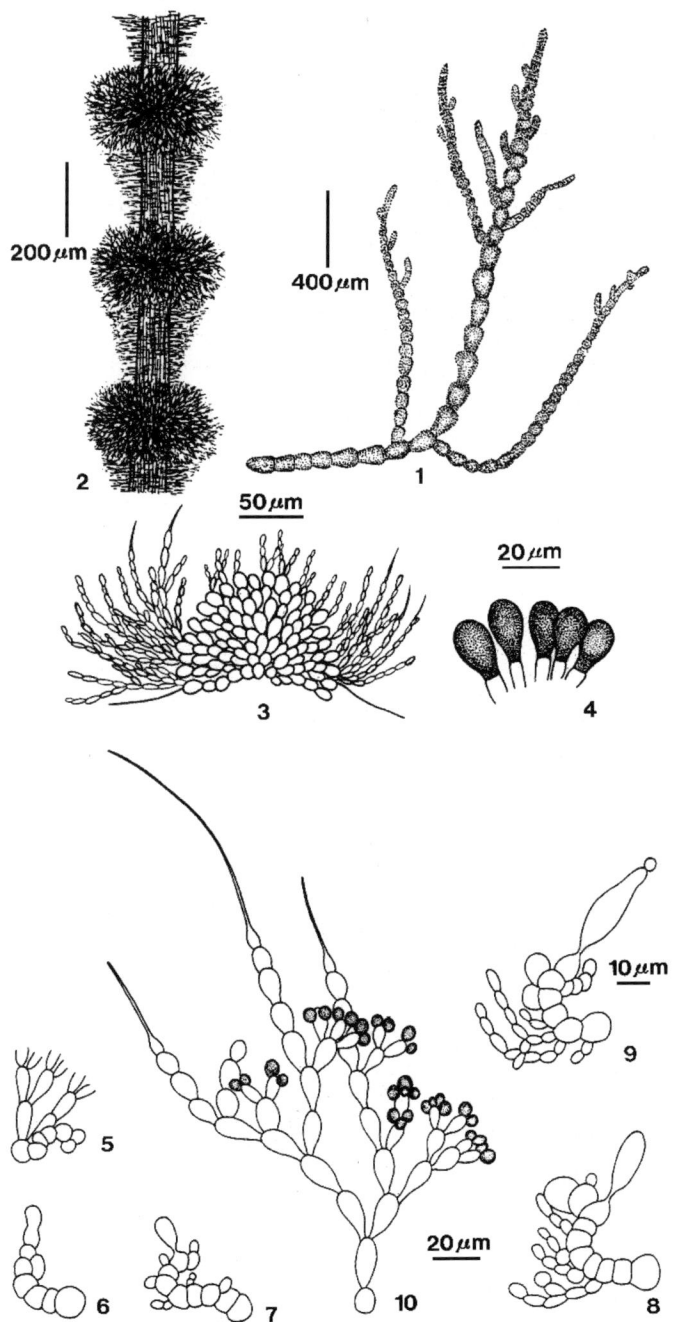

Plate 113
Batrachospermum curvatum, figs 1–10 (original author, Shi 1994)
1. a part of a plant showing pear-shaped or obconical whorls, 2. axial cells, whorls showing primary and secondary fascicles, 3. a carposporophyte, 4. carposporangia terminal on gonimoblast filaments, 5. a young carpogonium-bearing branch at the earlier stage in development, 6–7. a young carpogonium-bearing branch and an immature carpogonium, 8–9. a strongly curved mature carpogonium-bearing branch and a fertilized trichogyne with a attached spermatium, 10. spermatangia terminal or subterminal and terminal hairs on primary fascicle.

Batrachospermales

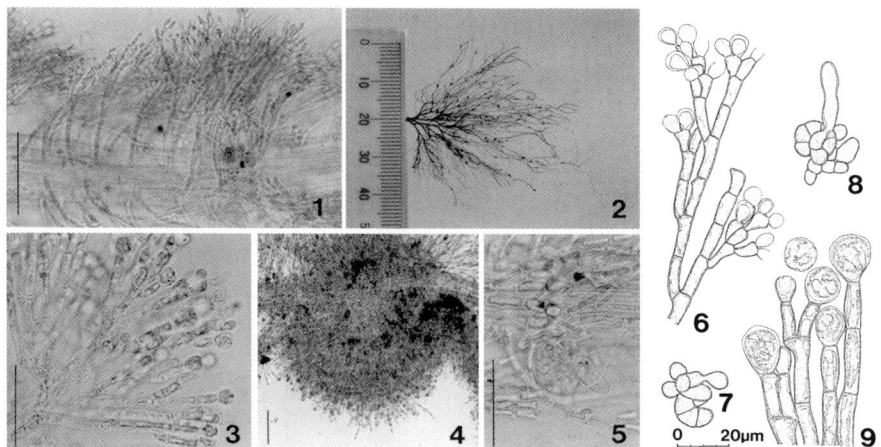

Plate 114 (a)
Batrachospermum globosporum as *B. cipoense,* figs 1–9 (original authors, Kumano & Necchi 1985, Kumano 1996h)
1. a portion of a plant showing well-developed cortical filaments and spermatangia terminal or lateral on primary and secondary fascicles, 2. a plant showing the ramification habit and several carposporophytes, 3. carposporangia terminal on gonimoblast filaments, 4. a hemispherical carposporophyte, 5. a carpogonium with a trichogyne, 6. spermatangia terminal on secondary fascicles, 7. the early stage in development of a coiled carpogonium-bearing branch with a young trichogyne, 8. a carpogonium-bearing branch with a mature trichogyne, 9. carposporangia terminal on gonimoblast filaments. Scale bars =100 μm (figs 1, 4); 50 μm (figs 3, 5); 20 μm (figs 6–9).

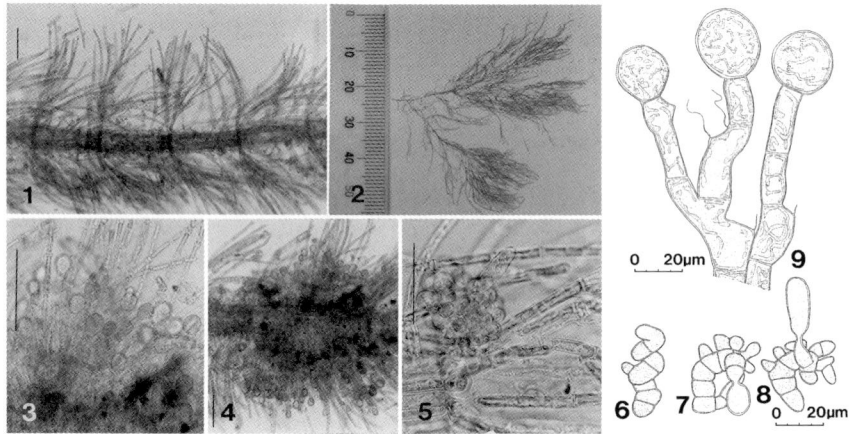

Plate 114 (b)
Batrachospermum equisetoideum, figs 1–9 (original authors, Kumano & Necchi 1985, Kumano 1996m)
1. a portion of a plant showing well-developed cortical filaments and many spermatangia terminal or lateral on primary and rarely secondary fascicles, 2. a plant showing the ramification habit and several carposporophytes, 3. carposporangia terminal on loosely agglomerated gonimoblast filaments, 4. an indefinite-shaped carposporophyte, 5. a portion of a plant showing primary fascicles, cortical filaments and a carpogonium-bearing branch surrounded by rounded involucral filaments, 6. a carpogonium-bearing branch at the very early stage in development, 7. the early stage in development of a twisted carpogonium-bearing branch with a young trichogyne, 8. a carpogonium-bearing branch with a mature trichogyne, 9. carposporangia terminal on gonimoblast filaments containing ribbon-shaped chloroplasts.
Scale bars =100 μm (figs 1, 3–4); 50 μm (figs 5); 20 μm (figs 6–9).

8–3. Procarpa

1) ***Batrachospermum equisetoideum*** Kumano et Necchi (1985:182, figs 1–16)

plate 114 (b), figs 1–9 (original authors, Kumano & Necchi 1985), (Kumano 1996m)

Plants monoecious, slightly mucilaginous, ca. 6 cm high, 300–800 μm in diameter, abundantly and irregularly branched, blackish-purple; whorls equisetum-like, separated or touching each other. Fascicles 1(–2), primary fascicles chantransia-like, curved, unilaterally, alternately or oppositely branched, 7–15 cell-storeys, cells cylindrical, 5–8.5 μm in diameter, 16–48 μm long, terminal hairs absent; cortical filaments well-developed; secondary fascicles rare; spermatangia spherical, obovoidal, 6–9 μm in diameter, lateral or terminal on primary (rarely on secondary) fascicles; carpogonium-bearing branches undulated, consisting of 5–7 disc-, barrel-shaped cells, arising from periaxial cells; involucral filaments numerous, short, consisting of rounded cells; carpogonia 7–8 μm in diameter at the base, 10–13 μm in diameter at the apex, 40–55 μm long, trichogyne ellipsoidal or urn-shaped, distinctly stalked; carposporophytes single, of indefinite shape, more or less diffuse, broader than whorls, 300–800 μm in diameter, gonimoblast filaments loosely agglomerated; carposporangia spherical, ovoid, 15–24 μm in diameter, 19–30 μm long.

Type Locality: Adolfo Ducke Forest Reserve, Município of Manaus, Amazonas, Brazil in South America. Epilithic in a clear, rapidly flowing rivulet in shaded places.

Holotype: SP 187177.

Distribution: Known from type locality in Brazil in South America and Tiria River, between La Laia and Base Camp, Chimantá Massif, Bolivar, Venezuela in Central America.

Note: Necchi (1990:90) included *Thorea trailii* Dickie (1881:123) as a synonym.

2) ***Batrachospermum procarpum*** Skuja var. ***procarpum*** (1931:81, tab. 1, figs 1–13)

plate 115 (a), figs 1–5 (original author, Skuja 1931)

Plants monoecious, slightly or moderately mucilaginous, 2–7 cm high, 150–400 μm in diameter, abundantly and irregularly branched, deep green. Whorls well-developed, lax, obconical or barrel-shaped, contiguous. Fascicles 1–2, primary fascicles chantransia-like, unilaterally branched, 6–10 cell-storeys, proximal cells cylindrical, 4–7 μm in diameter, 15–25 μm long, distal cells cylindrical or barrel-shaped, 4.5–6 μm in diameter, 10–18 μm long, terminal hairs numerous, long. Cortical filaments well-developed. Secondary fascicles numerous, reaching the length of primary ones, covering all internodes. Spermatangia spherical or obovoidal, 5–8.5 μm in diameter, terminal or subterminal on fascicles. Carpogonium-bearing branches spirally coiled, consisting of 6–9 disc- or barrel-shaped cells, arising from periaxial cells. Involucral filaments numerous and short. Carpogonia asymmetrical, 4–7 μm in diameter at the base, 8–11(–13) μm in diameter at the apex, 35–65 μm long, trichogyne cylindrical or club-shaped, sessile. Carposporophytes 1(–3), lax, hemispherical, higher than radius of whorls, 170–330 μm in diameter, 80–170 μm high. Carposporangia ovoidal, 7–10.5 μm in diameter, 10–15 μm long.

Type Locality: Santa Teresa, Espírito Santo, Brazil in South America.

Holotype: UPS, O. Conde, XI 1928.

Distribution: Occur in South America.

Batrachospermales

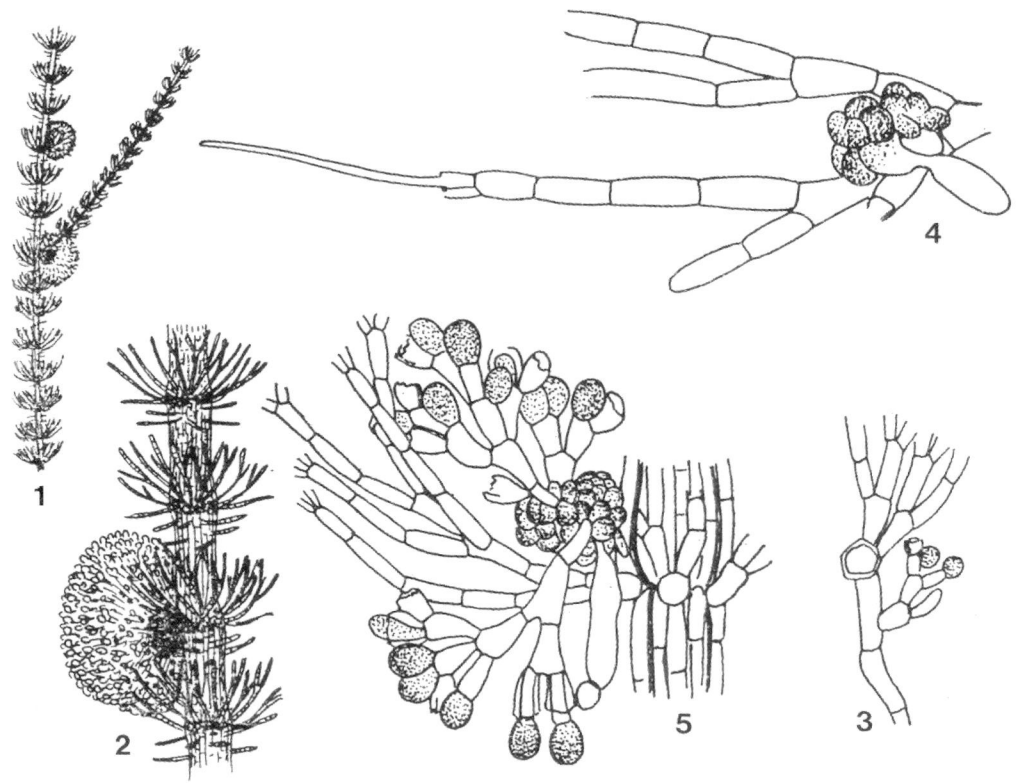

Plate 115 (a)
Batrachospermum procarpum var. *procarpum*, figs 1–5 (original author, Skuja 1931)
1–2. main axis with audouinelloid whorls and hemispherical axial carposporophytes, 3. spermatangia, 4. immature carpogonium with distinct stalk and twisted carpogonium-bearing branch, 5. fertilized carpogonium showing trichogyne with attached spermatium and obovoidal carposporangia terminal on gonimoblast filaments.

3) ***Batrachospermum procarpum*** Skuja var. ***americanum*** Sheath, Vis & Cole (1992:244, figs 27–32)

plate 115 (b), figs 1–6 (original authors, Sheath, Vis & Cole 1992)

Plants monoecious, green with deep blue. Whorls 191–363 µm in diameter, fascicles 5–11 cell-storeys, cells chantransia-like. Carpogonia 4.5–10.5 µm in diameter, 19.2–38.1 µm long, trichogyne cylindrical, stalked. Carposporophytes hemispherical, 89–214 µm in diameter, 56–131 µm high. Carposporangia obovoidal, 5.5–11.0 µm in diameter.
    Type Locality: Brushy Creek at Gunn Road, Citrus Park, Hillsborough, Florida, USA in North America.
    Holotype: UBC A8265.
    Distribution: Six populations from Florida, Louisiana, and Mississippi, USA in North America, have loose indistinct whorls, audouinelloid fascicle cells throughout the whorl, monoecious distribution of apical spermatangia and axial carpogonia, distinctly stalked trichogyne, and dense central carposporophytes contained within whorls. The whorls and carposporophytes size are significantly smaller than those of the type variety.

8-4. Kushiroense

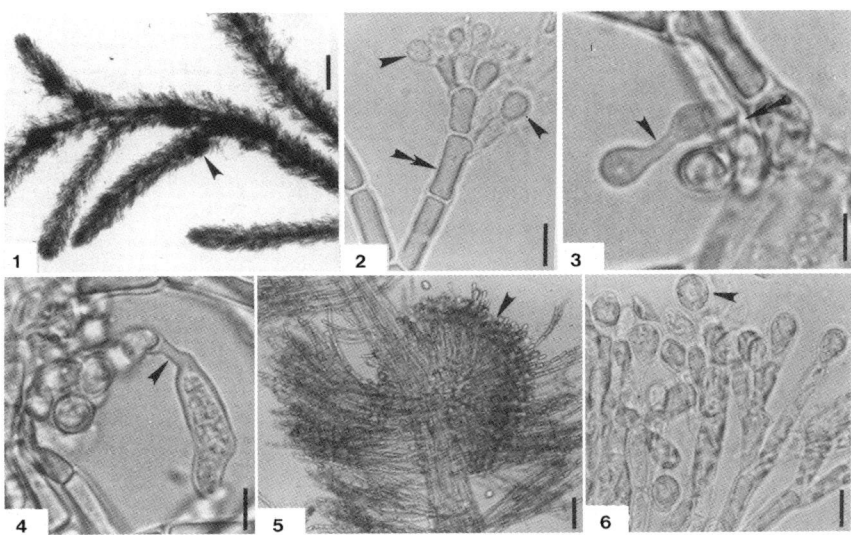

Plate 115 (b)
Batrachospermum procarpum var. *americanum*, figs 1–6 (original authors, Sheath, Vis & Cole 1992)
1. a main axis with branches showing distinct or confluent whorls and large carposporophytes (arrow), 2. a fascicle terminated with colourless spermatangia (arrows) and a audouinelloid fascicle (double arrow), 3. an immature carpogonium with a distinct stalk (arrow) and a twisted carpogonium-bearing branch (double arrow), 4. a mature carpogonium with a distinct stalk (arrow), 5. a dense axial carposporophyte (arrow), 6. Gonimoblast filaments with obovoidal carposporangia (arrow).
Scale bars = 250 µm (fig. 6); 10 µm (figs 7–11).

8-4. Subsection *Kushiroense* Kumano (1993:265)

Type: *Batrachospermum kushiroense* Kumano et Ohsaki

Plants irregularly branched. Carpogonium-bearing branches curved or strongly undulated, arising from periaxial cells. Trichogynes indistinctly stalked, often bent at the base. Carposporophytes spherical or hemispherical, inserted centrally. Gonimoblast filaments long, loosely agglomerated.

Key to the species of the subsection *Kushiroense*

1. Spermatiophore present..1) *B. spermatiophorum*
1. Spermatiophore absent..2
2. Carposporophytes < 300 µm in diameter..3
2. Carposporophyte > 300 µm in diameter..5
3. Carposporophytes < 130 µm in diameter...2) *B. kushiroense*
3. Carposporophytes > 130 µm in diameter...4
4. Carposporangia 12–14 µm long...3) *B. iriomotense*
4. Carposporangia 6.7–10.9 µm long...4) *B. louisianae*
5. Carposporophytes < 500 µm in diameter..6
5. Carposporophytes > 500 µm in diameter...11
6. Primary fascicles < 6 cell-storeys..5) *B. breviarticulatum*
6. Primary fascicles > 7 cell-storeys..7
7. Carpogonium > 50 µm long..6) *B. tabagatense*

Batrachospermales

7. Carpogonium < 50 μm long..8
8. Carpogonium > 30 μm long..9
8. Carpogonium < 30 μm long..10
9. Carposporangia 10–15 μm long..7) *B. guyanense*
9. Carposporangia 15–18 μm long..8) *B. nonocense*
10. Primary fascicles 7–11 cell-storeys..9) *B. globosum*
10. Primary fascicles 10–14 cell-storeys..10) *B. nechochoense*
11. Primary fascicles 8–15 cell-storeys..11) *B. capense*
11. Primary fascicles 13–20 cell-storeys..12) *B. skujanum*

1) ***Batrachospermum spermatiophorum*** Vis et Sheath in Vis, Sheath, Hambrook et Cole (1994:181, figs 3–11)

plate 116, figs 1–9 (original authors, Vis, Sheath, Hambrook & Cole 1994).

Plants monoecious, 548–1207 μm in diameter; whorls barrel-shaped, compressed; fascicles 10–14 cell-storeys; colourless spermatiophores formed at the tip of fascicles. Spermatangia formed directly from cells of fascicles, most spermatangia occurring at the apices of spermatiophores, which range from few-celled and unbranched complexes. Carpogonium-bearing branch curled; with maturation of carpogonium, the trichogyne changing from cylindrical to lanceolate, 6.4–11.4 μm in diameter, 29.3–43.1 μm long. Carposporophytes axial, hemispherical, 262–454 μm in diameter, 183–300 μm high. Gonimoblast filaments loosely agglomerated with 4–8 cells. Carposporangia spherical, 9.3–15.4 μm in diameter, 11.7–16.1 μm high.

Type Locality: tributary to Waiohue Gulch at Poa'aka'a State Wayside and Route 36, Hana District, Maui Island stream no 22, Hawaiian Islands in Pacific Ocean.Holotype: BISH, Herbarium Pacificum 628882.
Isotype: UBC A 80848
Distribution: Known from the type locality in Maui Island in Pacific Ocean.

2) ***Batrachospermum kushiroense*** Kumano et Ohsaki (1983:153, figs 1–22)

plate 117 (a), figs 1–4 (Kumano 1996y)
figs 5–7 (original authors, Kumano & Ohsaki 1983)

Plants monoecious, very mucilaginous, ca. 4.5 cm high, 300–350 μm in diameter, abundantly and irregularly branched, bluish green; whorls ellipsoidal, distant from each other, sometimes touching and compressed; primary fascicles abundantly branched, 7–12 cell-storeys, cells cylindrical or fusiform, 2–8 μm in diameter, 10–12 μm long, terminal hairs absent; secondary fascicles sparse; spermatangia rare, spherical, 4–7 μm in diameter, terminal on fascicles; carpogonium-bearing branches short, highly undulated, consisting of 3–7 disc-shaped to barrel-shaped cells, arising from periaxial cells; involucral filaments sparse, very short; carpogonia asymmetrical, 4–6 μm in diameter at the base, 4–7 μm in diameter at the apex, 17–34 μm long, trichogyne urn-shaped, indistinctly stalked; carposporophytes single or in pairs, spherical or semi-spherical, 80–190 μm in diameter, 40–130 μm high, inserted centrally, gonimoblast filaments loosely agglomerated; carposporangia spherical or ovoidal, 7–9 μm in diameter, 7–11 μm long.

Type Locality: Kushiro River, Kushiro Moor in Hokkaido, Japan in Eastern Asia. Occurs on submerged aquatic vascular plants and mollusks in stagnant pools in an oxbow pond.
Holotype: SAP 043462.
Distribution: Occurs the type locality only.

8–4. *Kushiroense*

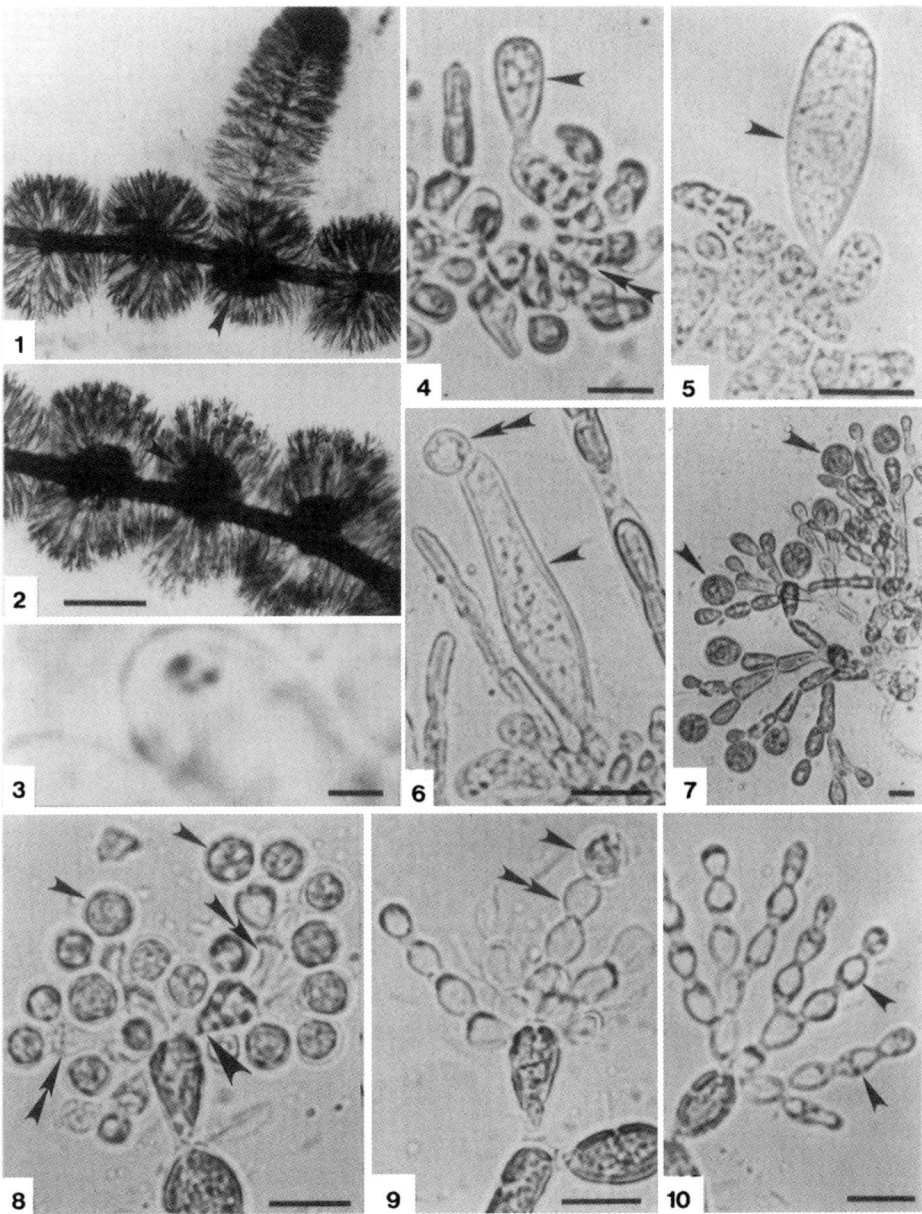

Plate 116
Batrachospermum spermatiophorum, figs 1–9 (original authors, Vis, Sheath, Hambrook & Cole 1994)
1. two branches with barrel-shaped, compressed mature whorls containing hemispherical, axial carposporophyte (arrowhead), 2. haematoxylin-stained fascicle cell showing n=3 chromosomes (double arrowhead), 3. immature carpogonium with inflated trichogyne (arrowhead) on curled carpogonium-bearing branch (double arrowhead), 4. developing carpogonium with cylindrical trichogyne (arrowhead), 5. mature carpogonium with lanceolate trichogyne (arrowhead) and attached spermatium (double arrowhead), 6. carposporophyte with numerous carposporangia (arrowhead) terminal on loosely agglomerated gonimoblast filaments, 7. spermatangia (arrowheads) attached directly to fascicle cells (large arrowhead) or to small colourless spermatiophores (double arrowheads), 8, spermatium (arrowhead) terminal on three-celled, colourless spermatiophore (double arrowhead), 9. well-developed cluster of colourless spermatiophores (arrowheads). Scale bars = 500 µm (fig. 1); 10 µm (figs 3–9); 2 µm (fig. 2).

Batrachospermales

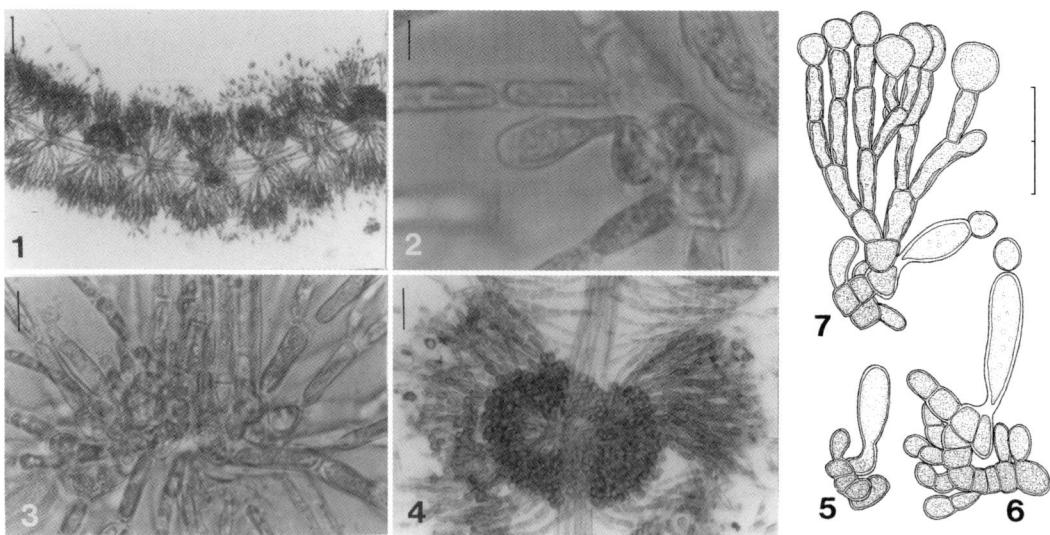

Plate 117 (a)
Batrachospermum kushiroense, figs 1–4 (, Kumano 1996y), figs 5–7 (original authors, Kumano & Ohsaki 1983)
1. the structure of whorls showing axial cells, primary fascicles, cortical filaments and axial carposporophytes, 2, 5. a young spirally coiled carpogonium-bearing branch, 3, 6. the early stage in development of gonimoblast filaments, 4. a portion of a plant showing two hemispherical carposporophytes, 7. carposporangia terminated on loosely agglomerated gonimoblast filaments.
Scale bars = 100 μm (fig. 1); 40 μm (fig. 4); 20 μm (figs 5–7); 10 μm (fig. 3); 5 μm (fig. 2).

3) **Batrachospermum iriomotense** Kumano (1982a:182, fig. 1, A–M, fig. 3, A–D)

plate 117 (b), figs 1–5 (Kumano 1996x)
figs 6–11 (original author, Kumano 1982a)

Plants monoecious, not very mucilaginous, 4–5 cm high, 150–240 μm in diameter, more or less abundantly and irregularly branched, reddish brown; whorls rounded pear-shaped, very frequently touching each other; primary fascicles abundantly branched, 8–10 cell-storeys, cells lanceolate-ellipsoidal or fusiform, terminal hairs rare; secondary fascicles numerous, covering all internodes; spermatangia spherical, 3–7 μm in diameter, terminal on fascicles; carpogonium-bearing branches very undulated, long, consisting of 8–12 barrel-shaped cells, arising from periaxial cells; involucral filaments numerous, short; carpogonia 5 μm in diameter at the base, 6–8 μm in diameter at the apex, 26–40 μm long, trichogyne club-shaped, indistinctly stalked, often bent at the base; carposporophytes single, hemispherical, big, 100–220 μm in diameter, 70–130 μm high, inserted centrally; carposporangia ellipsoidal or ovoidal, 12–14 μm in diameter, 16–19 μm long.
Type Locality: Sira-gawa River, Iriomote Island in Okinawa, Japan in Eastern Asia. Occurs on rocks and stones in a small mountain stream.
Holotype: Kobe University, Matsumoto, 23/III 1974.
Distribution: Known from the type locality and Kedah Peak, Kedah, Malaysia in Southeast Asia.

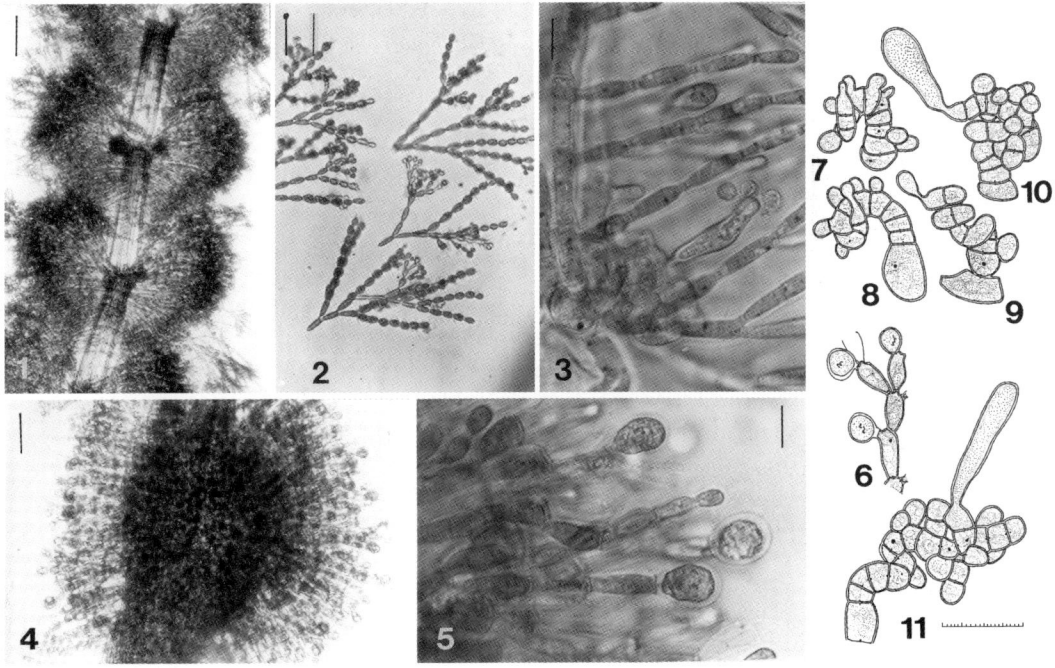

Plate 117 (b)
Batrachospermum iriomotense, figs 1–5 (Kumano 1996x), figs 6–11 (original author, Kumano 1982a)
1. a main axis with barrel-shaped whorls, 2, 6. spermatangia terminal or lateral on primary fascicles, 7–8. a terminal portion of a carpogonium swelling out, 9–11. a club-shaped trichogyne formed terminally on a carpogonium-bearing branch twisting twice or thrice, 3. a fertilized trichogyne with an attached spermatium, 4. an axial carposporophyte, 5. carposporangia terminal on gonimoblast filaments.
Scale bars = 40 μm (fig. 1); 20 μm (figs 2, 4, 7–11); 10 μm (figs 3, 5).

4) ***Batrachospermum louisianae*** Skuja in Flint (1949:549, figs 1–8)

        plate 111 (b). figs 1–8 (original author, Flint 1949)
        plate 111 (c). figs 1–6 (Sheath, Vis & Cole 1992)

Plants monoecious, up to 5 cm high, comprising bright blue-green, finely-textured tufts, centrally dark to black, resolving upon drying mostly into fan-shaped segments. In larger plants, these subordinate segments showing sub-dichotomous basal branching. Plants bearing a strong resemblance to *Sirodotia*, because carpogonia are sessile at the base of whorl elements. Carposporophytes flattened basely, axes undulated, with outline or silhouette of plants suggesting asymmetric development. Carposporophytes developing at adjoining nodes sometimes becoming confluent. Trichogynes small, simple, club-shaped structures. Spermatangia small, borne terminally on unmodified primary fascicles; also in small numbers on secondary fascicles. Vegetative monospores sometimes developing in profusion on old denuded stems.

    Type Locality: Mandeville, Louisiana, USA in North America
    Lectotype: US 56080, LA 101.
    Distribution: Known from the type locality only.

Batrachospermales

5) ***Batrachospermum breviarticulatum*** (Necchi et Kumano) Necchi (1990:69, figs 84, 86–94)
Basionym: *B. capense* Starmach ex Necchi et Kumano var. *breviarticulatum* Necchi et Kumano (1984:349, figs 2–11)

plate 118 (a), figs 3, 8–9 (Necchi 1988)
figs 1–2, 4–7 (original authors, Necchi & Kumano 1984)
as *B. capense* var. *breviarticulatum* Necchi et Kumano

Plants monoecious or dioecious, moderately mucilaginous, 3–7 cm high, 300–500 µm in diameter, abundantly and irregularly branched. Whorls reduced, dense, obconical or compressed, contiguous. Fascicles 2 or 3, primary fascicles straight, di-, rarely tri-chotomously branched, 4–6(–7) cell-storeys, proximal cells ellipsoidal or fusiform, 4–11 µm in diameter, 20–45 µm long, distal cells ellipsoidal or fusiform, 4.5–8 µm in diameter, 10–25 µm long, terminal hairs numerous, long or short. Cortical filaments well-developed. Secondary fascicles numerous, on all internodes, reaching the length of primary ones. Spermatangia spherical, 6–7 µm in diameter, terminal or subterminal on fascicles. Carpogonium-bearing branches spirally coiled, consisting of 5–8 barrel-shaped cells, arising from periaxial cells. Involucral filaments numerous, short. Carpogonia asymmetrical 4–6 µm in diameter at the base, 7–9 µm in diameter at the apex, 40–85 µm long, trichogyne cylindrical, club-shaped, sessile. Carposporophytes sessile, single, hemispherical, lax, higher than radius of whorls, 190–270 µm in diameter, 350–660 µm high. Carposporangia obovoidal or hemispherical, 8–12 µm in diameter, 11–16 µm long.

Type Locality: Itabaiana Mountains, Areia Branca, Sergipe, Brazil in South America.
Holotype: SP 187102.
Distribution: Known from the type locality and Sergipe, Matto Grosso and São Paulo, Brazil in South America.

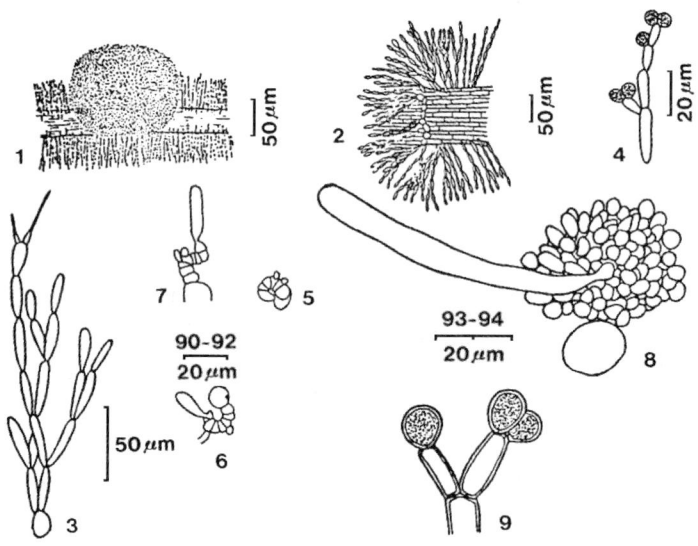

Plate 118 (a)
Batrachospermum breviarticulatum as *B. capense* var. *breviarticulatum*, figs 1–2, 4–7 (original authors, Necchi & Kumano 1984), figs 3, 8–9 (Necchi 1988)
1. the general view of whorls showing a carposporophyte, 2. details of a whorl showing cortical filaments, primary and secondary fascicles, 3. details of a primary fascicle, 4. distal cells of a primary fascicle with spermatangia, 5. a young carpogonium-bearing branch, 6–7. young carpogonia with developing trichogynes, 8. a mature carpogonium showing involucral filaments surrounding a carpogonium-bearing branch, 9. gonimoblast filaments with carposporangia.

6) *Batrachospermum tabagatense* Kumano et Bowden-Kerby (1986:117, figs 39–52)

plate 118 (b), figs 1–8 (original authors, Kumano & Bowden-Kerby 1986), (Kumano 1996zg)

Plants monoecious, very mucilaginous, ca. 3 cm high, 350–550 μm in diameter, sparsely and pseudodichotomously branched, grayish green; whorls cylindrical, touching each other; primary fascicles pseudodichotomously branched, 9–13 cell-storeys, cells lanceolate, club shaped, terminal hairs rare; cortical filaments well-developed; secondary fascicles numerous, dichotomously branched, 8–11 cell-storeys, covering all internodes, reaching the length of primary fascicles; spermatangia spherical, 4–5 μm in diameter, terminal or lateral on fascicles; carpogonium-bearing branches short, spirally coiled, consisting of 6–13 barrel-shaped cells, arising from periaxial cells; involucral filaments sparse, short; carpogonia 3–6 μm in diameter at the base, 8–10 μm in diameter at the apex, 50–65 μm long, trichogyne club-shaped, indistinctly stalked; carposporophytes single or in pairs, spherical or hemispherical, 180–300 μm in diameter, 130–250 μm high, inserted centrally, gonimoblast filaments more or less loosely agglomerated; carposporangia spherical or ovoidal, 10–14 μm in diameter, 12–16 μm long.

Type Locality: Tabagaten River, Nekking, Babeldaob Island, Palau, Western Caroline Islands, Micronesia in Pacific Ocean. Attached to rocks and free roots in a small rivulet of gentle current, arising from a leaf-clogged spring.

Holotype: Kobe University, Bowden-Kerby 10/V 1984.

Distribution: Known from the type locality and a seep-fed spring in Ibobang, Babeldaob Island, Palau, Western Caroline Islands in Pacific Ocean.

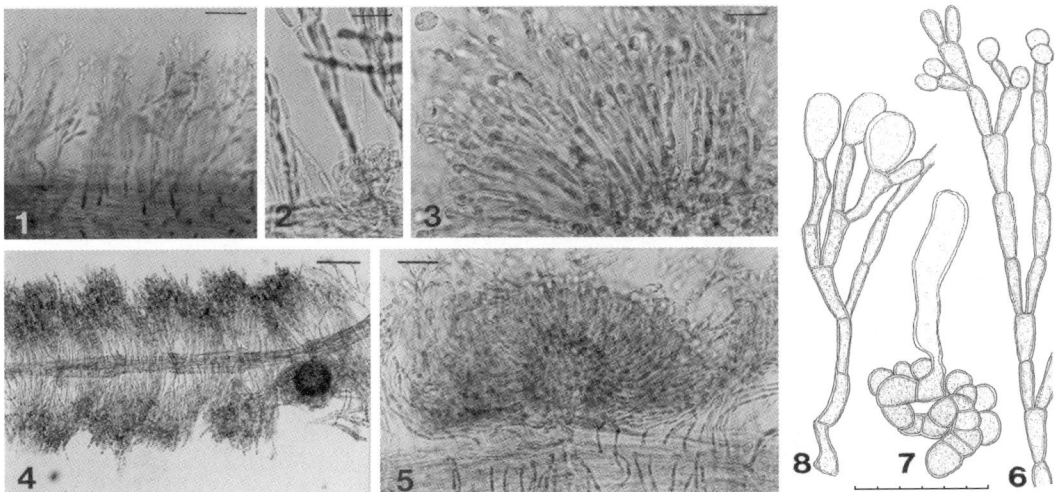

Plate 118 (b)
Batrachospermum tabagatense, figs 1–8 (original authors, Kumano & Bowden-Kerby 1986, Kumano 1996zg)
1. spermatangia terminal or lateral on secondary fascicles, 2. a carpogonium-bearing branch with a mature trichogyne, 3. carposporangia terminal on more or less loosely agglomerated gonimoblast filaments, 4. a portion of a plant showing cylindrical whorls and a carposporophyte, 5. a hemispherical carposporophyte, 6. spermatangia terminal or lateral on secondary fascicles, 7. a carpogonium-bearing branch with a mature trichogyne, 8. carposporangia terminal on more or less loosely agglomerated gonimoblast filaments.
Scale bars = 100 μm (fig. 4); 40 μm (figs 1, 5); 30 μm (figs 6–8); 20 μm (figs 2–3).

Batrachospermales

7) ***Batrachospermum guyanense*** (Montagne) Kumano (1990:284, figs 6–12)
Basionym: *B. vagum* var. *guyanense* Montagne (1850:295)

plate 119 (a), figs 1–4 (Kumano 1996q), figs 5–7 (Kumano 1990)

Plants monoecious, very mucilaginous, 9–10 cm high, ca. 500 μm in diameter, sparsely and irregularly branched; whorls cylindrical, touching each other; primary fascicles pseudo-dichotomously branched, 10–13 cell-storeys, cells lanceolate, club-shaped, terminal hairs short or long; cortical filaments very much developed; secondary fascicles rare, covering all internodes, reaching the length of primary fascicles in old plants; spermatangia spherical, 4–6 μm in diameter, terminal or subterminal on fascicles and on laterals around carpogonium-bearing branches; carpogonium-bearing branches undulated to spirally coiled, consisting of 6–11 disc-, barrel-shaped cells, arising from periaxial cells; involucral filaments numerous; carpogonia 7 μm in diameter at the base, 9–12 μm in diameter at the apex, 35–45 μm long, trichogyne club-shaped, more or less indistinctly stalked; carposporophytes spherical, 150–220 μm in diameter, inserted centrally, gonimoblast filaments long, consisting of 5–8 cylindrical cells, radially branched, loosely agglomerated; carposporangia spherical or ellipsoidal, 9–12 μm in diameter, 10–15 μm long, terminal on laterals of gonimoblast filaments.

Type Locality: near Cayenne, French Guiana in South America. In running freshwater.
Holotype: PC herb. Montagne, Le Prieur coll n° 1108.
Distribution: Known from the type locality only.

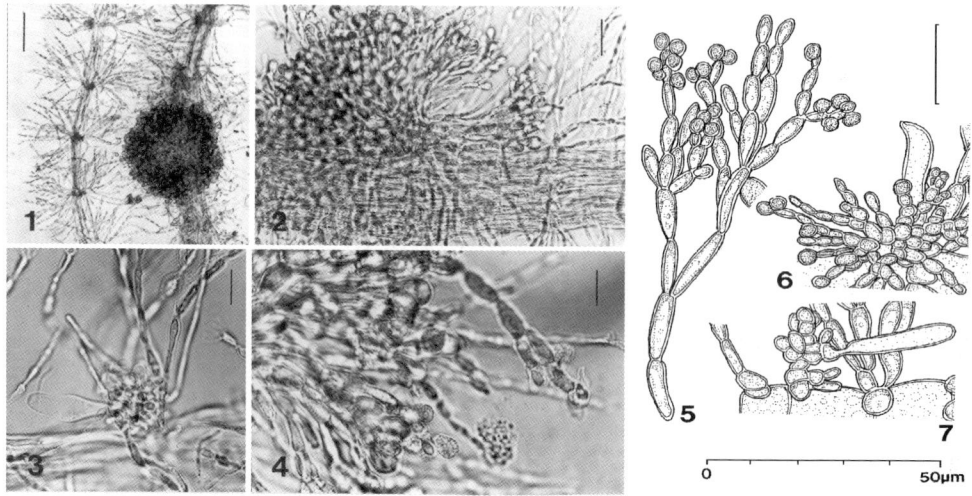

Plate 119 (a)
Batrachospermum guyanense, figs 1–4 (Kumano 1996q), figs 5–7 (Kumano 1990)
1. a main axis and whorls showing a carposporophyte centrally inserted, 2, 6. a carposporophyte and carposporangia terminal on gonimoblast filaments, 3, 7. strongly curved carpogonium-bearing branches, 4–5. spermatangia terminal on primary fascicles.
Scale bars = 100 μm (fig. 1); 40 μm (fig. 2); 20 μm (figs 4–7); 10 μm (fig. 3).

8) ***Batrachospermum nonocense*** Kumano et Liao (1987:101, figs 1–19)

plate 119 (b), figs 1–8 (original authors, Kumano & Liao 1987, Kumano 1996zd)

Plants monoecious, moderately mucilaginous, ca. 7 cm high, 300–370 μm in diameter, more or less dichotomously branched, bluish green; whorls pear-shaped or obconical in old plants;

primary fascicles di- or trichotomously branched, 6–10 cell-storeys, proximal cells club-shaped, 3–6 µm in diameter, 9–17 µm long, distal cells fusiform or ellipsoidal, 5–8 µm in diameter, 9–13 µm long, terminal hairs varying in length; cortical filaments well-developed; secondary fascicles 4–10 cell-storeys, covering all internodes; spermatangia spherical, 5–7 µm in diameter, terminal or lateral on fascicles; carpogonium-bearing branches short, spirally coiled, consisting of 9–13 barrel-shaped cells, arising from periaxial cells; involucral filaments numerous, very short; carpogonia 5–7 µm in diameter at the base, 5–8 µm in diameter at the apex, 25–40 µm long, trichogyne ellipsoidal, club-shaped, more or less indistinctly stalked; carposporophytes single, indistinguishable from whorls, equal in length, 150–250 µm high, gonimoblast filaments long, consisting of 6–10 fusiform cells, radially branched and loosely agglomerated; carposporangia spherical or ellipsoidal, 13–18 µm in diameter, 165–18 µm long.

Type Locality: Nonoc Island, near Surigao, Mindanao, the Philippines in Southeast Asia. Known from an unnamed stream near an open pit nickel mine.

Holotype: Kobe University, Largo & Liao, 29/III 1984.

Distribution: Known from the type locality only.

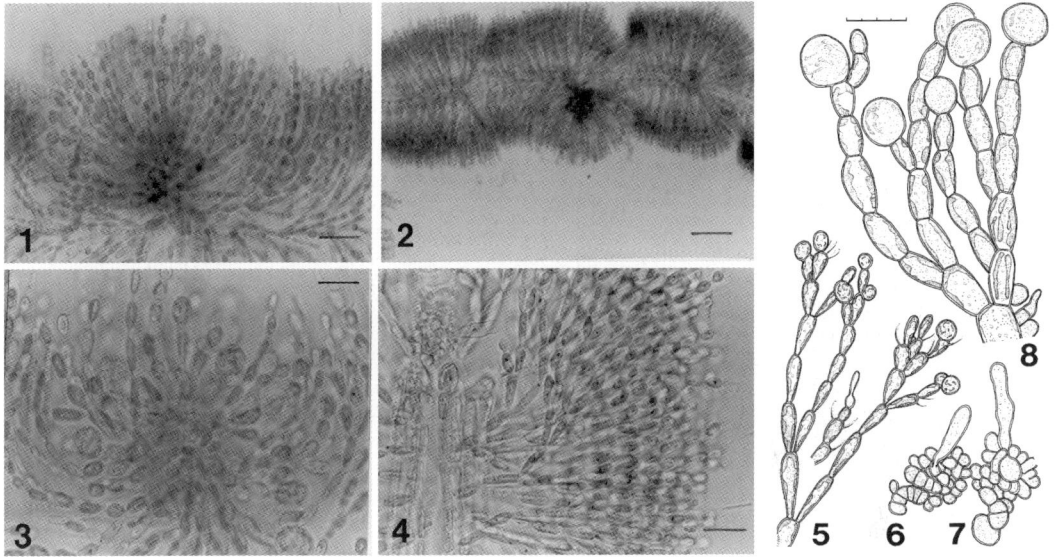

Plate 119 (b)
Batrachospermum nonocense, figs 1–8 (original authors, Kumano & Liao 1987, Kumano 1996zd)
1. loosely agglomerated gonimoblast filaments hardly distinguishable from primary fascicles, 2. a portion of a plant showing obconical whorls with a carposporophyte indistinguishable from the fascicle, 3. carposporangia terminal on loosely agglomerated gonimoblast filaments as long as primary fascicles, 4. a portion of a whorl showing a carpogonium-bearing branch with a mature trichogyne, cortical filaments and secondary fascicles, 5. spermatangia terminal or lateral on secondary fascicles, 6–7. spirally twisted carpogonium-bearing branches with mature trichogynes, 8. carposporangia terminal on gonimoblast filaments, consisting of fusiform cells, containing parietal chloroplasts.
Scale bars = 100 µm (fig. 2); 40 µm (fig. 1); 20 µm (figs 3–4, 5–8).

9) ***Batrachospermum globosporum*** Israelson (1942:44, pl. I, c–h, pl. II, b, d)
The synonymy was proposed by Sheath, Vis & Cole (1992).
Synonym: *B. cipoense* Kumano et Necchi (1985:183, figs 17–33); *B. jolyi* Necchi (1986:520, figs 13–20)

Batrachospermales

plate 114 (a), figs 1–9 (original authors, Kumano & Necchi 1985), Kumano 1996h as *B. cipoense*,
plate 120 (a), figs 1–6 (Sheath, Vis & Cole 1992)
plate 120 (b), figs 1–4 (original author, Israelson 1942)

Plants monoecious, moderately mucilaginous, 3–5 cm high, 350–500 μm in diameter, abundantly and irregularly branched. Whorls well-developed, obconical or compressed, contiguous. Fascicles 2 or 3, primary fascicles straight, dichotomously branched, 7–11 cell-storeys, proximal cells ellipsoidal or fusiform, 4.5–8 μm in diameter, 20–35 μm long, distal cells ellipsoidal or fusiform, 3.5–7 μm in diameter, 15–25 μm long, terminal hairs numerous, long. Secondary fascicles numerous, reaching the length of whorls. Spermatangia spherical or obovoidal, 6–7.5 μm in diameter, terminal or subterminal on fascicles. Carpogonium-bearing branches spirally coiled, consisting of 4–6 barrel-shaped cells, arising from periaxial cells. Involucral filaments numerous, short. Carpogonia asymmetrical 4–6 μm in diameter at the base, 4.5–7 μm in diameter at the apex, 20–30 μm long, trichogyne cylindrical, ellipsoidal, club-shaped, sessile. Carposporophytes sessile, single, hemispherical, lax, lower than whorls, 100–140 μm in diameter, 200–280 μm high. Carposporangia ovoidal or hemispherical, 7.5–10 μm in diameter, 9.5–13 μm long.

Type Locality: Lake Åsgarn, Folkärna, Dalarna, Sweden in Europe.
Holotype: UPS, (A–00174) 23438, G. Lohammar, 5/X 1941.
Distribution: Known from Europe; Paradise Road, Brisbane, Queensland in Australia; Adolfo Ducke Forest Reserve, Manaus, Amazonas (as *B. jolyi*), Cipo Mountains, Minas Gerais (as *B. cipoense*), Brazil in South America.

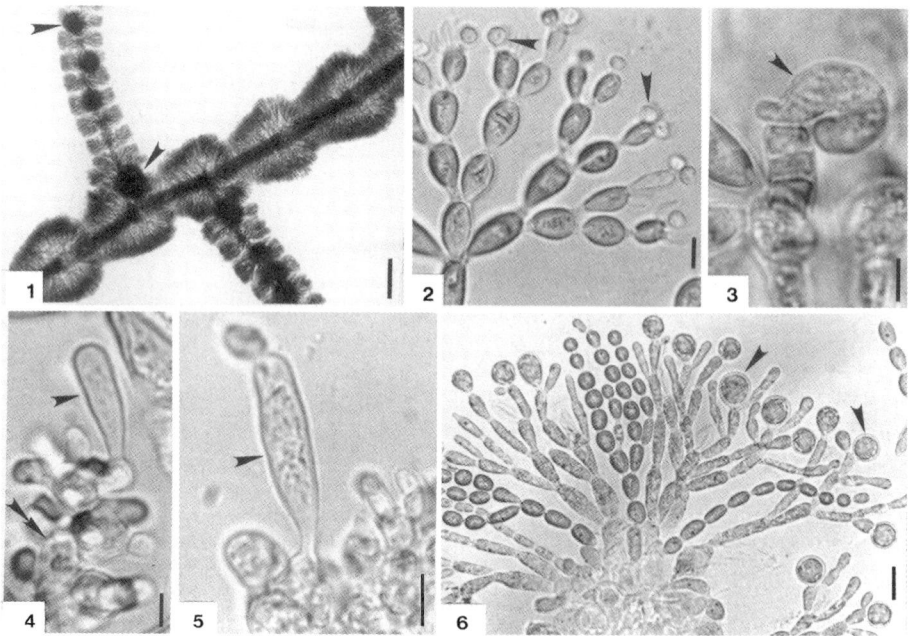

Plate 120 (a)
Batrachospermum globosporum, figs 1–6 (Sheath, Vis & Cole 1992)
1. a main axis and a branch showing barrel-shaped to obovoidal whorls and internal carposporophyte (arrowheads), 2. fascicle tips with colourless spermatangia (arrowheads), 3. an immature carpogonium-bearing branch showing curling (arrowhead), 4. an immature carpogonium (arrowhead) without a stalk and a highly twisted carpogonium-bearing branch (double arrowhead), 5. a fertilized carpogonium with slightly stalked trichogyne (arrowhead), 6. a mature carposporophyte with spherical carposporangia (arrowheads). Scale bars = 200 μm (fig. 1); 10 μm (figs 2–6).

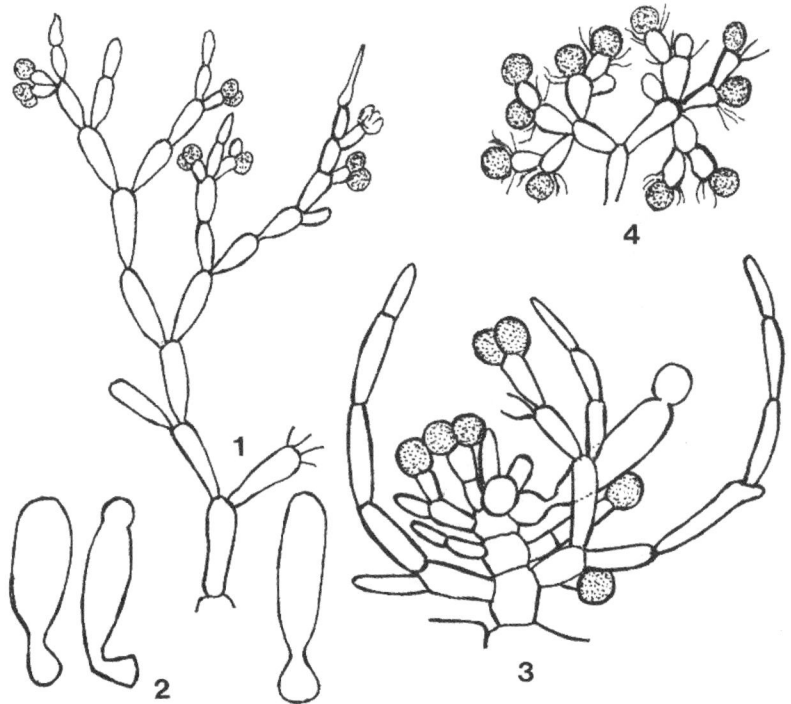

Plate 120 (b)
Batrachospermum globosporum, figs 1–4 (original author, Israelson 1942)
1. spermatangia terminal or subterminal on fascicles, 2. a carpogonium with a stalked trichogyne, 3. a curved carpogonium-bearing branch and spermatangia terminal on involucral filaments, 4. spherical carposporangia terminal on gonimoblast filaments.

10) ***Batrachospermum nechochoense*** Kumano et Bowden-Kerby (1986:120, figs 53–65)

plate 121 (a), figs 1–9 (original authors, Kumano & Bowden-Kerby 1986), Kumano 1996zb

Plants monoecious, mucilaginous, ca. 2 cm high, 350–550 μm in diameter, abundantly and irregularly branched, grayish green; whorls barrel-shaped, touching each other; primary fascicles di-, tri-, rarely tetra-chotomously branched, 11–14 cell-storeys, proximal cells lanceolate, club-shaped, distal cells obovoidal or pear-shaped, terminal hairs short; cortical filaments well-developed; secondary fascicles well-developed, dichotomously branched, 8–11 cell-storeys; spermatangia spherical, 5–7 μm in diameter, terminal or lateral on fascicles; carpogonium-bearing branches short, spirally coiled, consisting of 7–14 barrel-shaped cells, arising from periaxial cells; involucral filaments short; carpogonia 5–6 μm in diameter at the base, 7–12 μm in diameter at the apex, 25–30 μm long, trichogyne club- or urn-shaped, more or less indistinctly stalked; carposporophytes single, hemispherical, 150–220 μm in diameter, 140–180 μm high, inserted centrally, gonimoblast filaments loosely agglomerated; carposporangia spherical or ovoidal, 7–8 μm in diameter, 10–16 μm long.

Type Locality: Nechocho, Tol Island, Truk, Caroline Islands, Micronesia in Pacific Ocean. Known from a small spring-fed stream.

Holotype: Kobe University, Borden-Kerby, 14/III 1982.

Distribution: Known from the type locality and in Wichen River, Moen Island, Truk, Caroline Islands, Micronesia in Pacific Ocean.

Batrachospermales

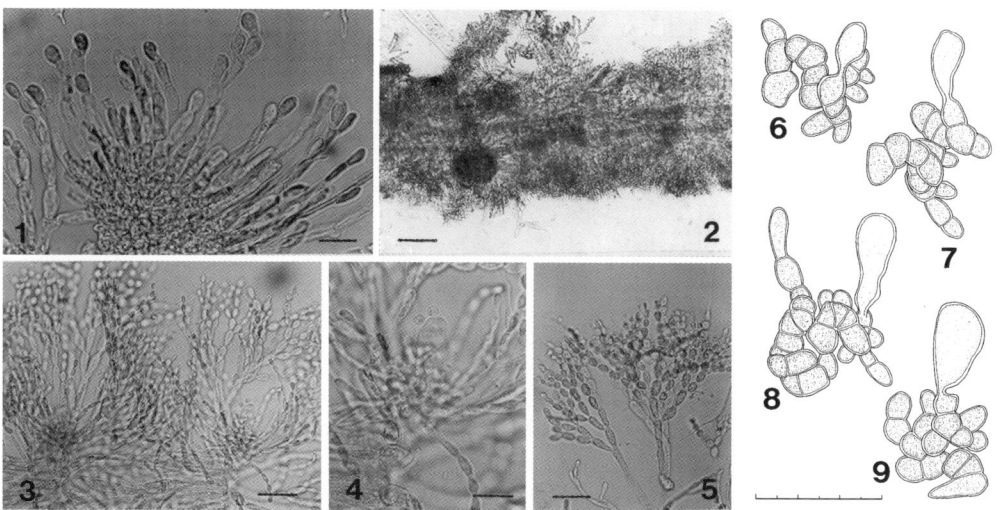

Plate 121 (a)
Batrachospermum nechochoense, figs 1–9 (original authors, Kumano & Bowden-Kerby 1986, Kumano 1996zb)
1. carposporangia terminal on loosely agglomerated gonimoblast filaments, 2. a portion of a plant showing barrel-shaped or cylindrical whorls, 3. a portion of a plant showing axial cells, cortical filaments, primary and secondary fascicles and a carpogonium-bearing branch, 4. a carpogonium-bearing branch with a mature trichogyne, 5. primary fascicles di- or trichotomously branched with terminal heirs, 6–9. coiled carpogonium-bearing branches with trichogynes.
Scale bars = 100 µm (fig. 2); 40 µm (figs 3, 5); 30 µm (figs 6–9); 20 µm (figs 1, 4).

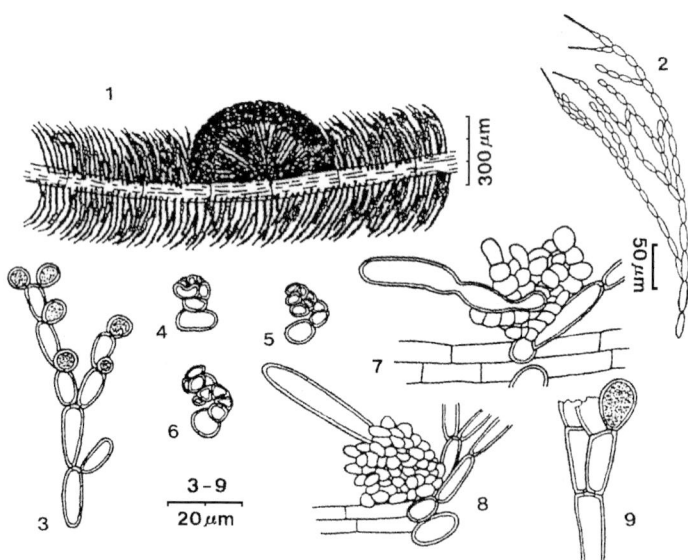

Plate 121 (b)
Batrachospermum skujanum, figs 1–9 (original author, Necchi 1990)
1. the general view of whorls with a carposporophyte, 2. details of a primary fascicle, 3. distal cells of primary fascicle with spermatangia, 4–6. the early stages in development of carpogonium-bearing branches, 7, 8. mature carpogonia showing trichogynes, involucral filaments surrounding carpogonium-bearing branches, pericentral and proximal cells, 9. gonimoblast filaments with carposporangia.

11) ***Batrachospermum capense*** Starmach ex Necchi et Kumano (1984:349)

Synonym: *B. capensis* Starmach (1975:210, pl. 4, a–e, pl. 5, figs 1–4, pl. 6, figs 5–10), *nom. illeg.*

plate 122, figs 1–7 (original author, Starmach 1976) as *B. capensis*

Plants dioecious, moderately mucilaginous, 3–8 cm high, 450–1000 µm in diameter, irregularly branched. Whorls well-developed, dense or lax, obconical or compressed, contiguous. Fascicles 2 or 3, primary fascicles straight, di-, rarely tri-chotomously branched, 8–15 cell-storeys, proximal cells ellipsoidal or fusiform, 4.5–8 µm in diameter, 17–35 µm long, distal cells ellipsoidal or fusiform, 3.5–6 µm in diameter, 5–10 µm long; cortical filaments well-developed. Secondary fascicles numerous, on all internodes, reaching the length of primary ones. Spermatangia spherical or obovoidal, 6–7 µm in diameter, terminal or subterminal on fascicles. Carpogonium-bearing branches spirally coiled, consisting of 6–9 barrel-shaped cells, arising from periaxial cells. Involucral filaments numerous, short. Carpogonia 5–7 µm in diameter at the base, 7–9 µm in diameter at the apex, 60–80 µm long, trichogyne cylindrical, club-shaped, sessile. Carposporophytes sessile, single, hemispherical, lax, higher than radius of whorls, 150–400 µm in diameter, 350–660 µm high. Carposporangia obovoidal or hemispherical, 8–11 µm in diameter, 10.5–16 µm long.

Type Locality: Du Cap stream, Mahé Island, Seychelles Archipelago in Indian Ocean.
Lectotype: KRA, J. Rzoska 1972, designated by Necchi & Kumano 1984.
Isotyoe: SP 187186.
Distribution: Known from the type locality and Le Nial stream on Mahé, Seychelles Archipelago in Indian Ocean; Manaus, Adolfo Ducke Forest Reserve, Amazonas, and Cipo Mountains, Minas Gerais, Brazil in South America.

12) ***Batrachospermum skujanum*** Necchi (1986:519, figs 2–12)

plate 121 (b), figs 1–9 (original author, Necchi 1990)

Plants monoecious, abundantly or moderately mucilaginous, 3–11 cm high, 500–1200 µm in diameter, abundantly and irregularly branched. Whorls well-developed, dense or lax, barrel-shaped, obconical or compressed, contiguous. Fascicles 2 or 3, primary fascicles straight, di-, rarely trichotomously branched, 13–20 cell-storeys, proximal cells cylindrical or ellipsoidal, 5–7 µm in diameter, 20–35 µm long, distal cells ellipsoidal, 4.5–8 µm in diameter, 10–20 µm long. Cortical filaments well-developed. Secondary fascicles numerous, reaching the length of primary ones. Spermatangia spherical or obovoidal, 5.5–7 µm in diameter, terminal or subterminal on fascicles. Carpogonium-bearing branches spirally coiled, consisting of 8–10 barrel-shaped cells, arising from periaxial cells. Involucral filaments numerous, short. Carpogonia asymmetrical, 5–7 µm in diameter at the base, 7–9 µm in diameter at the apex, 40–55 µm long, trichogyne cylindrical, club-shaped, sessile. Carposporophytes sessile, single, hemispherical, dense, higher than radius of whorls, 200–500 µm in diameter, 500–900 µm high. Carposporangia obovoidal or hemispherical, 8.5–12 µm in diameter, 10–15 µm long.

Type Locality: Evangelista de Aouza, Parenheiros, São Paulo, Brazil in South America.
Holotype: SP 176790.
Distribution: Known from the type locality and several localities in São Paulo, Brazil in South America.

Batrachospermales

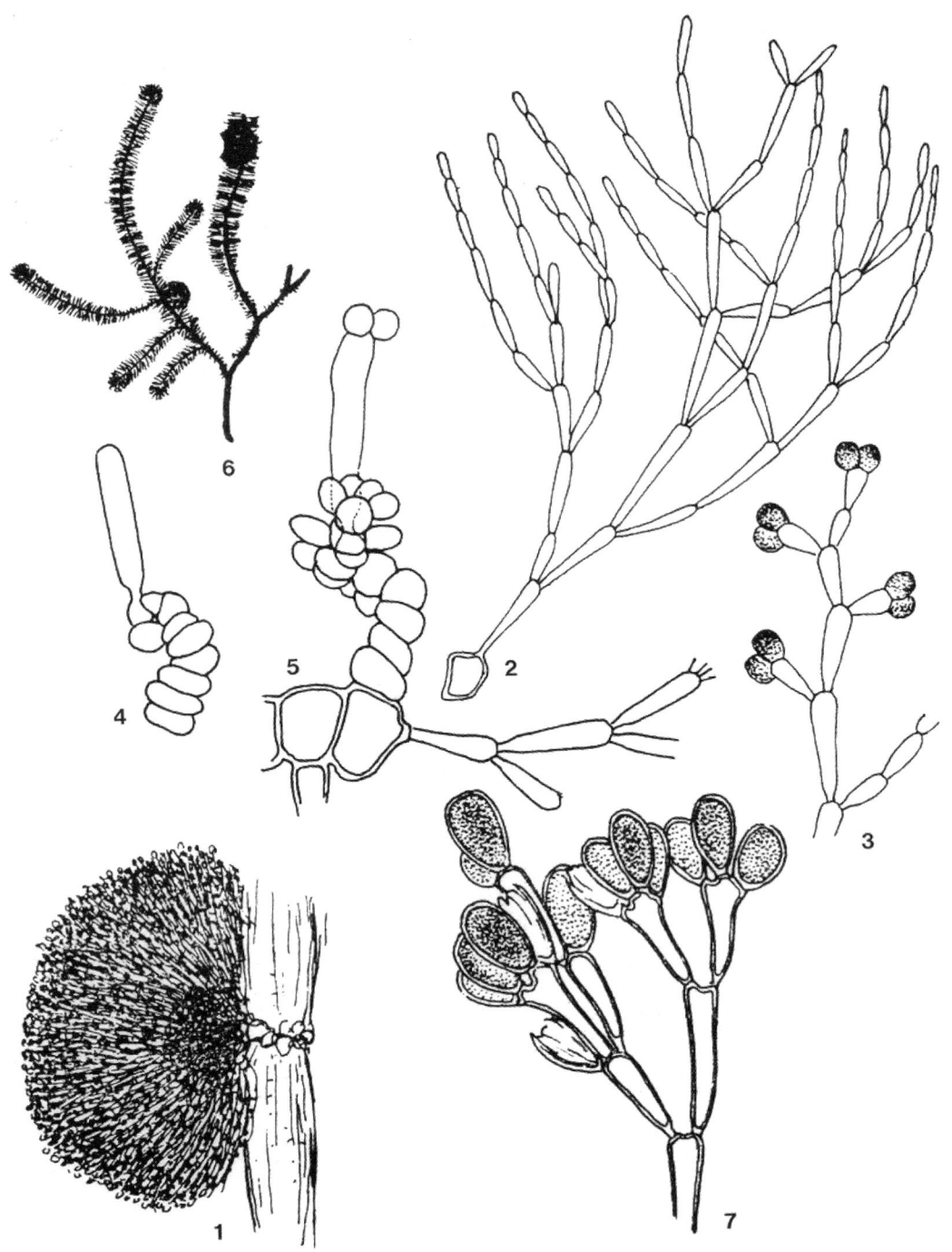

Plate 122
Batrachospermum capense as *B. capensis*, figs 1–7 (original author, Starmach 1976)
1. a hemispherical carposporophyte, 2. a primary fascicle, 3. spermatangia terminal on fascicles, 4–5. spirally twisted carpogonium-bearing branches, 6. a main axis and a branch showing large spherical or hemispherical axial carposporophytes, 7. carposporangia terminal on gonimoblast filaments.

8-5. Subsection *Ambigua* Kumano (1993:266)

Type: *Batrachospermum ambiguum* Montagne

Plants irregularly branched. Carpogonium-bearing branches strongly undulated. Trichogyne indistinctly stalked, often bent at the base. Carposporophytes spherical or hemispherical, inserted centrally. Gonimoblast filaments compactly agglomerated.

Key to the species of the subsection *Ambigua*

1. Primary fascicles pseudoparenchymatous..2
1. Primary fascicles not pseudoparenchymatous..3
2. Carpogonium-bearing branch 3–10 cells...1) *B. gibberosum*
2. Carpogonium-bearing branch 10–18 cells...2) *B. deminutum*
3. Spermatangia on middle of fascicles...3) *B. vittatum*
3. Spermatangia on terminal of fascicles...4
4. Primary fascicles up to 10 cell-storeys..5
4. Primary fascicles up to 15 cell-storeys..12
5. Carposporophytes up to 200 µm in diameter..6
5. Carposporophytes up to 500 µm in diameter..7
6. Trichogyne spathular-shaped...4) *B. hirosei*
6. Trichogyne ellipsoidal..8
7. Carpogonium-bearing branch 14–29 cells..5) *B. australicum*
7. Carpogonium-bearing branch 5–15 cells...6) *B. mahlacense*
8. Carpogonium < 25 µm long...7) *B. dasyphillum*
8. Carpogonium > 25 µm long...9
9. Trichogyne ellipsoidal...8) *B. nodiflorum*
9. Trichogyne cylindrical..10
10. Trichogyne up to 85 µm long...9) *B. gracillimum*
10. Trichogyne up to 50 µm long...11
11. Carposporangia < 14 µm long..10) *B. torsivum*
11. Carposporangia > 15 µm long..11) *B. iyengarii*
12. Carposporophyte < 200 µm in diameter...12) *B. tiomanense*
12. Carposporophyte > 200 µm in diameter...13
13. Trichogyne < 23 µm long...14
13. Trichogyne > 22 µm long..16
14. Trichogyne lanceolate..13) *B. zeylanicum*
14. Trichogyne ellipsoidal...15
15. Trichogyne 8–16 µm long..14) *B. kylinii*
15. Trichogyne 15–18 µm long.......................................15) *B. mahabaleshwarensis*
16. Carposporangia 8–11 µm long..16) *B. omobodense*
16. Carposporangia 10–17 µm long..17) *B. ambiguum*

1) ***Batrachospermum gibberosum*** (Kumano) Kumano (1986:24, figs 1–23)
Basionym: *Tuomeya gibberosa* Kumano (1978:105, figs 7, A–E).

plate 123 (a), figs 1–5 (Kumano 1996o)
figs 6–10 (original author, Kumano 1978) as *Tuomeya gibberosa*

Batrachospermales

Plants monoecious, mucilaginous, 1–4 cm high, 50–100 µm in diameter, sparsely and irregularly branched, dark green; whorls not recognizable, confluent; primary fascicles sparsely, sometimes dichotomously branched, consisting of 3–5 cells, proximal cells cylindrical, barrel-shaped, short, 5–10 µm in diameter, 6–8 µm long, distal cells ovoidal, terminal hairs rare, varying length; cortical filaments extremely numerous, pseudo-parenchymatously developed; secondary fascicles numerous, very short, sparsely, sometimes dichotomously branched, 2 or 3 cell-storeys, covering all internodes; spermatangia spherical, 4–5 µm in diameter, terminal on fascicles; carpogonium-bearing branches more or less undulated, coiled, short, 7–10 µm long, consisting of 3–10 disc-shaped cells, arising from periaxial cells; cell walls of carpogonium-bearing branches very thick, pit-connections between these cells clearly recognized; involucral filaments numerous, short; carpogonia somewhat asymmetrical, ca. 4 µm in diameter, 25–30 µm long, trichogyne cylindrical, stalked; carposporophytes single, swelling or wart-like, 100–140 µm in diameter, 70–100 µm high, protruding from whorls. Gonimoblast filaments radially branched, more or less compactly agglomerated; carposporangia spherical or ellipsoidal, 7–10 µm in diameter, 10–13 µm long.

Type Locality: Sungai Maron Kanan, Pasoh Forest Reserve, Negeri Sembilan, Malaysia in Southeast Asia. Growing on submerged woods and roots of vascular plants in a small rivulet running through a virgin forest.

Holotype: Kobe University, Kumano, 12/IX 1971.

Distribution: Known from the type locality and in Sungai Air Terjin, Pulau Langkawi, Malaysia in Southeast Asia.

2) ***Batrachospermum deminutum*** Entwisle et Foard (1999c: 627, fig. 6)

plate 124, figs 1–6 (original authors, Entwisle & Foard 1999c)

Plants monoecious, whorls conical, confluent, 100–160 µm in diameter, internodes 115–230 µm long. Axial cells 38–60 µm in diameter. Cortical filaments cylindrical, cells tending towards brick-like, 5–10 µm in diameter, 10–20 µm long. Primary fascicles 4 or 5 per periaxial cells, usually curved distally, 4–7 cell storeys, dichotomously 3–5 timed branched. Proximal cells cylindrical or obovoidal, 6–7 µm in diameter, 8–11 µm long, distal cells dome-shaped, ellipsoidal or spherical, 4–8 µm in diameter, 4–10 µm long. Terminal hairs uncommon, 7–18 µm long. Secondary fascicles common, some as long as primary fascicles. Spermatangia scattered, terminal on primary and secondary fascicles, spherical, ca. 6 µm in diameter. Carpogonium-bearing branch ca. 1-2 times spiralled or contorted, rising from periaxial cells, consisting of 10–18 modified, discoidal sequat cells, 6–11 µm in diameter, 2–4 µm long. Involucral filaments arising from a few cells of carpogonium-bearing branch, not extending beyond carpogonium. Carpo-gonium almost straight, 4–6 µm in diameter at base, 34–39 µm long. Trichogyne no stalked, club-shaped, ellipsoidal or fusiform, 4–8 µm in diameter at broadest portion. Carpo-sporophytes 1 or 2 per whorl, always exserted from whorl, semi-spherical, compact, 150–230 µm in diameter, 2–4 times whorl radius, inserted centrally. Gonimoblast filaments 4–9 cells long, carposporangia spherical or obovoidal, 7–11 µm in diameter, 7–13 µm long. No chantransia phase is observed.

Type locality: Little Wheeny, Mill Road Crossing, 2.5 km northwest of Kurrajong, New South Wales in Australia. Known from a slow-flowing stream through mixed farmland and remnant forest in warm-temperate Australia.

Holotype: NSW, Entwisle 2638, 24/vii 1996, Isotype: MEL

Distribution: Known from the type locality only.

8–5. *Ambigua*

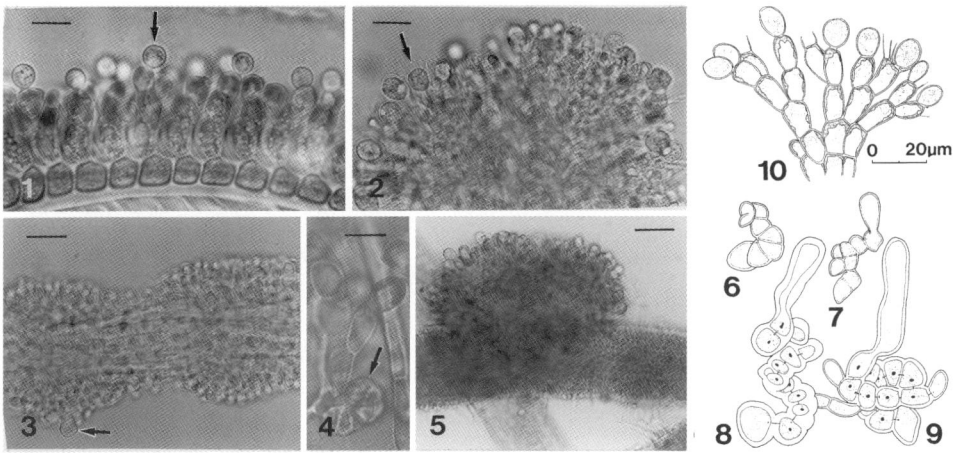

Plate 123 (a)
Batrachospermum gibberosum, figs 1–5 (Kumano 1996o),
figs 6–10 (original author, Kumano 1978 as *Tuomeya gibberosa*)
1. a cortical filament and spermatangia (arrow) terminal on secondary fascicles, 2. carposporangia (arrow) terminal on compactly agglomerated gonimoblast filaments, 3. a portion of a young plant showing cortical filaments, primary, secondary fascicles and a terminal portion of a trichogyne (arrow), 4. a primary fascicle and a carpogonium-bearing branch at the early stage in development (arrow), 5. a wart-like or hemispherical carposporophyte, 6–7. carpogonium-bearing branches at the early stages in development, 8–9. spirally twisted carpogonium-bearing branches with mature carpogonia, 10. carposporangia terminal on gonimoblast filaments. Scale bars = 10 μm (figs 1, 4); 20 μm (figs 2, 3, 6–10); 40 μm (fig. 5).

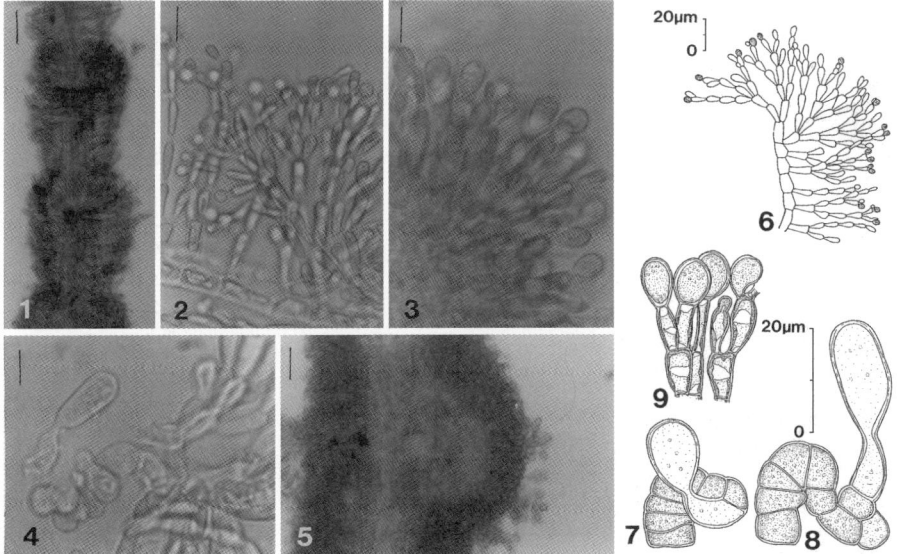

Plate 123 (b)
Batrachospermum hirosei, figs 1–5 (Kumano 1996r), figs 6–9 (original authors, Ratnasabapathy & Kumano 1982b)
1, 5. a portion of a plant with confluent whorls and axial carposporophytes, 2, 6. a portion of a whorl showing spermatangia terminal on primary, secondary fascicles and cortical filaments. 3, 9. carposporangia terminal on gonimoblast filaments, 4, 7–8. the development of a twisted carpogonium-bearing branch.
Scale bars = 100 μm (fig. 1); 40 μm (fig. 5); 20 μm (figs 6–9); 10 μm (figs 2–4).

Batrachospermales

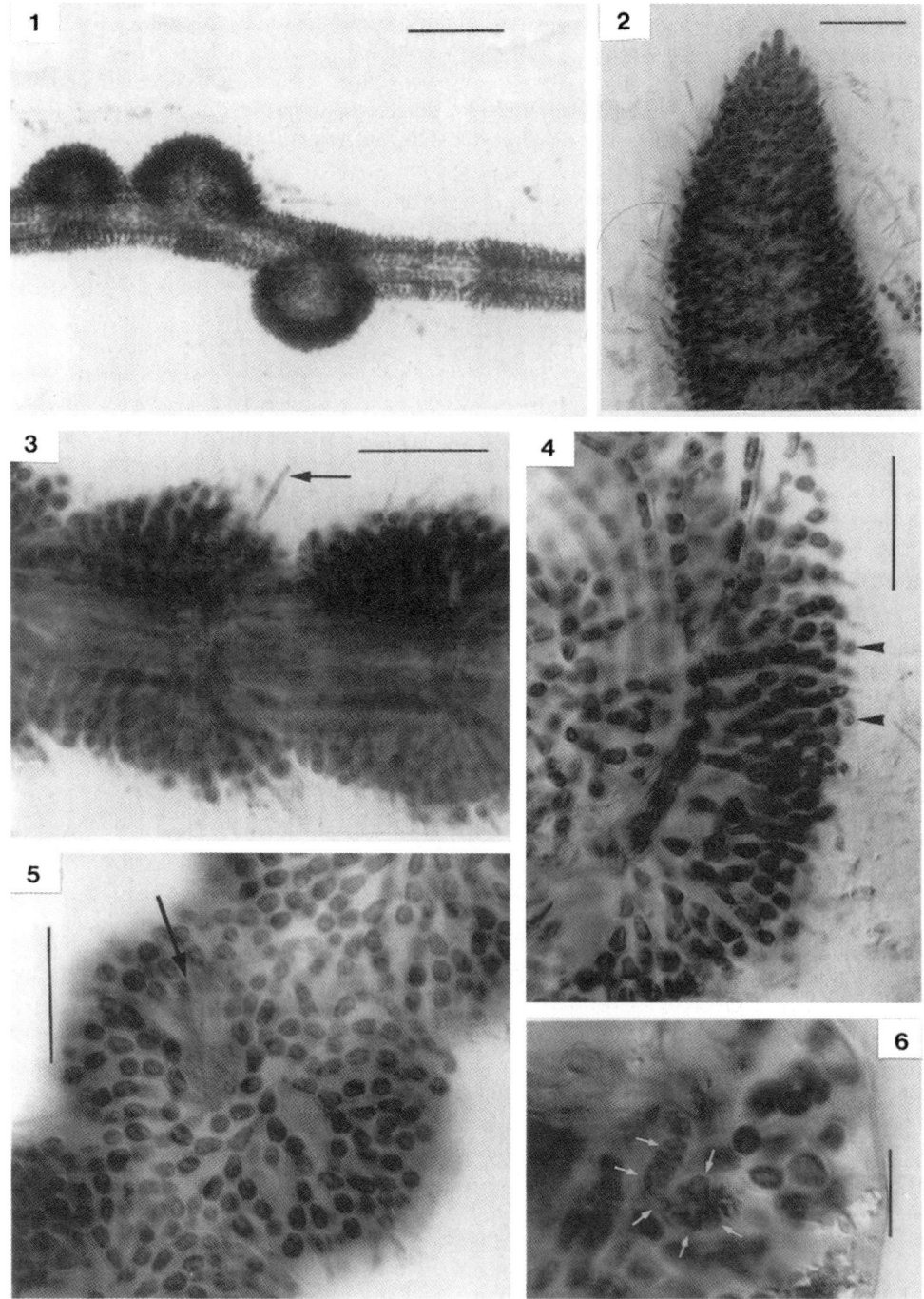

Plate 124
Batrachospermum deminutum, figs 1–6 (original authors, Entwisle & Foard 1999c)
1. plant showing reduced whorls and protuberant carposporophytes, 2. plant apex, 3. fascicles with protruding trichogyne (arrow), 4. fascicles with terminal spermatangia (arrowheads), 5. trichogyne (arrow) extending from carpogonium-bearing branch involucral filament mass to beyond whorl, 6, immature carpogonium-bearing branch (arrow).   Scale bars = 150 µm (fig. 1); 50 µm (figs 2–5); 15 µm (fig. 6).

3) ***Batrachospermum vittatum*** Entwisle et Foard (1999c: 619, fig. 3)

plate 125, figs 1–6 (original authors, Entwisle & Foard 1999c)

Plants monoecious, whorls conical, confluent, 390–480 µm in diameter, internodes 100–555 µm long. Axial cells 20–60 µm in diameter, cortical filaments cylindrical, 3–6 µm in diameter. Primary fascicles 2 or 3 per periaxial cell, more or less straight distally, 8–13 cell storeys, di- or trichotomously, 2–4 times (mostly restricted to proximal portion of fascicles) branched, proximal cell ellipsoidal, fusiform or obovoidal, 4–10 µm in diameter, 15–32 µm long, distal cells fusiform or obovoidal, 2.5–6 µm in diameter, 10–15 µm long. Spermatangia on primary or secondary fascicles, usually clustered on short lateral branch, in more or less distinct longitudinal band through the plant, also on some involucral filaments, spherical, 5–6 µm in diameter. Carpogonium-bearing branch once spiraled or contorted, arising from periaxial cell, consisting of 4–6 modified, discoidal cells, ca. 6 µm in diameter, 4 µm long. Involucral filaments spreading, (1–)3–4 cells long, not extending beyond carpogonium. Carpogonium obliquely attached, straight or curved, 3–4 µm in diameter at base, 32–40 µm long. Trichogyne not stalked, club-shaped or cylindrical, 4–9 µm in diameter at broadest portion. Carposporophytes 1 per whorl, in inner and outer portion of whorl, spherical to semi-spherical, compact, 110–260 µm in diameter, 0.5–1 times whorl radius, inserted centrally. Gonimoblast filaments compactly aggregated, 4–6 cells long. Carposporangia obovoidal, 7–10 µm in diameter, 13–15 µm long. No chantransia phase is observed.

Type locality: Gybara Pool, end of Gubara Walk, off Kakadu Highway between Cooinda and Jabiru, Northern Territory in Australia.

Holotype: DNA, coll. Entwisle, 6/VI, 1997. Isotype: MEL.

Distribution: Known from the type locality and Ji bal Creek, Amhem Land, in northern Territory in Australia.

4) ***Batrachospermum hirosei*** Kumano et Ratnasabapathy in Ratnasabapathy & Kumano (1982b:122, fig. 3, A–J)

plate 123 (b), figs 1–5 (Kumano 1996r)
figs 6–9 (original authors, Ratnasabapathy & Kumano 1982b)

Plants monoecious, very mucilaginous, 1–3 cm high, 100–320 µm in diameter, more or less irregularly branched, green or brown; whorls pear-shaped, obconical, touching each other in old plants; primary fascicles abundantly branched, 6–8 cell-storeys, cells fusiform or ovoidal, terminal hairs absent; cortical filaments well-developed; secondary fascicles numerous, covering all internodes; spermatangia spherical, 3–5 µm in diameter, terminal on fascicles; carpogonium-bearing branches strongly undulated, consisting of 6–13 barrel-shaped cells, arising from periaxial cells; involucral filaments short; carpogonia 4–10 µm in diameter at the base, 7–10 µm in diameter at the apex, 19–35 µm long, trichogyne ellipsoidal or irregularly spatula-shaped, distinctly stalked; carposporophytes single, spherical or hemispherical, 110–200 µm in diameter, 100–140 µm high, inserted centrally; carposporangia obovoidal or spherical, 7–8 µm in diameter, 10–15 µm long.

Type Locality: Sungai Pusu near Kampong Sungai Pusu, Selangor, Malaysia in Southeast Asia.

Holotype: Kobe University, Ratnasabapathy no. 1201b, 2/VI 1979.

Distribution: Known from type locality only.

Batrachospermales

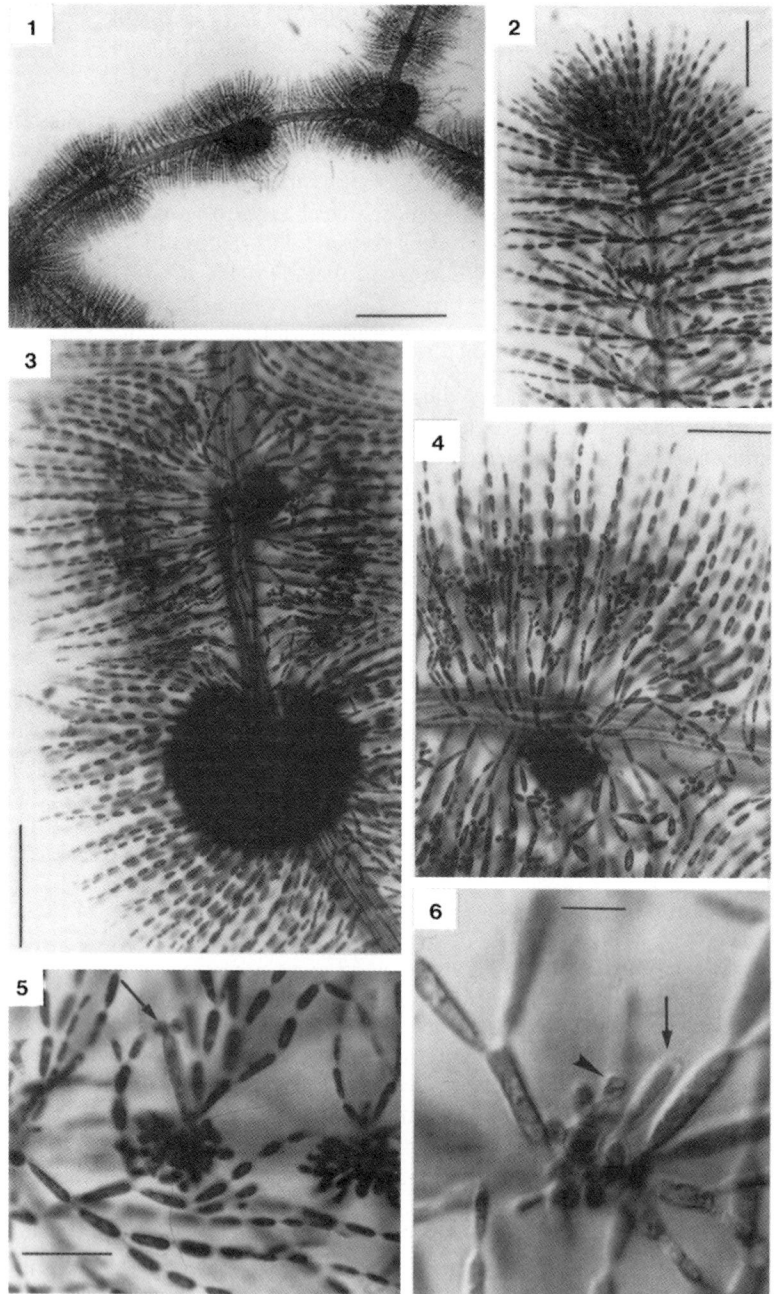

Plate 125
Batrachospermum vittatum, figs 1–6 (original authors, Entwisle et Foard 1999c)
1. plant with carposporophytes (dark masses), 2. plant apex, 3. plant with carposporophyte and spermatangia in a longitudinal mid-whorl band, 4. whorl with band of spermatangia (above) and developing carposporophyte (below), 5. fertilized carpogonium (arrow to attached spermatangium) protruding from relatively loosely arranged carpogonium-bearing branch involucral filament mass, 6, contorted carpogonium-bearing branch with immature carpogonium (arrow) and spermatangium borne on involucral filament (arrowhead).
Scale bars = 300 µm (fig. 1); 100 µm (fig. 3); 50 µm (figs 2, 4); 30 µm (fig. 5); 10 µm (fig. 6)

5) *Batrachospermum australicum* Entwisle et Foard (1999c: 624, fig. 5)

plate 126, figs 1–6 (original authors, Entwisle & Foard 1999c)

Plant monoecious, olive green, flaccid, whorls conical, confluent (rarely separate), 260–410 μm in diameter, internodes 160–740 μm long. Axial cells 40–85 μm in diameter, cortical filaments cylindrical, 2–5 μm in diameter. Primary fascicles 3 or 4 per periaxial cells, more or less straight distally, 6–9(–12) cell storeys, 2–4(–6) times (tri-) dichotomously branched. Proximal cells cylindrical or ellipsoidal (rarely ovoidal), 4–6 μm in diameter, 9–23 μm long. Distal cells ellipsoidal, 4–5 μm in diameter, 14–20 μm long. Terminal hairs 30–300 μm long. Secondary fascicles common, always shorter than primary fascicles. Spermatangia clustered, terminal or within on primary or secondary fascicles, spherical, 4–6 μm in diameter. Carpogonium-bearing branch spiraled, 2 or) 3 times, arising from periaxial cell of primary (rarely secondary) fascicles, consisting of 14–20(–25) modified, discoidal cells, 3–5 μm in diameter, 3–4 μm long. Involucral filaments arising from all cells of carpogonium-bearing branch, 1–3(–5) cells long, not extending beyond carpogonium. Carpogonium obliquely attached, straight or curved, 6–8 μm in diameter at base, 30–50 μm long. Trichogyne not stalked, club-shaped, cylindrical or fusiform, 4–8 μm in diameter at broadest portion. Carposporophytes 1 or 2 per whorl, in inner and outer whorl, semi-spherical, compact to loose, 115–200 μm in diameter, (0.3–)0.6–1.2 times whorl radius, centrally inserted. Gonimoblast filaments 2–5 cells long. Carposporangia obovoidal or ellipsoidal, 9–12 μm in diameter, 10–17 μm long. No chantransia phase is observed.

Type locality: Roper River, Elsey Station, Northern Territory in Australia.

Holotype: DNA 2709, Entwisle, 2/vi, 1997, Isotype: MEL.

Distribution: Known from the type locality in Northern Territory, and confluence of Waterhouse River, Thermal Spring outflow, Elsey National Park, Roynolds River, Daly River Road, southern perimeter of Litchfield National Park and Daley River, mango farm, Northern Territory in Australia. Attached to rocks or submerged vegetation in larger rivers in tropical Australia.

6) *Batrachospermum mahlacense* Kumano et Bowden-Kerby (1986:109, figs 1–12)

plate 127, figs 1–11 (original authors, Kumano & Bowden-Kerby 1986), Kumano 1996za

Plants monoecious, moderately mucilaginous, ca. 6 cm high, 250–400 μm in diameter, abundantly and irregularly branched, dark grayish green; whorls pear-shaped, primary fascicles dichotomously branched, 7–9 cell-storeys, cells ellipsoidal, terminal hairs more or less short; cortical filaments well-developed; secondary fascicles numerous, 6 or 7 cell-storeys, unbranched or dichotomously branched, covering all internodes; spermatangia spherical, 4–6 μm in diameter, terminal or lateral on fascicles; carpogonium-bearing branches strongly undulated, consisting of 5–15 barrel-shaped cells, arising from periaxial cells; involucral filaments numerous, short; carpogonia 4–5 μm in diameter at the base, 7–8 μm in diameter at the apex, 25–40 μm long, trichogyne ellipsoidal or urn-shaped, more or less distinctly stalked; carposporophytes sessile, single or paired, spherical or hemispherical, 140–170 μm in diameter, 80–160 μm high, inserted centrally; carposporangia obovoidal, 7–12 μm in diameter, 12–14 μm long.

Type Locality: Mahlac River, Talofofo, Guam Island, Mariana Islands, Micronesia in Pacific Ocean. Known from rocks in a perennial spring in upper reaches.

Holotype: Kobe University, Bowden-Kerby, 25/VIII 1983.

Distribution: Known from the type locality and Ibobang, Palau, Western Caroline Islands, Micronesia in Pacific Ocean.

Batrachospermales

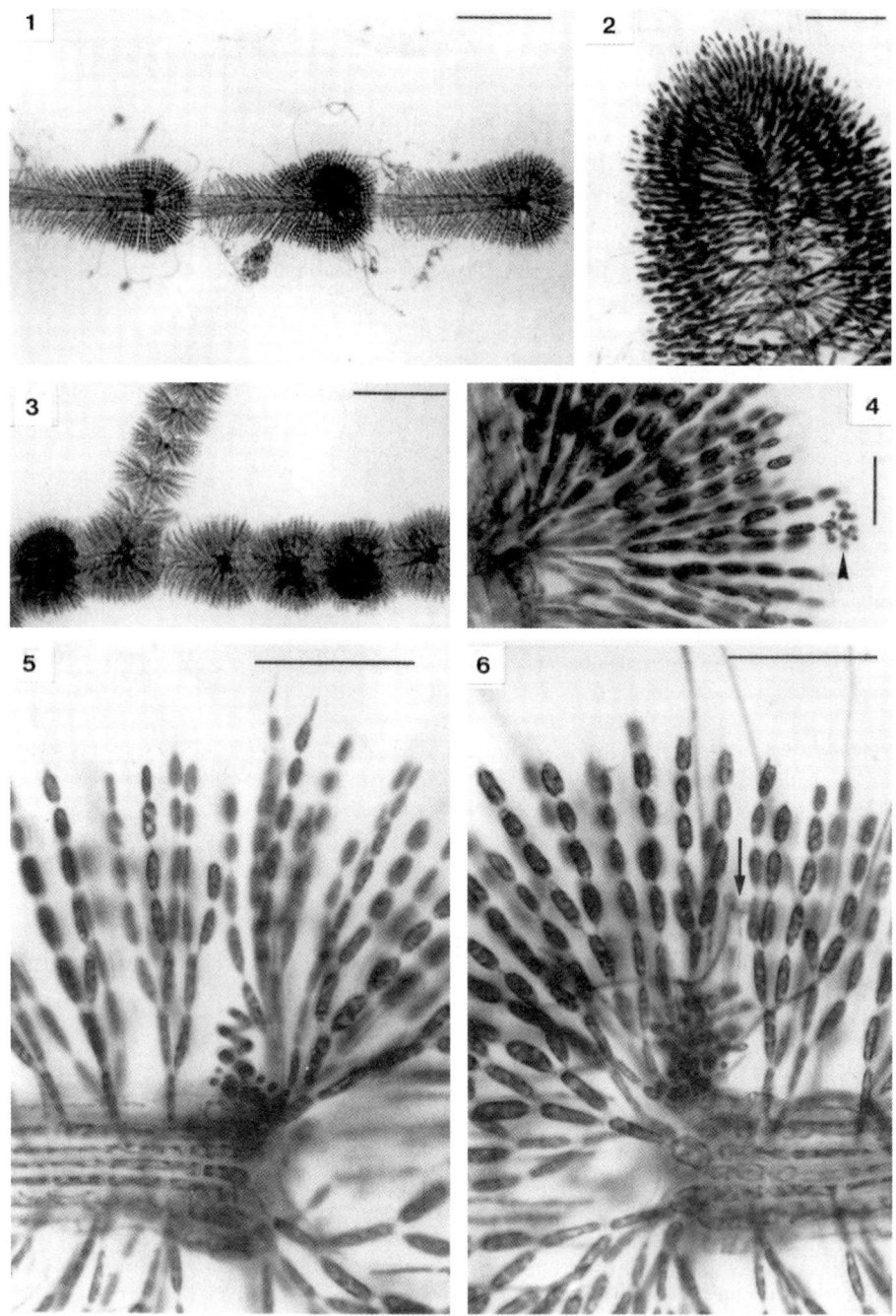

Plate 126
Batrachospermum australicum, figs 1–6 (original authors, Entwisle et Foard 1999c)
1. plant with elongate whorls, one with a carposporophyte (dark mass), 2. plant apex, 3. plant with relatively compact whorls, some with carposporophytes (dark masses), 4. Fascicles with terminal spermatangia (arrowhead), 5. fascicles with immature, cork-screw shaped carpogonium-bearing branch, 6. fascicles with fertilized carpogonium (arrow) extending beyond elongate carpogonium-bearing branch involucral filaments mass.   Scale bars = 300 µm (figs 1, 3); 50 µm (figs 2, 5, 6); 25 µm (fig. 4).

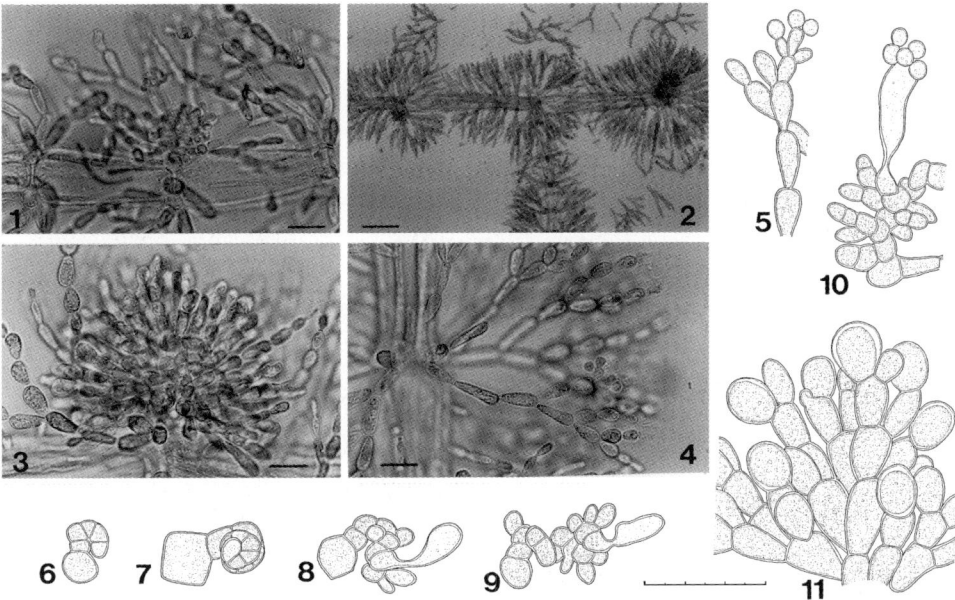

Plate 127
Batrachospermum mahlacense, figs 1–11 (original authors, Kumano & Bowden-Kerby 1986, Kumano 1996za)
1. a portion of a plant showing a carpogonium-bearing branch with a fertilized carpogonium, 2. a portion of a plant showing pear-shaped whorls, 3. a carposporophyte, 4–5. spermatangia terminal on primary fascicles, 6–7. coiled carpogonium-bearing branches at the very early stages in development, 8–9. the early stages in development of coiled carpogonium-bearing branches with young trichogynes, 10. a fertilized carpogonium with attached spermatia, 11. carposporangia terminal on gonimoblast filaments.
Scale bars = 100 µm (fig. 2); 30 µm (figs 5–11); 20 µm (figs 1, 3–4).

7) ***Batrachospermum dasyphillum*** Skuja in Balakrishnan et Chaugule (1980:230, figs 4–7)

plate 129, figs 4–7 (original authors, Balakrishnan & Chaugule 1980)

Plant monoecious, up to 6 cm high, colour of herbarium specimens bluish black, richly branched with blunt apex. Whorls disk-shaped, 95–159 µm in diameter, 63–79 µm high. Apical cell more or less dome-shaped, 8.5–10 µm in diameter, 6.5–9 µm long. Axial cells 145–316 µm long, cortical axes 47–316 µm in diameter. Periaxial cells hemispherical, 5–5.6 µm in diameter. Cortical filaments abundant, cells 6.5–8.5 µm in diameter, 11–69 µm long. Primary fascicles characteristically reduced and rarely conspicuous, 4–8 cell-storeys, cells short cylindrical to obovoidal, 4.5–6 µm in diameter, 8–13 µm long. Terminal hairs lacking. Spermatangia terminal on fascicles, spherical or semispherical, 6–5–8.5 µm in diameter. Carpogonium-bearing branch consisting of 10–15 cells, usually spirally coiled, generally single, rarely in pairs arising from periaxial cell. Involucral filaments short, simple or branched. Carpogonium hemispherical, 4.5–5 µm in diameter, 4–5 µm long; trichogyne conical, distinctly stalked, 4–5 µm in diameter, 13–21.5 µm long. Carposporophytes single or rarely geminate, spherical or hemispherical, 122–316 µm in diameter, situated in the axis. Gonimoblast filaments richly branched and densely aggregated; cells 11–16.5 µm in diameter, 13–22 µm long. Carposporangia ellipsoidal to obovoidal, 11–18 µm in diameter, 14–22 µm long.
Type Locality: Jog Falla, Karnata State, India in Asia. On rocky bed of stream.
Holotype: Iyengar no. 3 (Herbarium Uppsala, specimen no. 1, coll. Prof. Iyengar, 1930.

Batrachospermales

Distribution: Known from the type locality only.
Note: This species was mentioned by Skuja only in manuscript, Blakrishnan & Chaugule (1980) made the formal description and illustration for purpose of validation.

8) ***Batrachospermum nodiflorum*** Montagne (1850:294)
Synonym: *B. vagum* var. *nodiflorum* (Montagne) Sirodot (1884:266)

plate 128, figs 1–5 (Kumano 1996zc), fig. 6 (Kumano 1990)

Plants monoecious, moderately mucilaginous, 10–15 cm high, 100–240 µm in diameter, sparsely, more or less dichotomously branched; whorls dense, obconical, contiguous; primary fascicles short, 5 or 6 cell-storeys, cells barrel-shaped or obovoidal, terminal hairs short; cortical filaments very much developed; secondary fascicles, numerous short, consisting of 4 or 5 cell-storeys, cells obovoidal; spermatangia spherical, 5–8 µm in diameter, terminal on fascicles; carpogonium-bearing branches undulated, consisting of 3–9 disc- or barrel-shaped cells, arising from periaxial cells; involucral filaments very short; carpogonia 7–9 µm in diameter at the base, 9–13 µm in diameter at the apex, 30–50 µm long, trichogyne ellipsoidal or club-shaped, more or less indistinctly stalked; carposporophytes single, hemispherical or wart-like, 200–400 µm in diameter, 350–550 µm high, gonimoblast filaments long, consisting of 5–10 barrel-shaped cells, radially branched, compactly agglomerated; carposporangia obovoidal or ellipsoidal, 10–13 µm in diameter, 15–20 µm long, terminal on laterals of gonimoblast filaments.

Type Locality: near Tigres Mountains, Cayenne, French Guiana in South America. Occurs on rocks in quiet rivulets.
Holotype: PC herb Montagne, Le Prieur coll. n° 1107.
Distribution: Known from the type locality only.

9) ***Batrachospermum gracillimum*** W. West et G. S. West (1897:2) *emend.* Necchi (1989:69)

plate 130, figs 1–10 (Necchi 1990)

Plants monoecious, rarely dioecious, moderately mucilaginous, 2.5–12.5 cm high, 200–450 µm in diameter, abundantly and irregularly branched. Whorls reduced, dense or lax, pear-shaped, obconical or compressed, contiguous. Fascicles 2 or 3, primary fascicles straight, di- or trichotomously branched, 4–7 cell-storeys, proximal cells cylindrical or ellipsoidal, 3–10 µm in diameter, 20–45 µm long, distal cells barrel-shaped, ellipsoidal, 4–8 µm in diameter, 10–25 µm long, terminal hairs numerous, short. Cortical filaments poorly or well-developed. Secondary fascicles numerous, reaching the length of whorls. Spermatangia spherical, 5–7 µm in diameter, terminal, rarely subterminal on secondary fascicles, rarely primary ones. Carpogonium-bearing branches spirally coiled, consisting of 5–8 disc- or barrel-shaped cells, arising from periaxial cells. Involucral filaments numerous, short. Carpogonia asymmetric, 4–7 µm in diameter at the base, 6–10 µm in diameter at the apex, 35–85 µm long, trichogyne cylindrical, club-shaped, sessile or stalked. Carposporophytes sessile, single, dense, hemispherical, higher than whorls, 110–230 µm in diameter, 200–470 µm high. Carposporangia obovoidal, 8–12 µm in diameter, 12–17 µm long.

Type Locality: Andongo, Angola in Africa.
Lectotype: LISU, F. Welwitsch, 3/V 1857, designated by Necchi 1990. Isotype: BM.
Distribution: Known from the type locality and Amazonas, Matto Grosso and Paraná, Brazil in South America.

8–5. *Ambigua*

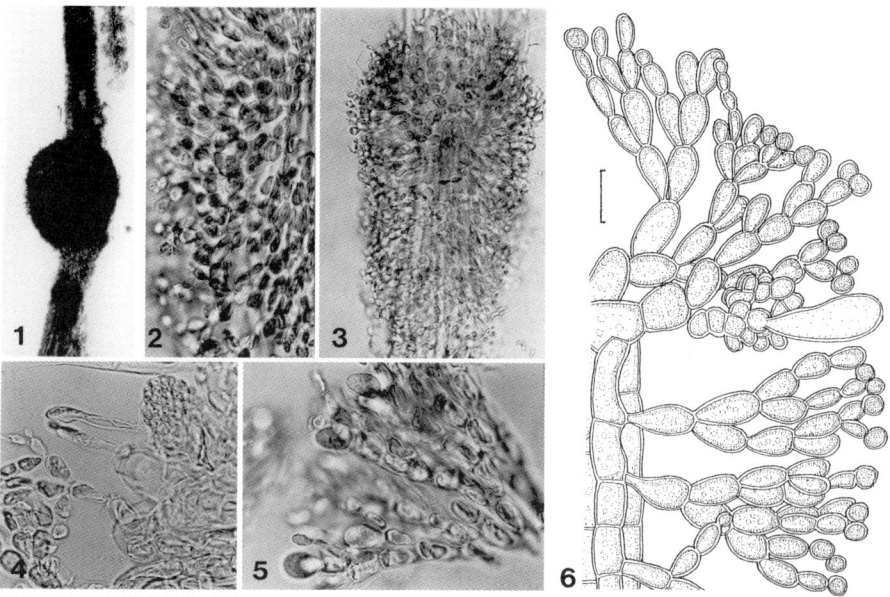

Plate 128
Batrachospermum nodiflorum, figs 1–5 (Kumano 1996zc), fig, 6 (Kumano 1990)
1. a main axis with cylindrical whorls and a hemispherical or wart-like carposporophyte, 2. spermatangia terminal on fascicles, 3. a subconical whorl, 4. an ellipsoidal or club-shaped trichogyne, 5. carposporangia terminal on compactly agglomerated gonimoblast filaments, 6. a portion of a plant showing spermatangia terminal on primary and secondary fascicles, cortical filaments and a twisted carpogonium-bearing branch with very short involucral filaments. Scale bars = 100 μm (fig. 1); 40 μm (fig. 3); 10 μm (figs 2, 4–6).

10) ***Batrachospermum torsivum*** Shi (1994:279, fig. 3, 1–4, fig. 4, 1–10)

plate 131, figs 1–14 (original author, Shi 1994)

Plants monoecious, 2.5–5 cm high, irregularly branched. Whorls more or less obconical or pear-shaped, hemispherical in young plants, compactly contiguous, 300–650 μm in diameter. Axial cells cylindrical, narrow in upper portion, wide in lower portion, 40–140 μm in diameter, 150–750 μm long. Cortical filaments conspicuous, upper portion of axial cells thick, lower portion of axial cells thin. Primary fascicles dichotomously branched, 7–9 cell-stories; proximal cells cylindrical or elongate obconical, 6.5–7.5 μm in diameter, 28–38 μm long, distal cells ellipsoidal or obovoidal, 5–7.5 μm in diameter, 11–25 μm long. Secondary fascicles very numerous, covering all internodes, unbranched or dichotomously branched, 5–8 cell-stories; cells 4–6.5 μm in diameter, 12–30 μm long. Terminal hairs few and short, only 13–28 μm long. Spermatangia spherical, 4.5–6.5 μm in diameter, terminal on primary and secondary fascicles. Carpogonium-bearing branch curved or undulated, arising from periaxial cells, 30–35 μm long, consisting of 5–8 disc- or barrel-shaped cells. Carpogonium 5–6 μm in diameter at base, trichogyne more or less cylindrical, club-shaped, narrowly fusiform or narrowly lanceolate, 8–12 μm in diameter, 28–36 μm long, with stalk conspicuously long, 5–9 μm long and sometimes undulated. Carposporophyte single, hemispherical, inserted centrally, 150–340 μm wide, 130–240 μm high. Carposporangia obovoidal, 8–13 μm in diameter, 15–23 μm long.
    Type Locality: Ruijin Xian, Jiangxi, China in Eastern Asia. Occurs on the sides of a well.
    Holotype: KSI 75215.
    Distribution: Known from the type locality only.

Batrachospermales

11) ***Batrachospermum iyengarii*** Skuja in Balakrishnan et Chaugule (1980c:232: figs 12–13)

plate 129, figs 1–2 (Balakrishnan & Chaugule 1980c)

Plants monoecious, up to 5 cm high, richly branched. Whorls more or less discoidal, separated or confluent due to growth of secondary fascicles, 159–280 µm in diameter. Apical cell dome-shaped, 6–7 µm in diameter, 9–10 µm long. Axial cells comparatively long, 238–480 µm long. Cortical axes 79–239 µm in diameter. Periaxial cells spherical to hemispherical, 6.5–11.5 µm in diameter. Cortical filaments quite prominent in older portion. Cortical cells 6.5–10 µm in diameter, 27–71.5 µm long. Primary fascicles 5–8 cell-storeys, not richly branched, parallel and slightly incurred, cells cylindrical to ellipsoidal, 4.5–6.5 µm in diameter, 16–26 µm long. Secondary fascicles frequent, rarely reaching the length of primary ones, 4–6 cell-storeys; cells 3–5 µm in diameter, 16–38 µm long. Terminal hairs hyaline and deciduous, 18–70 µm long. Spermatangia terminal on primary and secondary fascicles, characteristically in clusters, spherical, 4.5–6.5 µm in diameter. Carpogonium-bearing branch comparatively long, consisting of 13–17 barrel-shaped cells, spirally coiled, arising from periaxial cell. Involucral filaments short and mostly unbranched. Carpogonium hemispherical, 4–5 µm in diameter, 4–8 µm long. Trichogyne distinctly stalked, cylindrical, characteristically long, 4–7.5 µm in diameter, 26–48 µm long. Carposporophytes usually single, spherical, 143–240 µm in diameter. Gonimoblast filaments branched and compactly aggregated, cells 4.5–10 µm in diameter, 9.5–17.9 µm long. Carposporangia ellipsoidal to obovoidal, carpospores 8.5–11.5 µm in diameter, 9.5–13.5 µm long.

Type Locality: Mamandur, Andhra Pradesh, India in Asia. Attached to stones in streams.
Lectotype: Iyengar no. IB80, coll. Iyengar, 1930.
Distribution: Known from the type locality only.
Note: This species was mentioned by Skuja only in manuscript, Balakrishnan & Chaugule (1980) made the formal description and illustration for purpose of validation.

12) ***Batrachospermum tiomanense*** Kumano et Ratnasabapathy in Ratnasabapathy & Kumano (1982a:18, fig. 3, A–I)

plate 133, figs 1–6 (Kumano 1996zi)
figs 7–11 (original authors, Ratnasabapathy & Kumano 1982a)

Plants trioecious, not very mucilaginous, ca. 2 cm high, 150–300 µm in diameter, more or less dichotomously branched, olive green; whorls pear-shaped, obconical in aged plants; primary fascicles sparsely branched, 13–15 cell-storeys, cells cylindrical or fusiform, terminal hairs absent; secondary fascicles numerous, covering all internodes; spermatangia spherical, 3–5 µm in diameter, 4–6 µm long, unilateral on secondary fascicles; carpogonium-bearing branches strongly undulated, long, consisting of 6–10 barrel-shaped cells, arising from periaxial cells; involucral filaments numerous, very short; carpogonia asymmetrical, 5–9 µm in diameter at the base, 7–9 µm in diameter at the apex, 37–40 µm long, trichogyne urn-shaped, distinctly stalked, often bent at the base; carposporophytes single, hemispherical, 100–140 µm in diameter, inserted centrally; carposporangia hemispherical or ovoidal, 6–10 µm in diameter, 8–12 µm long.

Type Locality: Sungai Ayer Besar, Pulau Tioman, Malaysia in Southeast Asia. Attached to side of large submerged boulders on right bank of upper reaches.
Holotype: University of Malaya, Ratnasabapathy no. 15, 24/V 1974.
Distribution: Known from the type locality only.

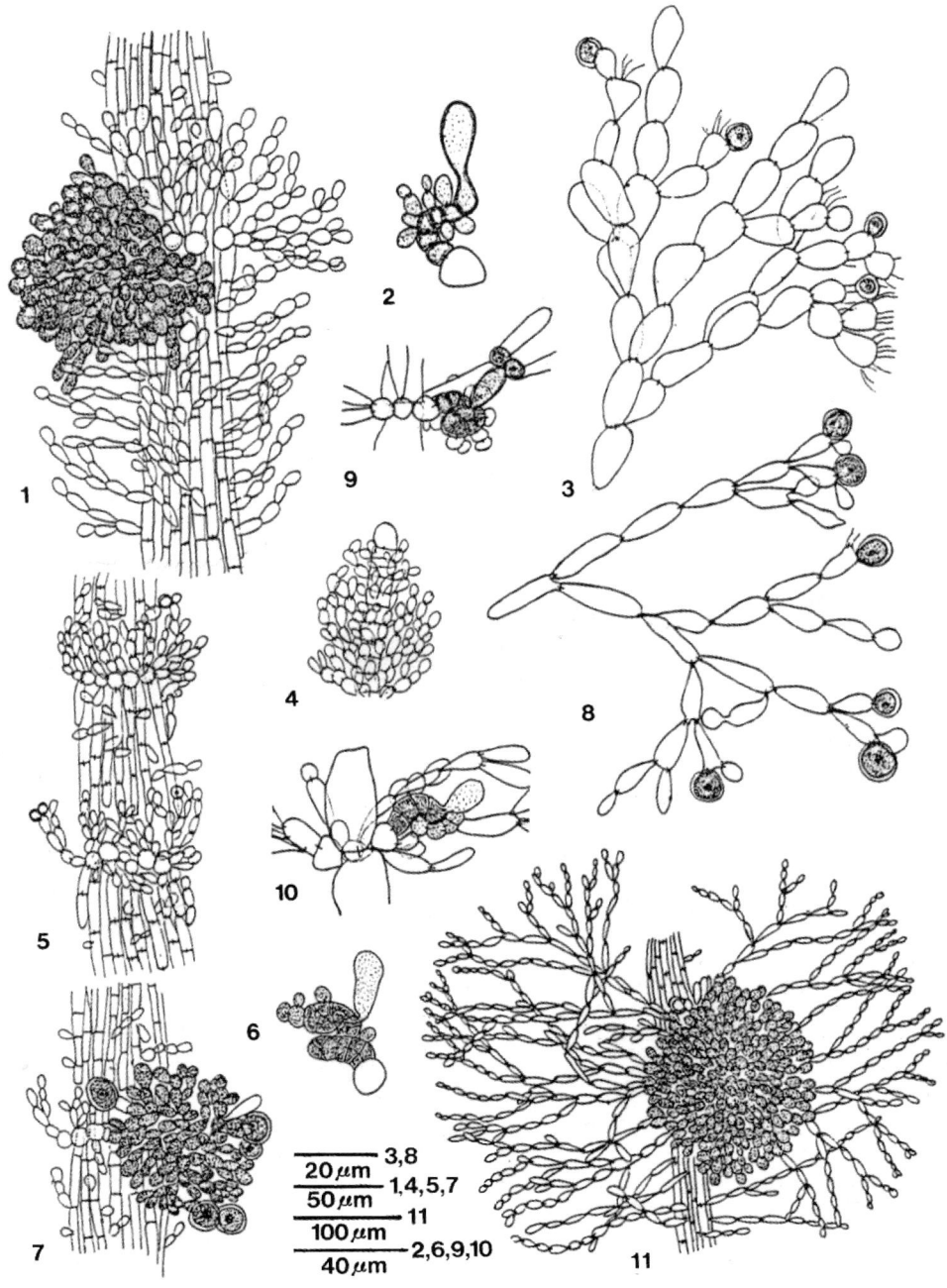

Plate 129

Batrachospermum turfosum Bory as *B. vagum* (Roth) C. Agardh, figs 1–3 (Balakrishnan & Chaugule 1980c)

Batrachospermum dasyphillum Skuja in Balakrishnan & Chaugule, figs 4–7 (Balakrishnan & Chaugule 1980c)

Batrachospermum mahabaleshwarensis Balakrishnan et Chaugule, figs 8–11 (original authors, Balakrishnan & Chaugule 1980c)

Batrachospermales

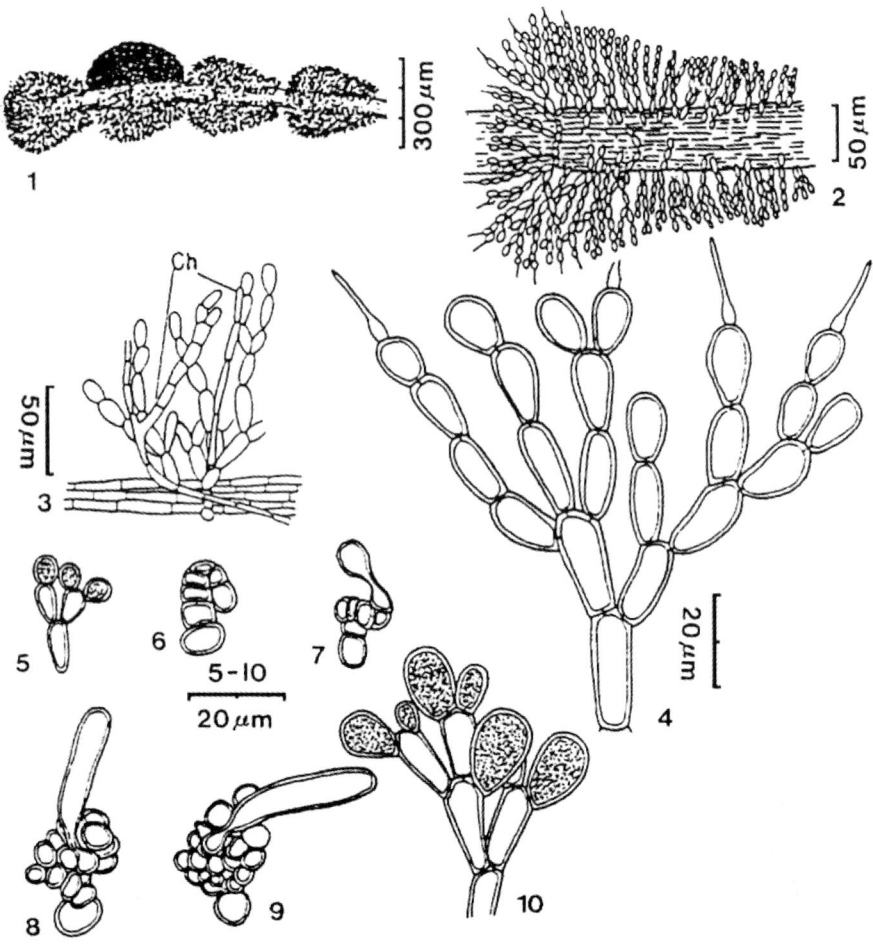

Plate 130
Batrachospermum gracillimum, figs 1–10 (Necchi 1990)
1. the general view of whorls with carposporophytes, 2. the structure of a whorl showing cortical filaments, primary and secondary fascicles, 3. filaments of '*chantransia* phase' (au) epiphytic on a gametophyte, 4. details of a primary fascicles, 5. a distal cell of a secondary fascicles with spermatangia, 6. a young carpogonium-bearing branch, 7. a young carpogonium with a developing trichogyne, 8–9. mature carpogonium-bearing branches, 10. gonimoblast filaments with carposporangia.

13) ***Batrachospermum zeylanicum*** Skuja in Balakrishnan et Chaugule (1980c:235:figs 14–16)

plate 129, figs 4–7, plate 132, figs 3–5 (Balakrishnan & Chaugule 1980c)

Plants monoecious, up to 9 cm high, brownish green to bluish black in herbarium specimens, profuse, somewhat irregularly branched with tapering apices. Whorls large, very conspicuous, ovoidal, 190–580 µm in diameter. Apical cells dome-shaped, 3–4 µm in diameter, 4.8–8.3 µm long. Axial cells comparatively very long, 600–800 µm long, corticated axes 60–115 µm in diameter. Periaxial cells 4–6 per axial cell, spherical to hemispherical, 4.5–11.5 µm in diameter. Primary fascicles 4–18 per periaxial cell, more or less richly branched, 10–15 cell-storeys; cells almost uniform, cylindrical to fusiform, 3–8 µm in diameter, 13–28 µm long. Terminal hairs often very thin hyaline, deciduous, 1–4 µm in diameter, about 26 µm long.

Cortical filaments abundant, with cells 2.4–3.3 μm in diameter, 16–54 μm long. Secondary fascicles short, branched or unbranched, 6–12 cell-storeys; cells sub-cylindrical, 1.5–3.5 μm in diameter, 11–12 μm long. Spermatangia terminal or subterminal on primary and secondary fascicles, spherical to hemispherical, 6.5–7.5 μm in diameter. Carpogonium-bearing branches spirally coiled or contorted, vary rarely curved, generally solitary, rarely in pairs, consisting of 3–13 cells, always arising from periaxial cell. Involucral filaments rather sparse, short, simple or once branched. Carpogonium hemispherical, 3–5.7 μm in diameter, 4.5–6.5 μm long. Trichogyne distinctly stalked, lanceolate, stout, 4.5–9 μm in diameter, 11–23 μm long. Carposporophytes generally single, sometimes geminate, 127–303 μm in diameter. Gonimoblast filaments densely branched, cells 3–3.5 μm in diameter, 14–25 μm long. Carposporangia terminal. ellipsoidal to obovoidal, carpospores obovoidal 9–14 μm in diameter, 14–22 μm long.

Type Locality: Jog falls, Karnataka state, India in Asia. On rocky bed of stream.
Lectotype: UPS, specimen no. 1, NS 238, Ferguson.
Distribution: Known from the type locality only.
Note: This species was mentioned by Skuja only in manuscript, Balakrishnan & Chaugule (1980) made the formal description and illustrations for purpose of validation.

14) ***Batrachospermum kylinii*** Balakrishnan et Chaugule (1980b:298: figs 1–22)

plate 135, figs 1–13 (original authors, Balakrishnan & Chaugule 1980b)

Plants monoecious, up to 6 cm high, highly mucilaginous, irregularly branched with tapering apices, light olive green to olive green becoming light yellowish green to bluish green when dry. Whorls discoid, 240–315 μm in diameter, becoming confluent in the very old portions due to growth of secondary fascicles from cortical filaments. Apical cells 5–6 μm in diameter, 6–7 μm long. Axial cells 250 μm in diameter, up to 700 μm long. Periaxial cells spherical to hemispherical, 8–12 μm in diameter. Corticated axes up to 300 μm in diameter; cortical filaments profuse, especially prominent in older portions; cells 3–6 μm in diameter, 7–21 μm long. Primary fascicles 6–12 cell-storeys, richly branched; cells ellipsoidal to ovoidal, 3–5 μm in diameter, 7–18 μm long. Secondary fascicles frequent, as long as primary one in older portion, 4–7 cell-storeys; cells 3–5 μm in diameter, 8–22.1 μm long. Terminal hairs hyaline and deciduous, 6–16 μm long. Spermatangia solitary or in pairs terminal or subterminal on primary and secondary fascicles, spherical to hemispherical, 6–9 μm in diameter. Carpogonium-bearing branch always spirally coiled, consisting of 9–14 cells, generally single, sometimes two at a node, usually arising from periaxial cells. Involucral filaments unbranched or branched and very well developed often covering the carpogonium. Carpogonium somewhat hemispherical, 14–16 μm in diameter, 15–18 μm long. Trichogyne rather short, sub-sessile, ellipsoidal, 8–11 μm in diameter, 17–19 μm long. Carposporophytes single, spherical to hemispherical, 185–280 μm in diameter. Gonimoblast filaments branched and aggregated compactly with cells 5–14 μm in diameter, 10–21 μm long. Carposporangia terminal and obovoidal, 12–17 μm in diameter, 20–28 μm long. Chantransia phase microscopic, and heterotrichous, cells of the prostrate system ovoidal to elongate, thick-walled, 6.5–11 μm in diameter, 8–16 μm long, erect filaments richly branched and uniseriate with cells 5–8.5 μm in diameter, 10–16 μm long,.

Type Locality: Matheran, a hill station, 21 km from Neral, on Poona, Bombay railroad, Maharashtra, India in Asia. Growing attached to rocks in shady bed of a stream.
Holotype: BBC no. 103, coll. Patel & Chaugule, 1972, deposited in Herbarium of University of Poona.
Distribution: Known from the type locality only.

Batrachospermales

15) ***Batrachospermum mahabaleshwarensis*** Balakrishnan et Chaugule (1980c: 231, figs 8–11)

plate 129, figs 8–11 (original authors, Balakrishnan & Chaugule 1980c)

Plants monoecious, 2–6.5 cm high, highly mucilaginous, richly branched with gradually tapering apices, light olive green to green, becoming olive green with a violet tinges when dry. Whorls conspicuous, separated and disk-shaped, 317–572 µm in diameter. Axial cells 47–112 µm in diameter, 158–635 µm long, corticated axes 63–238 µm in diameter. Periaxial cells spherical, semispherical or ellipsoidal, 5.3–8.5 µm in diameter. Cortical filaments enveloping the older

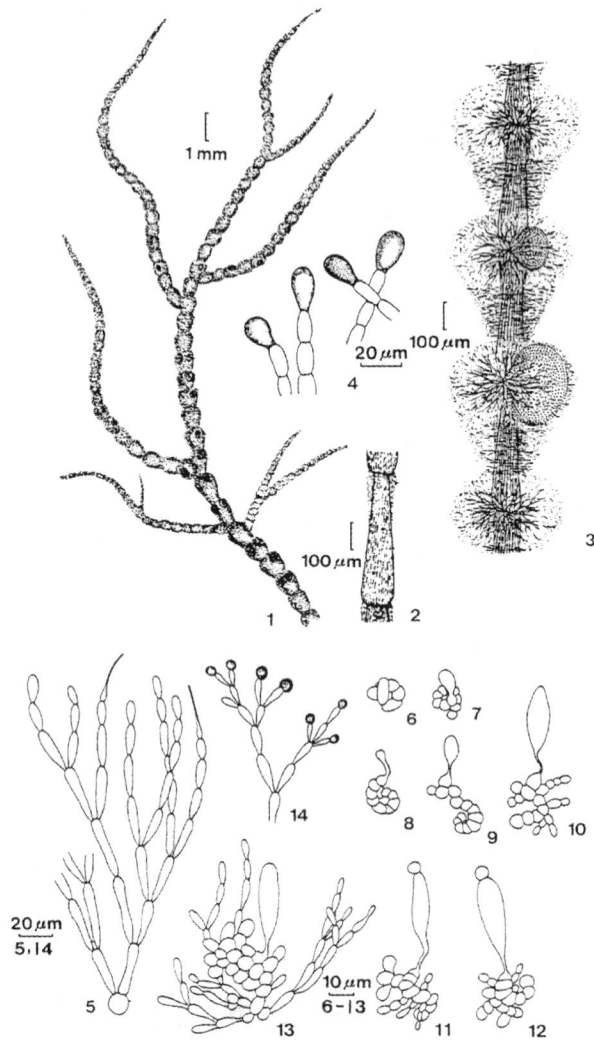

Plate 131
Batrachospermum torsivum, figs 1–14 (original author, Shi 1994)
1. a part of a plant showing obconical or pear-shaped whorls and carposporophytes, 2. axial cells and cortical filaments, 3. a part of a plant showing axial cells, whorls (primary and secondary fascicles) and carposporophytes inserted centrally, 4. carposporangia terminal on gonimoblast filaments, 5. primary fascicle with terminal hairs, 6. a carpogonium-bearing branch at the earlier stage in development, 7–9. young carpogonium-bearing branches with immature carpogonium, 10–13. mature carpogonium-bearing branches, 14. spermatangia terminal on primary fascicles.

8–5. Ambigua

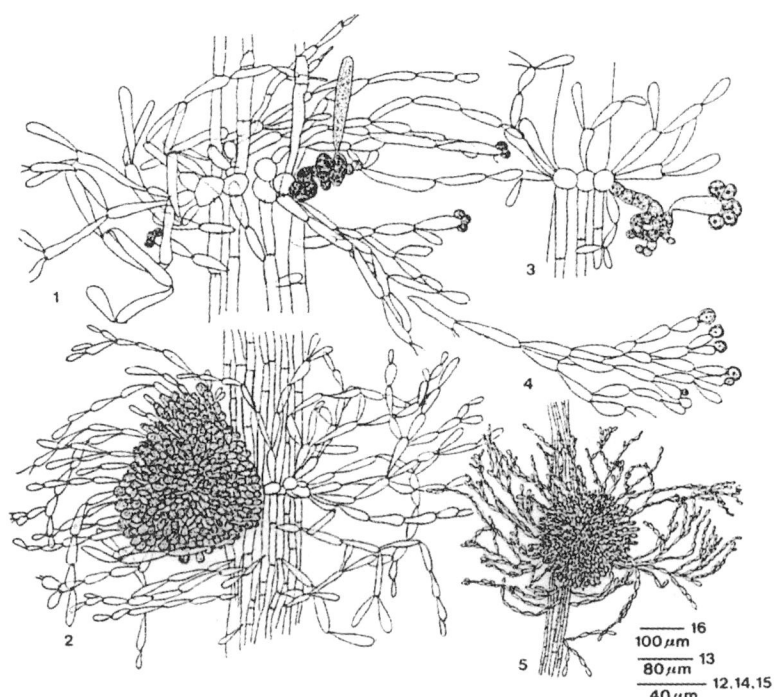

Plate 132
Batrachospermum iyengarii Skuja in Balakrishnan et Chaugule, figs 1–2 (Balakrishnan & Chaugule 1980c)

Batrachospermum zeylanicum Skuja in Balakrishnan et Chaugule, figs 3–5 (Balakrishnan & Chaugule 1980c)

portion, cells comparatively long, 3–5 µm in diameter, 14.5–64 µm long. Primary fascicles repeatedly formed, 4–14 cell-storeys, cells more or less uniform and almost cylindrical except for the terminal cells which are ellipsoidal. Secondary fascicles rather sparse, short constricted just below the nodes, consisting of thin filaments which are unbranched or branched once or twice, 3–10 cell-storeys; cells cylindrical to elongated, 3–4.1 µm in diameter, 16–20 µm long. Terminal hairs not prominent. Spermatangia terminal or subterminal on primary and rarely on secondary fascicles. solitary or in pairs, 4–10.6 µm in diameter. Carpogonium-bearing branches spirally coiled, sometimes curved, consisting of 3–38 cells, single, rarely in pairs, arising from periaxial cells. Involucral filaments very prominent and long, simple, or once branched. Carpogonium hemispherical, 4.5–8.5 µm in diameter, 4.5–9 µm long. Trichogyne distinctly stalked, mostly conical, sometimes ovoidal to ellipsoidal and compressed, 4.5–10 µm in diameter, 8–16.5 µm long. Carposporophytes somewhat large, spherical to hemispherical, generally single, rarely geminate, 158–260 µm in diameter. Gonimoblast filaments compactly aggregated, cells 3–10 µm in diameter, 16–25 µm long. Carposporangia terminal, ellipsoidal or obovoidal, 13–16 µm in diameter, 14.5–23 µm long. Chantransia phase filamentous and heterotrichous; cells of prostrate system thick walled, irregular in shape and 10–20 µm in diameter. Erect filaments submonopodially branched, upright branches 1–150 cell-storeys, elongated cells thick walled, 6–18 µm in diameter, 12–35 µm long. Monosporangia terminal or lateral on the upright filaments, semispherical, 7–12 µm in diameter.

Type Locality: Mahabaleshwar, India in Asia. Growing on rocky bed in streams.

Holotype: BBC 102, coll. Balakrishnan, 1962, deposited in the Herbarium of University of Poona.

Distribution: Known from the type locality only.

Batrachospermales

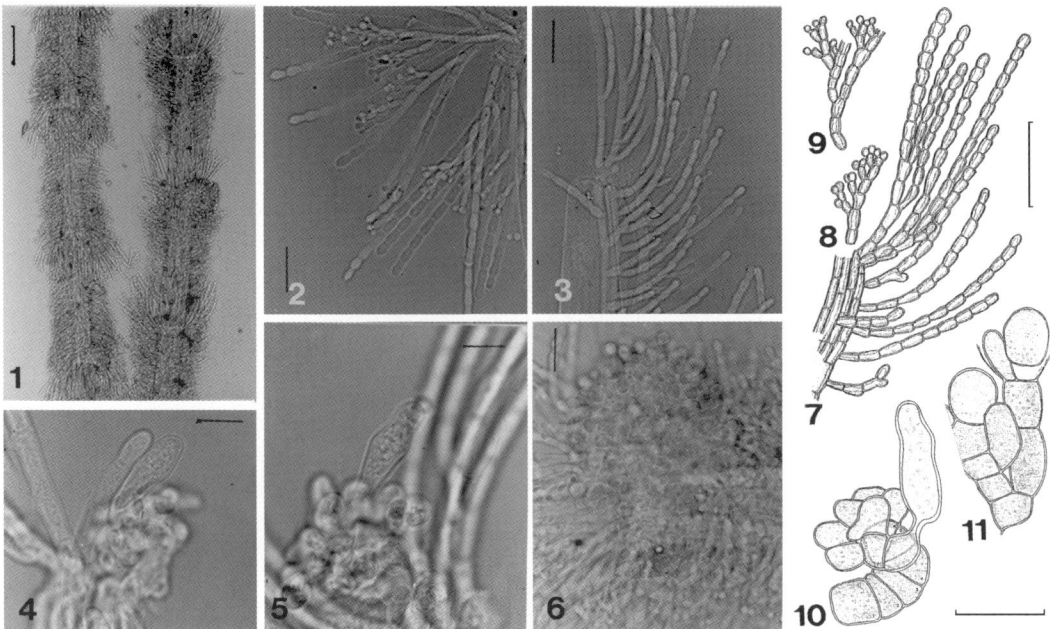

Plate 133
Batrachospermum tiomanense, figs 1–6 (Kumano 1996zi), figs 7–11 (original authors, Ratnasabapathy & Kumano 1982a)
1, 6. a main axis with subconical whorls showing primary, secondary fascicles and a spherical axial carposporophyte, 2, 8–9. spermatangia lateral on secondary fascicles, 3, 7. sparsely branched primary, secondary fascicles, 4–5, 10. a carpogonium with an urn-shaped trichogyne terminal on a twisted carpogonium-bearing branch, 11. ovoidal carposporangia terminal on gonimoblast filaments.
Scale bars = 200 μm (fig. 1); 50 μm (figs 7–9); 40 μm (figs 2–3, 6); 20 μm (figs 10–11); 10 μm (figs 4–5)

16) *Batrachospermum omobodense* Kumano et Bowden-Kerby (1986:114, figs 26–38)

plate 134, figs 1–9 (original authors, Kumano & Bowden-Kerby 1986), Kumano 1996zf

Plants monoecious, mucilaginous, ca. 4 cm high, 250–350 μm in diameter, very frequently pseudodichotomously branched, deep green; whorls barrel-shaped, touching each other in old plants; primary fascicles more or less unilaterally branched, 8–12 cell-storeys, cells ellipsoidal, terminal hairs absent; cortical filaments well-developed; secondary fascicles numerous, unbranched or dichotomously branched, 6–12 cell-storeys, covering all internodes; spermatangia spherical, 3–5 μm in diameter, terminal or lateral on secondary, rarely primary fascicles; carpogonium-bearing branches spirally coiled, consisting of 5–7 barrel-shaped cells, arising from periaxial cells; involucral filaments sparse, short; carpogonia 3–5 μm in diameter at the base, 7–8 μm in diameter at the apex, 35–40 μm long, trichogyne club-shaped; carposporophytes single or in pairs, spherical or hemispherical, 170–220 μm in diameter, 120–190 μm high, inserted centrally; carposporangia obovoidal, 8–11 μm in diameter, 10–14 μm long.

Type Locality: Omobodo stream arising from Ngerremlengui taro swamp, Ngerremlengui, Palau, Western Caroline Islands, Micronesia in Pacific Ocean. Occurs on rocks in slight current.
Holotype: Kobe University, Bowden-Kerby, 24/XII 1983.
Distribution: Known from the type locality only.

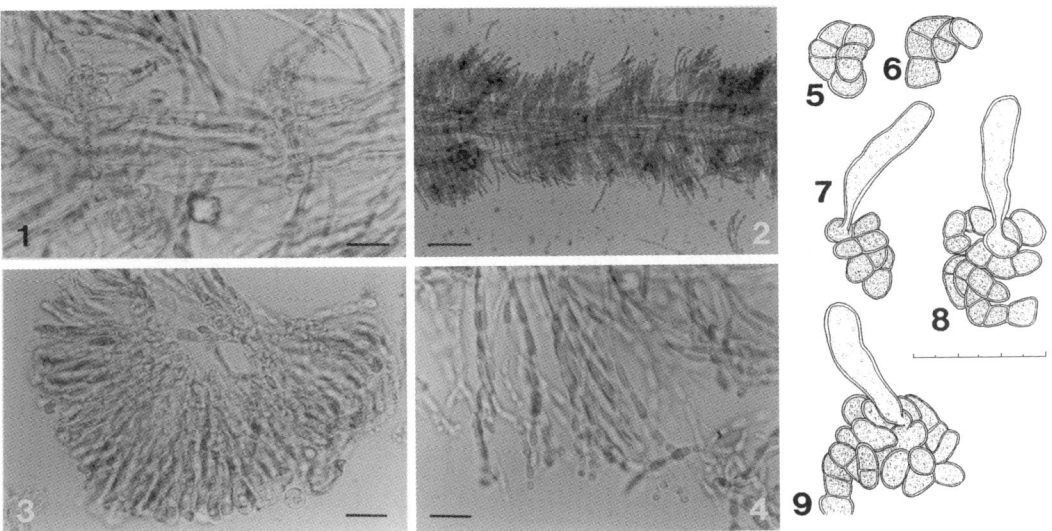

Plate 134
Batrachospermum omobodense, figs 1–9 (original authors, Kumano & Bowden-Kerby 1986, Kumano 1996zf)
1. a portion of a plant showing axial cells and primary fascicles with two carpogonium-bearing branches, 2. a portion of a plant showing barrel-shaped whorls, 3. carposporangia terminal on compactly agglomerated gonimoblast filaments, 4. spermatangia terminal on secondary fascicles, 5–6. coiled carpogonium-bearing branches at the very young stages in development, 7–9. coiled carpogonium-bearing branches with mature trichogynes. Scale bars = 100 μm (fig. 2); 30 μm (figs 5–9); 20 μm (figs 1, 3–4).

17) ***Batrachospermum ambiguum*** Montagne (1850:296)
The synonymy proposed by Vis *et al.* (1992) is as follows;
Synonym: *B. bicudoi* Necchi (1986:521, figs 21–19); *B. exsertum* Necchi (1986:523, figs 30–36); *B. basilare* Flint et Skuja in Flint (1953:10)

plate 136 (a), figs 1–4 (Kumano 1996b), figs 5–7 (Kumano 1990)
plate 136 (b), figs 1–6 (Sheath, Vis & Cole 1992)

Plants monoecious, moderately mucilaginous, 0.5–6 cm high, 300–600 μm in diameter, straight, abundantly and irregularly branched; whorls well-developed, dense or lax, obconical or barrel-shaped, contiguous, rarely separated; fascicles 3 or 4, primary fascicles di- or trichotomously branched, 7–12(–15) cell-storeys, proximal cells cylindrical or ellipsoidal, 4.5–8.5 μm in diameter, 20–45 μm long, distal cells ellipsoidal, obovoidal, pear-shaped, 4–7 μm in diameter, 7–15 μm long, terminal hairs numerous or few, short; cortical filaments well-developed; secondary fascicles numerous, reaching the length of whorls; spermatangia spherical or obovoidal, 5–7 μm in diameter, terminal or subterminal on fascicles; carpogonium-bearing branches spirally coiled, consisting of 4–8 disc- or barrel-shaped cells, arising from periaxial cells, rarely distal cells; involucral filaments numerous, short; carpogonia asymmetrical, 5–6 μm in diameter at the base, 7–10 μm in diameter at the apex, 22–65 μm long, trichogyne cylindrical, club-shaped, or elongate-conical, sessile or stalked; carposporophytes sessile, single, dense, hemispherical, shorter, rarely higher, than whorls, 100–250 μm in diameter, 120–450 μm high, inserted centrally; carposporangia obovoidal, 6.5–10.5 μm in diameter, 10–17.5 μm long.
Type Locality: in rivulet Oral, French Guiana in South America.
Holotype: PC Le Prieur coll. n° 1110.

Batrachospermales

Distribution: Known from the type locality and Harmyi Creek, Barran River, Davis Creek, Little Mulgave River, Mena Creek, Russell River, Bumbelta Creek, Coopers Creek in Queensland, Edith River, Gubera Pool in Northern Territory in Australia; several localities such as Cambriú Point, Cardoso Island, Minicípio of Canaéia, São Paulo, Brazil in South America.

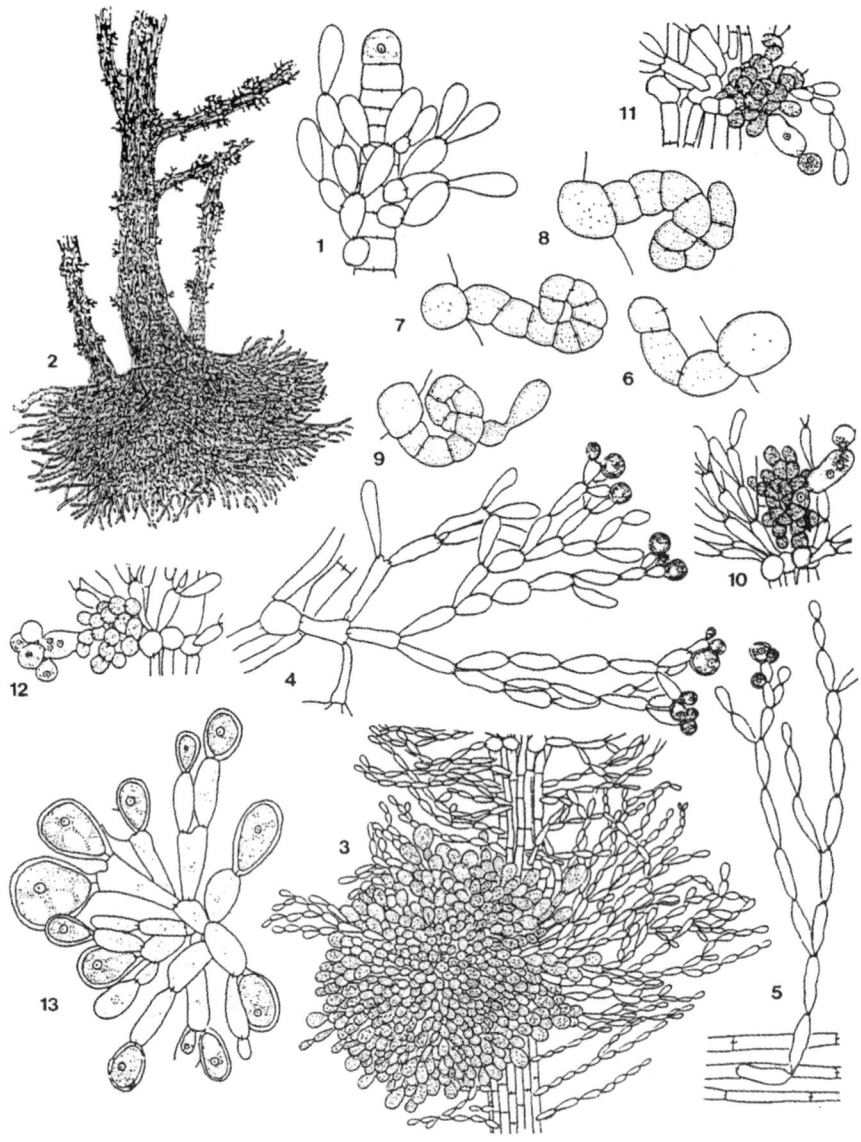

Plate 135
Batrachospermum kylinii Balakrishnan et Chaugule, figs 1–13 (original authors, Balakrishnan & Chaugule 1980b)
1. apical portion of thallus, 2. basal portion of thallus showing stupose rhizoidal growth, 3. portion of thallus showing primary and secondary fascicles as well as axial thallus, 4. spermatangia on a primary fascicle, 5. spermatangia on a secondary fascicle, 6–8. stages in development of a carpogonium-bearing branch, 9. young, spirally coiled carpogonium-bearing branch, distinctly showing the carpogonial and trichogyne nuclei, 10. node showing the spirally coiled carpogonium-bearing branch and an early stage in fertilization, 11–12, nodes enlarged showing the well-developed involucral filaments, 13. gonimoblast filaments with terminal carposporangia.

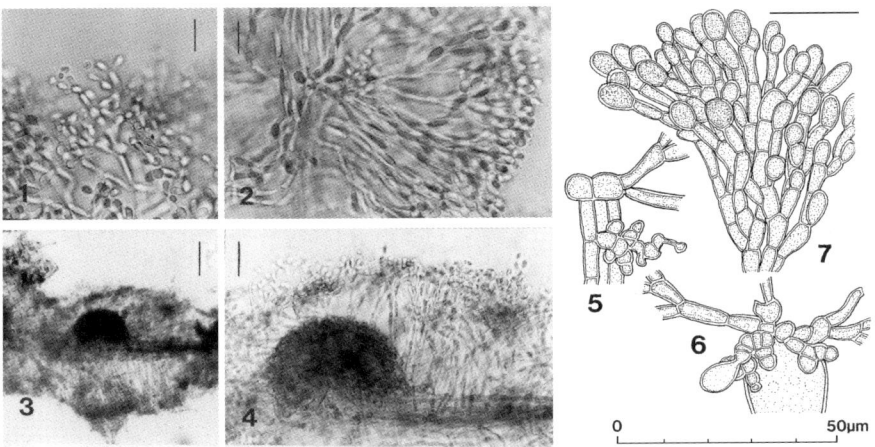

Plate 136 (a)
Batrachospermum ambiguum, figs 1–4 (Kumano 1996b), figs 5–7 (Kumano 1990)
1. spermatangia terminal on primary fascicles, 2, 6. a spirally coiled carpogonium-bearing branch arising from a periaxial cell with a carpogonium having an ellipsoidal trichogyne, 3–4. a hemispherical carposporophyte, 5. a carpogonium-bearing branch arising from a cell of cortical filaments, 7. carposporangia terminal on compactly agglomerated gonimoblast filaments.
Scale bars = 100 µm (fig. 3); 40 µm (fig. 4); 10 µm (figs 1–2); 20 µm (figs 5–7).

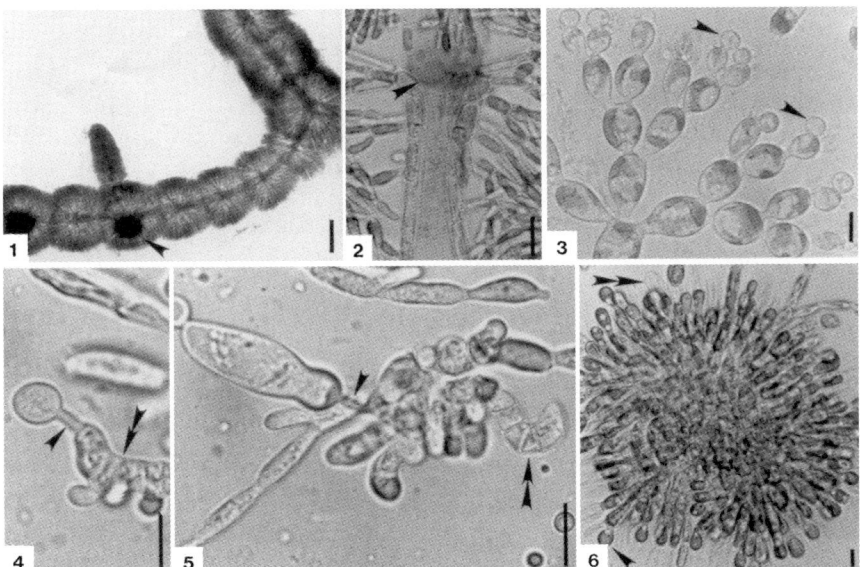

Plate 136 (b)
Batrachospermum ambiguum, figs 1–6 (Sheath, Vis & Cole 1992)
1. a main axis and a branch with barrel-shaped to obovoidal whorls and carposporophytes (arrowhead) inserted within a whorl, 2. a node showing a brown pigmented axial cell (arrowhead), 3. colourless spermatangia (arrowheads) terminal on a fascicle, 4. an immature carpogonium with a distinct stalk (arrowhead) and an upper portion of a carpogonium-bearing branch (double arrowhead), 5. a fertilized carpogonium with a distinct stalk (arrowhead) and an immature curved carpogonium-bearing branch (double arrowhead), 6. a mature carposporophyte with a carpospore (arrowhead) and empty carposporangia (double arrowhead).
Scale bars = 200 µm (fig. 1); 40 µm (figs 2, 6); 10 µm (figs 3–5).

Batrachospermales

Genus *Sirodotia* Kylin (1912:38)

Type: *Sirodotia suecica* Kylin
Synonym: *Batrachospermum* section *Sirodotia* (Kylin) Necchi et Entwisle (1990)

Plants irregularly branched. Carpogonium-bearing branches short, arising from periaxial cells or cells of fascicles or both. Carpogonia asymmetrical with elongate conical or club-shaped trichogynes. Carposporophytes indefinite in shape, extending over along cortical filaments. Gonimoblast filaments of a diffuse and indeterminate type, and developed directly from the fertilized carpogonia.

Note: Necchi & Entwisle (1990) proposed that *Sirodotia* should be reduced to a section of the genus *Batrachospermum*. Based on the examination of 25 North American populations and 10 type specimens of *Sirodotia*, Necchi *et al.* (1993) recognized the following species in North America: *Sirodotia huillensis* Welwitsch ex W. et G. S. West) Skuja (Synonym: *S. ateleia* Skuja), *S. suecica* Kylin (Synonym: *S. acuminata* Skuja ex Flint, *S. fennica* Skuja) and *S. tenuissima* (Collins) Skuja ex Flint.

Vis & Sheath (1998, 1999) stated that *S. suesica* and *S. tenuissima* are conspecific and morphological characters used to distinguished these species are not phylogenetically informative, because the sequence data for RuBisCo (rbcL, rbcL-S spacer, rbcS) gene revealed little variation among two species. On the other hand, *S. huillensis* and *S. suecica* are valid species, because the ITS sequences between these two species diverged significantly.

The following key is mainly based on the system of Necchi, Sheath & Cole (1993a).

Key to the species of the genus *Sirodotia*

1. Spermatangium-bearing branches specialized..2
1. Spermatangium-bearing branches not specialized..3
2. Spermatangia terminal on shortened or secondary fascicle....................................1) *S. yutakae*
2. Spermatangia in clusters on shortened involucral filaments around
   carpogonium-bearing branches, fascicles and cortical filaments........................2) *S. segawae*
3. Carpogonium-bearing branches arising from intercalary cells of
   primary branches and cortical filaments...3) *S. sinica*
3. Carpogonium-bearing branches arising from periaxial cells...4
4. Gonimoblast filaments arising from dorsal side of carpogonium...5
4. Gonimoblast filaments arising from ventral side of carpogonium...7
5. Carpogonium curved to deflexed...4) *S. goebelii*
5. Carpogonium not curved...6
6. Carpogonium 30–40 µm long..5) *S. suecica*
6. Carpogonium 20 µm long...6) *S. gardneri*
7. Carpogonium 37–53 µm long..7) *S. huillensis*
7. Carpogonium 25–35 µm long...8) *S. delicatula*

1) ***Sirodotia yutakae*** Kumano (1982:126, fig. 1, A–I)

   plate 137 (a), figs 1–5 (Kumano 1996zs)
   figs 6–13 (original author, Kumano 1982c)
   plate 141 (b), figs 1–4 (Segawa 1939) as *Sirodotia* sp.

Plants trioecious, richly mucilaginous, 2–10 cm high, 150–250 µm in diameter, abundantly, more or less irregularly branched, bluish deep green; whorls pear-shaped and distant, obconical and contiguous in female plants; primary fascicles abundantly branched, 5–7 cell-storeys, cells fusiform or ovoidal, terminal hairs rare; secondary fascicles numerous, covering all internodes; spermatangia spherical, 4–6 µm in diameter, terminal on shortened lateral of fascicles; carpogonium-bearing branches, 13–20 µm long, consisting of 2–6 barrel-shaped cells, arising from periaxial cells or cells of cortical filaments; involucral filaments short, more or less unilaterally issued; carpogonia with a hemispherical protuberance on one side of basal portion, 20–30 µm long, trichogyne irregular spatula-shaped, distinctly stalked; carposporophytes indefinite in shape, gonimoblast filament irregularly branched, creeping along cortical filaments; carposporangia ovoidal or obovoidal, 6–8 µm in diameter, 10–14 µm long.

Type Locality: Ohara, Hojo in Hyogo, Japan in Eastern Asia. Known from stones in a small stream.

Holotype: Kobe University, Kumano, 2/III 1961.

Distribution: Known from the type locality and in Ryu-ga-mori and Hariko-yachi in Iwate, Japan in Eastern Asia.

Note: Necchi, Sheath & Cole (1993a) could not confirm the presence of specialized spermatangial branches in the materials of this species examined, but retained this species pending further examination. The species was originally described and figured (as *Sirodotia* sp.) by Segawa (1939:2034, figs 1, 3).

2) *Sirodotia segawae* Kumano (1982:129: fig. 2, A–J)

plate 137 (b), figs 1–6 (Kumano 1996t)
figs 7–10 (original author, Kumano 1982c)
plate 141 (b), figs 5–9 (Segawa 1939) as *Sirodotia* sp.

Plants monoecious, richly mucilaginous, 3–6 cm high, 300–680 µm in diameter, abundantly, more or less irregularly branched, deep bluish green; whorls pear-shaped or obconical; primary fascicles abundantly branched, 8–13 cell-storeys, cells lanceolate, ellipsoidal or fusiform, 5–6 µm in diameter, 7–15 µm long, terminal hairs present; secondary fascicles numerous, covering all internodes; spermatangia ellipsoidal or ovoidal, 6–8 µm in diameter, 7–12 µm long, aggregated on short branchlets or on shortened involucral filaments of carpogonium-bearing branches; carpogonium-bearing branches, consisting of 3–9 barrel-shaped cells, arising from periaxial cells, or proximal cells of fascicles; involucral filaments numerous, embracing carpogonia; carpogonia with a hemispherical protuberance on one side of basal portion, 30–46 µm long, trichogyne spatula-shaped, distinctly stalked; carposporophytes indefinite in shape, gonimoblast filament irregularly branched, arising from the same side (ventral side) and the opposite side (dorsal side) of a basal hemispherical protuberance of carpogonium, creeping along cortical filaments; carposporangia ellipsoidal or ovoidal, 7–10 µm in diameter, 10–12 µm long.

Type Locality: Ruri-kei, Sonobe in Kyoto, Japan in Asia. Known from stones and roots of vascular plants in small streams in hillside swamps.

Holotype: Kobe University, Hirayama, 17/IV 1966.

Distribution: Known from the type locality and Takara-ga-ike in Kyoto, Japan in Asia.

Note: Necchi, Sheath & Cole (1993a) could not confirm the presence of specialized spermatangial branches in the materials of this species examined, but retained this species pending further examination. The species was originally described and figured (as *Sirodotia* sp.) by Segawa (1939:2035, figs 2, 4).

Batrachospermales

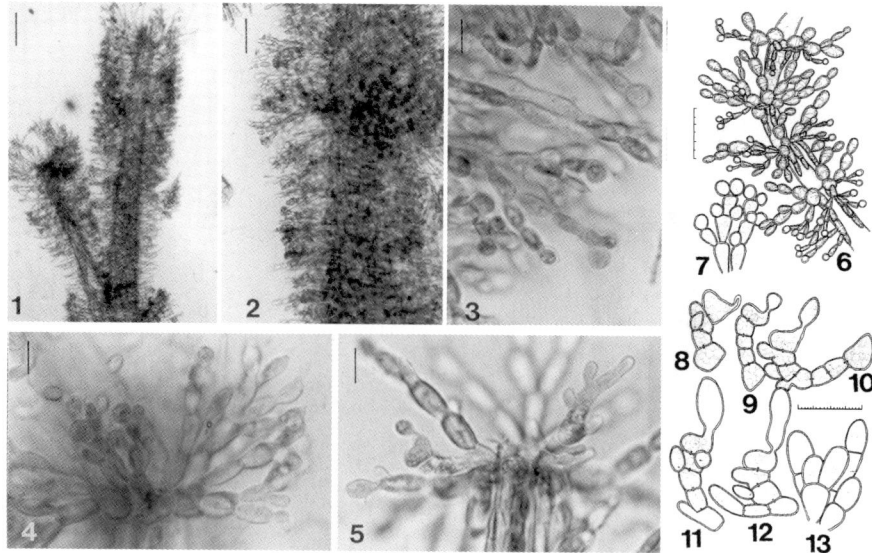

Plate 137 (a)
Sirodotia yutakae, figs 1–5 (Kumano 1996zs), figs 6–13 (original author, Kumano 1982c)
1–2. a main axis with cylindrical or subconical whorls, 3–4, 6–7. spermatangia terminal on shortened and slender laterals of fascicles, 5, 8–12. the successive stages in development of a carpogonium and a carpogonium-bearing branch arising from a periaxial cell, 13. carposporangia terminal on laterals of indeterminate gonimoblast filaments.
Scale bars = 100 μm (fig. 1); 40 μm (fig. 2); 30 μm (fig. 6); 20 μm (figs 7–13); 10 μm (figs 3–5).

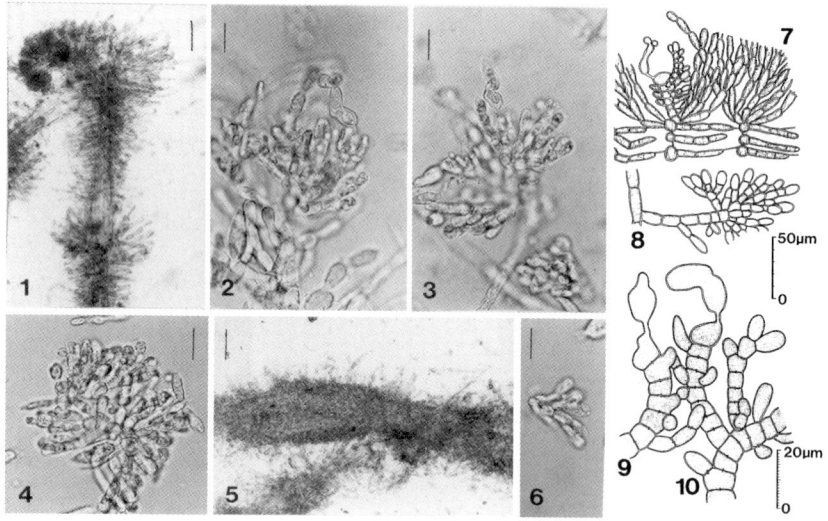

Plate 137 (b)
Sirodotia segawae, figs 1–6 (Kumano 1996t), figs 7–10 (original author, Kumano 1982c)
1. a main axis with obconical whorls, 2–4, 9–10. a mature carpogonium terminal on a carpogonium-bearing branch which also provides spermatagnia bearing involucral filaments, 5. a portion of a plant showing indeterminate and defused carposporophyte, 6. carposporangia terminal on gonimoblast filament, 7. a portion of a plant showing axial cells, primary fascicle, cortical filaments and a carpogonium-bearing branch, 8. spermatangia in clusters on the shortened branchlets arising from a cell of cortical filament.
Scale bars = 200 μm (fig. 5); 100 μm (fig. 1); 50 μm (figs 7–8); 20 μm (figs 9–10); 10 μm (figs 2–4, 6).

3) *Sirodotia sinica* Jao (1941:267, pl. V, figs 31–38)

plate 138 (a), figs 1–13 (Kumano 1996zr)
plate 138 (b), figs 1–5 (original author, Jao 1941)

Plants dioecious, richly mucilaginous, 3–4 cm high, 250–600 μm in diameter, abundantly, more or less irregularly branched, deep bluish green; whorls pear-shaped and separated when young, obconical in old plants; primary fascicles abundantly branched, 7–10 cell-storeys, cells lanceolate, ellipsoidal, fusiform or ovoidal, terminal hairs absent; secondary fascicles numerous, covering all internodes; spermatangia spherical, 5–6 μm in diameter, terminal on primary fascicles; carpogonium-bearing branches more or less curved, 10–20 μm long, consisting of 2–5 barrel-shaped cells, arising from proximal cells of fascicles and cells of cortical filaments; involucral filaments short; carpogonia with a hemispherical protuberance on one side of basal portion, 20–28 μm long, trichogyne spatula-shaped, distinctly stalked; carposporophytes indefinite in shape, gonimoblast filament irregularly branched, arising from the same side (ventral side) and the opposite side (dorsal side) of a basal hemispherical protuberance of carpogonium, creeping along cortical filaments; carposporangia ovoidal or obovoidal, 8–9 μm in diameter, 14–17 μm long.

Type Locality: in pond, Po-sang, Mien-ning, Yunnan, China in Eastern Asia.
Holotype: YN 2, no 1341.
Distribution: Known from the type locality and Yamada-gawa River, Ono in Hyogo, Japan in Eastern Asia.

Note: According to Jao (1941), this species differs from *S. suecica*, in having different form of primary fascicles and secondary fascicles, and trichogynes with an irregular shape and not distinctly pedicellate. In general appearance, it seems to be closely related to *S. huilensis* which, however, has much shorter carpogonium-bearing branches, only 1–3 cells long, and small carpogonia.

4) *Sirodotia geobelii* Entwisle et Foard (1999b:610, fig. 3)

plate 139, figs 1–8 (original authors, Entwisle & Foard 1999b)

Plant monoecious, whorls conical, confluent or separated, 230–379 μm in diameter, internodes 368–529 μm long. Axial cells 11–90 μm in diameter, cortical filament cells cylindrical to somewhat barrel-shaped, 4–7 μm in diameter, 7 μm or more long. Primary fascicles curved, 9–12 cell storeys, di- or trichotomously, 4–5 times branched; proximal cells ellipsoidal, 8–10 μm in diameter, 18–32 μm long, distal cells variable in shape, 3–7 μm in diameter, ca. 14 μm long. Secondary fascicles common, some as long as primary fascicles. Spermatangia scattered on primary fascicles, spherical to ellipsoidal, 5–9 μm in diameter. Carpogonium-bearing branch curved, arising from 2–6 cells from periaxial cells on primary (and rarely secondary) fascicles, consisting of 2–3 modified, barrel-shaped to discoidal, 6–8 μm in diameter, 4–8 μm long. Involucral filaments 1–4 cells long, arising from all cells of carpogonium-bearing branch, lower ones extending beyond carpogonium. Carpogonium curved to deflexed, 11–16 μm in diameter at base, 26–38 μm long, trichogyne stalked (rarely less than 5 μm long), radially symmetrical, ellipsoidal to cylindrical, 5–7 μm in diameter at broadest portion. Gonimoblast initial arising from the opposite side (dorsal side) of a basal hemispherical protuberance of carpogonium, erect gonimoblast filaments extending into outer portion of whorl, with up to 9 cell storeys, cells to 20 μm long. Carposporangia ellipsoidal, oblong or dome-shaped, 6–9 μm in diameter, 13–21 μm long. No chantransia phase are observed.

Batrachospermales

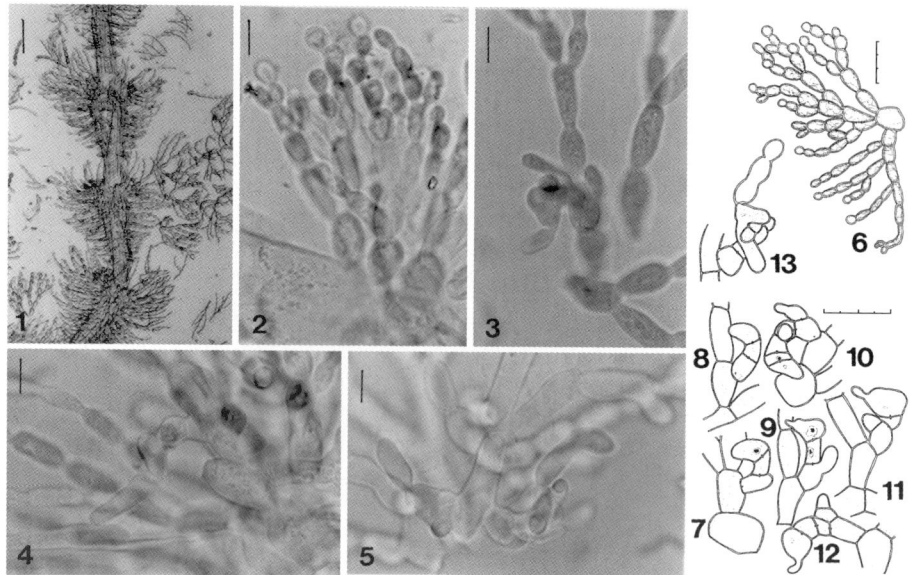

Plate 138 (a)
Sirodotia sinica, figs 1–13 (Kumano 1996zr)
1. a main axis with subconical whorls, 2, 6. spermatangia terminal on fascicles, 3–5, 7–12. the various stages in development of carpogonium-bearing branches, 13. a fertilized carpogonium with an attached spermatium, a slightly stalked, cylindrical trichogyne, a basal protuberance, and a gonimoblast initial on the opposite side (dorsal side) of a basal hemispherical protuberance of carpogonium.
Scale bars = 100 μm (fig. 1); 30 μm (fig. 6); 20 μm (figs 7–13); 10 μm (figs 2–5).

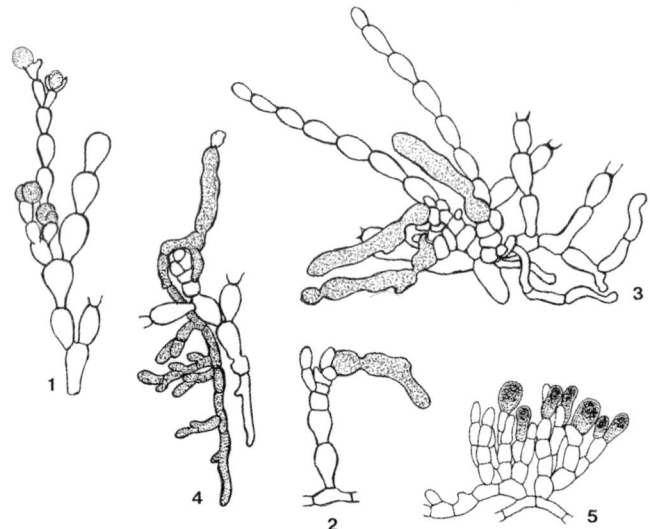

Plate 138 (b)
Sirodotia sinica, figs 1–5 (original author, Jao 1941)
1. spermatangia terminal on primary fascicle, 2. a young carpogonium-bearing branch arising from a cell of a cortical filament, 3. two carpogonium-bearing branches arising from proximal cells of fascicles, and one fertilized carpogonium beginning to produce a gonimoblast initial, 4. a fertilized carpogonium and gonimoblast filaments arising from the same side (ventral side) of a basal hemispherical protuberance of carpogonium, 5. well-developed indeterminate gonimoblast filaments showing its ramification and carposporangia.

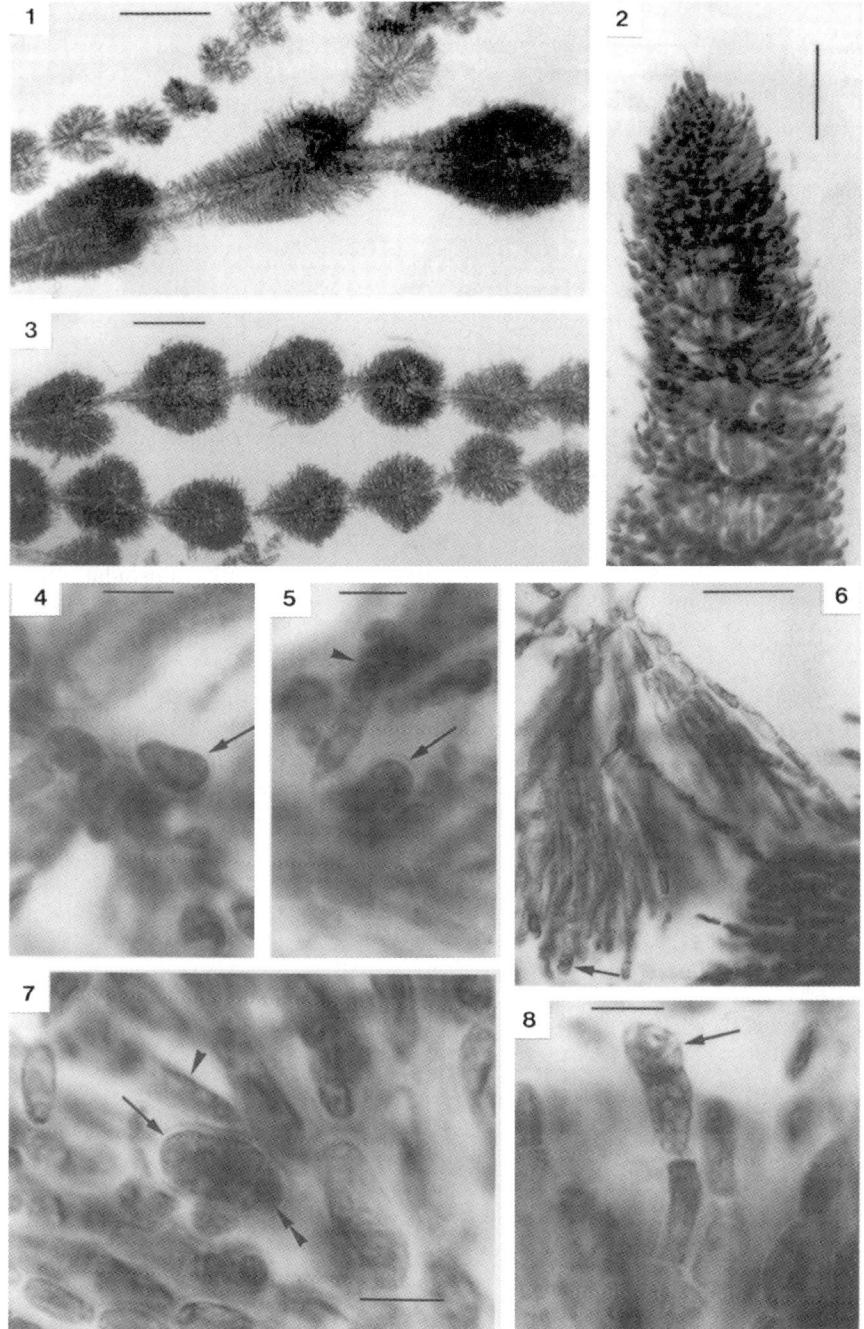

Plate 139
Sirodotia goebelii, figs 1–8 (original authors, Entwisle & Foard 1999b)
1. plant with carposporophytes (dark masses) occupying most of whorl, 2. plant apex, 3. plant portion with shorter whorls, 4, 5, 7. carpogonia showing carpogonium base (arrows), trichogyne (arrowhead) and gonimoblast initial (double arrowhead), 6. long erect gonimoblast filaments with carposporangium (arrow), 8. carposporangium (arrow) and elongate subtending cell.
Scale bars = 250 µm (figs 1, 3); 50 µm (figs 2, 4); 10 µm (figs 4–6, 8).

Batrachospermales

Type locality: the site of the type locality is uncertain, The type specimen is labeled Hermitage, Western Australia, a place name that is no longer recorded in gazetteers. Presumably, Hermitage is somewhere in the south-west of West Australia (Entwisle & Foard 1999).
Holotype: UPS, Hermitage, Western Australia, 1898, coll. Geobel.
Distribution: Known from the type locality only.

5) *Sirodotia suecica* Kylin (1912:38, fig. 16 a–f)
Synonym: *Sirodotia fennica* Skuja (1931:297, tab. 6, figs 1–14, tab. 7, figs 1–11); *S. acuminata* Skuja in Flint (1950:755, figs 7–12); *S. tenuissima* (Collins) Skuja ex Flint (1948:431, figs 25–28); *Batrachospermum vagum* (Roth) C. Agardh f. *tenuissimum* Collins (1895:110);

plate 140, figs 1–5 (Necchi, Sheath & Cole 1993a)
figs 6–9 (Necchi, Sheath & Cole 1993a) as *S. tenuissima*
plate 141 (a), figs 1–7 (Kumano 1997m), figs 8–10 (Kumano 1982c)
plate 142, figs 1–6 (original author, Kylin 1912)
figs 7–12 (original author, Skuja 1931) as *Sirodotia fennica*

Plants trioecious, richly mucilaginous, 2–10 cm high, 200–300 µm in diameter, more or less abundantly and irregularly branched, yellowish green. Whorls small, ellipsoidal, separated each other in male plants, finally pear-shaped or confluent. Primary fascicles 8–10 cell-storeys, cells cylindrical or barrel-shaped, ellipsoidal or obovoidal, terminal hairs present. Secondary fascicles numerous, finally reaching the length of primary ones. Spermatangia spherical, 5–6 µm in diameter, terminal on fascicles. Carpogonium-bearing branches, short, consisting of 2–5 barrel-shaped more or less iso-diametrical cells, arising from periaxial cells. Involucral filaments short, consisting of ellipsoidal cells. Carpogonia with a hemispherical protuberance on one side of basal portion, 30–40 µm long, trichogyne ellipsoidal, finally cylindrical or spatula-shaped more or less distinctly stalked. Carposporophytes indefinite in shape, gonimoblast filament irregularly branched, arising from the opposite side (dorsal side) of a basal hemispherical protuberance of carpogonium, creeping along cortical filaments. Carposporangia obovoidal to elongated pear-shaped, 5–8 µm in diameter, 10–12 µm long.
Type Locality: Osby, Skåne, Sweden in Europe.
Lectotype: LD, Kylin, 3/VIII 1909.
Distribution: Known from the type locality in Europe; Japan in Eastern Asia; South Australia, Queensland, New South Wales in Australia, North Island, South Island in New Zealand; USA and Canada in North America and widely distributed in the world. In Japan, Sawano-ike in Kyoto. In North America, an outlet of an oxbow lake, near Lake Piseco, Adirondack Mountains, Hamilton, New York, Louisiana, Maine, and Rhode Island in USA, Quebec (as *S. tenuissima*), in Canada.
Note: Vis & Sheath (1998, 1999) stated that *S. suesica* and *S. tenuissima* are conspecific and morphological characters used to distinguished these species are not phylogenetically informative, because the sequence data for RuBisCo (rbcL, rbcL-S spacer, rbc-S) gene revealed little variation among two species. On the other hand, *S. huillensis* and *S. suecica* are valid species, because the its sequences between these two species diverged significantly.

Sirodotia

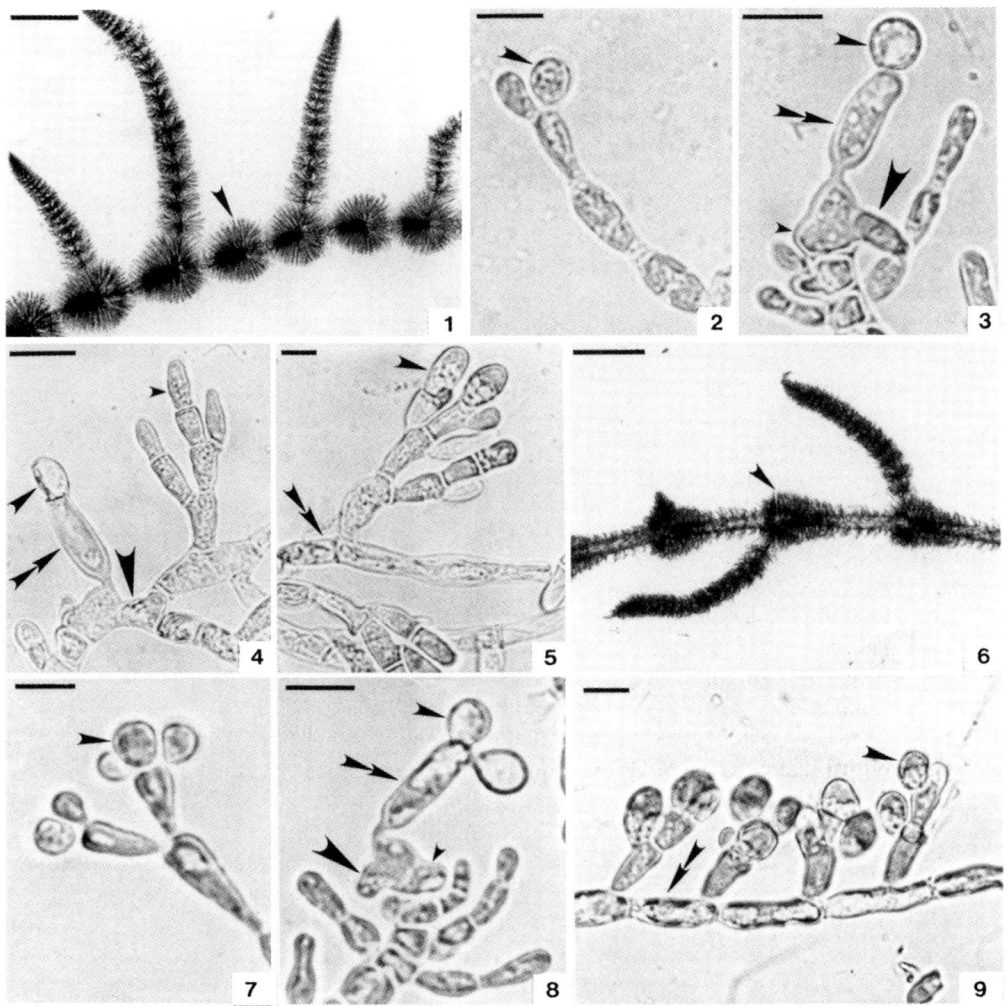

Plate 140
Sirodotia suecica, figs 1–5 (Necchi, Sheath & Cole 1993a)
1. a main axis with obconical whorls (arrowhead) and young branches with dense whorls, 2. spermatangia (arrowhead) terminal on fascicle, 3. a fertilized carpogonium with an attached spermatium (arrowhead), a slightly stalked, cylindrical trichogyne (double arrowhead), a basal hemispherical protuberance (smallest arrowhead), and a gonimoblast initial on the opposite side, dorsal side (largest arrowhead) of a basal hemispherical protuberance of carpogonium, 4. a fertilized carpogonium with an attached spermatium (arrowhead), a slightly stalked trichogyne (double arrowhead), a branched gonimoblast filament (largest arrowhead), and an immature carposporangium (smallest arrow), 5. a mature carposporophyte with carposporangium (arrowhead) and indeterminate prostrate gonimoblast filaments (double arrowhead).
Scale bars = 500 μm (fig. 1); 10 μm (figs 2–5).

Sirodotia suecica as *S. tenuissima*, figs 6–9 (Necchi, Sheath & Cole 1993a).
6. a main axis with truncate-pyramidal whorls (arrowhead) and young branches with dense whorls, 7. a spermatangium (arrowhead) terminal on fascicle, 8. a fertilized carpogonium with an attached spermatium (arrowhead), a slightly stalked, cylindrical trichogyne (double arrowhead), a basal hemispherical protuberance (smallest arrowhead), and a gonimoblast initial on the opposite side, dorsal side (largest arrowhead) a basal hemispherical protuberance of carpogonium, 9. a mature carposporophyte with carposporangia (arrowhead) and an indeterminate prostrate gonimoblast filament (double arrowhead).
Scale bars = 400 μm (fig. 6); 10 μm (figs 7–9).

Batrachospermales

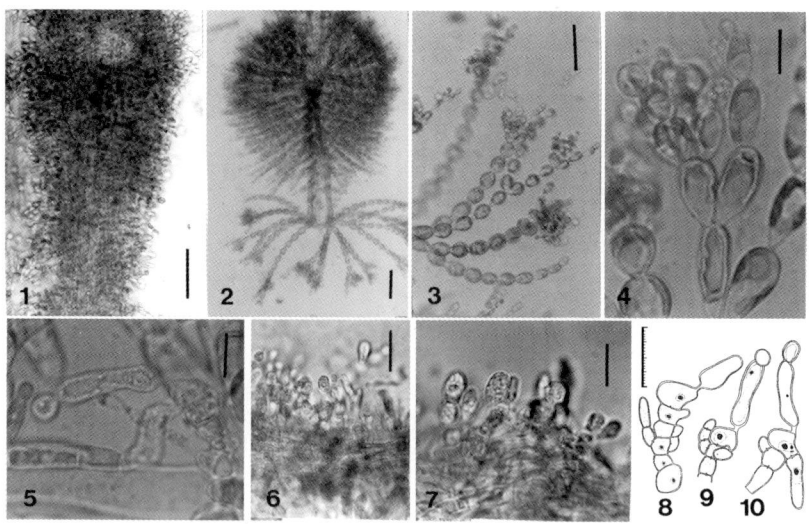

Plate 141 (a)
Sirodotia suecica, figs 1–7 (Kumano 1997m), figs 8–10 (Kumano 1982c)
1. a subconical whorl, 2, a subconical whorl and spermatangia in clusters terminal on primary fascicles, 3–4. spermatangia in clusters terminal on primary fascicles, 5. a fertilized trichogyne with an attached spermatium and gonimoblast filaments, 6–7. carposporangia terminal on lateral of prostrate indeterminate gonimoblast filaments, 8. a carpogonium-bearing branch arising from a periaxial cell, 9. a carpogonium having a trichogyne with an attached spermatium, 10. a fertilized carpogonium with an attached spermatium, a slightly stalked cylindrical trichogyne, a basal hemispherical protuberance and a gonimoblast filament arising from the opposite side (dorsal side) of a basal hemispherical protuberance of carpogonium.
Scale bar = 50 μm (figs 1–3); 20 μm (figs 6, 8–10); 10 μm (figs 4–5, 7).

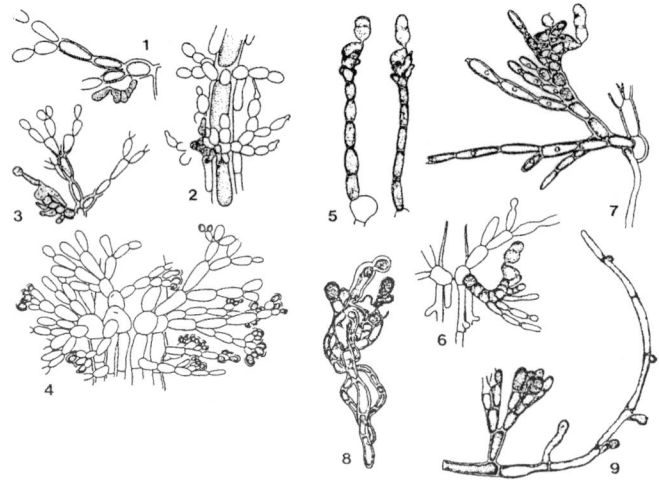

Plate 141 (b)
Sirodotia yutakae as *Sirodotia* sp. figs 1–4 (Segawa 1939)
1, 2. carpogonium-bearing branches, 3. gonimoblast filament arising from the opposite side (dorsal side) of a basal hemispherical protuberance of fertile carpogonium, 4. spermatangia in clusters on the shortened branchlets arising from a cell of cortical filament.

Sirodotia segawae as *Sirodotia* sp., figs 5–9 (Segawa 1939)
5, 6, 7 carpogonium-bearing branches, 8. gonimoblast filament arising from dorsal side of fertile carpogonium base, 9. gonimoblast filament with carposporangia.

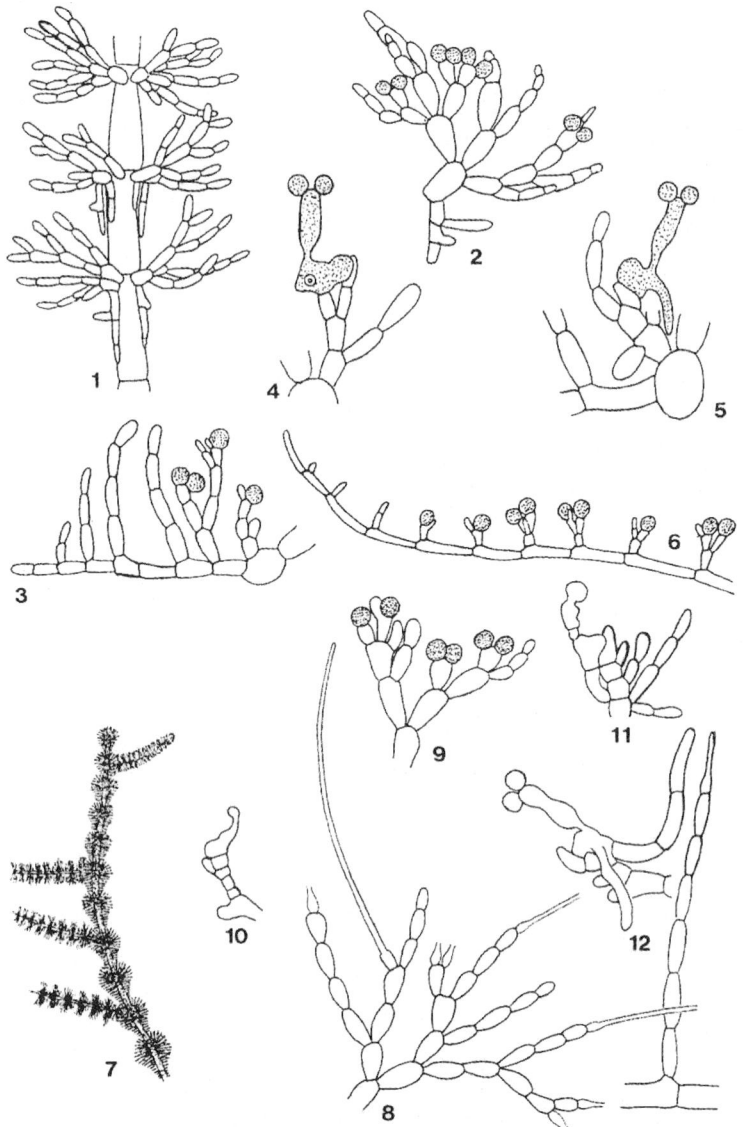

Plate 142
Sirodotia suecica, figs 1–6 (original author, Kylin 1912)
1. a young main axis with whorls, 2. spermatangia terminal on primary fascicles, 3. spermatangia terminal on secondary fascicle, 4. a fertilized carpogonium with attached two spermatia, a slightly stalked cylindrical trichogyne, a basal protuberance and a gonimoblast initial on the opposite side (dorsal side) of a basal hemispherical protuberance of carpogonium , 5. a fertilized carpogonium with attached two spermatia, a slightly stalked cylindrical trichogyne, a basal hemispherical protuberance and a gonimoblast filament arising from the opposite side (dorsal side) of a basal hemispherical protuberance of carpogonium, 6. a carposporophyte with an indeterminate prostrate gonimoblast filament and carposporangia.

Sirodotia suecica as *S. fennica*, figs 7–12 (original author, Skuja 1931)
7. a main axis with truncate-pyramidal whorls (arrow) and young branches with dense whorls, 8. primary fascicle with terminal hairs, 9. spermatangium terminal on fascicle, 10. an immature carpogonium and a carpogonium-bearing branch, 11–12. a fertilized carpogonium with attached spermatia, a slightly stalked, cylindrical trichogyne, a basal hemispherical protuberance of carpogonium, and a gonimoblast filament on the opposite side (dorsal side) of a basal hemispherical protuberance of carpogonium.

Batrachospermales

6) *Sirodotia gardneri* Skuja ex Flint (1950:754, figs 1–6)

plate 145, figs 5–10 (original author, Flint 1950)

Plants monoecious, 3.2–10 cm high, 288–463 μm in diameter, blue-green, grayish or reddish gray; whorls obconical, barrel-shaped and distant; primary fascicles abundantly branched, 6–10 cell-storeys; spermatangia spherical, 5.3–8.0 μm in diameter, mostly in pairs, terminal on fascicles; carpogonia sessile at the nodes, but in some instances, developed from the periaxial cells, about 20 μm long, trichogyne ovoidal, slightly stalked; carposporophytes indefinite in shape, gonimoblast filament irregularly branched, creeping along cortical filaments; carposporangia in individual cluster in the general nodal region.

Type Locality: Arroyo Seco, Los Angeles, California, USA in North America.
Lectotype: UC 395470, Gardner, 31/V 1908, designated by Necchi, Sheath & Cole 1993a.
Distribution: Known from the type locality and from a stream in Verduga Canyon, north of Glendale, Los Angeles, California, USA in North America.
Note: Necchi *et al.* (1993b), did not have materials of this species suitable for comparison with the other species of the genus. It appears to be similar to *S. tenuissima* in having long, well-separated whorls.

7) *Sirodotia huillensis* (Welwitsch ex W. et G. S. West) Skuja (1931:304, tab. 8, figs 1–20)
Basionym: *Batrachospermum huillense* Welwitsch ex W. et G. S. West (1897:3).
Synonym: *Sirodotia ateleia* Skuja (1938:617, tab. XXXII, figs 1–16); *S. cirrhosa* Skuja in Balakrishnan et Chaugule (1980:242, figs 17–21).

plate 143, figs 1–7 (Kumano 1997e), figs 8–19 (Skuja 1931)
plate 144, figs 1–5 (Necchi, Sheath & Cole 1993a)
plate 145, figs 1–4 (original author, Skuja 1931)

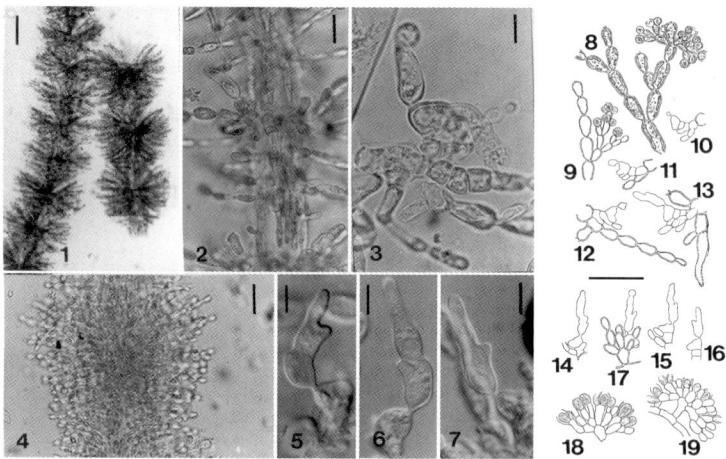

Plate 143
Sirodotia huillensis, figs 1–7 (Kumano 1997e), figs 8–19 (original author, Skuja 1931)
1, 4. a main axis with subconical whorls, 2. a portion of a plant showing a carpogonium-bearing branch, 3. a fertilized carpogonium with an attached spermatium, a trichogyne and a gonimoblast initial arising from the same side (ventral side) of a basal hemispherical protuberance of carpogonium, 5–7, 14–17. a carpogonium with a trichogyne irregular in shape, 8–9. spermatangia in clusters subterminal on fascicles, 10–13. a carpogonium in the various stages in development, 18–19. carposporangia in clusters terminal on laterals of indeterminate gonimoblast filaments.
Scale bars = 100 μm (fig. 1); 40 μm (fig. 4); 20 μm (fig. 2); 10 μm (figs 3, 5–7); 50 μm (figs 8–19).

Sirodotia

Plants dioecious, richly mucilaginous, 3–7 cm high, 250–400 μm in diameter, abundantly branched, bluish green; whorls ellipsoidal, separated from each other when young, afterwards pear-shaped and confluent due to development of secondary fascicles; primary fascicles abundantly branched, 3–10 cell-storeys, cells ovoidal, ellipsoidal, lanceolate, obovoidal, 5–6 μm in diameter, 7–15 μm long, terminal hairs rare; secondary fascicles numerous, finally reaching the length of primary fascicles; spermatangia spherical, 5–7 μm in diameter, terminal on primary and secondary fascicles; carpogonium-bearing branches short, consisting of 2–5 barrel-shaped cells, arising from periaxial cells; involucral filaments short, rare; carpogonia without a distinct hemispherical protuberance on one side of basal portion, 37–53 μm long, trichogyne elongate conical, club-shaped, more or less indistinctly stalked; carposporophytes indefinite in shape, gonimoblast filament irregularly branched, arising from the same side (ventral side) of a basal hemispherical protuberance of carpogonium, creeping along cortical filaments; carposporangia obovoidal, 6–8 μm in diameter, 10–13 μm long.

Type Locality: Lopollo, Huilla, Angola in Africa.

Isotype: LISU, Welwitsch, V 1680.

Distribution: Known from the type locality and India in Asia, Malaysia, Indonesia in Southeast Asia, Arizona, USA in North America.

Note: *S. ateleia* was regarded as a synonym of *S. delicatula* by Umezaki (1960) and Necchi (1991). In contrast, based on the examination of North American populations of *S. ateleia* and *S. huillensis,* Necchi *et al.* (1993) considered these two taxa to be synonymous, and the oldest epithet is *huilensis*. Vis & Sheath (1998, 1999) stated that, *S. huillensis* and *S. suecica* are valid species, because the ITS sequences between these two species diverged significantly.

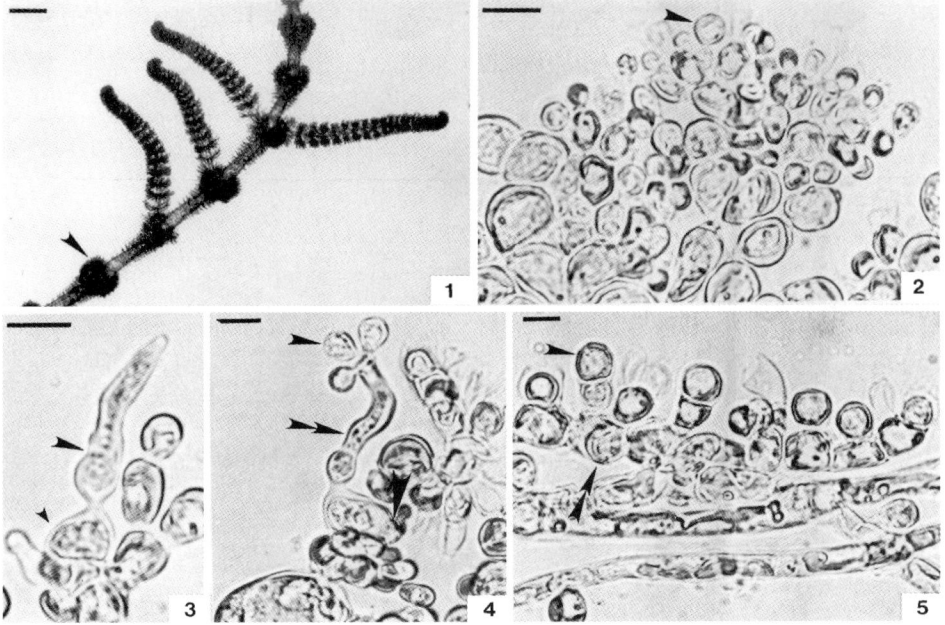

Plate 144
Sirodotia huillensis, figs 1–5 (Necchi, Sheath & Cole 1993a)
1. densely clustered spermatangia (arrow) terminal on fascicle, 2. a carpogonium with an irregular trichogyne (arrow) and a basal hemispherical protuberance (small arrow) of carpogonium, 3. a fertilized carpogonium with an attached spermatium (arrow), 4. an irregular trichogyne (double arrow), and a gonimoblast initial (largest arrow) arising from the same side (ventral side) of a basal hemispherical protuberance of carpogonium, 5. a mature carposporophyte with carposporangia (arrow) and an indeterminate prostrate gonimoblast filament (double arrow). Scale bar = 10 μm.

Batrachospermales

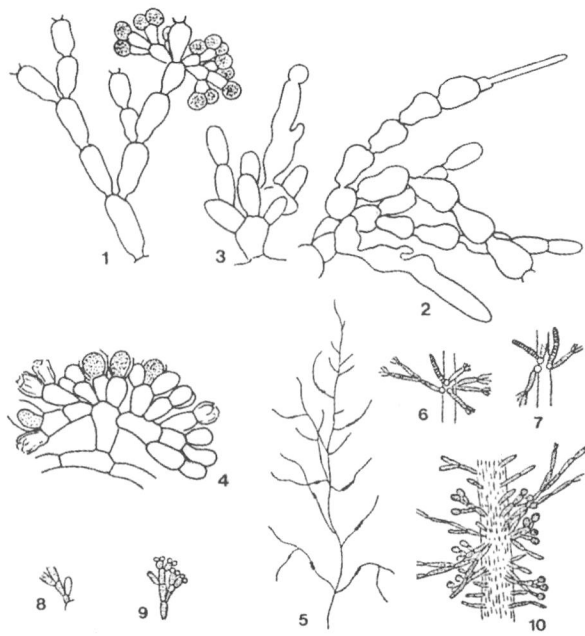

Plate 145
Sirodotia huillensis, figs 1–4 (original author, Skuja 1931)
1. densely clustered spermatangia terminal on fascicle, 2–3. a carpogonium with an irregular trichogyne and a basal hemispherical protuberance, 4. a mature carposporophyte with carposporangia and an indeterminate prostrate gonimoblast filament.

Sirodotia gardneri, figs 5–10 (original author, Flint 1950)
5. plant showing ramification habit, 6. vegetative bud at node, 7. two bud at node, 8. carpogonium, 9. spermatangia terminal on unmodified fascicles, 10. portion of plant showing carposporangia.

8) ***Sirodotia delicatula*** Skuja (1938:614, tab. XXXI, figs 1–15)

   plate 146, figs 1–17 (original author, Skuja 1938)

  Plants dioecious, richly mucilaginous, 3–9 cm high, 140–300 µm in diameter, abundantly branched, bluish green. Whorls ellipsoidal, distant from each other when young, afterwards pear-shaped and confluent due to development of secondary fascicles. Primary fascicles abundantly branched, 4–6 cell-storeys, cells ellipsoidal or ovoidal, 5–6 µm in diameter, 7–15 µm long, terminal hairs rare. Secondary fascicles numerous. Spermatangia spherical, 4–5 µm in diameter, terminal on primary fascicles. Carpogonium-bearing branches short, 8–15 µm long, consisting of 3–5 barrel-shaped cells, arising from periaxial cells. Involucral filaments short, consisting of ellipsoidal cells. Carpogonia with a hemispherical protuberance on one side of basal portion, 25–30 µm long, trichogyne cylindrical, distinctly stalked. Carposporophytes indefinite in shape, gonimoblast filament irregularly branched, arising from the same side (ventral side) of a basal hemispherical protuberance of carpogonium, creeping along cortical filaments. Carposporangia spherical, 6–8 µm in diameter.
  Type Locality: Tjilinwong, Bogor, Java Island, Indonesia in Southeast Asia.
  Holotype: UPS, 19/IX 1928.
  Distribution: Known from Indonesia, Malaysia in Southeast Asia.
  Note: The observation that gonimoblast initials develop from the protuberant, ventral side of the carpogonium (Kumano 1982 and Necchi, Sheath & Cole 1993a) is in contrast with the observations of Umezaki (1960) and Necchi (1991).

Sirodotia

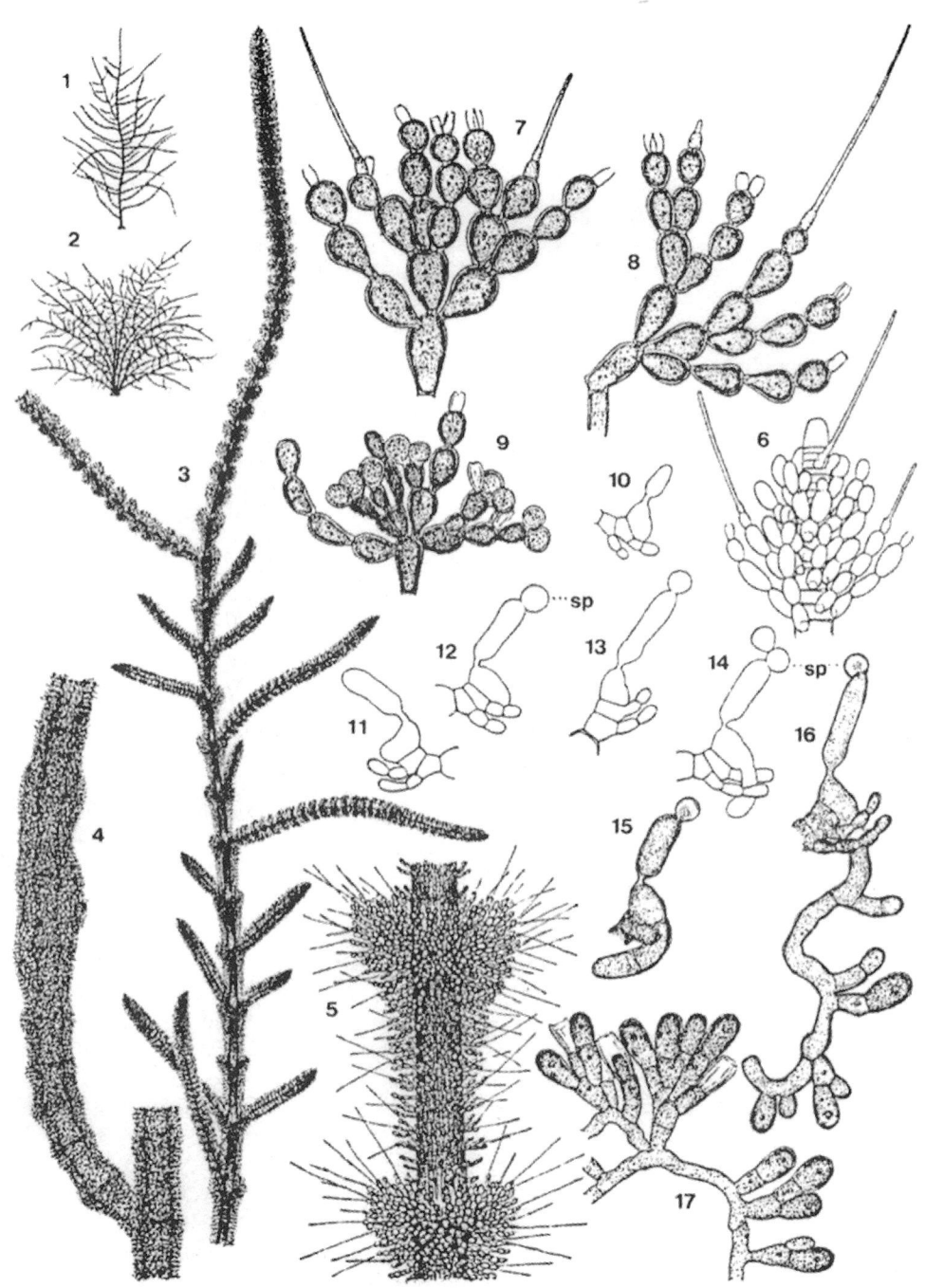

Plate 146
Sirodotia delicatula, figs 1–17 (original author, Skuja 1938)
1–4. habit of plants, 5. primary and secondary fascicles, , 6. apical portion of plant, 7–8. terminal portion of fascicles, 9. spermatangia terminal on fascicle, 10–13. various stages of development of carpogonium with cylindrical trichogyne and basal hemispherical protuberance, 14–16. development of gonimoblast filament arising from the same side (ventral side) of a basal hemispherical protuberance of fertilized carpogonium. 17. carposporangia terminal on laterals of gonimoblast filament.

Batrachospermales

Genus ***Tuomeya*** Harvey (1858:64) in NCU–3 (1993:1184)

Type: *Tuomeya fluviatilis* (Harvey) Harvey
= *Tuomeya americana* (Kützing) Papenfuss
Synonym: *Batrachospermum* section *Tuomeya* (Harvey) Necchi et Entwisle (1990)

Plants irregularly branched, cartilaginous, continuous, solid, at first transversely banded, afterwards annulately constricted. Plants composed of longitudinal axial cell filament with cortical filaments and whorls. Spermatangia ovoid, terminating on lateral branch. Carpogonium-bearing branches short, arising from periaxial cells and or cells of fascicles or both. Carpogonia asymmetrical with elongate-conical or club-shaped trichogynes. Carposporophytes definite in shape. Gonimoblast filaments of the radially branched type, and arising from the dense mass of fusion cells (gonimoblast placenta), which includes the original involucral cells.

1) ***Tuomeya americana*** (Kützing) Papenfuss (1958:104)
Basionym: *Baileya americana* Kützing (1857:35, pl. 87, III, f, f', g)
Synonym: *Tuomeya fluviatilis* (Harvey) Harvey (1858:64)

plate 147, figs 1–5 (Setchell 1890) as *T. fluviatilis*,
plate 148, figs 1–12 (Skuja 1944) *T. fluviatilis*,
plate 149, figs 1–6 (Sheath 1984) as *T. americana*,
plate 150, figs 1–9 (Kaczmarczyk, Sheath & Cole 1993) ("1992") as *T. americana*

Plants bushy, about 2.5–5 cm high, about 2 mm in diameter, reaching 5 mm in diameter at the base, branched irregularly. Axial cell filament covered with compact cortical cell filaments of parenchymatous type. Outer cortex not distinct. Whorl branches abundantly branched at the ends, their branching in close contact with branching of neighboring whorls. Spermatangia, small, ovoid, 1 or 2 terminating on lateral branch, 3.5–4.0(–5.5–6) µm long. Carpogonium-bearing branches short, arising from periaxial cells and or cells of fascicles. Carpogonia 38–43 µm long, asymmetrical with elongate conical or club-shaped trichogynes, 9–10 µm in diameter. Carposporophytes definite in shape. Carposporangia terminating on the gonimoblast filaments growing radially from gonimoblast placenta, 9–10 µm in diameter, 15–16 µm long.

Type Locality: near Fredericksburg, Virginia, USA in North America
Lectotype: TCD, designated by Kaczmarczyk, Sheath & Cole 1993 ("1992").
Distribution: Known from the type locality and in about 16 localities from Newfoundland, Canada, to Louisiana, USA in North America (Kaczmarczyk, Sheath & Cole 1993 ("1992")), and in South Africa (Borge 1928).

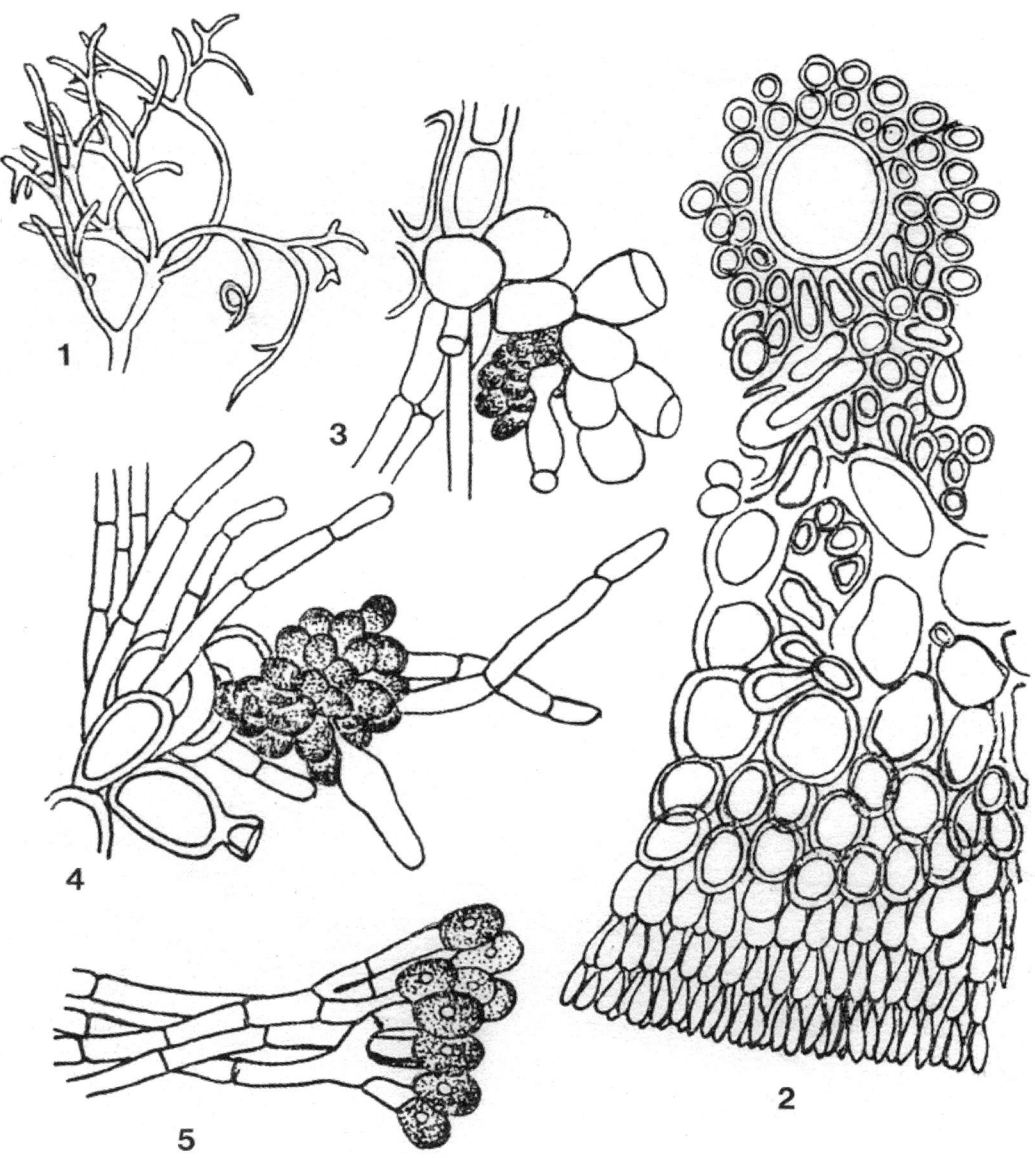

Plate 147
Tuomeya americana as *T. fluviatilis*, figs 1–5 (Setchell 1890).
1. the macroscopic view of a plant with dense branches, 2. the cross section showing an axial cell filament, cortical filaments and fascicles, 3–4. a carpogonium with a trichogyne and a young carposporophyte, 5. carposporangia terminal on gonimoblast filaments.

Batrachospermales

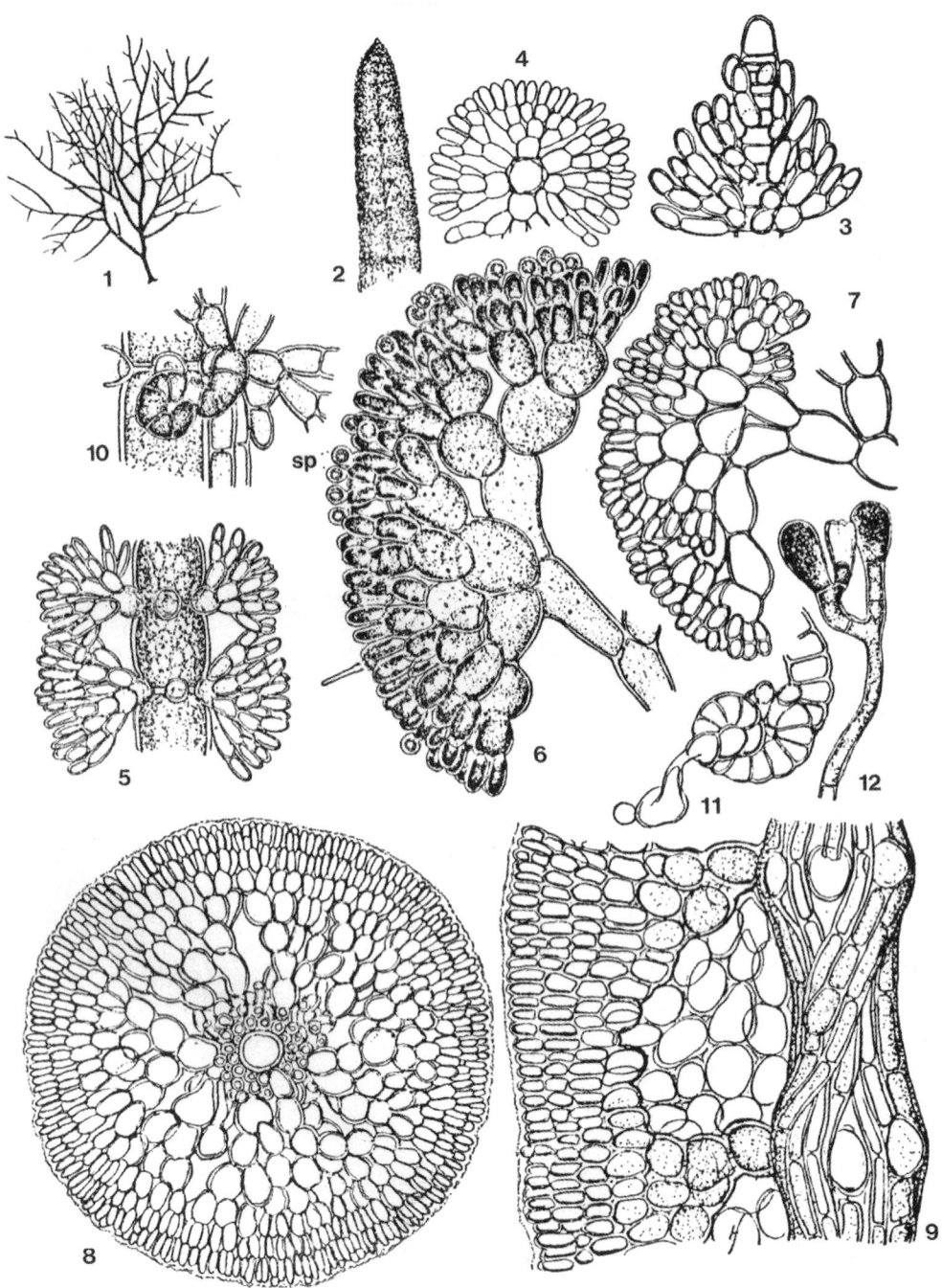

Plate 148
Tuomeya americana as *T. fluviatilis*, figs 1–12 (Skuja 1944)
1. habit of plant, 2. apical portion of branch, 3. apical portion of branch with apical cell, 4. a young whorl observed from axis, 5. two young whorls observed from side, 6–7. distal portion of primary fascicles with a terminal hair and spermatangia (sp), 8. cross section of thallus, 9. longitudinal section of thallus, 10. a young carpogonium-bearing branch, 11. spirally coiled carpogonium-bearing branch, 12. carposporangia terminal on gonimoblast filaments.

Tuomeya

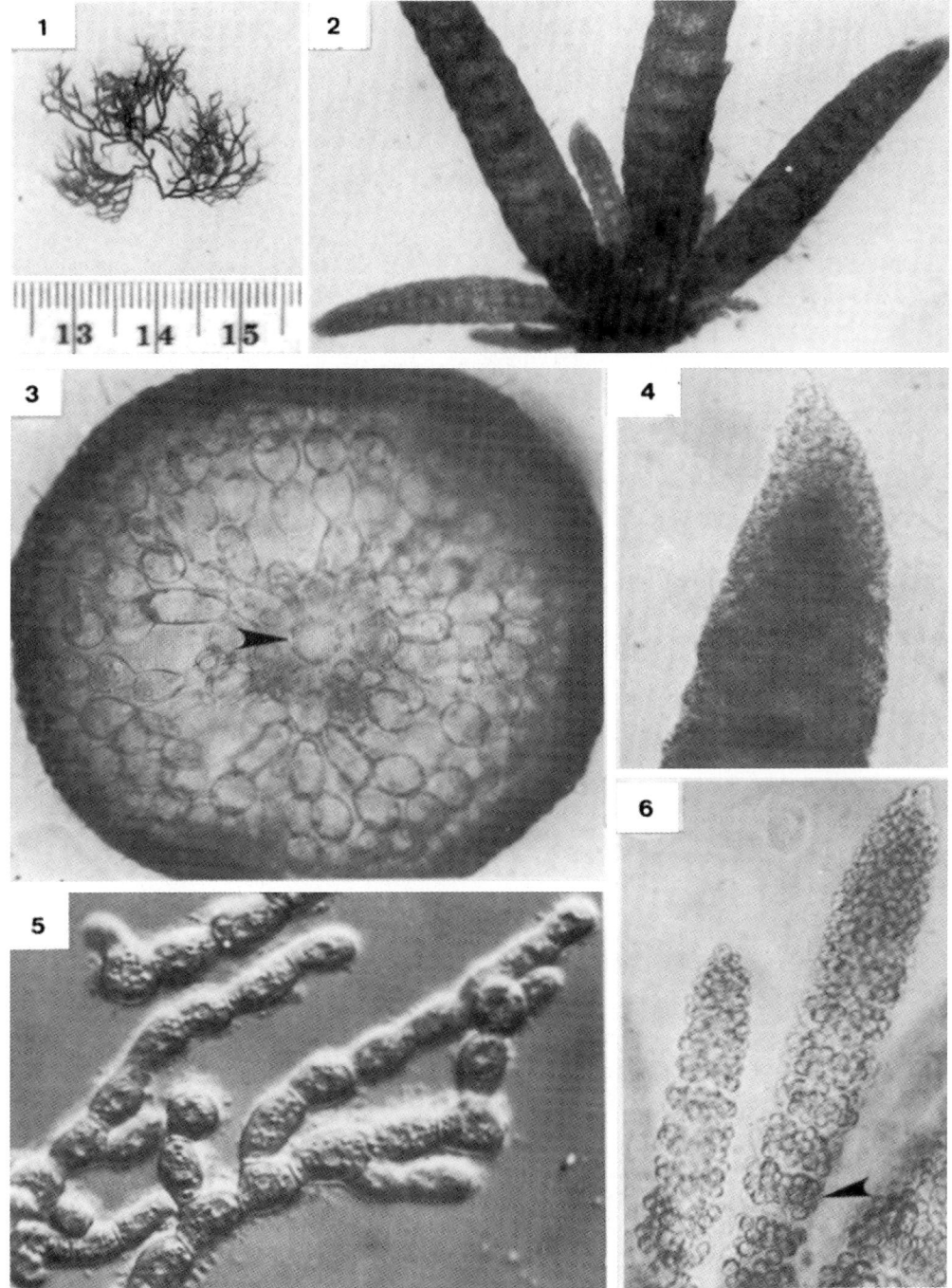

Plate 149
Tuomeya americana, figs 1–6 (Sheath 1984)
1. the macroscopic view of a mature plant with dense ramification, 2. mature plants and juvenile plants attached to a common 'chantransia stage', 3. the cross section through a node showing an axial cell filament (arrow) and fascicles, 4. an apex with a prominent apical cell, 5. 'chantransia stage' with irregular ramification, 6. juvenile gametophyte showing the compact whorls (arrow).

Batrachospermales

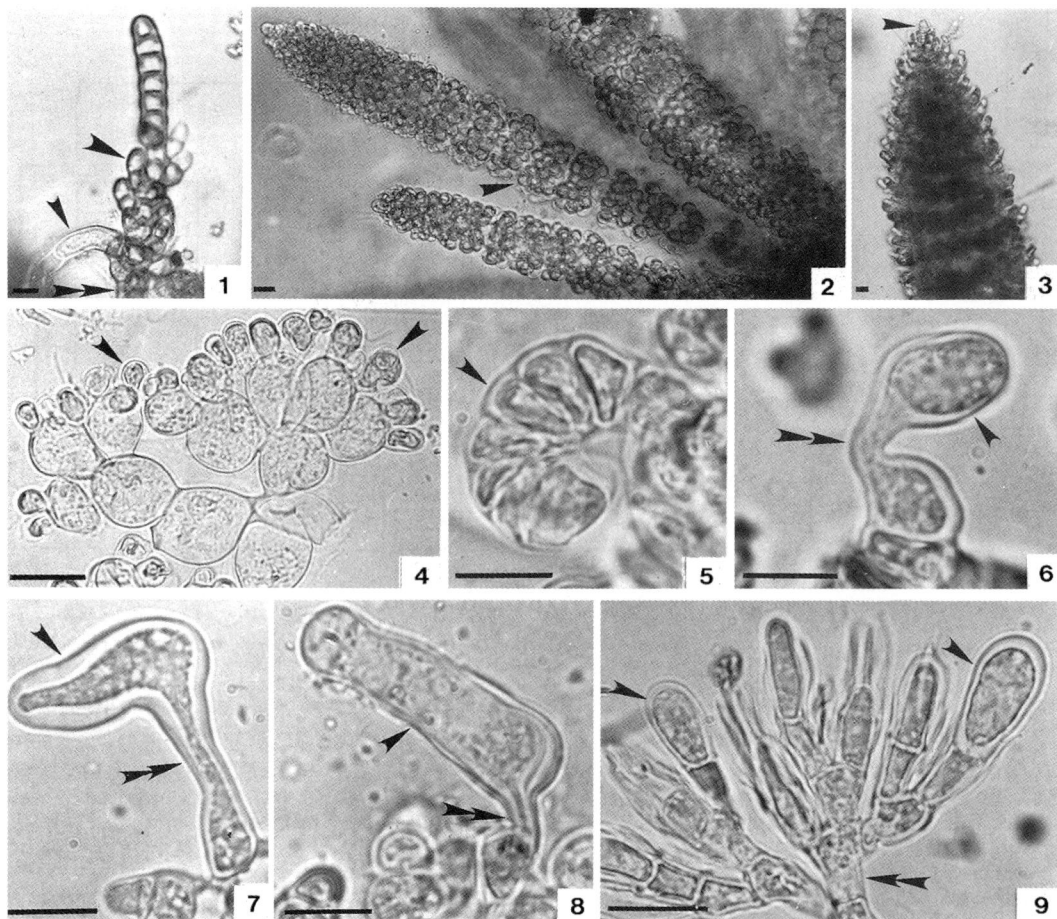

Plate 150
Tuomeya americana, figs 1–9 (Kaczmarczyk, Sheath & Cole 1993 ("1992")).
1. a juvenile gametophyte attached to a basal mass of cells (double arrowhead) by a rhizoidal filament (small arrowhead). Fascicles arise at an acute angle and adhere to the main axis (large arrowhead), 2. a mature gametophyte with determinate fascicles apressing to those above when they become four to five cells long (arrowhead), 3. an apex of mature gametophyte with a prominent dome-shaped apical cell (arrowhead), 4. spermatangia (arrowhead) terminal on determinate fascicles, 5. an immature carpogonium-bearing branch (arrowhead) showing distinct curving, 6. an immature carpogonium with a trichogyne (arrowhead) perpendicular to a prominent stalk (double arrowhead), 7. a mature carpogonium with a trichogyne (arrowhead) perpendicularly attached to a long stalk (double arrowhead), 8. a mature carpogonium with a long trichogyne (arrowhead) attached obliquely to a short stalk (double arrowhead), 9. a mature carposporophyte with ovoidal carposporangia (arrowhead) and cylindrical cells of gonimoblast filaments (double arrowhead). Scale bar = 10 μm.

Genus *Nothocladus* Skuja (1934:186).

Type: *Nothocladus nodosus* Skuja
Synonym: *Batrachospermum* section *Nothocladus* (Skuja) Necchi et Entwisle (1990:485)

Plants irregularly branched. Carpogonium-bearing branches somewhat long, sometimes curved, arising from periaxial cells and or cells of fascicles. Carpogonia symmetrical with elongate-conical or club-shaped trichogynes. Carposporophytes indefinite in shape, extending over along outer cortical filaments. Gonimoblast filaments of an indeterminate diffused type. This genus is restricted to New Zealand, Australia and Madagascar.

Key to the species of the genus *Nothocladus*

1. Carpogonium-bearing branch consisting of up to 12 cells..................................1) *N. afroaustralis*
1. Carpogonium-bearing branch consisting of up to 9 cells..2
2. Internodes < 400 µm long in mature axis..2) *N. lindaueri*
2. Internodes > 500 µm long in mature axis..3) *N. nodosus*

1) *Nothocladus afroaustralis* Skuja (1964:310, abb. 1, figs 1–15)

plate 151, figs 1–6 (original author, Skuja 1964)

Plants richly cartilaginous, long, 3.5–6 cm (and more) long, 200–350 µm, rarely 450 µm in diameter, bristle-like, slightly, but regularly nodulose or nearly cylindrical, monopodial, more or less irregularly, alternately or rarely oppositely branched, with the branches short to moderately long, often unilaterally encurved or arched. Cortical layer extremely dense on main axis of adult plants, 80–90 µm in thick, formed from very coherent cortical filaments. Whorls scarcely evident, primary fascicles pseudo-dichotomously branched and consisting of more or less cylindrical or club-shaped cells. Secondary fascicles abundant, mostly similar. Monoecious. Spermatangia sparse, spherical or rounded pear-shaped, 3.5–4.5 µm in diameter, occurring on primary and secondary fascicles. Carpogonium-bearing branch, more or less twisted, consisting of 12 or more cells, simple, terminal with several involucral filaments consisting of a few cell. Trichogyne elongate-club-shaped, 30–40 µm, rarely up to 60 µm long, 5–6 µm in diameter. Carposporophytes indefinite, gonimoblast filaments long, creeping irregularly along fascicles, producing carposporangia at periphery of whorls. Carposporangia obovoidal or obovoidal, pear-shaped, 12–15 µm in diameter, 15–27.5 µm long.
   Type Locality: Mamery River near Ebakika, Fort-Dauphin, Madagascar Island in Africa.
   Holotype: UPS, 12/VII, 1932, col. Decary
   Distribution: Known from the type locality only.

2) *Nothocladus lindaueri* Skuja (1944:20, pl. II, figs 1–10, pl. III, figs 1–10)
Synonym: *Batrachospermum lindaueri* (Skuja) Necchi et Entwisle (1990:485)

plate 151, figs 7–12 (original author, Skuja 1944)
plate 152, figs 1–6 (Entwisle & Kraft 1984)

Plants tufted, monoecious, mucilaginous, 0.5–7 cm high, 100–600 µm in diameter, profusely or sparingly branched, brown-red to olive-green; whorls usually slightly separated in

Batrachospermales

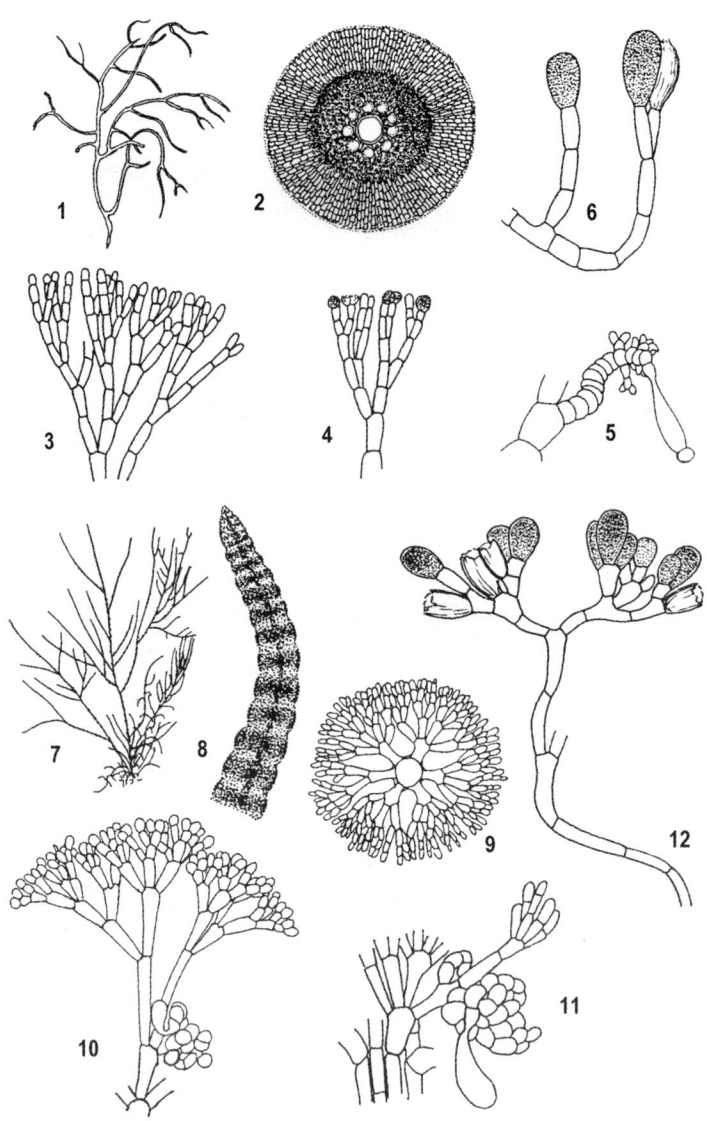

Plate 151
Nothocladus afroaustralis, figs 1–6 (original author, Skuja 1964)
1. the habit of a well-branched plant, 2. the cross section of a mature axis showing a broad, pseudo-parenchymatous axial cell filaments, cortical cell filaments and a relatively dense, narrow fascicles, 3. a cylindrical to ellipsoidal cells of primary fascicle, 4. spermatangia terminal on fascicles, 5. a carpogonium-bearing branch slightly twisted and arising from a proximal cell of fascicle, 6. carposporangia terminal on ultimate ramification of a gonimoblast filament.

Nothocladus lindaueri, figs 7–12 (original author, Skuja 1944)
7. the habit of a plant, 8. an apex of a mature plant with axial cell filaments and cortical cell filaments visible through surface of whorls, 9. the cross section of a young axial portion showing an axial cell and a relatively lax fascicles, 10. a primary fascicle composed of cylindrical proximal cells and ellipsoidal distal cells, a spirally twisted carpogonium-bearing branch arising from a proximal cell of a primary fascicle, 11. a spirally twisted carpogonium-bearing branch arising from a proximal cell of a primary fascicle, 12. carposporangia terminal on ultimate ramification of gonimoblast filament.

Nothocladus

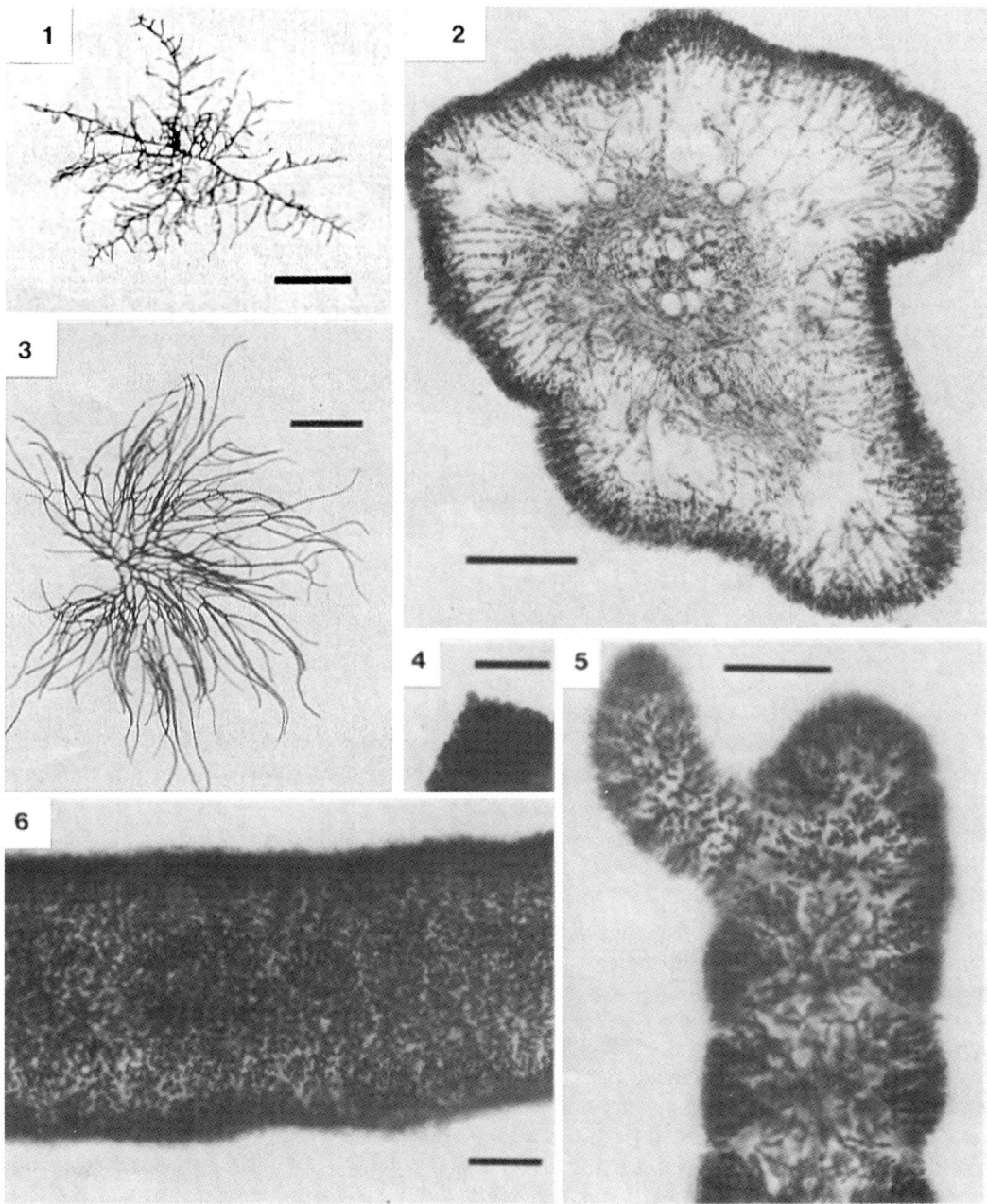

Plate 152
Nothocladus lindauri, figs 1–6 (Entwisle & Kraft 1984)
1. the habit of a dried, relatively short-branched plant, 2. the cross section of a mature axis showing a broad, pseudo-parenchymatous axial cell filaments, cortical cell filaments and a relatively dense, narrow fascicles, 3. the habit of a dried, relatively long branched plant, 4. an apex of a mature axis showing protruding apical cell, 5. an apex of a lateral branch, 6. a surface view of a lower main axial portion showing a dense fascicles which completely obscured axial cell filaments and cortical cell filaments.
Scale bar = 1 mm (figs 1, 3); 60 µm (figs 2, 6); 40 µm (figs 4–5).

young plants but confluent below. Cortical filaments and secondary fascicles very abundant. Primary fascicles di- tri- and tetrachotomously branched; secondary fascicles of similar structure and abundant; cells lunate, ellipsoidal to globular; proximal cells cylindrical to obovoidal; distal cells forming densely consolidated outer layer over markedly laxer inner cortex, terminal hairs common on fascicle, at times absent; spermatangia globular, 3–4 µm in diameter, terminal on specialized laterals arising from periaxial cells; carpogonium-bearing branches curved, consisting of 4–9 cells, arising from periaxial cells and cortical filaments but not from cells of fascicles; involucral filaments short; carpogonia 4–6 µm in diameter at the base, 20–40 µm long, trichogyne cylindrical or elongate clavate; carposporophytes indefinite in shape, consisting of branched gonimoblast filament of unlimited growth spreading through fascicles, producing carposporangium-bearing laterals in cymose clusters at whorl surface; carposporangia obovoidal or rarely ellipsoidal, 6–10 µm in diameter, 8–14 µm long, produced in profusion at surface of whorls.

Type Locality: Waitangi Falls, North Island, New Zealand in Oceania.
Lectotype: UPS, designated by Entwisle (1989).
Isotype: MELU, Lindauer, 1/XII 1937.
Distribution: Known from Bay of Islands area, North Island, New Zealand and Victoria (east of Melbourne to central Gippsland), and Tasmania (a single collection in northern Tasmania) in Australia.

3) ***Nothocladus nodosus*** Skuja (1934:186, abb. 2, figs 1–13)
Synonym: *N. tasmanicus* Skuja (1934:187, abb. 3, figs 1–13); *Batrachospermum nodusum* (Skuja) Necchi et Entwisle (1990:485)

plate 153, figs 1–4 (Entwisle & Kraft 1984)
plate 154, figs 1–5 (original author, Skuja 1934)
figs 6–9 (original author, Skuja 1934) as *N. tasmanicus*

Plants tufted, monoecious, mucilaginous, 1–4 cm high, 250–1000 µm in diameter, irregularly, radially alternate, rarely oppositely or pseudodichotomously branched, olive-green to brown; whorls confluent to apex of plants. Cortical filaments and secondary fascicles very abundant. Fascicles cells cylindrical, fusiform to ellipsoidal, terminal hairs rare or common; proximal cells often with swollen ends, distal cells ellipsoidal to obovoidal, no differentiation of cell size and shape of cortical filaments; secondary fascicles numerous, covering all internodes; spermatangia globular, 2–3 µm in diameter, terminal on specialized laterals arising from periaxial cells; carpogonium-bearing branches curved, consisting of 4–9 cells, arising from periaxial cells and all but distal cells of fascicles; involucral filaments short; carpogonia 4–6 µm in diameter at the base, 35–55 µm long, trichogyne elongate, clavate, often protruding through fascicles; carposporophytes indefinite in shape, consisting of branched gonimoblast filaments of indeterminate growth spreading through fascicles, producing carposporangium-bearing laterals in cymose clusters at whorls surface; carposporangia ellipsoidal or obovoidal, 7–12 µm in diameter, 10–15 µm long. (after Entwisle & Kraft 1984)

Type Locality: Yarra River, Collingwood, Melbourne, Victoria, Australia.
Lectotype: UPS, designated by Entwisle (1989).
Distribution: Known from Australia; in Victoria (east of Melbourne to central Gippsland, similar to *N. lindaueri*) and Tasmania (one collection only, from near Launceston in northern Tasmania).
Note: Entwisle & Kraft (1984) gave reasons for tentative merging. *N. tasmanicus* was also described by Skuja (1934) from Tasmania with *N. nodosus*. It was said to have a thinner, pseudodichotomously branched axis, a higher frequency of dichotomously branched fascicles, no terminal hairs, and ellipsoidal rather than ovoidal carposporangia.

Nothocladus

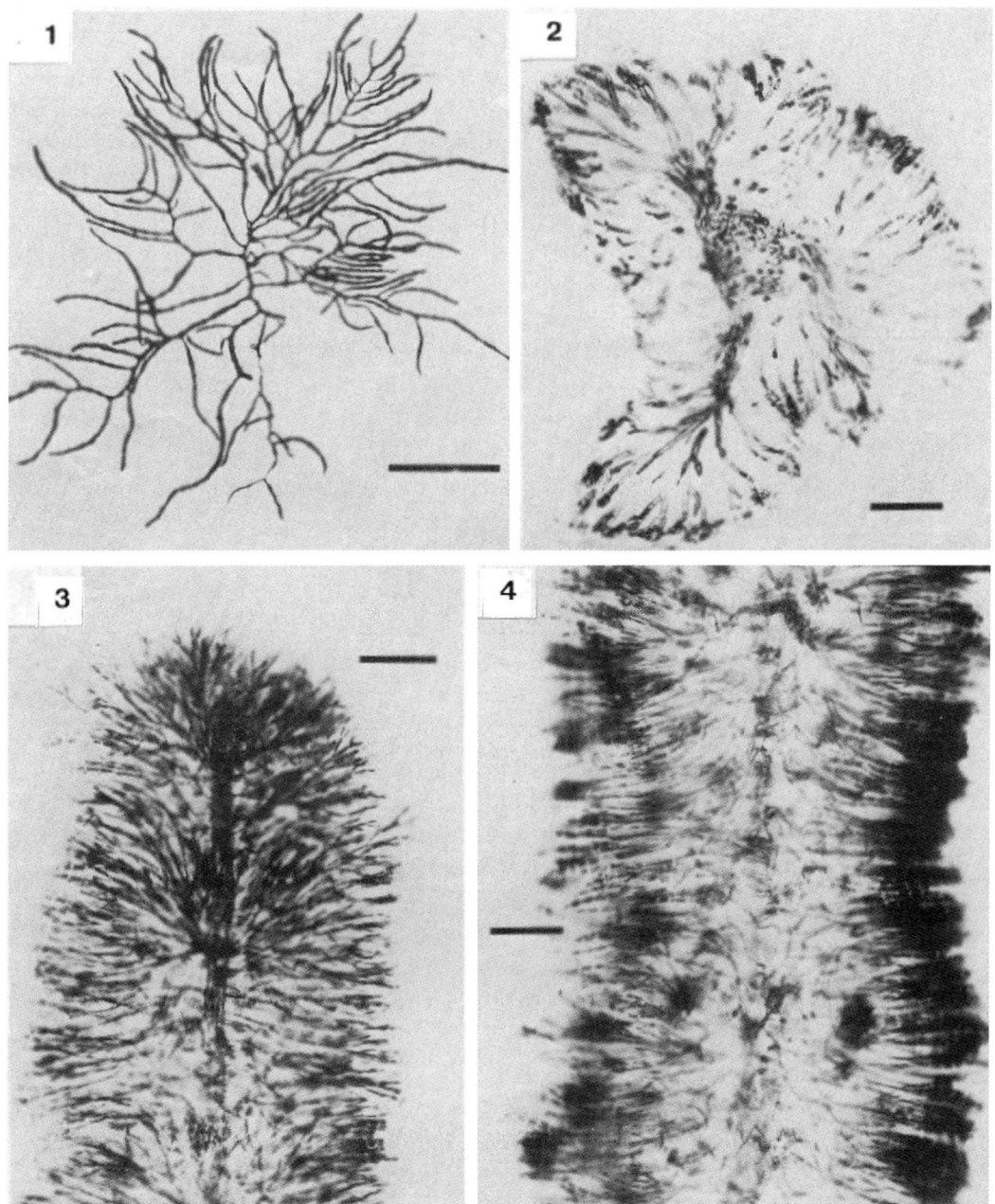

Plate 153
Nothocladus nodosus, figs 1–4 (Entwisle & Kraft 1984)
1. the habit of a dried plant, 2. the cross section of a mature axial portion showing dense, relatively narrow axial cell filaments, cortical cell filaments and a broad, lax fascicles, 3. an apex of a mature axial portion with axial cell filaments, cortical cell filaments visible through surface of fascicles, 4. the surface view of a lower main axial portion showing a relatively stable proportion of fascicles to axial cell filaments and cortical cell filaments throughout a plant.
Scale bars = 8 mm (fig. 1); 70 µm (figs 2–4).

Batrachospermales

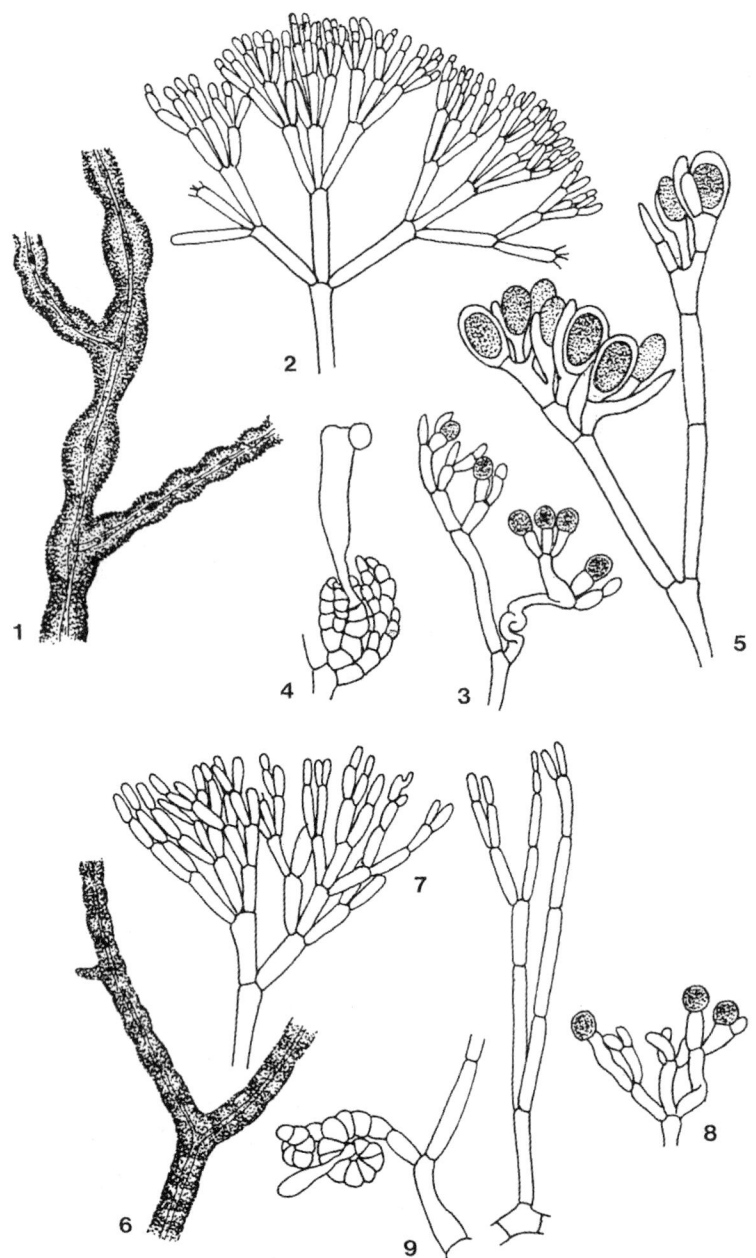

Plate 154
Nothocladus nodosus, figs 1–5 (original author, Skuja 1934)
1. a modulating mature plant with axial portion visible through surface of fascicles, 2. a primary fascicle composed of elongate cylindrical proximal and intercalary cells and short cylindrical distal cells, 3. spermatangia terminal of a fascicle, 4. a carpogonium-bearing branch arising from a proximal cell of fascicle, 5. carposporangia terminal on gonimoblast filament.

Nothocladus nodosus as *N. tasmanicus*, figs 6–9 (original author, Skuja 1934)
6. a mature plant with axial portion visible through surface of fascicles, 7. a primary fascicle composed of elongate cylindrical proximal and intercalary cells and short cylindrical distal cells, 8. spermatangia terminal of fascicle, 9. a spirally twisted carpogonium-bearing branch arising from a proximal cell of a fascicle.

Family **Psilosiphonaceae** Entwisle, Sheath, Müller et Vis

in Sheath, Müller, Vis & Entwisle (1996:245)

Type: genus *Psilosiphon* Entwisle

Plants pseudoparenchymatous with outer cortex cells that are uniform in composition; vacuoles few or absent. Medullar area with numerous filaments which are connected to the outer cortex, medullar cells with numerous chloroplasts, largely without vacuoles. Reproduction by adventitious filaments and spores; asexual spores cleaved off obliquely. Putative spermatangia dispersed on plants.

Note: The genus *Psilosiphon* comprises Australian species having cortical filaments throughout the interior of the plant and no obvious ray cells or large, outer cortex cells (Entwisle 1989b). Sheath, Müller, Vis & Entwisle (1996) confirmed the distinctiveness of the three genera, *Psilosiphon, Lemanea* and *Paralemanea*, then classified in the Lemaneaceae. Based on the ultrastructural features, they concluded that *Lemanea* and *Paralemanea*, while they are distinct from each other, are closely related in their morphological construction, ultrastructure and reproduction. In contrast, *Psilosiphon* is quite different from these two genera. Analysis of the rbcL gene and 18S rRNA gene (Vis, Saunders, Sheath, Dunse & Entwisle 1998) supports the separation of *Psilosiphon* from *Lemanea* and *Paralemanea*.

Genus *Psilosiphon* Entwisle (1989b:469)

Type: *Psilosiphon scoparium* Entwisle

Plants caespitose, erect, narrowly fusiform and undulate, uniaxial, arising from basal cushions of filaments; medulla lax, with profuse filaments; cortex crowded with short, similarly shaped cells which finally terminate into moniliform chains.

1) ***Psilosiphon scoparium*** Entwisle (1989b:470, figs 1–15)

> plate 155 (a), figs 1–5 (original author, Entwisle 1989)
> plate 155 (b), figs 1–5 (original author, Entwisle 1989)

Plants occurring in clumps of some 10–20 erect axes subtended by a firm basal cushion of uniseriate filaments. Erect plants are olive-green, cylindrical, and 14–33 mm long by 0.5–1.5 mm in diameter. Axes producing short lateral branches only rarely and having smooth surfaces without nodulation, 0.5–1.5 mm in diameter. Each axial cell produces four periaxial cells and successive periaxial cells are orthostichous. The periaxial cells further divide and differentiate to produce longitudinal medullar filaments which then give rise to radial cortical filaments. Attachment rhizoidal filaments arise from the periaxial cells in the basal portion. Ascending and descending medullar filaments derived from the axial cells do not densely surrounding the axial cell. Ultimately, the medulla becoming a mass of tangled medullar filaments, up to several hundred in cross-sections of mature axes, each 7–15 μm in diameter, enveloping a thick-walled (to 5 μm), axial cell, ca. 23 μm in diameter. Medullar filaments producing lateral, initially short, fascicles from an internodal portion of the cell. The cortex appearing to be mostly adventitious and is relatively cohesive and firm, consisting of filaments 4 or 5 cells long and pseudo- di-trichotomously branching. Cortical filaments are sometimes terminated with a unicellular hair up to 60 μm long. In older plants, continuing growth of the cortical filaments resulting in numerous free-standing series, up to 12 cells long, of ellipsoidal cells. Inflated cells, 12–20 μm in diameter

Batrachospermales

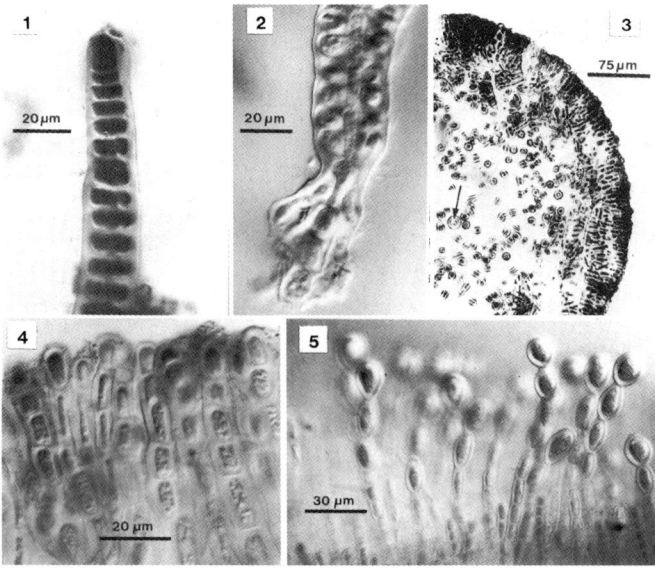

Plate 155 (a)
Psilosiphon scoparium, figs 1–5 (original author, Entwisle 1989)
1. an apex of a juvenile plant showing monosiphonous construction, 2. a base of a plant showing axial cell filament and rhizoidal filament extensions from basal cells, 3. the transverse section through a plant, showing cross-section of axial cell filament (arrow) and subsidiary medullar filaments, 4. the transverse section of a plant, showing a typical appearance of cortical structure, 5. series of inflated cells produced exogenously on an older plant, apparently functioning in vegetative propagation.

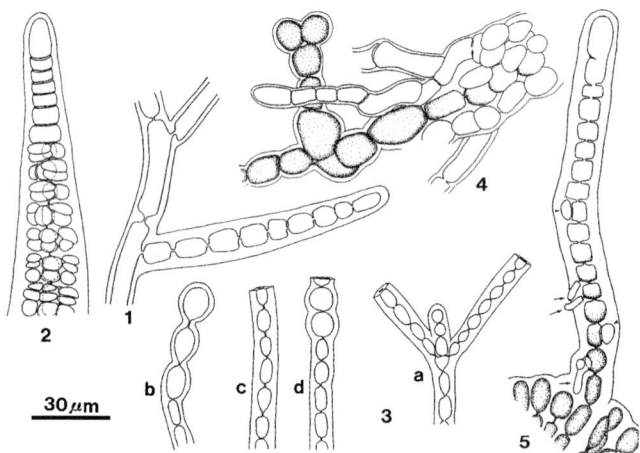

Plate 155 (b)
Psilosiphon scoparium, figs 1-5 (original author, Entwisle 1989)
1. prostrate, rhizoidal filaments of a basal cushion giving rise to an erect filament, 2. an apex of a plant showing development of periaxial cells from axial cell filaments (dotted line), 3. exogenous filaments extending from the outer cortical layer in an older plant, a: a branched filament, b: swollen cells, c: an unbranched filament of non-inflated cells, d: a transition from non-inflated to inflated cells, 4. a plantlet developing from a detached exogenous filament (shaded), 5. a cortical filament at the early stage of conversion into a new plantlet in situ by the initiation of basal rhizoids (arrows) and the first periaxial cells (arrows).

and 16–28 μm long often forming at the end of the external filaments and resembling chains of sporangia, but evidence of spore release not present. Adventitious plantlets, however, arising directly from some of these exogenous filaments, both as a direct extension of attached filaments and on those which broken off in culture. In older plants, clumps of new multiseriate plants sometimes arising from the thallus surface.

Type Locality: Lamond Creek, 'Natural Stone Bridge', Barren Grounds Nature Reserve, Mossvale-Kiama Road, between Robertson and Jamberoo, New South Wales in Australia.

Holotype: MELU TJE 1565.

Distribution: Known from the type locality and a rivulets, Frankland Range, Southwest Tasmania in Australia and North Island in New Zealand as mentioned bellow.

Note: Entwisle, Vis & Foard (2000) proposed that three genetic races be recognized and sufficient independent molecular data for complete congruence, these races are not given formal taxonomic recognition, because the phylogeny of this species is strongly supported by the *rbc*L data, and the *rbc*L base-pair differences between collections from the three regions are large. However the phylogeny supports an origin for these genetic races prior to the separation of New Zealand from Gondwana.

1. New South Wales genetic race

Remarks: Further collecting in humic streams through the sandstone region around Sydney has not resulted in any new localities for this species.

Representative specimen (Holotype of *Psilosiphon scoparium*) : MELU TJE 1565.

Distribution: A single reach of a stream 100 km south of Sydney, New South Wales, Australia.

2. Tasmania genetic race

Remarks: The thalli are variable in size, even within the one stream.

Representative specimen: MEL 2033445, Entwisle 2533 & Walsh TAS3, ca. 1 km north of Harrys Bluff, stream from Harrys Bluff into Old River, South West National Park, Tasmania, Australia.

Distribution: Widespread but uncommon in south-west Tasmania, Australia.

3. New Zealand genetic race

Remarks: Although the thalli are substantially smaller than most Australian specimens, the anatomy is smaller.

Representative specimen: MEL 2036082, N. G. Walsh 4686, Waikotahu Stream, Waipoua, Northland, North Island, New Zealand.

Distribution: Apparently rare in Northland, New Zealand. A second collection from New Zealand was also from Kauri forest in Northland.

Family **Lemaneaceae** C. Agardh (1828:1)

as Ordo Lemanieae

Type: genus *Lemanea* Bory

As mentioned previously, plants with a *Lemanea*-type life history. Namely, carpospores in *Lemanea* (Magne 1967), germinating into small plants called the chantransia phase with meiosis taking place in an apical cell, then, gives rise to a new plant (gametophyte).

Plants pseudoparenchymatous with fascicles of cells that are progressively smaller towards the exterior, vacuoles large, occupying most of cell volume in cells towards interior. Medullar area with scattered ray cells (periaxial cells) not obviously continuous with cortical filaments; ray cell and inner fascicle cells with small chloroplasts and considerable vacuolation. Reproduction similar to other Batrachospermales but with carposporangia produced in chains, sporangia and spermatangia cleaved apical, spermatangia in discrete patches or rings.

Note: In this family the plants are a branched, solid or hollow cylinders and does not show the axial plan of the Batrachospermaceae. Other essential characters are found in the behavior of the fertilized carpogonium and the method by which carpospores are formed by all cells of gonimoblast filaments instead of only the terminal cells. Within the Lemaneaceae there are two plant morphologies which differ in internal structure. One has cortical filaments localized around an axial cell with simple ray cells (periaxial cells) that do not abut the large cells of the outer fascicles. A second one has no cortical filaments and has T- or L-shaped ray cells (periaxial cells) closely applied to the outer fascicles (Sirodot 1872). The two, originally included in the genus *Lemanea* Bory (1808), were recognized as distinct genera, *Lemanea* Bory and *Paralemanea* (Silva) Vis et Sheath (=*Lemanea* subgenus *Paralemanea* Silva 1959) were recognized by Vis & Sheath (1992). Following analysis of the rbcL gene and 18S rRNA gene (Vis, Saunders, Sheath, Dunse & Entwisle 1998) retained the two genera pending further investigation, but noted the possibility that *Paralemanea* is paraphyletic.

Species of family Lemaneaceae occur in the Northern hemisphere, namely, from Asia (Jao 1941, Khan 1973), Europe (Sirodot 1872, Hamel 1925, Israelson 1942), North America (Atkinson 1890 etc., Palmer 1941, Vis & Sheath 1992). There are a few records from southern South America (Necchi & Zucchi 1995).

Based on multivariate morphometrics and image analysis, Vis & Sheath (1992) recognized 5 species and 1 variety from North America, however, the revised descriptions with few photomicrographs for these taxa were incomplete.

Blum described 6 new species of *Paralemanea* from Indiana (1993) and California (1994) in western USA. He questioned the practice of applying European names to the American materials and advocated using additional characters in the taxonomy of the genus.

In the following treatment descriptions of 6 species by Blum are followed by species reported from North America and Europe and China.

Key to the genera *Lemanea* and *Paralemanea*

1. Axial cells without cortical filaments; ray cells (periaxial cells)
T- or L-shaped, closely applied to the outer fascicles; spermatangial
papillae in patches on node; plants stalked or unstalked....................................genus *Lemanea*
1. Axial cells with cortical filaments; ray cells (periaxial cells) simple,
not abutting the outer fascicles; spermatangial papillae in rings on
node; plants unstalked...genus *Paralemanea*

Genus *Lemanea* Bory (1808:178) *nom. cons.*

Type: *Lemanea corallina* Bory, *nom. illeg.* =*Lemanea fluviatilis* (Linnaeus) C. Agardh
Synonym: genus *Lemanea* subgenus *Sacheria* (Sirodot) Atkinson (1931:225);
genus *Sacheria* Sirodot (1872:70)

Plants consisting of tufts of tubular bristle-like axes with regularly placed swellings (nodes). Each axis with an axial cell filament of very long cells that may be naked or clothed with cortical cell filaments. Each axial cell having four cells, T- or L-shaped ray cells (periaxial cells), representing the basal cell of a whorl branchlet perpendicular to it. Ray cells (periaxial cells) closely applied to the outer fascicles. Whorl branchlets repeatedly branched toward their outer face, with the distal portion of fascicles so closed to one another forming a pseudoparenchymatous tissue. Spermatangia elevated or flush on the frond surface at the nodes, forming either circular patches or in a continuous transverse ring. The carpogonium-bearing branches 3 or 4 cells long borne either in the spermatangial zone or midway between two spermatangial zones. Hypogynous cells of carpogonium-bearing branch naked. The gonimoblast filaments growing from the base of the carpogonium and extend inward. All cells of gonimoblast filaments developing into carposporangia.

Chantransia phase only slightly branched and very fugitive. Prostrate form of protonema mainly cellular. In very turbulent water.

Key to the species of the genus *Lemanea*

1. Plants unstalked, seldom branched, diameter < 0.4 mm..2
1. Plants stalked, few to much branched, diameter > 0.3 mm...3
2. Spermatangial papillae patch like...1) *Lemanea borealis*
2. Spermatangial papillae flat...2) *L. simplex*
2. Spermatangial papillae distinctly pronounced....................................3) *L. condensata*
3. Plants narrowing abruptly towards the base, if branched
    usually ≤ 4 branches per plants..4) *L. fluviatilis*
3. Plants gradually narrowing towards the base...4
4. Spermatangial papillae are irregularly distributed on the branches
    with indistinct nodes ....5) *L. sudetica*
4. Spermatangial papillae are distributed more or less regularly, node
    distinct or indistinct...5
5. Spermatangial papillae 2–8 in number, node indistinct..............................6) *L. ciliata*
5. Spermatangial papillae protruding...6
6. Plants thick, olive-green or yellowish..7
6. Plants bright or dark green, grows in bunches or forming tuft concentrations............8
7. Spermatangial papillae fusing into rings only rarely broken, base of
    plant elongated into a long stalk..7) *L. sinica*
7. Spermatangial papillae not fusing into rings...8) *L. rigida*
8. Plants do not form compact tuft concentrations..9
8. Plants at the base coalescent, forming compact tufts, 4–10 cm high,
    spermatangial papillae 3–4 in whorls..9) *L. mamillosa*
9. Plants 12.6–20 cm long..10) *L. fucina*
9. Plants 2.7–7.1 cm long..11) *L. fucina* var. *parva*

1) ***Lemanea borealis*** Atkinson (1904:26)

Plants evenly tufted, slender, little branched, 130–350 µm in diameter, 0.6–4.1 cm long; sterile basal portion 0.5–1 cm long, slender, unstalked, gradually tapering into the fertile portion, the transition very rarely abrupt. Spermatangial zone when young prominently tubercuate with 2–5 spermatangial papillae, these disappearing in age so that the older plants are plane. Carposporophyte zone usually cylindrical, rarely constricted in the middle, sometimes slightly so near the apex, the result being that in age, with the disappearance of the spermatangial papillae, the plants are nearly or quite cylindrical, the younger and middle-aged ones appearing slightly nodes. Carposporophytes arising in both the spermatangial zone and carposporophyte zone, but not quite reaching the middle of the carposporophyte zone. Carpospores in tuft throughout the entire length of the plant, ellipsoidal to oblong, 18–25 μm in diameter, 25–45 μm long. (Atkinson 1904)

Chantransia phase, dull green when dry, represent only by fragments at season when collected, but filaments 18–25 μm in diameter; cells 35–45 μm long, often slightly constricted at the septa. the spores sometimes blue and darkening.

Type Locality: on rocks in a waterfall, Bay of Island, Newfoundland, Canada in North America.

Lectotype: NY, no. 1108, coll. Howe & Long, 9/VIII, 1901.

Distribution: Known primarily from the north, ranging from the north slope of Alaska to Oregon, USA, in the west, and from western Greenland (Denmark) to central Quebec, Canada, in the east (Vis & Sheath 1992), from Colourado to Utah, USA (Palmer 1945) in North America.

Note: Atkinson (1931) merged *L. borealis* with *L. rigida* (as *L. fucina* var. *rigida*). Vis & Sheath (1992), however, retained both species, pointing out they differ in presence of stalks, in size and branching.

2) ***Lemanea simplex*** Jao (1941:272, pl. VI, fig. 46, pl. VII, figs 47–48)

plate 156, figs 1–3 (original author, Jao 1941)

Plants densely caespitose, red-violet, brownish, olive-green or dark green, thick, turf, up to 2 cm high, up to 350 μm in diameter, curved or straight, unbranched, irregularly undulately constrict. Spermatangial papillae flat, irregular in outline, 2–4 in a whorl, are distributed more or less regularly, protruding, not fusing into rings. Carpogonia in the middle parts of the carpogonial zones. Carpogonium-bearing branches 4-celled. Mature carpospores large, 50–70 μm in diameter, 110–140 μm long, cylindrical-obovoid, single, not comb-like. The carpospore formation of this species is similar to that of *L. sinica*, but the stalks of the carpospores are always unbranched. After carpospores discharged, some stalks sometimes divided again and producing two or more carpospores arranged into a chain. (after Jao 1941). Chantransia phase unknown.

Type Locality: on rocks in a rapid of Chialing River near Pehpei, Szechwan, China in Eastern Asia.

Holotype: SC 1128.

Distribution: Known from the type locality only.

3) ***Lemanea condensata*** Israelson (1942:20, pl.2, a)

Plants greenish or brownish, blackish- or brownish-olive, unbranched, curved, sometimes gradually narrowing towards the base, growing in bunches or forming loose tuft concentrations, 1–2(–3) cm high, in middle parts 100–250 μm in diameter, internodal zone concave. Hairs absent or present, sometimes abundant. Spermatangial papillae usually 3, distributed more or less regularly, protruding, distinctly pronounced, or, more or less prominent. Spermatangial papillae zone 220–350 μm in diameter, the papillae included. Carposporophyte zones hour-glass-shaped or towards the proximal parts of plants cylindrical, 100–280 μm in diameter at the middle. Internodes 320–830 μm long, 1.5–3.5 times longer than the diameter at the middle portion. Carpogonium-bearing branches developed both along the entire carposporophyte zone and the spermatangial papillae zone. (Israelson 1942)

Chantransia phase composed of irregularly branched filaments interwoven into a disc and of upright, more or less richly branched filaments of varying resistance. Upright filaments of cylindrical cells, 18–25 μm in diameter and usually 3–4 times longer than diameter. Terminal hairs absent or present and of varying length. Monosporangia unknown.

Type Locality: Vallen, near Östmark, Värmland, Sweden in Europe. Known together with *L. fluviatilis*.

Holotype: Mus. Bot. Upsaliensis

Distribution: Known from the type locality only.

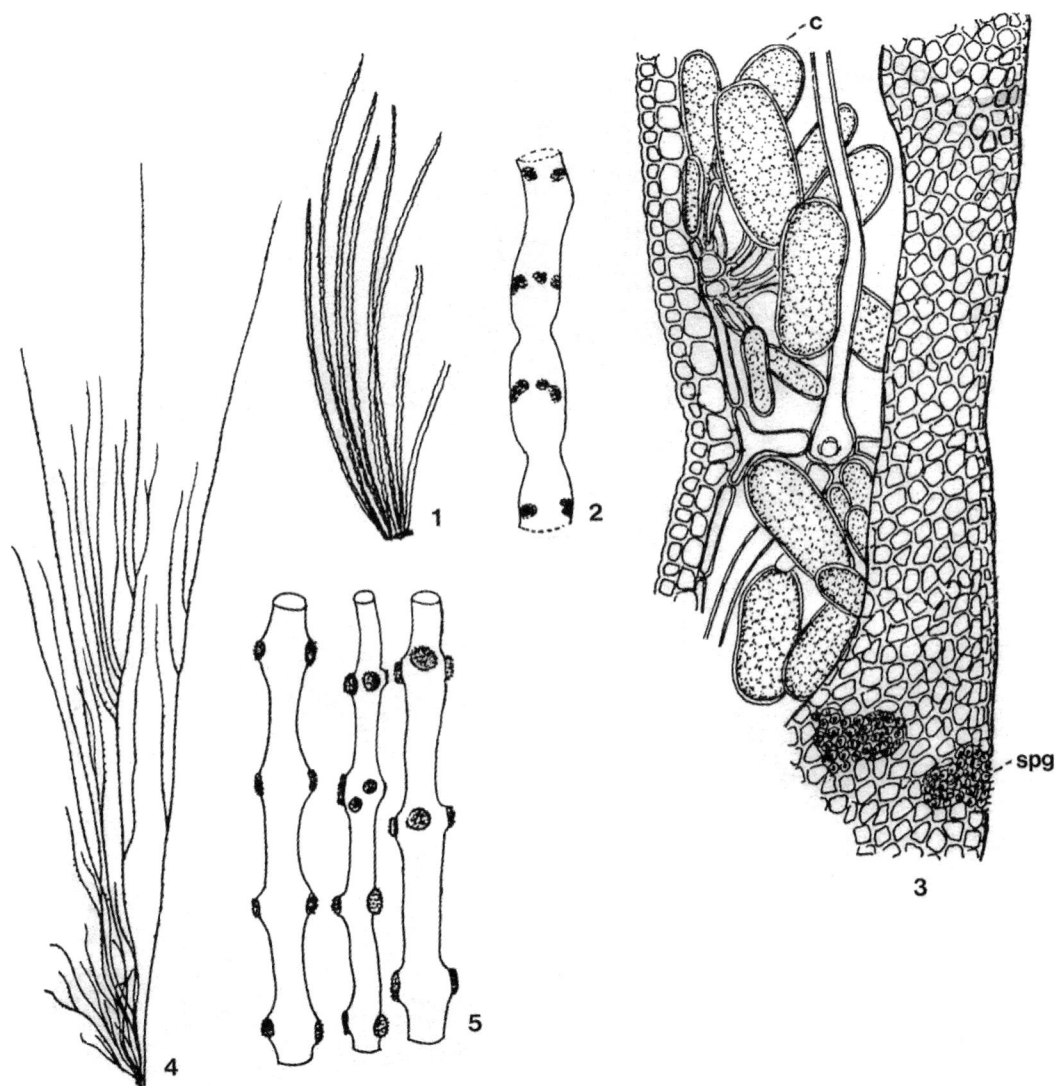

Plate 156
Lemanea simplex, figs 1–3 (original author, Jao 1941)
1. the macroscopic view of a plant showing ramification habit, 2. a plant showing spermatangial papillae, 3. the surface and sectional views of a portion of gametophytes showing the structure of a plant, spermatangial zone, and mature gonimoblast filaments with carposporangia. sgp; spermatangial papillae, c; carposporangia.

Lemanea mamillosa as *Sacheria mamillosa*, figs 4–5 (original author, Sirodot 1872)
4. the macroscopic view of a plant, showing ramification habit, 5. a portion of a plant showing spermatangial papillae.

Batrachospermales

4) *Lemanea fluviatilis* (Linnaeous) C. Agardh (1811:25)
Basionym: *Conferva fluviatilis* Linnaeous (1753:1165).
Synonym: *Sacheria fluviatilis* (Linnaeous) Sirodot (1872:70, pl. 1, fig. 7, pl. 2, fig. 14, pl. 3, figs 15–19, 21, 23, pl. 6, figs 47–50, pl. 7, figs 62–63, pl. 8, fig. 81); Atkinson (1890:221, pl. 9, figs 52, 58): *Lemanea fucina* (Bory) Atkinson var. *subtilis* Atkinson (1890:225)

plate 157, figs 1–7 (original author, Sirodot 1872) as *Sacheria fluviatilis*

Plants confined to lower cells of chantransia phase, or in the middle branching. Fertile plants usually 5–30 cm high, irregularly and usually very often simple, frequently branched, branches often fasciculate, arcuate, long, slender-pedicelled, narrowed abruptly towards the base and transformed into a thin, cylindrical stalk; nodes fairly distinct. Terminal hairs usually absent, sometimes abundant. Colour blackish-, brownish- or yellowish olive or blackish violet, usually reddening the water in which they stand, after drying black. Spermatangial papillae usually 3 or 4 in number, more or less prominent, in mature filaments usually less so, well separated and of about the same shape and size, occasionally more irregular and partly confluent. Spermatangial zone 300–1200(–1600) µm in diameter, the papillae included. Carposporophyte zones cylindrical, slightly thinner initially than spermatangial zones, at maturity slightly broader at the middle, 250–750(–1100) µm in diameter at the middle. Internodes 1000–3000(–4500) µm long, 3.3–8 times longer than the diameter at the middle portion. Carpogonium-bearing branches developed all along the generative filaments from the first cell as well as from the cells in the spermatangial zone. (after Israelson 1942).

Chantransia phase composed of irregularly branched threads interweaving to a disc and upright filaments of limited resistance. Upright filaments in dense tuft, more or less sparsely and irregularly branched, alternate, final branching sometimes opposite or fasciculate, hairs, composed of cylindrical cells. Cells 15–22 µm in diameter and 2.5–4 times longer than diameter. Terminal hairs of varying length may occur. Monosporangia unknown.

Type Locality: Ruisseau de Beaufort, France in Europe ?
Holotype: REN 2086?, Sirodot, 18/IV 1869, for *Sacheria fluviatilis* Sirodot.
Distribution: Known from the type locality, France, Belgium in Europe; China in Eastern Asia; from south-central Alaska to northern California in the west, extending into the range of *L. borealis*, from southern Ontario to North California (Vis & Sheath 1992); from New Jersey, South California, Alabama and California (Wolle 1997), Ontario (Palmer 1945), Oregon (Atkinson), USA in North America. Grows in swift running streams.

5) *Lemanea sudetica* Kützing (1857:33, pl. 85, I, a,a', b,b',b'',b'')
Synonym: *Lemanea kalchenbenneri* Rabenhorst (1863) ?; *Lemanea daldinii* Rabenhorst (1863:421) ?

plate 160, fig. 1 (original author, Kützing, 1857)

Plants red violet, brownish, olive-green or dark green, gradually narrow towards the base. Spermatangial papillae are irregularly distributed on the branches, these 2–9 cm high, about 1 mm in diameter, with not very pronounced nodes (Starmach 1976).

Type Locality: Mts. Sudete, northern Italy in Europe.
Holotype:
Distribution: Known from the type locality, Mts. Sudete, northern Italy and Switzerland in Europe.

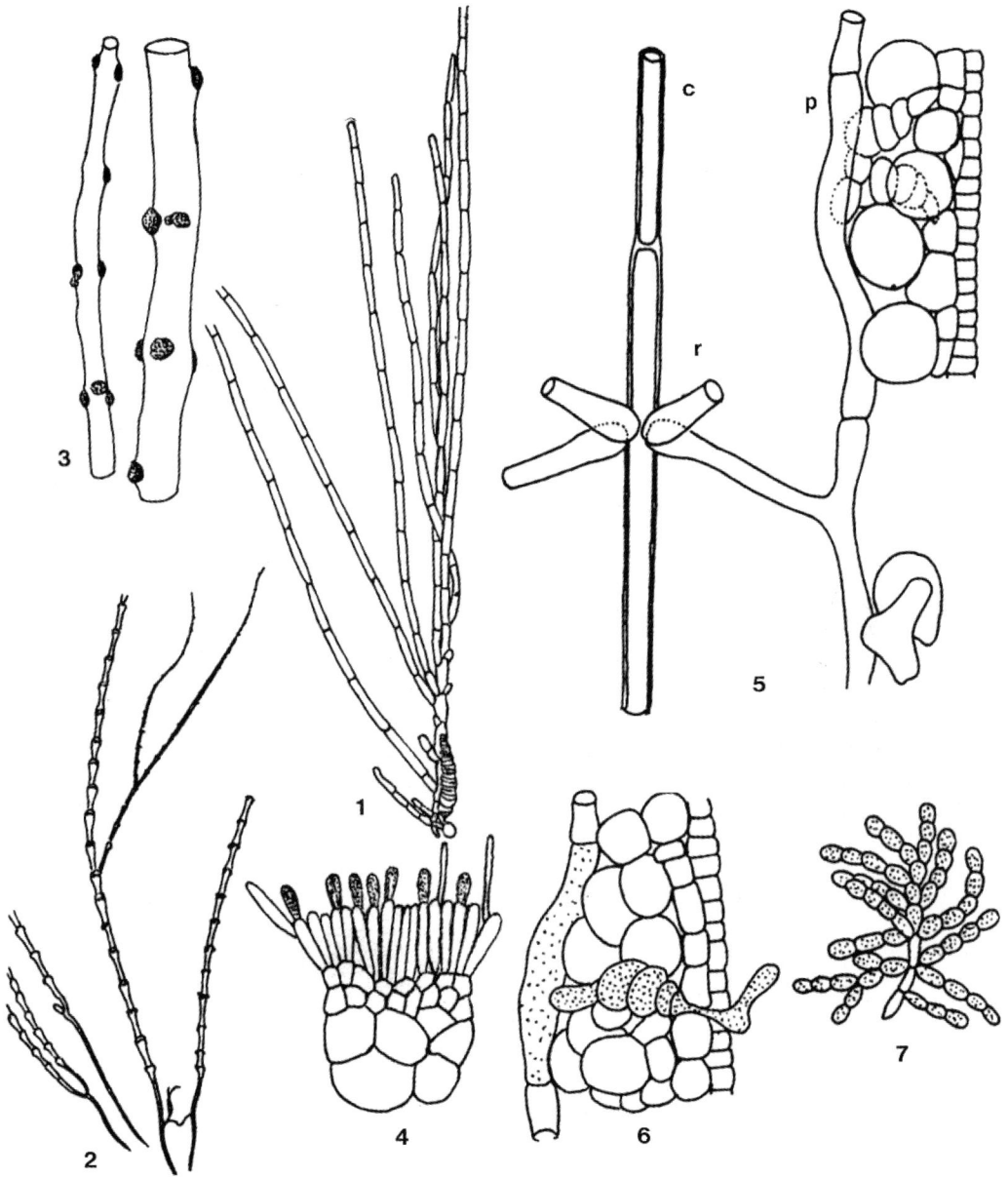

Plate 157
Lemanea fluviatilis as *Sacheria fluviatilis*, figs 1–7 (original author, Sirodot 1872)
1. a chantransia phase with a base of a young gametophyte, 2. the macroscopic view of mature plants showing ramification habit, 3. plants showing spermatangia papillae, 4. the section of spermatangial papilla at time spermatangial development, 5. the longitudinal section showing ramification of sexual filaments and an origin of carpogonium-bearing branches, 6. the section of carposporophyte zone showing a carpogonium-bearing branch, 7. the early development of gonimoblast filaments, a carposporophyte. a: ascending generative filaments, c: axial cells, p: a procarp, d: descending generative filaments, r: a ray-cell, T-shaped cell (periaxial cell), t: a tie-cell, tr: a trichogyne.

Batrachospermales

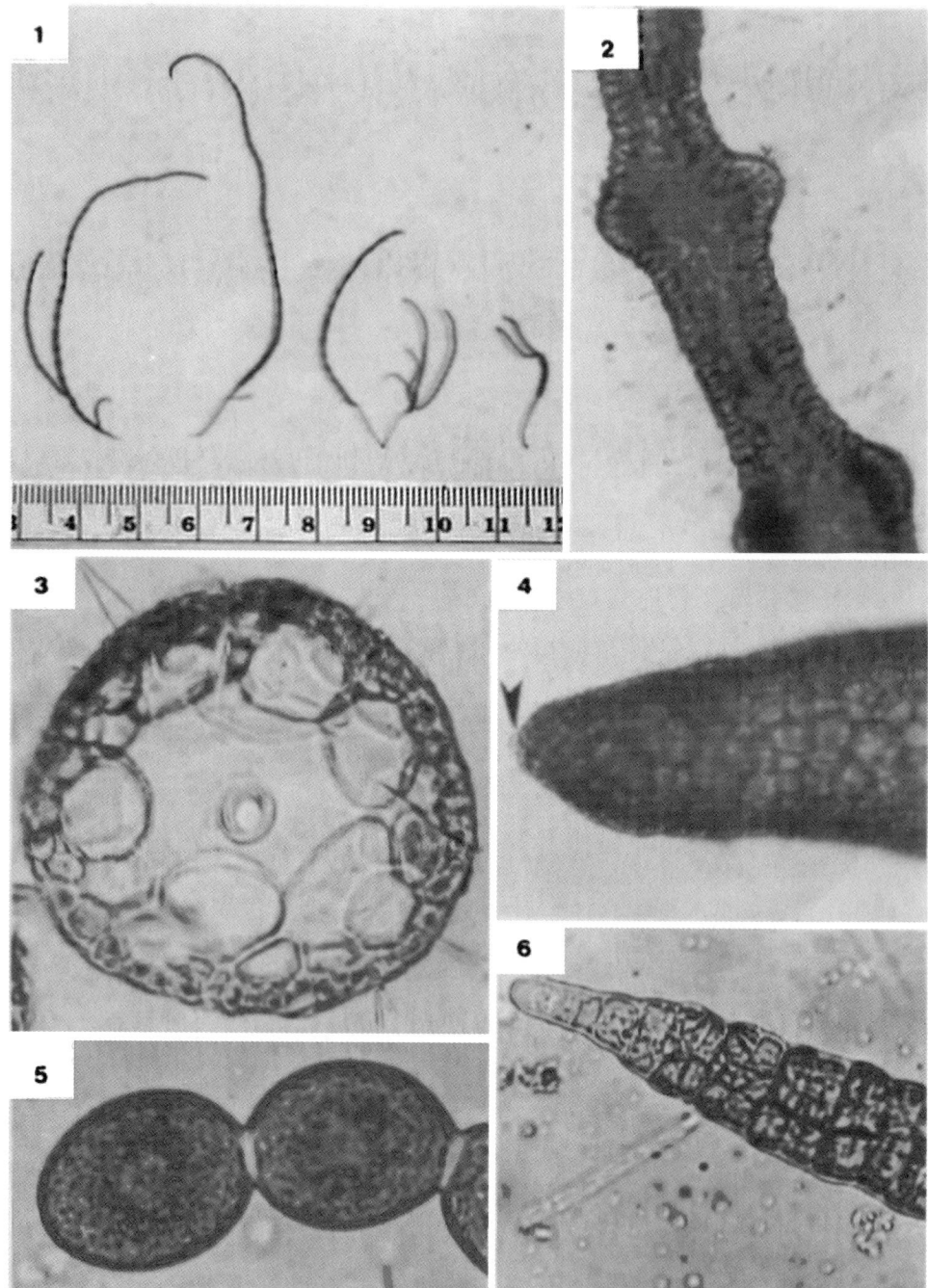

Plate 158
Lemanea fucina, figs 1–6 (Sheath 1984)
1. the macroscopic view of mature plants showing ramification habit, 2. nodes obvious as a series of swellings, 3. the transverse section through an internode of a young branch in which there are no cortical cell filaments surrounding axial cell filament, 4. an apex of a mature plant with a prominent apical cell (arrow), 5. carposporangia in a series, 6. an apex of a juvenile gametophyte showing axial cell filament producing fascicles.

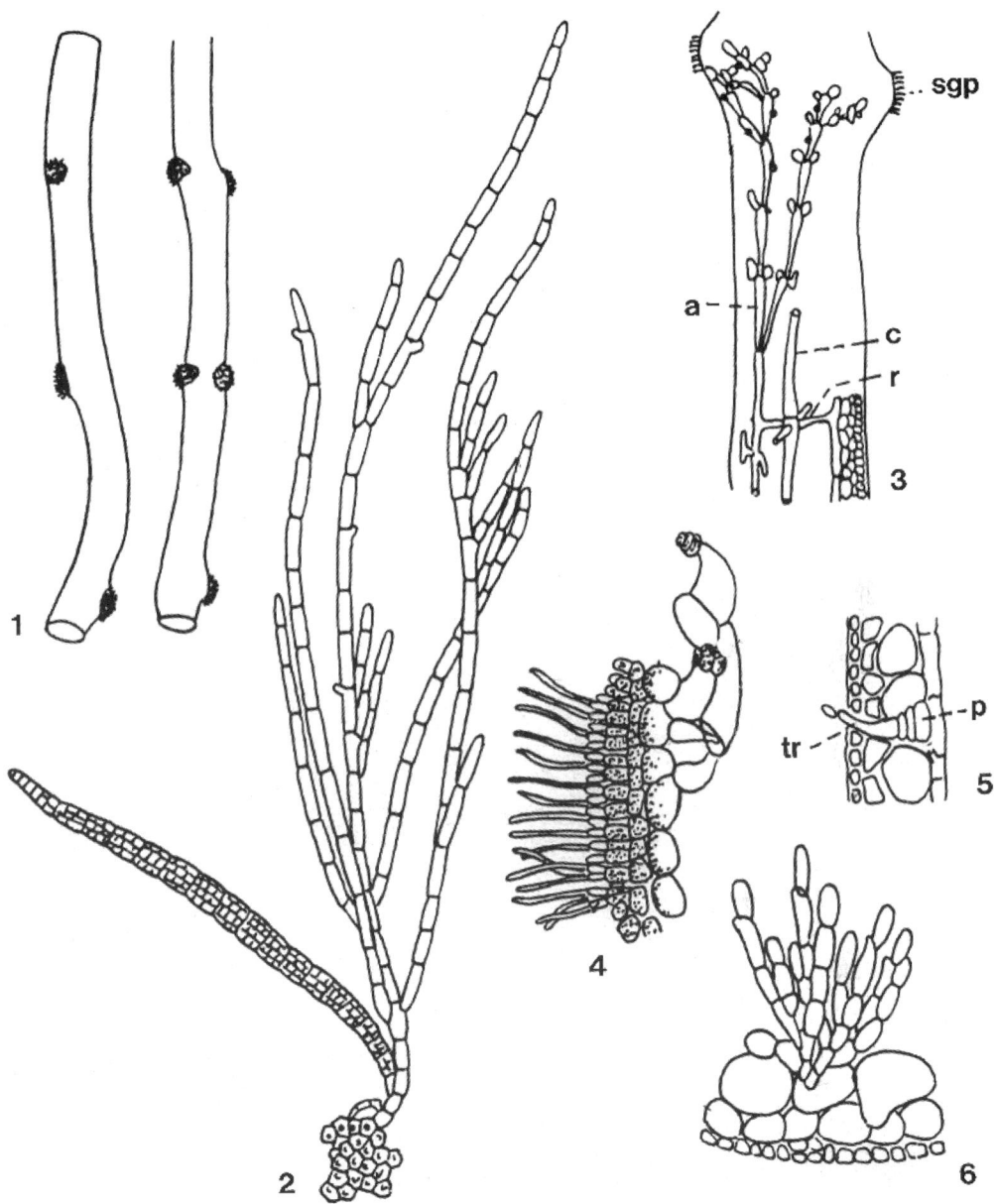

Plate 159
Lemanea fucina, figs 1–2 (Sirodot 1872 as *Sacheria fucina*), *L. mamillosa,* figs 3–6 (Atkinson 1890 as *L. fucina* var. *mamillosa*)
1, a plant showing spermatangial papillae, 2. chantransia phase with a base of a young gametophyte, 3. the longitudinal section of a portion of a sexual segment at time of fertilization, showing ramification of generative filaments and origin of carpogonium-bearing branches, 4. the section of spermatangial papilla at time spermatangial development, 5. the section of carposporophyte zone showing a portion of a wall and a carpogonium-bearing branch at time of fertilization, 6. the early development of a carposporophyte. a: ascending generative filaments, c: axial cell filament, p: a procarp, d: descending generative filaments, r: a ray-cell, T-shaped cell (periaxial cell), sgp: spermatangial papilla, sg: spermatangia, sp: spermatium, tr: a trichogyne.

Batrachospermales

6) *Lemanea ciliata* (Sirodot) De Toni (1897:42)
Basionym: *Sacheria ciliata* Sirodot (1872:71, pl. 2, figs 8–11, pl. 3, figs 24–25, pl. 7, figs 51–61, pl. 8, fig. 82)

plate 160, figs 2–7 (original author, Sirodot 1872) as *Sacheria ciliata*

Plants red-violet, brownish, olive-green or dark green, branching frequent, gradually narrowing towards the base, with an indistinct pronounced stalk at the base. Branches terminating in terminal hairs. Spermatangial papillae are distributed more or less regularly, 2–8 in number, not very pronounced (Starmach 1977).

Type Locality: Betton, écluse du Haut-Chalet (or Vau-Chalais), canal d'Ille-et-Rance, France in Europe.
Holotype:
Distribution: Known from streams, France in Europe.

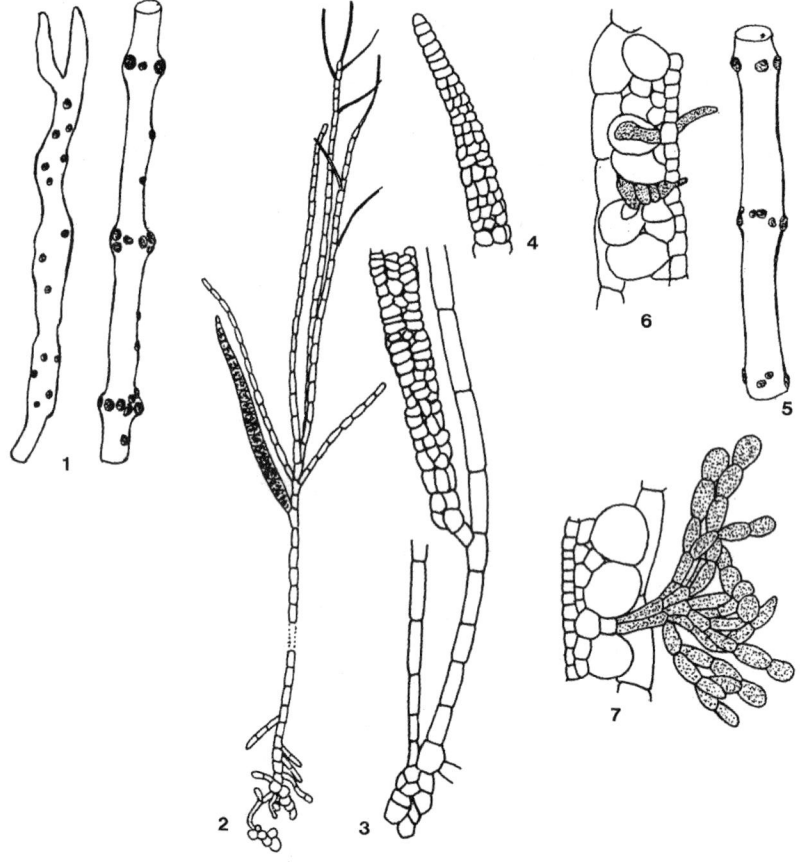

Plate 160
Lemanea sudetica, figs 1 (original author, Kützing 1857)
1. plants showing spermatangial papillae.

Lemanea ciliata as *Sacheria ciliata*, figs 2–7 (original author, Sirodot 1872)
2. chantransia phase producing a gametophyte, in middle ramification, 3–4. a gametophyte arising from a middle portion of chantransia phase, 5. a plant showing spermatangial papillae, 6. the section of a carposporophyte zone, 7. the early development of gonimoblast filaments, a carposporophyte.

7) *Lemanea sinica* Jao (1941:270, pl. 7, figs 49–57)

plate 161, figs 1–4 (original author, Jao 1941)

Plants rigid, red-violet, brownish, olive-green or dark green, olive-green, elongate, up to 16 cm long, attenuated downwards at the base into a long stalk, usually branched at the basal portion. Branches opposite, alternate or sometimes forked, end usually capillary. Carpogonial zone cylindrical in upper parts of the plants and slightly tumid (up to 1 mm in diameter below. Carpogonium-bearing branches usually 4 cells long, very rarely 3, numerous, scattered throughout the entire length of the carposporophytes zone and even into the spermatangial zone. Mature carpospores large, cylindrical-obovoid, 30–50 µm in diameter, 75–95 µm long, solitary, never chained. Spermatangial zones conspicuously inflated at upper parts of the filaments, not prominent below; upper spermatangial bands widely annulate, lower ones occasionally interrupted, the interruption probably due to the hypertrophy of the tissue beneath the spermatangial bands. Growing in streams.
Chantransia phase unknown.
Type Locality: Lung-tan, Chun-liang-shan, Chun-tien, Yunnan, China in Eastern Asia. Known from running water.
Holotype: YN no. 1276.
Distribution: Known from the type locality only.

8) *Lemanea rigida* (Sirodot) De Toni (1897:42)
Basionym: *Sacheria rigida* Sirodot (1872:72, pl. 2, figs 12–13. pl. 8, fig. 86)
Synonym: *Lemanea fucina* Bory var. *rigida* Atkinson (1890:225, pl. VII, figs 6–7); *Lemanea torulosa* sensu Kützing (1843: vol. VII, pl. 84, fig. 2)

plate 161, figs 5 (original author, Sirodot 1872) as *Sacheria rigida*,
plate 165, figs 1–2 (Sirodot 1872), figs 3–4 (Schmitz) as *Lemanea torulosa*

Plants simple or branched at various lengths in chantransia phase, red-violet, brownish, olive-green or dark green, thick, rough, curved, of various size, branching at the base, gradually narrowing towards the base. Carposporophyte zones nearly cylindrical or strongly constricted. Spermatangial papillae distributed more or less regularly, nearly or completely plane, often confluent, 3–7 in a whorl, sometimes not apparent, sometimes protruding, not fusing into rings. In age usually drying yellowish, sometimes darkened by the greenish or dark colour of the spores clustered in the spermatangial zone, rarely fading so as to be nearly colourless (after Atkinson 1890).
Type Locality: stream in the valley of Saint-Lazare, near Montfort, France in Europe.
Holotype: REN 2078, Sirodot, 22/V 1878, for *Sacheria rigida*.
Distribution: Occurs Belgium, France in Europe (Starmach, 1977) and North America. North American records derive from Atkinson (1931); they need to be reevaluated. Vis & Sheath (1992) did not analyze this species, although unlike Atkinson, they regarded it as separate from *L. borealis*.

Batrachospermales

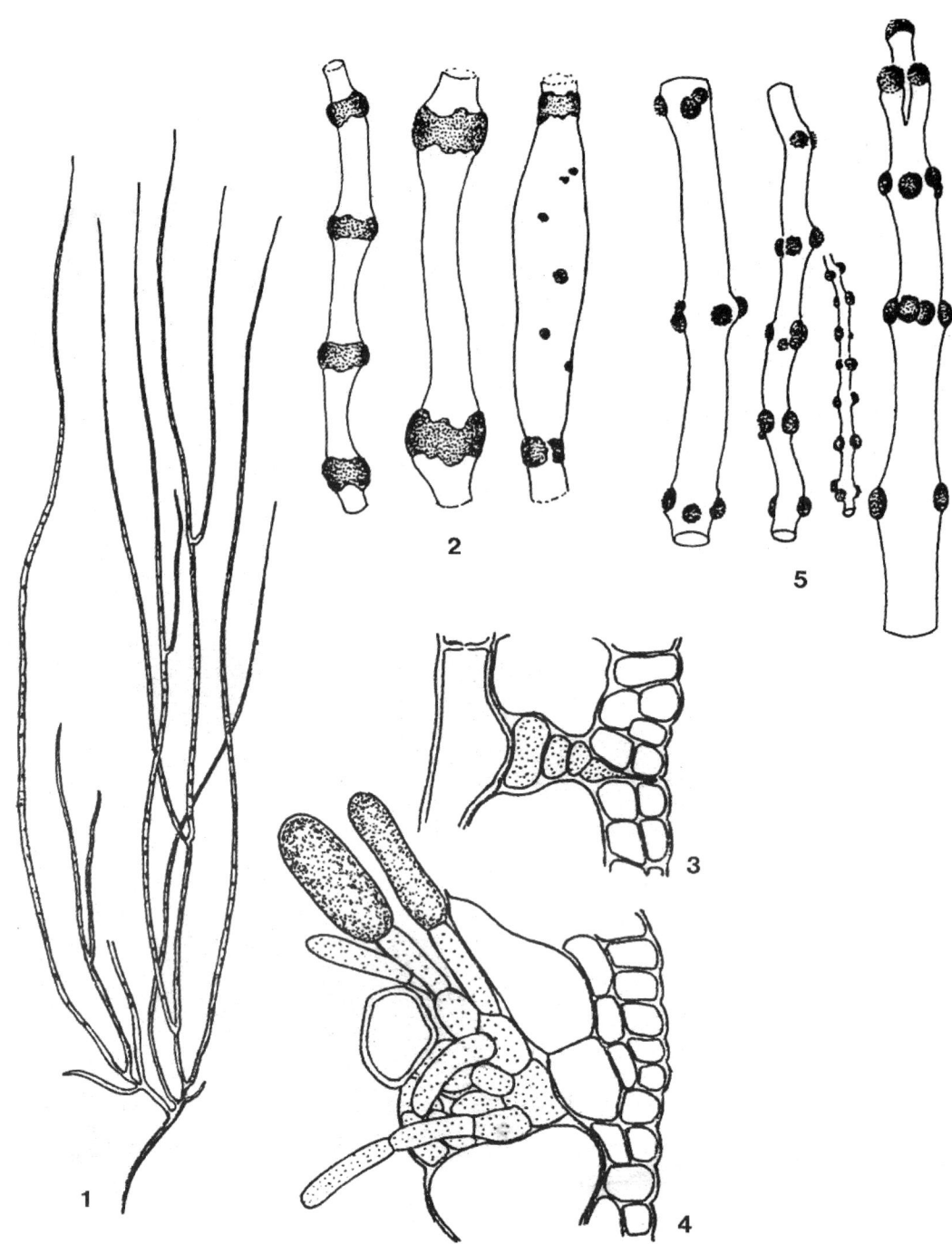

Plate 161
Lemanea sinica, figs 1–4 (original author, Jao 1941)
1. the macroscopic view of a plant showing ramification habit, 2. a plant showing spermatangia papillae, 3. the section of a carposporophyte zone, 4. ellipsoidal carposporangia terminal on gonimoblast filaments.

Lemanea rigida as *Sacheria rigida*, fig. 5 (original author, Sirodot 1872)
5. plants showing spermatangia papillae.

9) *Lemanea mamillosa* Kützing (1845:261)
Synonym: *Sacheria mamillosa* (Kützing) Sirodot (1872:75, p. 1, fig. 7, pl. 8, figs 84–85): *Lemanea fucina* var. *mamillosa* (Kützing) Atkinson (1890:225, pl. VII, figs 8–18)

plate 156, figs 4–5 (original author, Sirodot 1872) as *Sacheria mamillosa*
plate 159, figs 3–6 (Atkinson 1890) as *L. fucina var. mamillosa.*

Plants growing in bunches or forming tuft concentrations, stout, red-violet, brownish, olive-green or dark green, young Plants of a dark violet colour, reddening the water in which they stand for several days, in age drying yellow, gradually narrowing towards the base, at the base coalescent, forming compact tufts, 4–10 cm high, confluent with the basal-cells of the chantransia phase. Carposporophyte zones cylindrical or nearly so. Spermatangial papillae very prominent at time of fertilization, sometimes more so in age by hypertrophy of the tissue at base of spermatangia, sometimes confluent, 3 or 4 in whorls, distributed more or less regularly, protruding, growing in bunches or forming tuft concentrations (after Atkinson 1890, Starmach 1976).

Type Locality: a locality in Germany in Europe.
Holotype: To be sought among the Kützing collections at Leiden
Distribution: Known from France, Germany, Sweden? (Israelson, 1942) in Europe; North Carolina, Alabama, USA in North America.

10) *Lemanea fucina* Bory (1808:185, pl.21, fig.3) var. *fucina*
Synonym: *Lemanea mamillosa* var. *fucina* Kützing (1843:261); *Sacheria fucina* (Bory) Sirodot (1872:74, pl. 3, fig. 20, pl. 8, fig. 83)

plate 158, figs 1–6 (Sheath 1984)
plate 159, figs 1–2 (original author, Sirodot 1872) as *Sacheria fucina* Sirodot

Plants before maturity olivaceous, or yellowish-green, sometimes dark violet, sometimes reddening the water in which they stand, confined to the basal cell of the chantransia phase or arising from the middle branching. When young usually preserving colour in drying. In age drying yellowish, greenish, or blackish; from 2–40 cm long, very delicate or stout, with an abrupt contraction at the beginning of the fertile portion, marked in stout specimens; simple or very much branched; branches distributed all along the main axis of the Plants; main axis reaching beyond the branches and easily traced, or indistinguishable from them; branches unilateral, or fasciculate or both; when profusely branched the final branches slender, often capillary, becoming stouter in age by the breaking away of the capillary apices. Carposporophyte zone nearly cylindrical, or constricted in the middle; spermatangial papillae plane or prominent, so that the Plants varies from cylindrical to torulose, or with regularly recurring whorls of prominent papillae. Papillae in verticils of two to seven, sometimes irregular, often confluent, sometimes increasing after fertilization by hypertrophy of the tissue beneath the spermatangia, so that they are very prominent in age, sometimes less prominent after fertilization; carposporophyte zone in some specimens strongly constricted just above the spermatangial zone, so that with the next spermatangial zone it appears nearly clavate, more strongly so in age and toward the distal end of the plants. Spermatangial zones sometimes distant, sometimes rather near each other. Carpogonium-bearing branch developed in and near the spermatangial zone, never in the middle of carposporophyte zone, so that at maturity the clusters of spores alternate with the sterile middle portion of carposporophyte zone; spores when mature giving a darker colour to the plants. (After Atkinson 1890).

Batrachospermales

Chantransia phase producing extensive mats or individual tufts, greenish or bluish-green, 1–2 mm long, primary branches usually alternate, final branching unilateral, alternate, or sometimes opposite or slightly fasciculate, sometimes pilose, filaments not much attenuated at base, 15–35 μm in diameter.

Type Locality: rapid streams between Vitré and Fougères, France in Europe.

Holotype: PC herb. Bory.

Distribution: Known from France, Belgium in Europe; from Vermont and New Hampshire (Flint 1947), Wisconsin (Prescott 1962), North Carolina and Massachusetts (Atkinson 1931), USA and Canada in North America. Known from fast flowing streams. However, it is unclear from the descriptions whether these populations are *L. fucina*, *L. fucina* var. *parva* or *L. fluviatilis* (Vis & Sheath 1992).

11) ***Lemanea fucina*** Bory var. ***parva*** Vis et Sheath (1992:177)

Plants 2.9–7.1 cm long, 0.34–0.67 mm in diameter, stalked and very branched (0–18). Spermatangia in circular patches on nodes. Internal structure consisting of a central axis without cortical filaments and T- or L-shaped ray cells applied to the outer cortex.

Type Locality: Concheco River crossing Route 16, 200 m north of Rochester, New Hampshire, USA in North America.

Holotype: UBC A8264.

Distribution: Known from the type locality and in a river at the junction of Route 71 and 270, 4.6 km east of Polk, in a river crossing Route 59 east of Page, 4.8 km east of Route 259, Arkansas, USA in North America.

Genus ***Paralemanea*** (Silva) Vis et Sheath (1992:177)

Type: *Paralemanea catenata* (Kützing) Vis et Sheath
Synonym: *Lemanea* subgenus *Paralemanea* Silva (1959:62);
subgenus *Eulemanea* (*Paralemanea*) Atkinson (1931:225);
genus *Lemanea* Sirodot (1872:77)

Plants cylindrical, bristle-like, usually not branched. Axial filaments surrounded by slender, longitudinal medullar filaments arising from the proximal ends of the ray-cells and forming a loose cortex. Ray-cells (periaxial cells) simple, not reaching the outer cortex. Central axis soon surrounded by slender filaments developed from the lower face of the ray-cells. Generative filaments widely separated from the wall except in the spermatangial zone, normal number at first six above and eight below the ray-cells, soon becoming eight above by the branching of two opposite first cell. Cortex sometimes comprising 2 or 3 layers of cells developed serially. Spermatangial sori in rings around nodes; internal structure of cortical filaments around central axis and simple ray cells not abutting the outer cortex. Carposporophyte zone hour-glass shaped, the spermatangia forming regular or interrupted confluent bands around the spermatangial zone. Carpogonium-bearing branch of 5–10 cells, always developed in the middle of the carposporophyte zone, never as near the spermatangial zone as in *Sacheria*. The basal cell of gonimoblast filaments short, oval. Hypogynous cells of the carpogonium-bearing branch producing short celled involucral filaments at time of fertilization. Prostrate form of the protonema mainly confervoid.

Chantransia phase well developed, richly branched and persistent. The genus *Paralemanea* could be found in turbulent water and in slow-running water, but if it was slow running water, the streams were deep and strongly flowing. (After Atkinson 1890).

Key to the species of the genus *Paralemanea*

1. Plants with many branches usually in rebranching whorls...................1) *Paralemanea mexicana*
1. Plants unbranched...2
2 Plants ≥ 8 cm long, ≥ 0.7 mm in diameter..2) *P. catenata*
2. Plants < 6 cm long, < 0.7 mm in diameter..3
3. Carposporangia produced in chains..3) *P. annulata*
3. Carposporangia produced solitary...4) *P. grandis*

1) ***Paralemanea mexicana*** (Kützing) Vis et Sheath (1992:177)
Basionym: *Lemanea mexicana* Kützing (1857:34)
Synonym: *Lemanea feldmannii* Sánchez-Rodríguez et Huerta (1969:27)

Plants bristle-like, cylindrical with bulges, unstalked, unbranched, whorled branched and rebranched, branches not be confused, 290–620 μm in diameter, 3.6–11.4 cm long. Papillae greatly aggregated. Cellular structure, 4 celled layers, outer cortex smaller, inner gradually larger. Medullar layer, axial cortical filaments densely reticulate without gelatinous matrix. Carpospores in chain.
Type Locality: rivers in mountains of Mexico in Central America.
Holotype: L herb. Lugd. Bat. 941. 96.41
Distribution: Known from central Mexico, geographically disjunct from other species.

2) ***Paralemanea catenata*** (Kützing) Vis et Sheath (1992:177)
Basionym: *Lemanea catenata* Kützing (1845:261, 1867: pl. 87, fig. 1)
Synonym: *Lemanea nodosa* Kützing (1867: pl. 87, fig. 2; 1859: 528); *Lemanea pleocarpa* Atkinson (1931:236)

plate 162, figs 1–7 (Sirodot 1872) as *Lemanea catenata*,
plate 163, figs 3–4 (Sirodot 1872) as *Lemanea nodosa*

Plants violet colour, becoming very dark, when dry, unstalked, unbranched, (5–) 8.0–14.8 cm long, 0.69–0.93 (–1.5) mm in diameter (Vis & Sheath 1992)., distinctly narrowed at the base, straight or slightly arcuate. Spermatangial zones of fertile portions prominent, papillae in nodal rings, broad, irregular, sometimes interrupted near base. Carposporophyte zones nearly cylindrical. Carpogonium-bearing branch in middle of carposporophyte zone (after Atkinson 1890)
Chantransia phase in tufts; branches attenuated at base, alternate, sometimes unilateral above, shaped like an artist's brush.
Type Locality: in rivers of Rhine region of western Germany in Europe.
Holotype: L herb. Lugd. Bat. 10, 941.149, -343.
Distribution: Known from Belgium, France, Germany in Europe; Kentucky (type of *L. pleocarpa*), Indiana (Vis & Sheath 1992 as *L. catenata*), possibly California, USA in North America.

3) ***Paralemanea annulata*** (Kützing) Vis et Sheath (1992:177)
Basionym: *Lemanea annulata* Kützing (1845:261, 1867: pl. 84, fig. 1)

plate 163, figs 1–2 (Sirodot 1872) as *Lemanea annulata*,
plate 164, figs 1–6 (Necchi & Zucchi 1995)

Batrachospermales

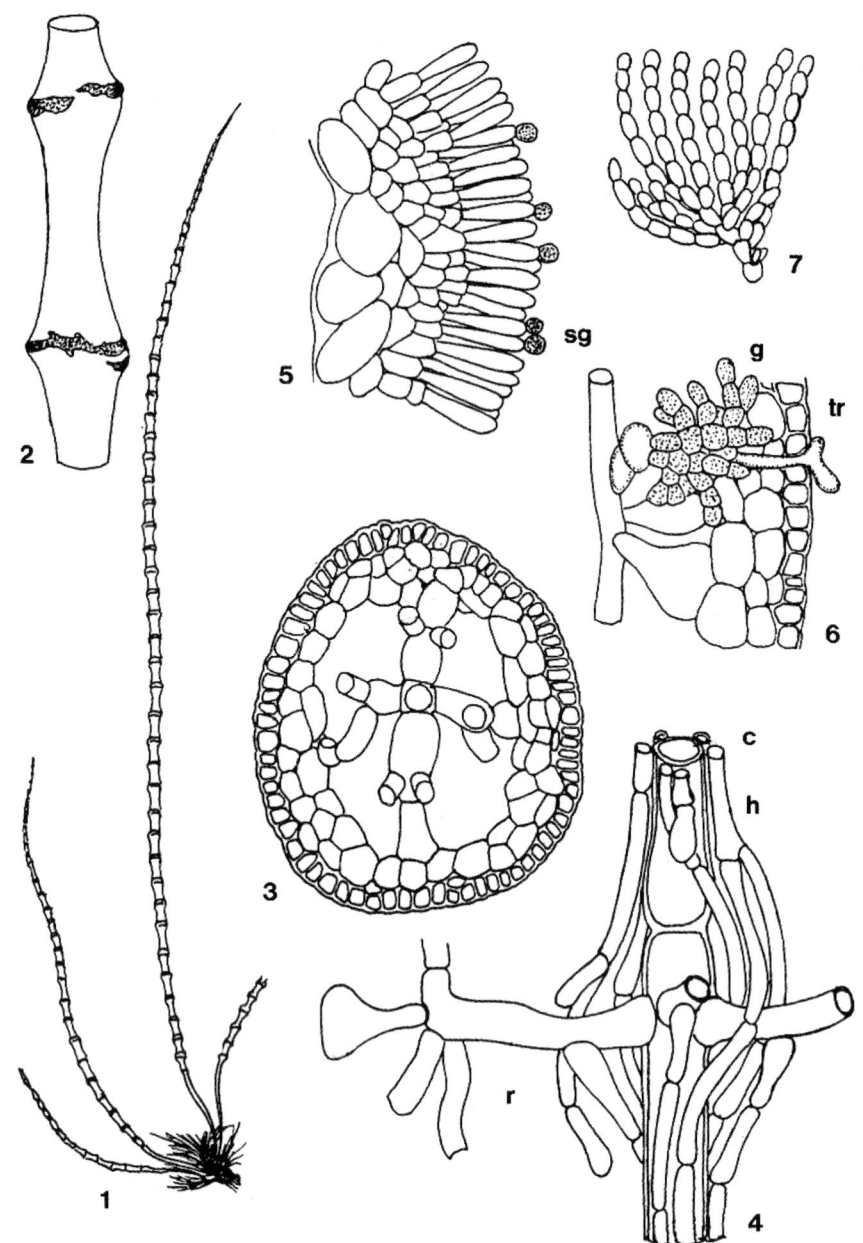

Plate 162
Paralemanea catenata as *Lemanea catenata*, figs 1–7 (Sirodot 1872)
1. a gametophyte arising from a basal portion of chantransia phase, 2. a portion of a plant showing spermatangial papillae, 3. the transverse section through an internode of a young branch in which there are cortical cell filaments surrounding an axial cell filament, 4. the longitudinal section of a portion of a nodal segment, showing cortical cell filaments localized around an axial cell filament with simple ray cells (periaxial cells) that do not abut the large cells of the outer fascicles, 5. the cross section of a spermatangial zone, 6. the longitudinal section of carposporophyte zone showing a trichogyne and gonimoblast filaments in the early stage of development, 7. a carposporophyte, gonimoblast filaments in the early stage of development. c; an axial cell filament, g; gonimoblast filaments, h; cortical cell filaments, r; ray cell (periaxial cell), sg; spermatangia, tr; trichogyne.

Paralemanea

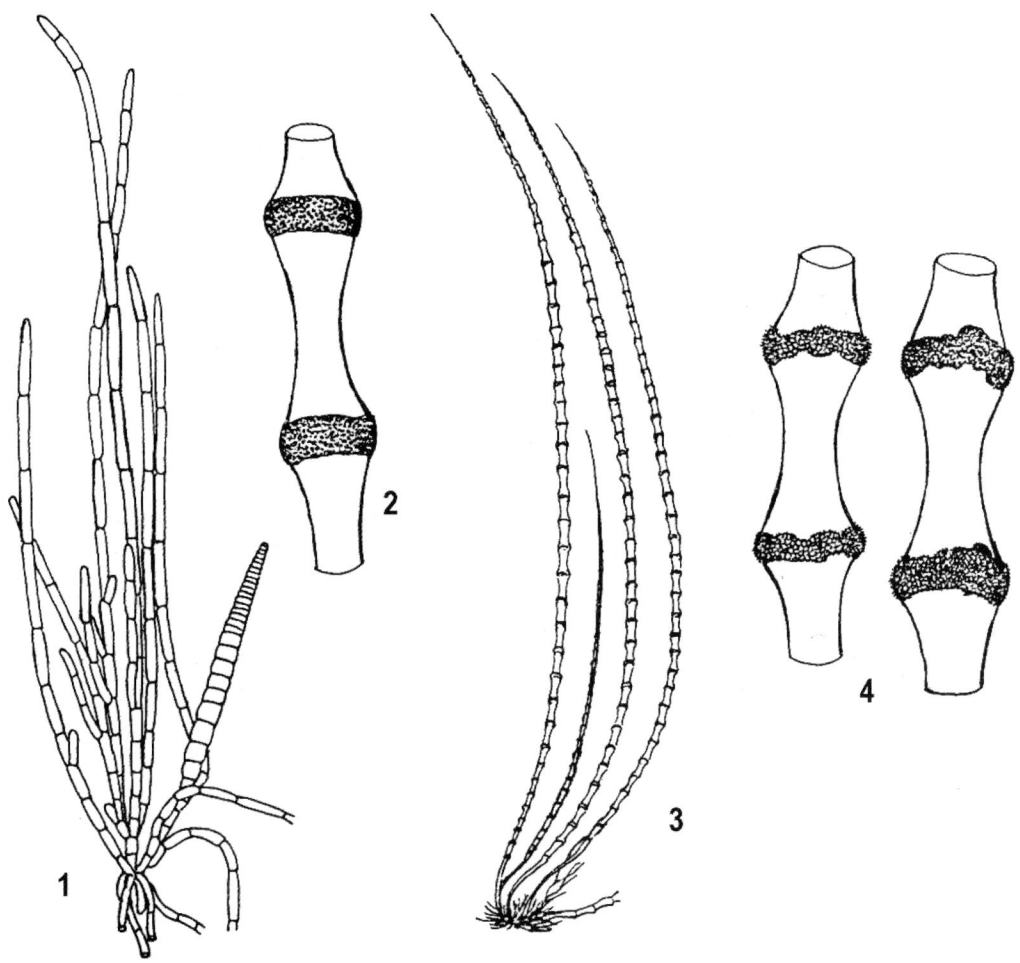

Plate 163
Paralemanea annulata as *Lemanea annulata*, figs 1–2. (Sirodot 1872)
1. a chantransia phase with a gametophyte in a basal ramification, 2. a portion of a plant showing spermatangial papillae.

Paralemanea catenata as *Lemanea nodosa*, figs 3–4 (Sirodot 1872)
3. a gametophyte arising from a basal portion of chantransia phase, 4. a portion of a plant showing spermatangial papillae.

Plants arising at the ends of a principal axis of the chantransia phase, or on a short branch, restricted to the base, of a violet colour when young, sometimes fading out in age, young ones becoming black when dry. Plants 8–15 cm high, unstalked, unbranched or very rarely branched, with the branches fasciculate and developing from carpogonium-bearing branches in the middle of the carposporophyte zone; up to 2 mm in diameter at nodes, concave in internodes. Sexual segments regularly constricted. Papillae in nodal rings, at the time of fertilization of a lighter colour than the adjacent parts when in water, broad, usually regular, sometimes interrupted near the base, forming large rings. Spermatangial zone sometimes enlarging after fertilization by hypertrophy of the tissue at the base of the spermatangia. Carposporophytes developed in the middle of carposporophyte zone. (After Atkinson 1890).

Batrachospermales

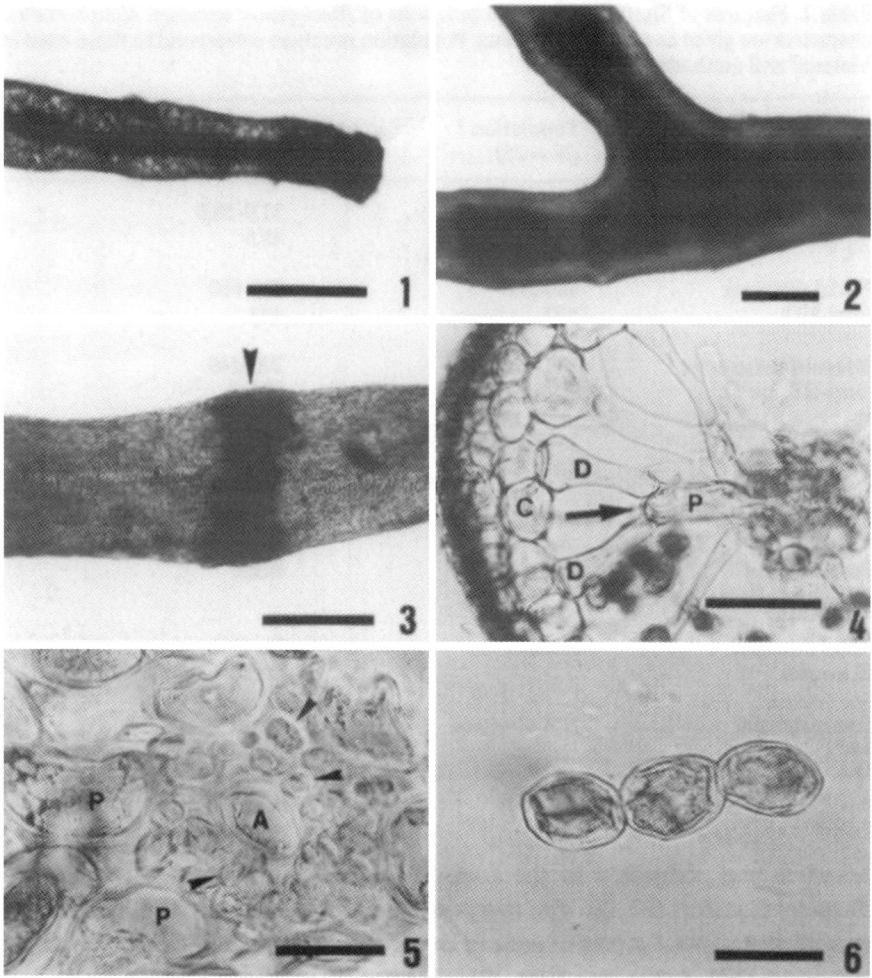

Plate 164
Paralemanea annulata, figs 1–6. (Necchi & Zucchi 1995)
1. an unstalked basal portion of a plant, 2. a detail of a branching portion, 3. details of a nodal region showing spermatangial papillae in ring (arrowhead), 4. a cross section of a nodal region showing two layers of ray cells (P; a periaxial cell, and D; distal cells, lower cells of fascicles), the "Y" branched distal cells (arrow) and touching the fascicle cells (C), 5. a cross section of a nodal region showing axial cell (A) wrapped by cortical cell filaments (arrowheads) and proximal ray cells (P; periaxial cell), 6. carposporangia in chain.
Scale bars = 500 µm (fig. 1); 250 µm (fig. 2); 200 µm (fig. 3); 100 µm (fig. 4); 50 µm (figs 5–6).

Chantransia phase 2–3 mm long of a dark violet colour unless faded with age, cells 30–40 µm in diameter, filaments of nearly the same diameter throughout, branching below alternate, then unilateral, alternate or rarely opposite.
Type Locality: Saale River at Halle, Germany in Europe.
Holotype: L herb. Lugd. Bat. 10, 941.149. 336.
Distribution: Known from France, Germany in Europe; Arkansas, California ?, Indiana, Nevada, North Carolina, Oregon, Washington, West Virginia, USA in North America; Brazil, Argentina in South America. Occurs on stones in fast-flowing streams.

4) ***Paralemanea grandis*** (Wolle) Kumano *comb. nov.*
Basionym: *Entothrix grandis* Wolle (1877:183)
Synonym: *Tuomeya grandis* (Wolle) Wolle (1887: pl. 66, figs 2–8); *Lemanea grandis* (Wolle) Atkinson (1889: 292); *Lemanea australis* Atkinson (1890:218, pl. IX, figs 43–44, 47)

plate 165, figs 5–9 (original author, Atkinson 1890) as *Lemanea australis*,
figs 10–11 (Atkinson 1890) as *Lemanea grandis*

Plants green, when young not blackening upon drying, in age blackening or darkening when dry, confined to the base of chantransia phase, either from a short principal axis, or a short branch; stout; sterile base gradually enlarging to fertile portion. Fertile segment nearly plane at base, strongly constricted at time of fertilization in the middle region and nearly plane at the distal end; fascicles of mature specimens in two layers of cells; spermatangial band narrow and interrupted near the base, perfect and irregular in middle region, and broad and regular at the distal end. At maturity spermatangial band sometimes constricted because the tissue here ceases to grow, sometimes prominent by hypertrophy, when a prominent ring is formed. Carpogonium-bearing branch in the middle of the carposporophyte zone. Carpospores 35–45 µm in diameter (after Atkinson 1890).

Chantransia phase in dense tufts or in unbroken patches, yellowish or bluish-green, of two forms, sterile form 4–5 mm, fertile 2–3 mm. Branching alternate, or rarely, unilateral, or opposite at base, penicillate when crowded, corymbiform when completely free, slender at base, gradually increasing in size to middle region, where the cells are very large, sometimes nearly spherical.

Type Locality: Bethlehem, Pennsylvania, USA in North America.
Holotype: NY, vi.1877
Distribution: Known from type locality and Vermont, Alabama and New Hampshire seems to be common in the south-eastern USA, and less common in the north-eastern USA (as *L. australis,* Flint 1947), from North Carolina, South Carolina, Maryland, West Virginia, Georgia and Mississippi (as *L. australis*, Atkinson 1931), from Pennsylvania and Maryland (as *L. grandis*) Broad River, Columbia, South Carolina, USA in North America.

According to Blum (1997 ('1994')), the existing descriptions and keys to the species of the genus *Paralemanea* rely largely on the size and shape of the plants, which represent only one stage of the development. They are imprecise when describing nodal shape and other features, which vary from season to season depending on stages in the development. The following five species are described by Blum (1993b, 1997('1994')) based on the different and (according to Blum) more reliable parameters than those conventionally used.

Key to the species of the genus ***Paralemanea***
from Indiana (Blum 1993b) and California (Blum 1997('1994'))

1. Carpospores > 60 µm in diameter..1) *Paralemanea deamii*
1. Carpospores < 30 µm in diameter...2
2. Complete annular bands starting at node ≤ 15..2) *P. gardneri*
2. Complete annular bands starting at node > 9..3
3. Cortical filaments < 22..3) *P. parishii*
3. Cortical filaments > 20...4
4. Internodes < 2 mm long...4) *P. tulensis*
4. Internodes > 2 mm long..5
5. Complete annular bands starting at node 2–3..5) *P. californica*
5. Complete annular bands starting at node 6–9..6) *P. brandegeei*

Batrachospermales

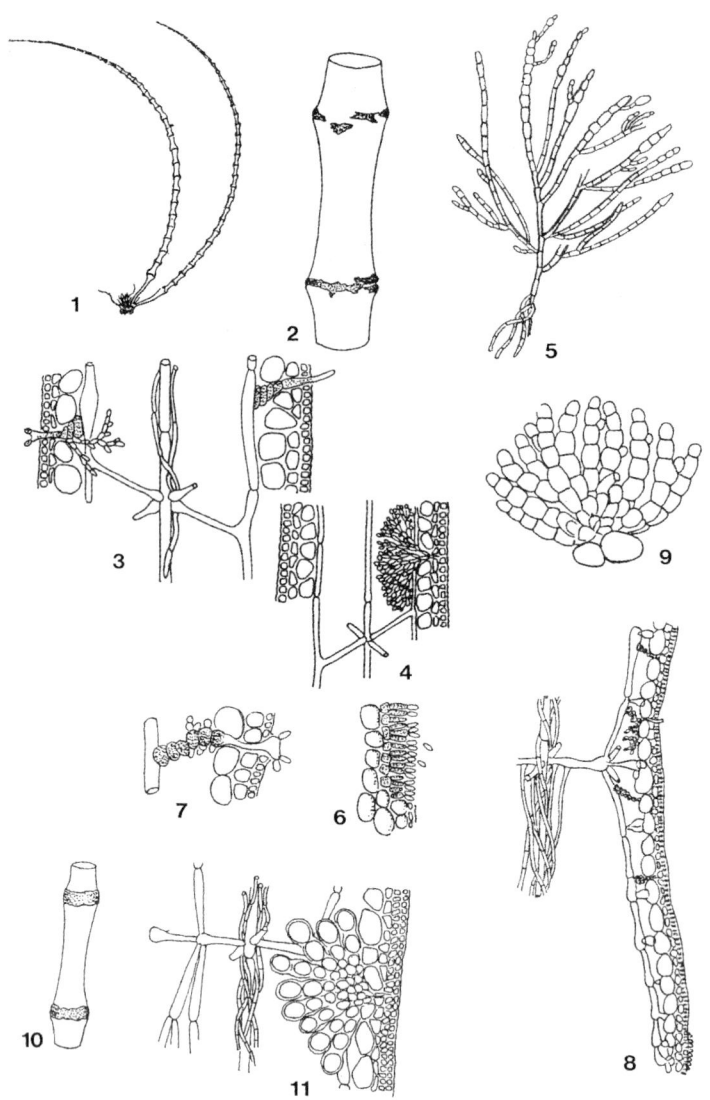

Plate 165

Lemanea rigida as *L. torulosa*, figs 1–2 (Sirodot 1872), figs 3–4 (Schmitz)
1. a gametophyte arising from a basal portion of chantransia phase, 2. a portion of a plant showing spermatangial papillae, 3. the section of a carposporophyte zone showing axial cell filaments, ray cells (periaxial cells), cortical cell filaments (enveloping filaments), generative filaments, carpogonium-bearing branches, one of which with a fertilized trichogyne, 4. gonimoblast filaments.

Paralemanea grandis as *Lemanea australis*, figs 5–9 (original author, Atkinson 1890).
5. gametophytes arising from the middle of chantransia phase, 6. the section of a spermatangial zone, 7. a carpogonium-bearing branch with a fertilized trichogyne, 8. the section of a carposporophyte zone showing axial cell filaments, ray cells (periaxial cells), cortical cell filaments, generative filaments, carpogonium-bearing branches and spermatangia, 9. gonimoblast filaments.

Paralemanea australis as *Lemanea grandis*, figs 10–11 (Atkinson 1890).
10. a portion of a plant showing spermatangial papillae, 11. the section of a procarp zone showing cortical filaments dissected away to show cruciate ramification of rays (periaxial cells) from a central axis, a carposporophyte reaching from generative filament into tissue of a wall, from which point the basidia (bases of gonimoblast filaments) radiate and bear carposporangia.

1) *Paralemanea deamii* Blum (1993b:4, figs 3–4)

> plate 166, figs 1–2 (original author, Blum 1993b)

Plants axis 0.8–1.6 mm in diameter at the node, to 13 cm overall length, with ca. 45–(58) internodes, sometimes abruptly narrowed to a basal stipe. Carpospores subspherical, very large, 66–68 μm in diameter at maturity.

Chantransia phase to 2.7 mm in length, 17–24 μm in diameter at the crosswalls, sometimes narrowed toward the base, but not typically so; sometimes branching from the base or just above the base, not much branched above.

Type Locality: Sec. 6, T2S, R2E, near Marengo, Whiskey Run, Crewford County, Indiana, USA in North America.

Holotype: UC 4971, Blum 18/VIII, 1989

Distribution: Known from the type locality and Indian River, near Corydon, Harrison, Indiana, USA in North America.

2) *Paralemanea gardneri* Blum (1997 ('1994'):11, figs 5–6, 20–21)
Synonym: *Lemanea annulata* var. *franciscana* Atkinson (1931:232)

> plate 167, figs 1–2 (original author, Blum 1997) ('1994'),
> plate 169, figs 2–3 (original author, Blum 1997) ('1994')

Plants axis slender, often cylindrical; internodes mostly 1.0–1.9 mm long, sometimes with gently swollen nodes; central strand in the mesothallus of 10–30 cortical filaments. Lowest spermatangial zone (at node 1) 4.3–7.2 mm above the rhizoidal base but with ostensibly complete annular bands or rings starting only at node 15 or higher; the annular bands 25–90 μm wide, not increasing in overall width on superior nodes; many nodes showing shallow annular through approximately coinciding in position with the spermatangial bands. Carposporophytes usually 2–6 per internode; banded appearance typical, if carpospores are present. Carpospores 13–31 μm in diameter, 19–42 μm long.

Chantransia phase without markedly conical filaments, but sometimes with an identifiable main filaments wider than the primary branches; the basal filaments 13–26 μm in diameter at the crosswalls, branch filaments 13–17 μm in diameter.

Type locality: San Francisquito Creek, Santa Clara, California, USA in North America.

Holotype: UC 27167, N.L. Gardner 2846.

Distribution: Known from many localities such as Palm Canyon, Borrego Valley, San Diego, Mill Valley, Mt. Tamalpais, and Mill Creek, Marin, California, USA in North America.

3) *Paralemanea parishii* Blum (1997('1994'):17, figs 13–14, 23)

> plate 168, figs 3–4 (original author, Blum 1997) ('1994')
> plate 169, fig. 5 (original author, Blum 1997) ('1994')

Plants axis narrow and nodose, often with cylindrical internodes and slightly enlarged nodes. Internodes mostly 1.0–1.5 mm long, the central strands in the mesothallus comprising 5–22 cortical filaments. Carposporophytes 2–7 per internode. Mature carpospore mass filling the internode. Spermatangial zones beginning (node 1) with one or more rounded and raised papillae at (4.8)–6–7–9 mm above the rhizoidal base, this pattern of papillae sometimes continuing distally for 15 or more nodes; the remaining spermatangial bands, 50–100 μm in diameter, not increasing materially in width distally. carpospores 11–20 μm in diameter, 15–31 μm long.

Batrachospermales

Plate 166
Paralemanea deamii, figs 1–2 (original author, Blum 1993b)
1. a chantransia phase showing two juvenile gametophytes, a broken one (a) and perfect one (b),
2. the development of a carposporophyte showing larger carposporangia of this species (right) and smaller ones of another species (left).

Chantransia phase without markedly conical filaments but sometimes with an identifiable main filament wider than the principal branches. Main filament 14–19 μm in diameter; apical cell cylindrical.

Paralemanea

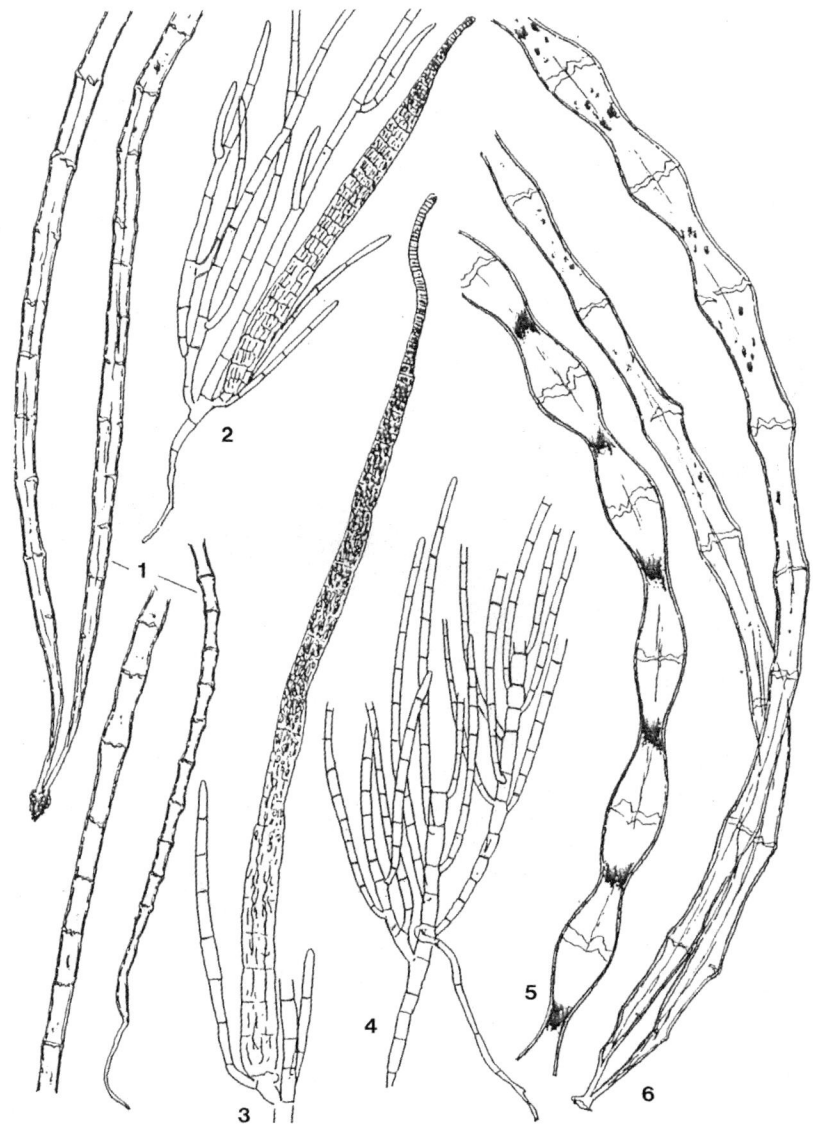

Plate 167
Paralemanea gardneri, figs 1–2 (original author, Blum 1997) ('1994')
1. gametophytes (type), 2. a chantransia phase with a juvenile gametophyte axis.

Paralemanea californica, figs 3–6 (original author, Blum 1997) ('1994')
3. a young gametophyte axis showing a proboscoid tip, 4. a chantransia phase (type), 5. a gametophyte axis (blastulae coll.), 6. gametophyte axes (type).

Type locality: Tahquitz Creek, about 0.6 km below Tahquitz Fall, Palm Springs, Riverside, California, USA in North America.
Holotype: UC 5175, J.L Blum 9/XII, 1993.
Distribution: Known from the type locality and a historical collection from hills 2 miles southwest of Bloomington in the vicinity of San Bernardino, San Bernardino, California, USA in North America.

Batrachospermales

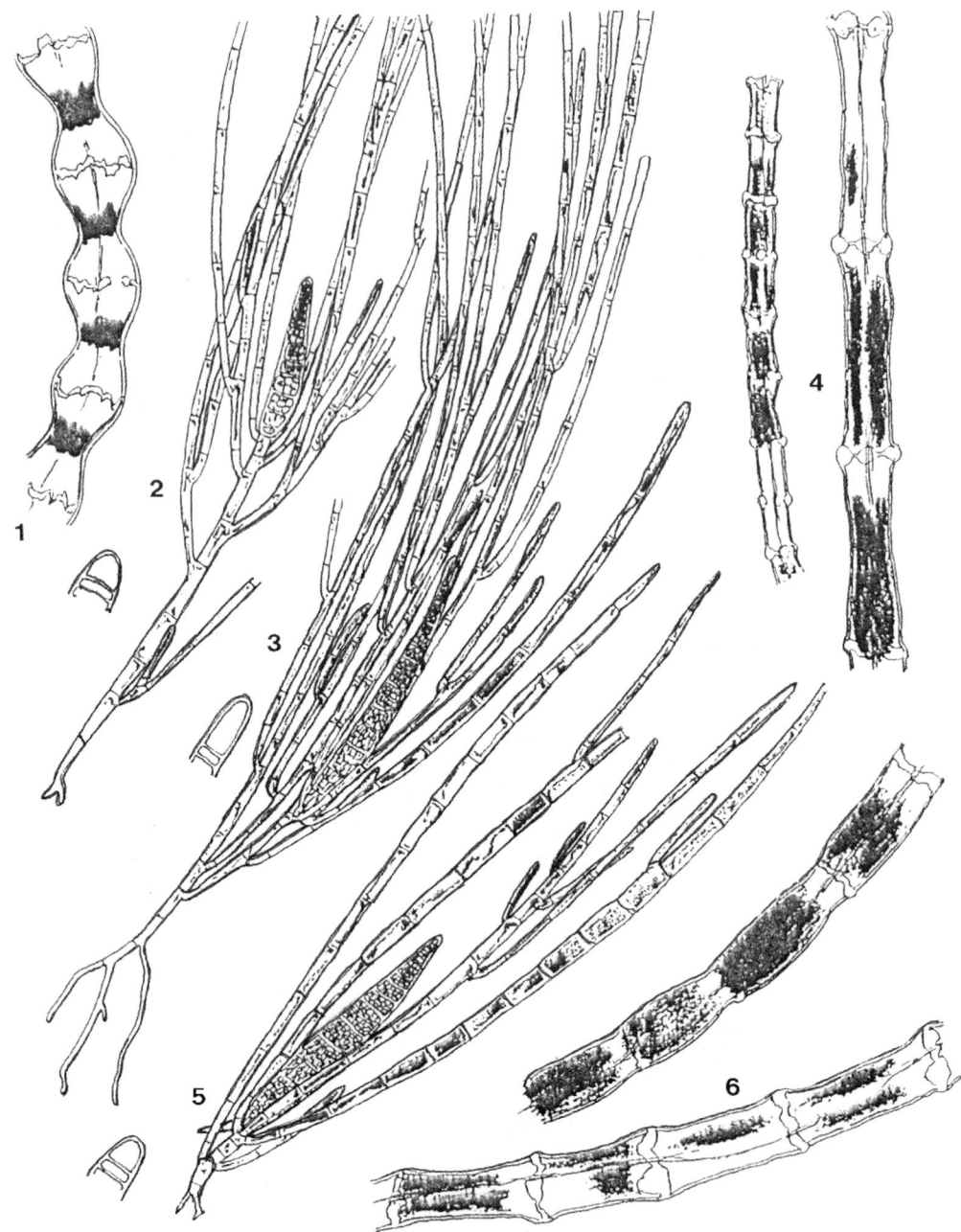

Plate 168
Paralemanea brandegeei, figs 1–2 (original author, Blum 1997) ('1994'
1. a mature gametophytes axis (type), 2. a young gametophyte arising from a chantransia phase.

Paralemanea parishii, figs 3–4 (original author, Blum 1997) ('1994')
3. a young gametophyte arising from chantransia phase, 4. mature gametophyte axes (type).

Paralemanea tulensis, figs 5–6 (original author, Blum 1997) ('1994')
5. a young gametophyte arising from a chantransia phase (type), 6. mature gametophyte axes (type).

4) *Paralemanea tulensis* Blum (1997 ('1994'):19, figs 15–16)

plate 168, figs 5–6 (original author, Blum 1997) ('1994')

Plants axis with relatively small epidermal cells of which 170–210 occur on a 1 mm transect; slightly nodose with internodes, where carpospore-filled, frequently exceeding in diameter the nodal diameter. Internodes ca. 1.0–1.8 mm long. Central strand in the mesothallus comprising 20–30 cortical filaments. Lowest spermatangial zone (node 1) 4.3–7.2 mm above the rhizoidal base but with ostensibly complete annular bands or rings at the nodes distally from nodes 3–5; the annular bands showing a substantial increase in width from node 3 distally. Mature carpospore mass approximately filling the internode, Carpospores 8–24 μm in diameter, 15–31 μm long.

Chantransia phase with markedly conical filaments which enlarge distally from 13–33 μm; apical cell dome-shaped.

Type locality: North Fork of Tule River on large rocks at the crossing of Highway 190, Springville, Tulare, California, USA in North America.

Holotype: UC 5176, J. L. Blum, 11/XII, 1993.

Distribution: Known from the type locality only.

5) *Paralemanea californica* Blum (1997) ('1994'):15, figs 7–10, 19)

plate 167, figs 3–6 (original author, Blum 1997) ('1994')
plate 169, fig. 1 (original author, Blum 1997) ('1994')

Plants axis robust; internodes mostly 2–4 mm long, often with clearly swollen nodes; central strand in the mesothallus comprising 20–50 cortical filaments. Lowest spermatangial zone (at node 1) ca. 2.9–5.4 mm above the rhizoidal base and consisting of one or more spermatophorous papillae, but with complete annular bands or rings starting at node 2 or 3(–8), the annular bands 110–300 μm in diameter, increasing distally in width after the initial series (nodes 1–7) of spermatangial nodes; many nodes showing annular troughs approximately coinciding in position with the spermatangial bands. Carposporophytes usually 5–8 per internode; mature carpospore mass not filling the internode. Banded appearance typical, if carpospores are present. Carpospores ca. 13.2–28.6 μm in diameter, 15.4–33 μm long.

Chantransia phase in places heavily tufted; the basal filaments 11–20 μm in diameter at the crosswalls; branch filaments mostly 11–15 μm in diameter at the crosswalls. Young axis initials terminating distally in an elongate cylindrical portion.

Type locality: Four miles north of Ben Hur Post Office, Mariposa, California, USA in North America.

Holotype: UC 502456, R. S. Ferris & R. Bacigalupi, 29/V, 1941.

Distribution: Known from several localities in California and one in eastern Oregon, USA in North America.

6) *Paralemanea brandegeei* Blum (1997('1994'):16, figs 11–12, 22)

plate 168, figs 1–1 (original author, Blum 1997) ('1994'),
plate 169, fig. 4 (original author, Blum 1997) ('1994'

Plants axis robust, internodes mostly 2–3 mm long with swollen, frequently inflated nodes; central strand in the mesothallus comprising 30–40 cortical filaments; lowest spermatangial zone (at node 1) 6–8.5 mm above the rhizoidal base and consisting of one or more spermatangial

Batrachospermales

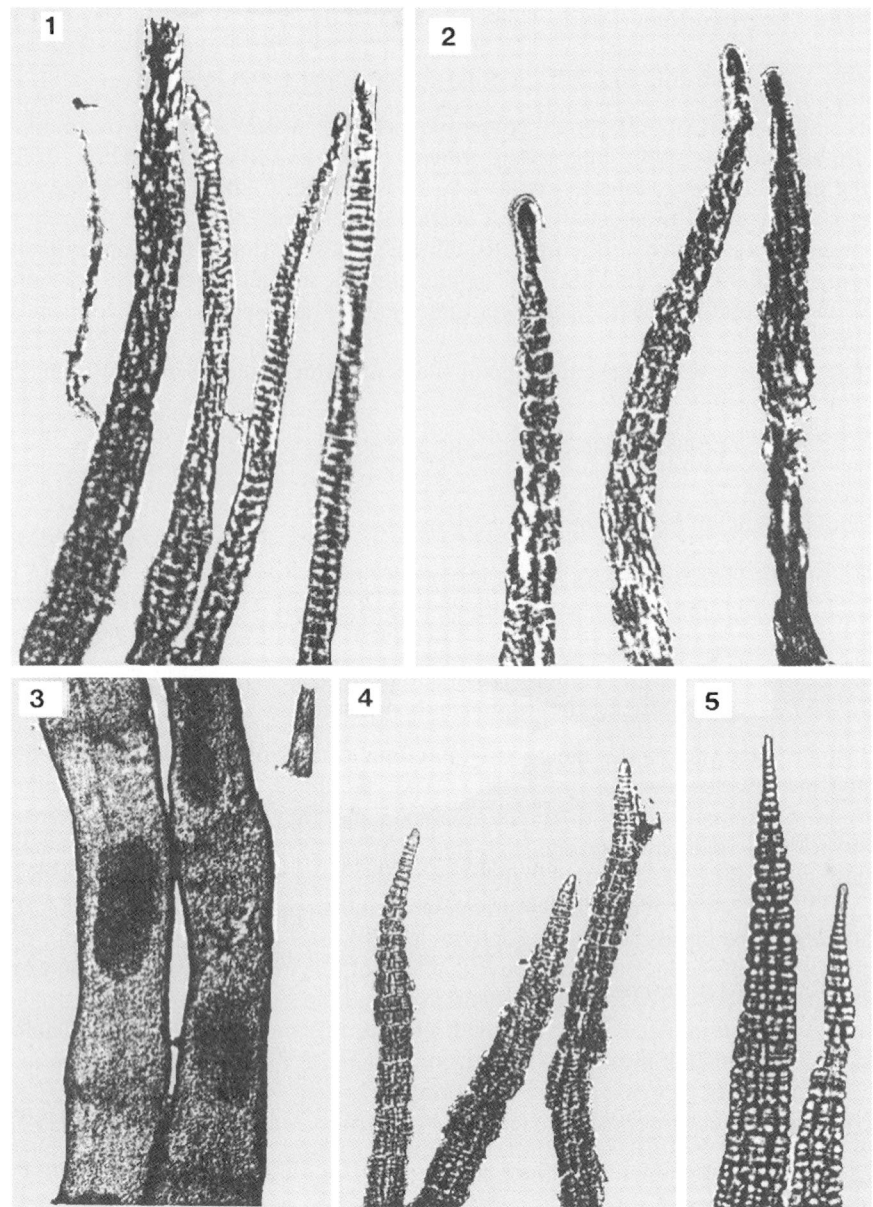

Plate 169
Paralemanea californica, fig. 1 (original author, Blum 1997) ('1994')
1. juvenile gametophyte axis (tips) at about the 800 cell stage; the tip of the left axis in the photomicrograph is broken off.

Paralemanea gardneri, figs 2–3 (original author, Blum 1997) ('1994')
2. juvenile gametophyte axis tip, 3. mature gametophyte axis.

Paralemanea brandegeei, fig. 4 (original author, Blum 1997) ('1994')
4. juvenile gametophyte axis tips.

Paralemanea parishii, fig. 5 (original author, Blum 1997) ('1994')
5. juvenile gametophyte axis tips.

papillae, but with complete annular bands or rings at about nodes 6–9; the annular bands 80–150 μm in diameter, increasing distally in width after the initial series (nodes 1–7) of spermatangial nodes. Carposporophytes usually 5–8 per internode; mature carpospore mass not filling the internode; banded appearance is typical, if carpospores are present. Carpospores 15.4–24.2 μm in diameter, 17.6–30 μm long.

Chantransia phase tufts without markedly conical filaments when reproductive, but sometimes with a main filament wider than the primary branches. Young axis initials without an elongate cylindrical portion; apical cell dome-shaped.

Type locality: New York River, Eldorado, California, USA in North America.
Holotype: UC 277581, Mrs. Katherine Brandegee, 10/V, 1945.
Distribution: Known from the type locality only.

Order **Thoreales** Sheath, Müller & Sherwood (2000) nom. illeg.

Sequences of the rbcL gene and 18S rRNA genes of *Thorea violacea* imply that the Thoreaceae is not closely related to Batrachospermales (Vis, Saunders, Sheath, Dunse & Entwisle, 1998). They regarded the family as *incertae sedis* pending analysis of additional species. Sheath, Müller & Sherwood (2000) and Hanyuda *et al.* (2001) invalidly proposed the order Thoreales to accommodate the two recognized genera, *Thorea* and *Nemalionopsis*.

Family **Thoreaceae** (Reichenbach) Hassall (1845:64)

Type: genus *Thorea* Bory

Plants mucilaginous, usually abundantly branched, multiaxial in organization, lateral assimilatory filaments disposed along the entire axis. Alternation of three phases, Plants, carposporophyte and chantransia phase. Sexual structures reported for three species of *Thorea*, *T. okadae* by Yoshizaki (1986), *T. bachmannii* by Necchi (1987) and *T. hispida* (carpogonium only) by Sheath, Vis & Cole 1993a; only asexual reproduction by monospores is known for the genus *Nemalionopsis*.

Note: This family originally included only the genus *Thorea*. *Nemalionopsis* was added by Kylin (1956). Based on the molecular phylogenetic analyses, significant genetic differences were found among *T. okadae*, *T. gaudicahudii* and *T. violacea*, so that these three species had better to be regarded as distinct species (Hanyuda *et al.* 2001).

Based on the examination of American populations and type specimens of the family Thoreaceae, Sheath, Vis & Cole 1993a recognized worldwide 4 species of *Thorea* ; *T. hispida* (Synonym: *T. andina, T. lehmannii, T. ramosissima*), *T. violacea* (Synonym: *T. bachmannii, T. brodensis, T. gaudichaudii, T. okadae, T. prowsei* and *T. riekei*), *T. clavata*, and *T. zollingeri*, and 2 species of *Nemalionopsis*; *N. shawii* and *N. tortuosa* (Synonym: *N. shawii* f. *caroliniana*).

According to Necchi & Zucchi (1997), it is particularly important to distinguish spermatangia from carposporangia. The former are almost colourless and with sparse contents, whereas the latter are strongly coloured and have dense contents. In addition, spermatangia are smaller (8.0–13.0 μm long, 4.0–7.0 μm in diameter) than carposporangia (17.0–26.0 μm long, 8.5–16.0 μm in diameter), according to Yoshizaki (1986) for *T. okadae* and Necchi (1987) for *T. bachmannii*. Sheath, Vis & Cole (1993a) followed the traditional interpretation and treated the reproductive structures generally as monosporangia, including a wide morphometric range (8.7–25.8 μm long, 4.2–11.5 μm in diameter) for *T. violacea* sensu Sheath *et al*. Some of the monosporangia (e.g., fig. 9) illustrated by Sheath, Vis & Cole (1993a) are very similar in shape to the clusters of carposporangia (fig. 15) shown by Yoshizaki (1986).

Necchi (1987) stated that the structures described as 'monosporangia' by several authors for different species of *Thorea* are possibly spermatangia or carposporangia and the occurrence of monosporangia in the genus has to be carefully studied in the light of a new information concerning sexual reproduction.

Thus, characters of sexual reproductive structures as well as of carposporophytes should be described in detail for every species of the genus *Thorea*.

A tentative key to the species of the Thoreaceae, including some species in which sexual reproductive organs are unknown, is shown here.

Key to the genera of the family **Thoreaceae**

1. Sporangium (carposporangium, spermatangium and monosporangium)-bearing branches short, sparsely branched and assimilatory filaments loosely arranged ....genus *Thorea*
1. Sporangium (probably monosporangium)-bearing branches long and much-branched, assimilatory filaments compressed........................genus *Nemalionopsis*

Genus *Thorea* Bory (1808:127) in NCU–3 (1993:1146)

Type: *Thorea ramosissima* Bory [= *Thorea hispida* (Thore) Desvaux]

Plants gelatinous, much branched, multiaxial in organization, lateral assimilatory filaments disposed along the entire axis. Sexual reproductive organs (spermatangia and carpogonia) and alternation of three phases (plant, carposporophyte and chantransia phase) have been reported for *T. okadae* and *T. bachmannii*. Asexual reproductive organs reported as monosporangia for some species may be spermatangia (those of smaller size) or carposporangia (those of larger size). Sporangium-bearing branches short, sparsely branched and assimilatory filaments loosely arranged.

Key to the species of the genus *Thorea*

1. Assimilatory filaments clavate, unbranched..2
1. Assimilatory filaments not clavate, variably branched..3
2. Monosporangia in clusters of 1–3...1) *Thorea clavata*
2. Monosporangia in clusters of 5–8..2)*T. zollingeri*
3. Secondary branching sparse......................3) *T. violacea sensu* Sheath *et al*. (part).................4
3. Secondary branching copious..5
4. Plants 500–1890 μm in diameter...4) *T. hispida*
4. Plants 180–460 μm in diameter...5) *T. conturba*
5. Sexual reproduction organs known..6
5. Sexual reproduction organs unknown..7
6. Plants 2100–4000 μm in diameter, up to 3 m long...6) *T. okadae*
6. Plants 800–1300 μm in diameter, 10–15 cm long..7) *T. bachmannii*
7. Plants up to 200 cm high..8) *T. reikei*
7. Plants < 60 cm high...8
8. Plants < 15 cm high...9) *T. prowsei*

8. Plants up to 6 cm high..9
9. Assimilatory filaments 300–800 μm long..10) *T. gaudichaudii*
9. Assimilatory filaments up to 1415–1750 μm long...11) *T. brodensis*

1) ***Thorea clavata*** Seto et Ratnasabapathy in Ratnasabapathy & Seto (1981:248, fig. 3, a–f, fig. 5, a–h)

plate 170, figs 1–6 (Seto 1998a)
plate 171, fig. 9 (Sheath, Vis & Cole 1993a)

Plants small, rather slender, tufted, very mucilaginous, 480–1425 μm in diameter, 4.5–12 cm long, dull brown, sparsely or abundantly branched, multiaxial, consisting of medullar filaments and cortical assimilatory filaments, attached to substrata with discoid holdfasts, 380–1350 μm in diameter. Medullar portion 115–420 μm in diameter; medullar filaments interlacing. Assimilatory filaments 130–840 μm long, consisting of 8–40 cells; apical cells clavate with rounded apices. Distal portion of assimilatory filaments unbranched or sparsely branched, clavate, gradually tapered from apex toward proximal portion. Sexual reproduction unknown. Monosporangia (carposporangia?) solitary or in clusters derived from the base of assimilatory filaments, ovoidal, obovoidal or pear-shaped, 5.5–14 μm in diameter, 8–20 μm long.

Type Locality: Gombak River, Selangor, Malaysia in Southeast Asia. Known from rocks in shaded place in a clean river, 5–40 cm depth, 2–3 m wide, running through tropical rain forest.

Holotype: Herb. Ratnasabapathy, RS 490, Seto, 6/V 1978.

Distribution: Known from the type locality only.

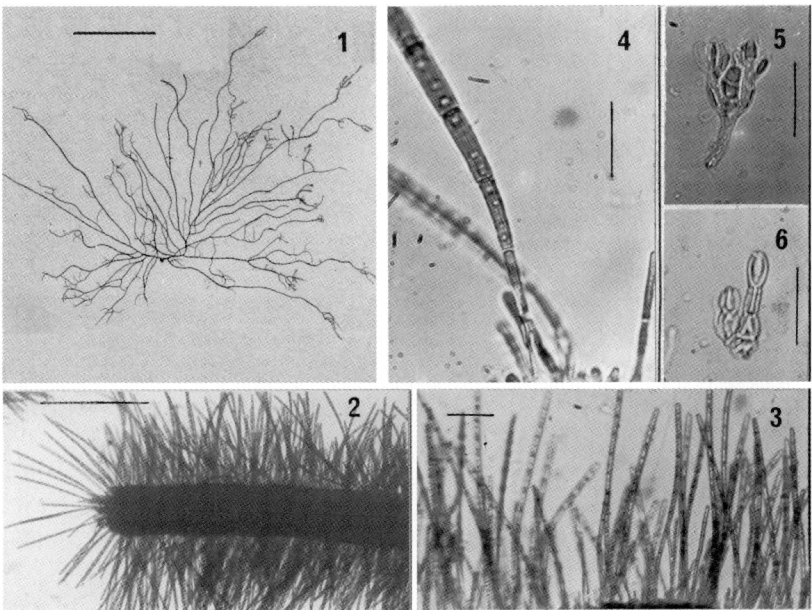

Plate 170
Thorea clavata, figs 1–6 (Seto 1998a)
1. a plant showing ramification habit, 2. assimilatory filaments and medullar layer of a main branch, 3. a few branched assimilatory filaments at a distal portion, 4. assimilatory filaments gradually tapering from an apex to the base, 5–7. monosporangia.
Scale bars = 2 cm (fig. 1); 400 μm (figs 2, 4); 100 μm (fig. 3); 30 μm (figs 4–6).

Thoreales

2) ***Thorea zollingeri*** Schmitz (1892:134) *emend.* Sheath, Vis et Cole (1993a:240, fig. 12)

plate 171, fig. 10 (Sheath, Vis & Cole 1993a)

Secondary branches 0.6 per 30 mm, plants 799–1188 μm in diameter, medulla 105–471 μm in diameter, assimilatory filaments clavate and unbranched, and monosporangia 9.4–15.6 μm long, in cluster of 5–8.

Type Locality: Java, Indonesia in Southeast Asia.
Holotype: L 941.182.160, herb. Weber-van Bosse, Zollinger coll. No. 3692.
Distribution: Known from the type locality only.
Note: Sheath, Vis & Cole (1993a) added two distinguishing features, clavate assimilatory filaments and large clusters of monosporangia to the original description by Schmitz (1892).

3) ***Thorea violacea*** Bory (1808:133, pl.18, fig. 2) *sensu* Sheath, Vis & Cole (1993a:238, figs 8–9)

plate 171, figs 6–7 (Sheath, Vis & Cole 1993a)

Secondary branches 2–9 per 30 mm, plant 540–1878 μm in diameter, medulla 126–304 μm in diameter, assimilatory filaments non-clavate with variable branching, and monosporangia (carposporangia?) 8.7–25–8 μm long, in clusters of up to three sporangia.

Type Locality: Rivière des Ramparts, la Réunion, Mascarene Islands in Indian Ocean.
Lectotype: BM, Bory 1801–1802, designated by Sheath, Vis & Cole (1993a).
Distribution: Known from the type locality only.
Note: Sheath, Vis & Cole (1993a), included *Thorea gaudichaudii* C. Agardh, *Thorea okadae* Yamada, *Thorea bachmannii* Pujals, and *Thorea prowsei* Ratnasabapathy et Seto in this species. Because sexual reproductive organs are unknown in *Thorea violacea* I think that such a merger is premature and Hanyuda *et al.* (2001) treated all of these species separately.

4) ***Thorea hispida*** (Thore) Desvaux (1818:16) *emend.* Sheath, Vis et Cole (1993a:238, figs 3–7)
Basionym: *Conferva hispida* Thore (1799:398, fig. A)
Synonym: *Conferva flexuosa* Bory (1804:336), *nom. illeg.*; *Thorea ramosissima* Bory (1808:128), illeg.; *Thorea lehmannii* Hornemann (1818: pl. 1594, fig, 1); *Thorea andina* Lagerheim et Möbius in Möbius (1891:338, figs 1–6).

plate 171, figs 1–5 (Sheath, Vis & Cole 1993a)

Plants 100 cm or more high, abundantly branched, secondary branches 11–41 per 30 mm, plant diameter 513–1890 μm, medulla 88–611 μm, in diameter. Assimilatory filaments sparsely or variably branched, non-clavate, 700–1400 μm long, consisting of 18–20 cells, 6–10 μm in diameter, 18–40 μm long. Chloroplasts laminate, greenish or violet. A mucilaginous envelope surrounding the axial part of plants, enveloping bases of assimilatory filaments. Monosporangia relatively rare, 8.6–30.0 μm long, in clusters of up to 8 sporangia. Monospores initially spherical, becoming pear-shaped.

Type Locality: L'Adour Rivière at Dax, France in Europe.
Lectotype: P–JU 375–D, Thore, 1801.
Distribution: Known from France, Belgium in Europe; China (as *Thorea ramosissima* Bory) in Eastern Asia. In Europe grows in rocky rivers and ditches in scattered sites, sometimes found in 2–3 m depth, by August.
Note: When transferring *Conferva hispida* Thore into his new genus *Thorea*, Bory (1808) unnecessarily changed the epithet, *Thorea ramosissima*. The conspecificity of *T. lehmannii* and

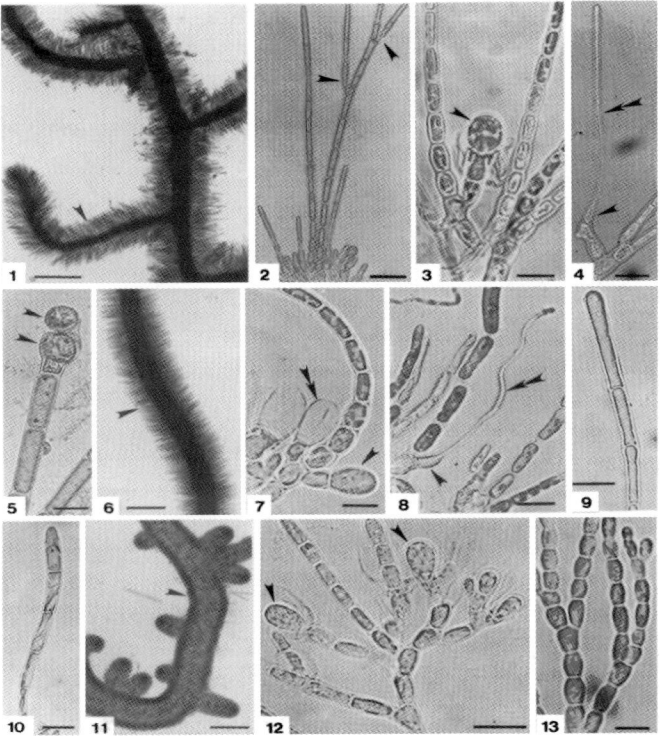

Plate 171
Thorea hispida, figs 1–5 (Sheath, Vis & Cole 1993a).
1. a plant with a dense, central medulla, loosely arranged assimilatory filaments (arrow) and numerous secondary branches, 2. non-clavate assimilatory filaments with variable lengths and degree of ramification (arrows), 3. a monosporangium (arrow) on a 2-celled sporangium bearing branch arising from the basal cell of the assimilatory filament, 4. a carpogonium with a swollen base (arrow) and a narrow, elongate trichogyne (double arrow) on a 2-celled carpogonium-bearing branch arising from the base of the assimilatory filaments, 5. putative seirospores (arrows) at the tip of the assimilatory filaments.
Scale bars = 500 μm (fig. 1); 10 μm (figs 2–5).

Thorea violacea, figs 6–7 (Sheath, Vis & Cole 1993a).
6. a plant with a dense medulla, loosely arranged assimilatory filaments (arrow) and no secondary ramification, 7. a monosporangium (arrow) and empty monosporangium (double arrow) on a 1-celled spore bearing branches arising from the base of the non-clavate assimilatory filaments,

Thorea violacea as *T. okadae,* fig. 8 (Sheath, Vis & Cole 1993a).
8. a carpogonium with a swollen base (arrow) and a narrow, wavy, elongate trichogyne (double arrow) on a 1-celled carpogonium-bearing branch arising from the base of the assimilatory filament.
Scale bars = 100 μm (fig. 6); 10 μm (figs 7–8).

Thorea clavata, fig. 9 (Sheath, Vis & Cole 1993a).
9. a clavate unbranched assimilatory filament. Scale bar =10 μm.

Thorea zollingeri, fig. 10 (Sheath, Vis & Cole 1993a).
10. clavate, unbranched assimilatory filaments. Scale bar =10 μm.

Nemalionopsis tortuosa, figs 11–12 (Sheath, Vis & Cole 1993a).
11. a plant with a central medulla and compacted assimilatory filaments (arrow), 12. monosporangia (arrows) terminal on cylindrical assimilatory branches. Scale bar = 300 μm (fig. 11); 10 μm (fig. 12).

Nemalionopsis shawii, fig. 13 (Sheath, Vis & Cole 1993a).
13. a barrel-shaped cells of assimilatory filament. Scale bar =10 μm.

Thoreales

T. hispida was proposed by Hassal (1845). Although *Thorea andina* has been separated on the basis of having shorter assimilatory filaments and smaller assimilatory cell diameter (Bischoff 1965), Sheath, Vis & Cole (1993a) found that type specimens overlapped slightly in both of these features. They illustrated a carpogonium with a long trichogyne but did not discuss it.

5) ***Thorea conturba*** Entwisle et Foard (1999a:49, figs 12–19)

plate 172, figs 1–8 (original authors, Entwisle et Foard 1999a)

Plant dioecious, branching about 40 times in 30 mm of plant length, 180–400(–460) μm in diameter, medulla 48–79 μm in diameter. Medullar filament cells cylindrical, longer and thinner than assimilatory filaments cells, 3–6 μm in diameter, 15–44 μm long; cells giving rise to assimilatory filaments broader and irregular in shape. Assimilatory filaments audouinelloid, rarely somewhat clavate, 65–160(–260) μm long, 8–18(30) cell-storeys, branching 0–1(–3) times, mostly in lower portion of filaments; cells mostly cylindrical (rarely discoidal or somewhat barrel-shaped), 4–7 μm in diameter, 4–10(–15) μm long; terminal cells occasionally swollen to almost twice the subtending cell diameter, possibly seirosporangia; chloroplast 1 per cell, parietal, mostly extending the length of the cell. 'Rhizoidal filaments' produced in lower portion of cortex, growing into central medulla. Spermatangium-bearing filaments 0.3–1 times the length of vegetative assimilatory filaments, branched, spermatangia clustered, groups of 1–3 subtended by 1- or 2-celled filaments, ellipsoidal, 3.3–4 μm in diameter, 10–13 μm long. Male plants with some elongate cells (function unknown; possibly aborted carpogonia), 4–6 μm in diameter, 23–40 μm long, borne on 1- or 2-celled filaments. Carpogonia lateral on assimilatory filaments (i.e. carpogonium-bearing branch 1-celled), usually arising from proximal cell, linear to fusiform, base slightly swollen, 3–5 μm in diameter, 60–140 μm long; trichogyne 2–3 μm in diameter. Carposporangium-bearing filaments 0.5–0.8 times the length of vegetative assimilatory filaments; carposporangia obovoidal to dome-shaped, 6–9 μm in diameter, 11–15 μm long.

Chantransia phase consisting of uniseriate filaments; prostrate system profusely branched, cells somewhat irregularly shaped; erect filaments forming compact tuft, richly and irregularly branched; cells discoidal, cylindrical, or somewhat barrel-shaped, 6–10(–12) μm in diameter, 6–12 μm long; chloroplasts parietal, apparently 1 or 2 per cell; apical cells sometimes elongate, up to 25 μm long. Monosporangia spherical, about 10 μm in diameter, about 11 μm long. Filaments resembling assimilatory filaments sometimes arising directly from chantransia phase, 150–330 μm long, 17–33 cell-storeys; cells 3–6 μm in diameter, 6–10(–16) μm long. (after Entwisle & Foard 1999)

Type Locality: Byrangery Creek, tributary of Cooper Creek, 25 km northeast of Lismore, New South Wales in Australia.

Holotype: MEL 2045617, 15/VII, 1997.

Distribution: Known from type locality only.

6) ***Thorea okadae*** Yamada (1949:158, figs 1–3)

plate 171, fig. 8 (Sheath, Vis & Cole 1993a)
plate 174, figs 1–11 (Yoshizaki 1993a)

Plants monoecious, mucilaginous, up to 2.1–4 mm in diameter, up to 3 m long, dark purple with red, irregularly and densely branched, multiaxial, consisting of medullar filaments and cortical assimilatory filaments, attached to substrata with discoid holdfasts, 2–3 mm in diameter. Medullar portion forming core; medullar filaments interlacing. Assimilatory filaments dichotomously, two to three times branched, issuing two type of assimilatory branches: long assimilatory branches 8–12 μm in diameter, up to 400 μm long, consisting of 10–21 cells and short assimilatory branches 5–18 μm in diameter, less than 150 μm long, consisting of 3–6 cells. Spermatangia terminal in small clusters on short assimilatory branches, ovoidal, 5–6 μm in diameter, 10–13 μm long. Carpogonia 6–7 μm in diameter at the base, 12–15 μm long, trichogyne straight, cylindrical, 3–4 μm in diameter, 160 to 350 μm long. Carposporophytes indefinite in shape, gonimoblast filaments irregularly branched, creeping downwards toward the medulla and penetrating among assimilatory filaments; carposporangia terminal on short branchlets of gonimoblast filaments, obovoidal, 7–18 μm in diameter, 10–26 μm long. Monosporangia terminal, 2 or 3 per cluster on assimilatory filaments, obovoidal, 6–12 μm in diameter, 8–16 μm long.

Type Locality: Hishikari along Sendai-gawa River in Kagoshima, Japan in Eastern Asia.
Holotype: SAP 046883, Okada, 28/III 1939.

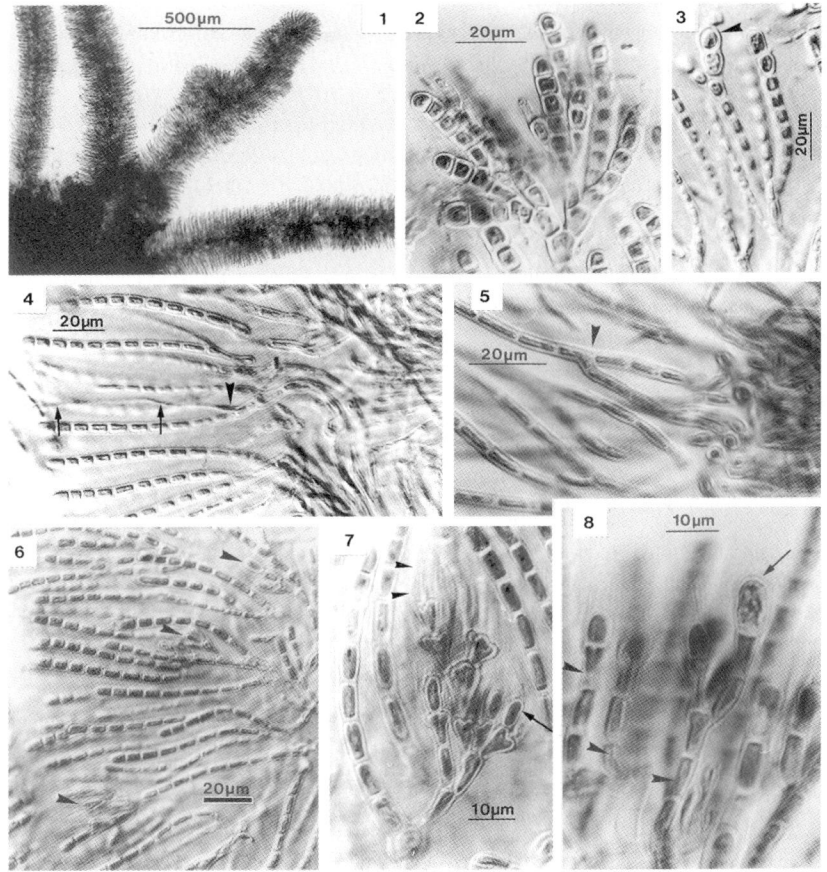

Plate 172
Thorea conturba, figs 1–8 (original authors, Entwisle & Foard 1999a)
1. habit of gametophyte with chantransia phase in bottom-left corner, 2. erect filaments of chantransia phase, 3. apparent seirosporangium (arrowhead) terminal on assimilatory filament, 4. carpogonium (arrowhead to base, arrows to trichogyne) extending from the base of the assimilatory filament to the outer cortex. 5. rhizoidal filament arising from assimilatory filament (arrowhead) and extending into medullar region, 6. spermatangia (arrowheads) at various position in the cortex, 7. clusters of spermatangia (arrow) and empty (arrowhead) spermatangia, 8. gonimoblast filaments (arrowheads) in mid-cortex with carposporangium (arrow).

Thoreales

Distribution: in Japan, occurs near Tanushimaru along Chikugo-gawa River in Fukuoka, near Yamaga along Kikuchi-gawa River, near Nishiki along Kuma-gawa River in Kumamoto, from Ebino in Miyazaki to Hishikari in Kagoshima along Sendai-gawa River, Japan in Eastern Asia.

Note: Sheath, Vis & Cole (1993a) did not consider morphological differences significant and hence regarded *Thorea okadae* as conspecific with *Thorea violacea*. However, *Thorea okadae* is distinguished on the genetic base (Hanyuda *et al*. 2001).

7) ***Thorea bachmannii*** Pujals (1967:1–2)

plate 175, figs 1–8 (Necchi 1989) as *Thorea bachmannii*,
plate 176, figs 1–4 (Necchi & Zucchi 1997) as *T. violacea*

Plants dioecious, moderately mucilaginous, up to 800–1300 µm in diameter, up to 10–15 cm long. Main branch 800–1200 µm in diameter, medullar portion 250–400 µm in diameter. Assimilatory filaments 1–4(–5) times, alternately or oppositely branched, 300–550 µm long, 10–18(–21) cell-storeys, proximal cells barrel-shaped, distal cells cylindrical, elongate, 6–11 µm in diameter, 15–30 µm long. Spermatangia terminal or sub-terminal in pairs on short specialized assimilatory branches near the base of assimilatory filaments, ellipsoidal or ovoidal, 4–7 µm in diameter, 8–10 µm long. Carpogonium-bearing branch arising from short branches, consisting of 1 or 2 cylindrical cells. Carpogonia 5–7 µm in diameter at the base, 12–15 µm long, trichogyne straight or slightly curved, elongate-filiform, 2–4 µm in diameter, 100–300 µm long. Carposporophyte a simple cluster of gonimoblast filaments, of indefinite shape; gonimoblast filaments short and sparsely branched, consisting of cylindrical cells, compactly arranged. Carposporangia terminal or subterminal on gonimoblast filaments, solitary or in pairs, club-shaped or obovoidal, 8.5–13 µm in diameter, 17–25 µm long. Monosporangia solitary or in pairs, obovoidal, 7–10 µm in diameter, 12–16 µm long.

Type Locality: La Plata, Arroya del Gato, Buenos Aires, Argentina in South America.
Holotype: BA 12709, Bachmann 27/X 1965.
Distribution: Known from the type locality and Conchas River, Marechal Rondon Highway (SP–300), Minicipio Conchas, São Paulo, Urumbeba River, Mato Grosso do Sul, Brazil in South America.

Note: *Thorea bachmannii* Pujals has been distinguished on the basis of the dimensions of assimilatory cells and diameter of clustering monosporangia. Sheath, Vis & Cole (1993a) did not consider these differences significant and hence regarded *Thorea bachmannii* as conspecific with *Thorea violacea*.

8) ***Thorea riekei*** Bischoff (1965:111, figs 1–18)

plate 173, figs 1–4 (Sheath 1984)

Plants up to 200 cm long, unbranched or very sparsely branched, up to 1.7 mm in diameter, central axis 275 µm in diameter at the base, decreasing slightly toward the apex, dark wine-red. Gelatinous matrix extending centrifugal from medullar portion to apical cell of assimilatory filaments. Assimilatory filaments always unbranched, up to 775 µm long, consisting of 4 or 5 cells; cells uninucleate, 5–7 µm in diameter, 12–22 µm long; apical cells up to 40 µm long with pointed apex. Chloroplast parietal, laminate, cinnamon-coloured, brown, bluish or reddish. Monosporangia single or 3–5 in cluster. Monospores pear-shaped or oblong when young, rounded to spherical when released, averaging 15 µm in diameter (after Bischoff 1965).

Type Locality: Landa Park, New Braunfels, Comal, Texas, USA in North America. Growing on rocks, 0.5–1.0 m below the surface of a rapid in spring-fed stream.

Thorea

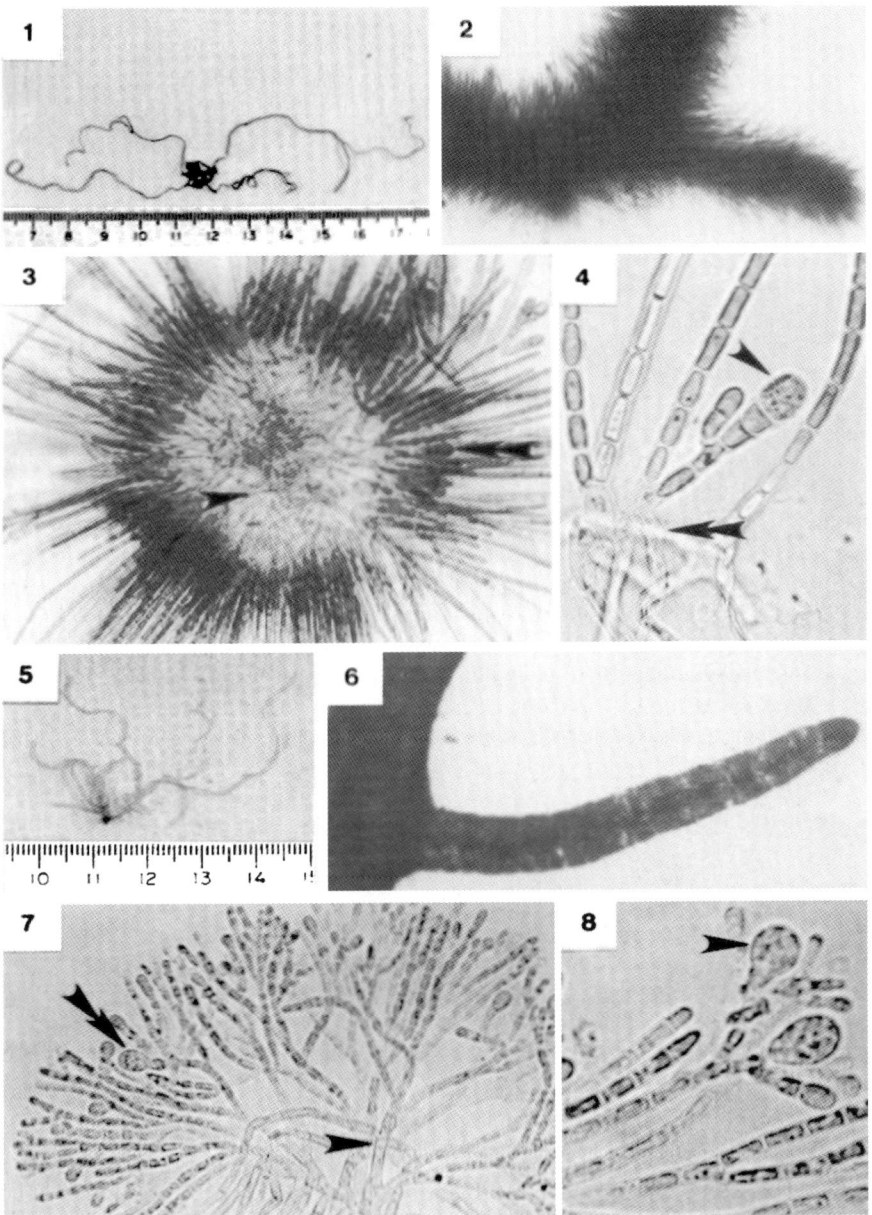

Plate 173
Thorea violacea as *T. riekei*, figs 1–4 (Sheath 1984)
1. the macroscopic view of a plant exhibiting sparse ramification, 2. hairy appearance due to copious assimilatory filaments which are not enclosed in a common mucilage, 3. the cross section of a plant showing a central medulla (arrow) and assimilatory filaments with basal monosporangium, 4. a monosporangium (arrow) attached to colourless medullar filaments (double arrow).

Nemalionopsis tortuosa as *N. shawii* f. *caroliniana*, figs 5–8 (Sheath 1984)
5. the macroscopic view of a plant with scattered ramification, 6. a smooth appearance due to copious assimilatory filaments enclosed in a common mucilage, 7. the transverse section with a colourless medulla (arrow) and a cortex with apical monosporangia (double arrow), 8. a monosporangium terminal on assimilatory filament (arrow).

Thoreales

Holotype: unknown.
Distribution: Known from the type locality only.
Note: *Thorea riekei* Bischoff has been distinguished on the basis of diameter of medulla and length of assimilatory filaments and cells. Sheath, Vis & Cole (1993a) did not consider these differences significant and hence regarded *Thorea riekei* as conspecific with *Thorea violacea.*

9) ***Thorea prowsei*** Ratnasabapathy et Seto (1981:246, fig. 2, a–g, fig. 4, a–k)

plate 177, figs 1–7 (Seto 1998b)

Plants tufted, mucilaginous, 5–15 cm high, deep brown, sparsely branched, lateral branches alternate, usually near the base of the main branches. Main axis 540–915–1500 μm in diameter, consisting of medullar layer and assimilatory filaments. Medullar layer with interlacing filaments, 262–395 μm in diameter. Assimilatory filaments 115–342(–605) μm long, consisting of a proximal portion and distal portion with 13–18(–30) cell stories. Distal portion of assimilatory filaments densely branched, attenuated toward apex, apical cells rounded. Monosporangia (carposporangia?) in dense clusters or solitary, produced on the basal cells of assimilatory filaments, obovoidal or ellipsoidal, 7–11(–20) μm in diameter, 10–15(–24) μm long.

Type Locality: Sungai Tahan, Pahang, Malaysia in Southeast Asia. Attached to rocks in sunny place in a clean stream, 10–20 cm depth, running through tropical rain forest.
Holotype: Herb. Ratnasabapathy RS 494, Seto, 5/VIII 1971.
Distribution: Known from the type locality only.
Note: *Thorea prowsei* Ratnasabapathy and Seto has been distinguished on the basis of diameter of medulla and dimensions of monosporangium clusters. Sheath, Vis & Cole (1993a) did not consider these differences significant and hence regarded *Thorea prowsei* as conspecific with *Thorea violacea.*

10) ***Thorea gaudichaudii*** C. Agardh (1824:56)

Plants, tufted, very mucilaginous, 1000–2000 μm in diameter, 12–58 cm long, dark or reddish brown, 2 or 3 times alternately branched, with branches originating near the base, with short patent ultimate branchlets, multiaxial, consisting of medullar filaments and cortical assimilatory filaments, attached to substrata with discoid holdfasts, 1–4.3 mm in diameter. Medullar filaments interlacing. Assimilatory filaments 300–800 μm long, consisting of 18–36 cells; apical hair cell cylindrical, 5–6 μm in diameter, 16–36 μm long, basal cells barrel-shaped, 6–11 μm in diameter, 7–15 μm long. Sexual reproduction unknown. Monosporangia solitary or in clusters derived from the base of assimilatory filaments, ovoidal, obovoidal or pear-shaped, 7–16 μm in diameter, 11–28.5 μm long.

Type Locality: Pago River on Guam Island, Mariana Islands, Micronesia in Pacific Ocean.
Lectotype: LD herb. Agardh 17811, found by Gaudichaud, designated by Sheath, Vis & Cole (1993a).
Distribution: Known from the type locality and in Ylig River, a small stream located just south of Pago River in Guam Island, Mariana Islands, Micronesia in Pacific Ocean, on rocks in rather stagnant water; Amisu-gah and Yahu-gah in Okinawa Island in Okinawa, Japan in Eastern Asia, in small springs.
Note: Sheath, Vis & Cole (1993a) did not consider morphological differences significant and hence regarded *Thorea gaudichaudii* as conspecific with *Thorea violacea*. However, *Thorea gaudichaudii* is distinguished on the genetic base (Hanyuda *et al.* 2001).

Thorea

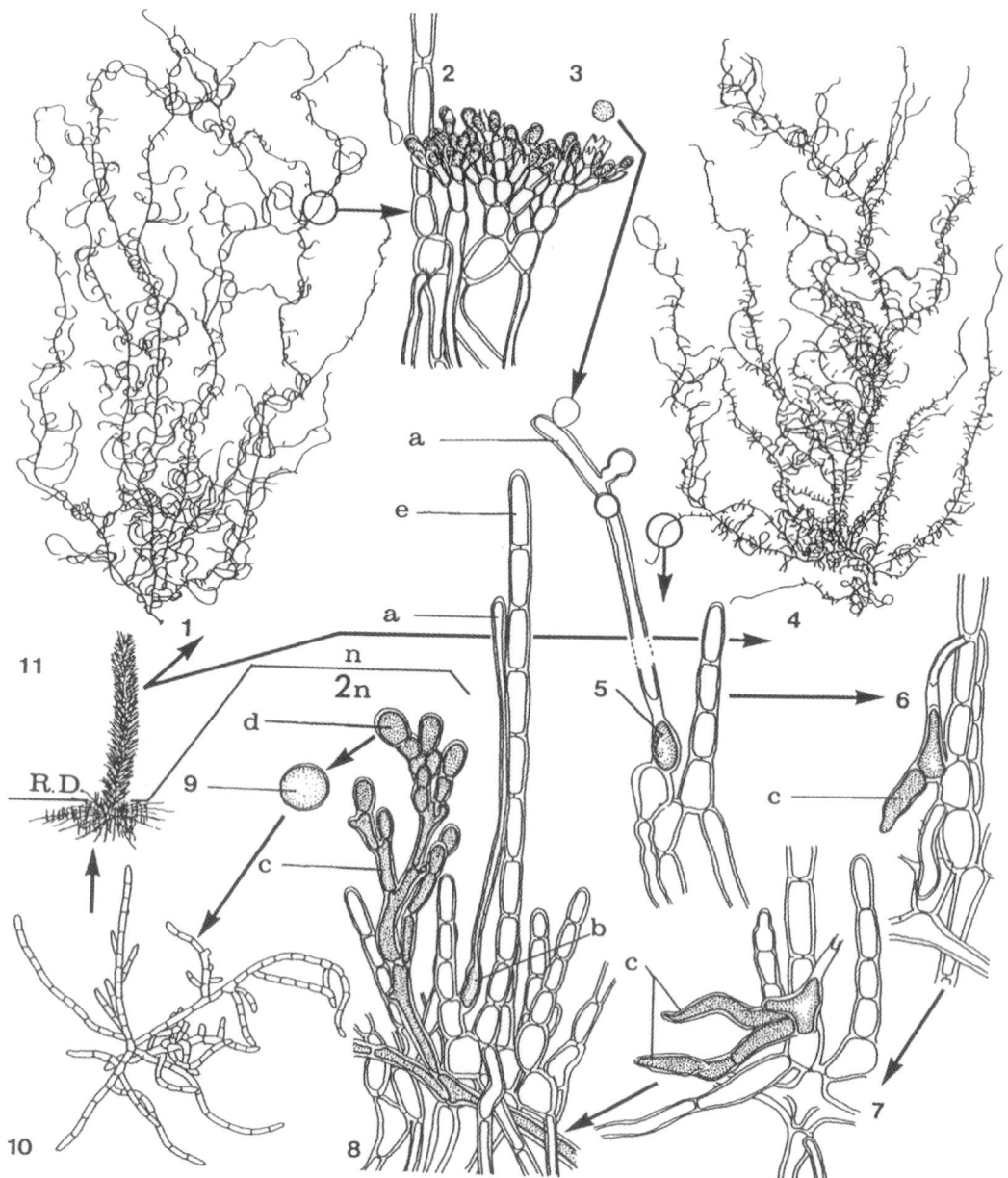

Plate 174
Thorea okadae, figs 1–11 (Yoshizaki 1993a)
1. a male gametophyte, 2. spermatangia, 3. a released spermatium, 4. a female gametophyte, 5. spermatia attached on the top of a trichogyne and the cytoplasm of zygote separate from the trichogyne, 6. gonimoblast filaments developing from the lower portion of the zygote, 7. the development of gonimoblast filaments, 8. the well-developed stages of gonimoblast filaments forming clusters of carposporangia, 9. a released carpospore, 10. a germling of a carpospore (chantransia phase), 11. a young multiaxial gametophyte arising from the chantransia phase after experience the reduction division in a cell of chantransia phase. 1: a trichogyne, b: a carpogonium, c: gonimoblast filaments, d: a carposporangium, e: an assimilatory filament.

Thoreales

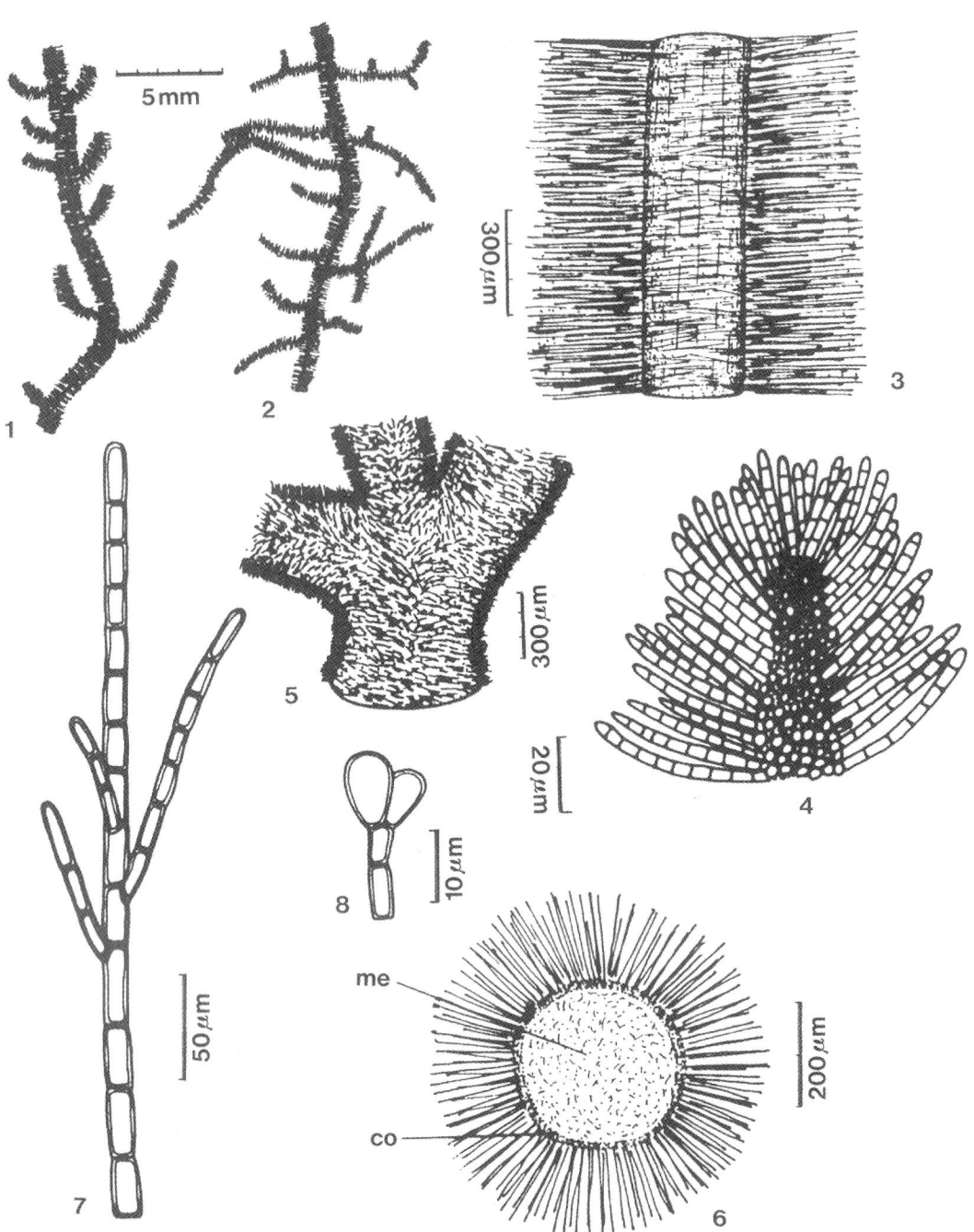

Plate 175
Thorea bachmannii, figs 1–8 (Necchi 1989)
1–2. an aspects of a middle portion of a plant, 3. details of a middle portion, 4. a details of an apex showing multiaxial structure, 5. a basal portion of a plant, 6. the cross section of a middle portion showing a cortex (co) and a medulla (me), 7. an assimilatory filament, 8. carposporangia.

11) ***Thorea brodensis*** Klas (1936:283, Ab. 2–7, Taf. V, fig. 1, Taf. VI, figs 3–4)

Plants whip-like, robust, olive-green to violet-brown, 20–50 cm long, sparsely branched or not branched at all, central axes 2000 µm or more in diameter. Main axes evident, assimilatory filaments lacking, 570–750 µm in diameter. Secondary axes about 200 µm in diameter, with much-branched assimilatory filaments; assimilatory filaments in summer 750–1415 µm in length, in winter only 400–750 µm in length, consisting of cells in summer 5–16 µm in diameter, in winter 8–16 µm in diameter, 30–55 µm long. Sporangia ellipsoidal or pear-shaped, 13–20 µm long, single on small stalk or in clusters on fascicles, inserted alternately or oppositely on main axis or on secondary axes.

Type Locality: Sava River near Slavonski Brod, Croatia in Europe.

Holotype: Deposited in Herbarium, Institute of Botany, University of Zagreb, Croatia, coll. Zora Klas.

Distribution: Known from the type locality only.

Note: *Thorea brodensis* Klas was distinguished on the basis of diameter of medulla, lengths of assimilatory filaments and cells. Sheath, Vis & Cole (1993a) regarded *Thorea brodensis* as conspecific with *Thorea violacea*.

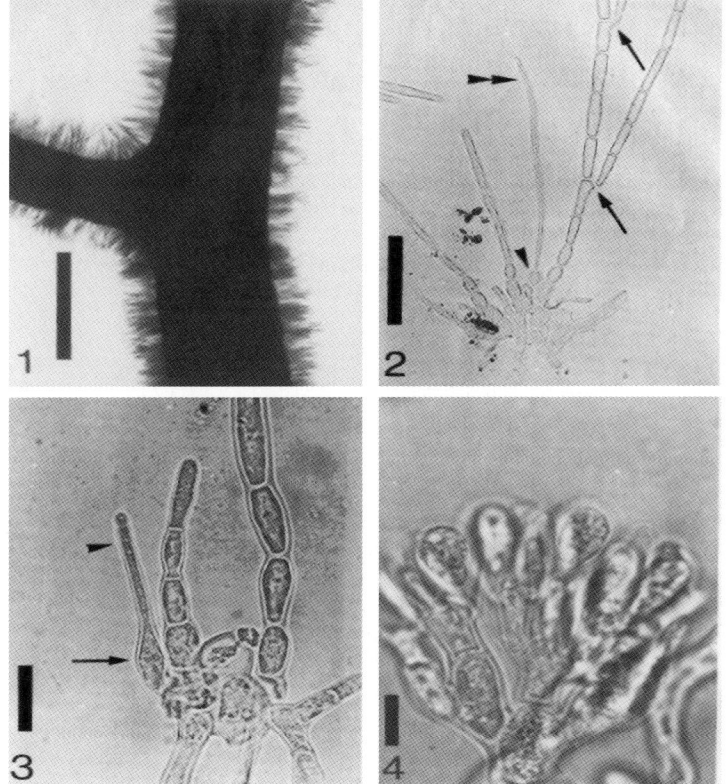

Plate 176
Thorea bachmannii as *T. violace,* figs 1–4 (Necchi & Zucchi 1997)
1. details of a female plant with secondary branching, 2. non-clavate assimilatory filaments with sparse branching (arrow) and a carpogonium with a swollen base (arrowhead) and an elongate trichogyne (double arrow), 3. details of a young carpogonium with a swollen base (arrow) and a developing trichogyne (arrowhead), 4. carposporangia terminal on gonimoblast filaments.
Scale bars = 500 µm (fig. 1); 100 µm (fig. 2); 10 µm (figs 3–4).

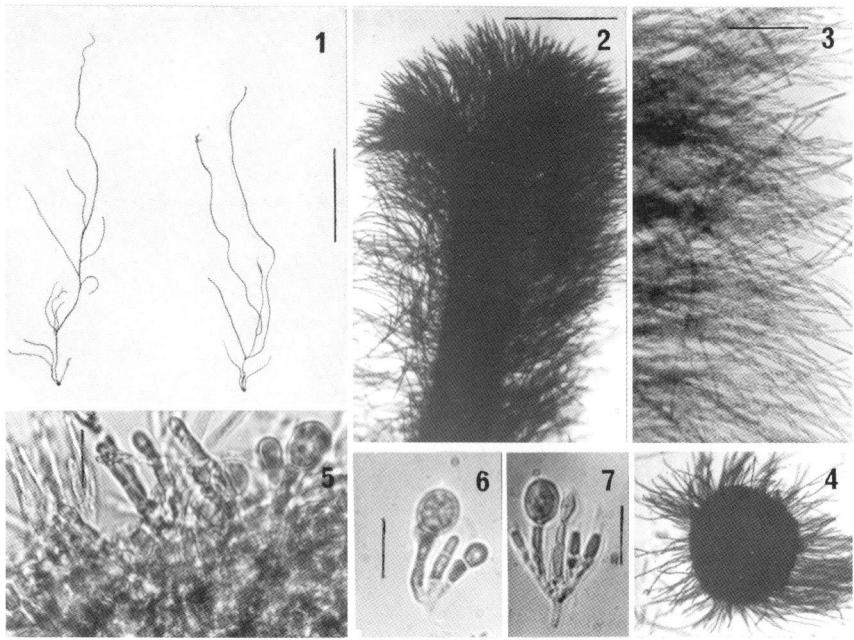

Plate 177
Thorea prowsei, figs 1–7 (Seto 1998b)
1. a plant showing ramification habit, 2. assimilatory filaments and medullar layer of a main branch, 3. frequently branched assimilatory filaments at distal portion, 4. the cross section of a main branch, 5–7. monosporangia.
Scale bars = 2 cm (fig. 1); 400 μm (figs 2, 4); 100 μm (fig. 3); 30 μm (figs 5–7).

Genus *Nemalionopsis* Skuja (1934:191)

Type: *Nemalionopsis shawii* Skuja

Plants mucilaginous, much branched, multiaxial in organization, lateral assimilatory filaments disposed along the entire axis. Sexual reproductive organs unknown. Sporangium (probably monosporangium) -bearing branches long and well-branched, assimilatory filaments compressed.

Key to the species of the genus *Nemalionopsis*

1. Assimilatory filaments short, composed of cylindrical cells.....................................1) *N. tortuosa*
1. Assimilatory filaments long, composed of barrel-shaped cells...................................2) *N. shawii*

1) ***Nemalionopsis tortuosa*** Yoneda et Yagi in Yagi & Yoneda (1940:83, fig. 1, fig. 2, 1–8)
Synonym: *Nemalionopsis shawii* Skuja f. *caroliniana* Howard et Parker (1979:333, figs 1–13)

plate 171, figs 11–12 (Sheath, Vis & Cole 1993a)
plate 173, figs 5–8 (Sheath 1984) as *N. shawii* f. *caroliniana*,
plate 178, figs 1–8 (original author, Yagi & Yoneda 1940)

Plants mucilaginous, 400–840(–900) μm in diameter, 5–30 cm long, dull reddish brown, main axis with sparse lateral branches, multiaxial, consisting of medullar filaments and cortical assimilatory filaments, attached to substrata with discoid holdfasts. Medullar portion up to 450 μm in diameter, medullar filaments long and interlacing, cells 3–7 μm in diameter, 13–100 μm long. Assimilatory filaments 60–290 μm long, 6–12 cell-storeys, basal cells cylindrical, 3.6–9 μm in diameter, 8–36 μm long, distal cells ovoidal or pear-shaped, 3–7 μm in diameter, 3–14 μm long. Sexual reproduction unknown. Monosporangia solitary or in clusters derived terminally from assimilatory filaments, ovoidal, ellipsoidal, 5–12 μm in diameter, 6.5–18 μm long.

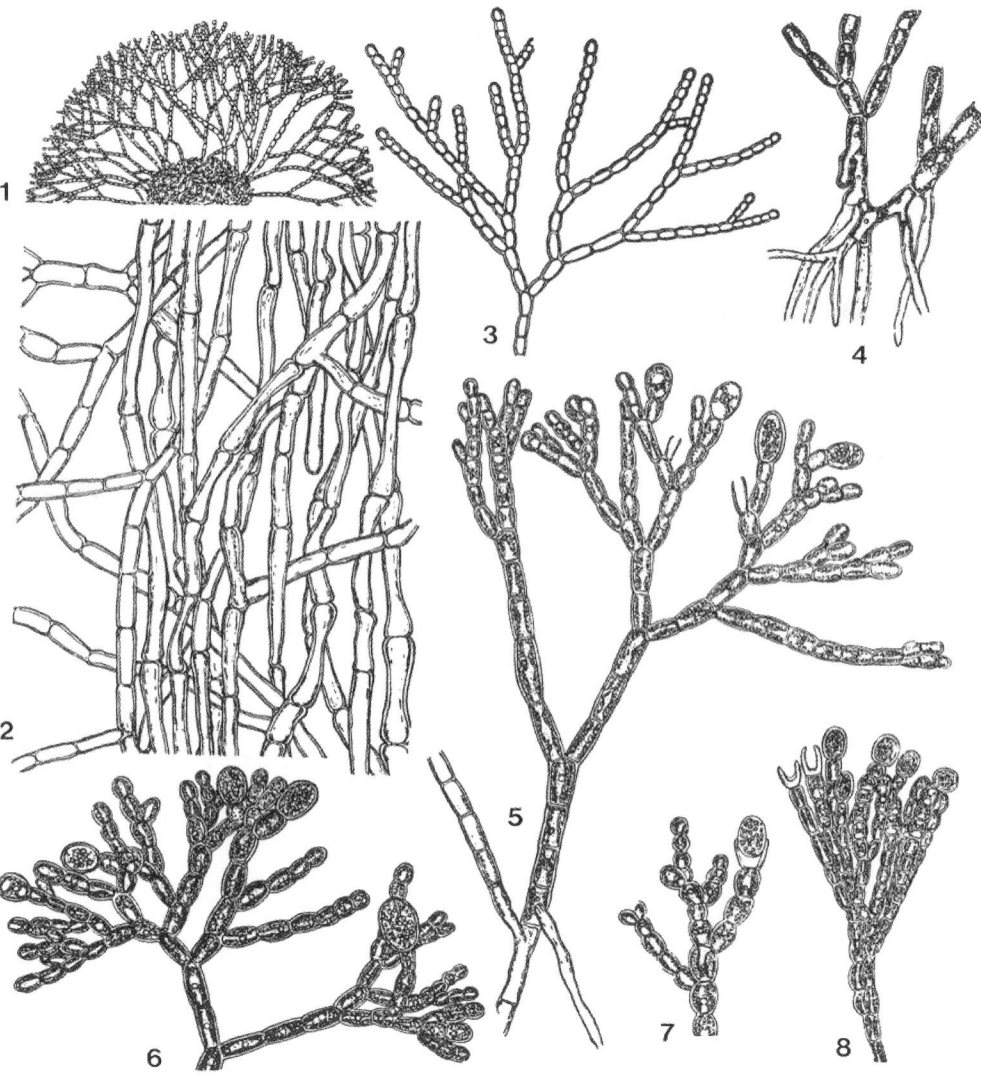

Plate 178
Nemalionopsis tortuosa, figs 1–8 (original authors, Yagi & Yoneda 1940)
1. cross section of plant, 2. longitudinal section of medullar portion, 3. habit of ramification of assimilatory filaments, 4. proximal portion of assimilatory filaments showing rhizoids, 5. assimilatory filaments with monosporangia, 6–8. distal portion of assimilatory filaments showing monosporangia terminal on short branchlets.

Hildenbrandiales

Type Locality: Okichi-izumi, near Matsuyama in Ehime, Japan in Eastern Asia. Occurs on stones in a shallow stream running in shady place.

Holotype: not designated.

Distribution: Known from the type locality, and Japan in Eastern Asia and USA in North America. In Japan, near Minami-oguni along Shitsu-gawa River, near Kikuchi along Kikouji-gawa River, tributaries of Chikugo-gawa River in Kumamoto, near Kunimi along Koujiro-gawa River in Nagasaki. In North America, Lower Barton Creek, 1–5 km down stream a bridge of Route 1005, Wake, North Carolina.

2) *Nemalionopsis shawii* Skuja (1934:191, Ab. 4, figs 1–9)

plate 171, fig. 13 (Sheath, Vis & Cole 1993a)

Plants mucilaginous, ca. 700 µm in diameter, ca. 6.5 cm long, gray-brown or violet, main axis with a few lateral branches, multiaxial, consisting of medullar filaments and cortical assimilatory filaments, attached to substrata with discoid holdfasts. Medullar portion up to 270–280 µm in diameter, medullar filaments long and interlacing, cells 4–5 µm in diameter, 16–25 µm long. Assimilatory filaments 145–400 µm long, 14–32 cell-storeys, basal cells cylindrical, 4–16 µm long, distal cells ellipsoidal or barrel-shaped, 3–5.5 µm in diameter, 5–8.3 µm long. Sexual reproduction unknown. Monosporangia solitary or in clusters derived terminally from assimilatory filaments, ovoidal, ellipsoidal, 9.5 µm in diameter, 11–13 µm long.

Type Locality: Llama Forest Reserve, Bataan Prov., Luzon, the Philippines in Southeast Asia.

Holotype: NY, Shaw & Day coll. #490, 28/IV 1907.

Distribution: Known from the type locality only.

Order **Hildenbrandiales** Pueschel et Cole (1982:718)

Type: family Hildenbrandiaceae Rosenvinge

Plants crustaceous with smooth surface or wart-like protuberances, composed of a basal layer and usually erect filaments. Basal layer prostrate, lacking rhizoids, composed of branched filaments consisting of cells densely aggregated laterally, each cell producing an erect filament upwards. Erect filaments coherent, consisting of cylindrical cells. Pit-plug type 4, with core enveloped by a cap layer and a cap membrane. Sexual reproduction unknown. Tetrasporangia unknown in freshwater representatives, borne in conceptacles in marine representatives.

Family **Hildenbrandiaceae** Rosenvinge (1917:202)

Type: genus *Hildenbrandia* Nardo

Description as for order.

Genus *Hildenbrandia* Nardo (1834:676) in NUC-3 (1993:529) nom. et orth. cons.

Type: *Hildenbrandia prototypus* Nardo = *Hildenbrandia rubra* (Sommerfelt) Meneghini (1841:426).

Plants crustaceous, forming red patches on stone in shaded streams and in lakes. Vegetative reproduction by fragments of plants, and by means of offshoots and gemmae.

Note: Based on morphometric analysis and phylogenetic analysis of sequences of the rbcL gene and 18S rRNA gene, three groups were distinguished: group 1 with freshwater species, *H. angolensis* (North America, Europe) and *H. rivularis* (Europe), group 2 with parallel tetrasporangial division, *H. occidentalis* and group 3 with non-parallel tetrasporangial division, *H. rubra*, but it is clear that they are closely related ones on the type of pit plug structure (Sherwood & Sheath 1998, 1999, 2000, Saunders & Bailey 1999).

1) ***Hildenbrandia rivularis*** (Liebmann) J. Agardh (1851:379)
Basionym: *Erythroclathrus rivularis* Liebmann (1839:174)

plate 179, figs 1–9 (Seto 1993b)

Plants bright crimson, forming more or less regularly circular discs, attaining a diameter about 2.5 cm; neighbouring discs often fusing into extensive crusts. Discs usually about 50–100 μm thick, consisting of a basal layer and erect filaments. A basal layer composed of radiating rows of somewhat elongate cells. Erect filaments densely aggregated, simple or more rarely dichotomously branched and composed of almost isodiametric cells, usually 7–10(–15) cells and 4–12 μm in diameter. Asexual propagation by formation of stolons, gemmae, and fragmentation of erect filaments.

Type Locality: in stream at Kongens Moøller, Sjaelland, Denmark in Europe.
Holotype: C, collected by Hornemann, S., VI, 1826.
Distribution: Known from Denmark, Belgium, Poland and in whole of Europe; River Congo in Africa; China, Japan in Eastern Asia; Sungai Gombal, Selangor, Malysia, Java, Sumatra, Indonesia in Southeast Asia; Australia; Pennsylvania, USA in North America; Venezuela, Jamaica in Central America.

In China, on limestone in a lake near Yungning in Sikang. In Japan, Benten-ike, Sano in Tochigi, Takagami, Chosi in Chiba, Kikusui-sen in Gifu, Uriwari Spring Park, Uenaka in Fukui, Iwashimizu-hachiman in Kyoto, Aotani-gawa River, Kobe in Hyogo, Ritsurin Park in Kagawa, Kohara, Kobayashi in Miyazaki.

2) ***Hildenbrandia angolensis*** Welwitsch ex W. West et G. S. West (1897:3)

plate 180, figs 1–2 (Sheath & Cole 1996)

Discs usually about 21.2–162 μm (4–17 cells) thick, 2.0–9.9 μm in diameter, 3.9–12.4 μm long.
Type. Loc. Golungo Alto, Angola in Africa.
Lectotype: BM 3435, Welwitsch Collection no. 150.
Distribution: Known from Angola in Africa, Spain in Europe; Kauai Island, Oahu Island, Hawaii Island (Vis, Sheath, Hambrook & Cole 1994), Fiji Island (Sheath & Cole 1996) in Pacific Ocean; in Pennsylvania (Wolle 1887), Texas (Flint 1955), USA in North America; through Central America to the Caribbean Islands (Sheath *et al.* 1993).

Hildenbrandiales

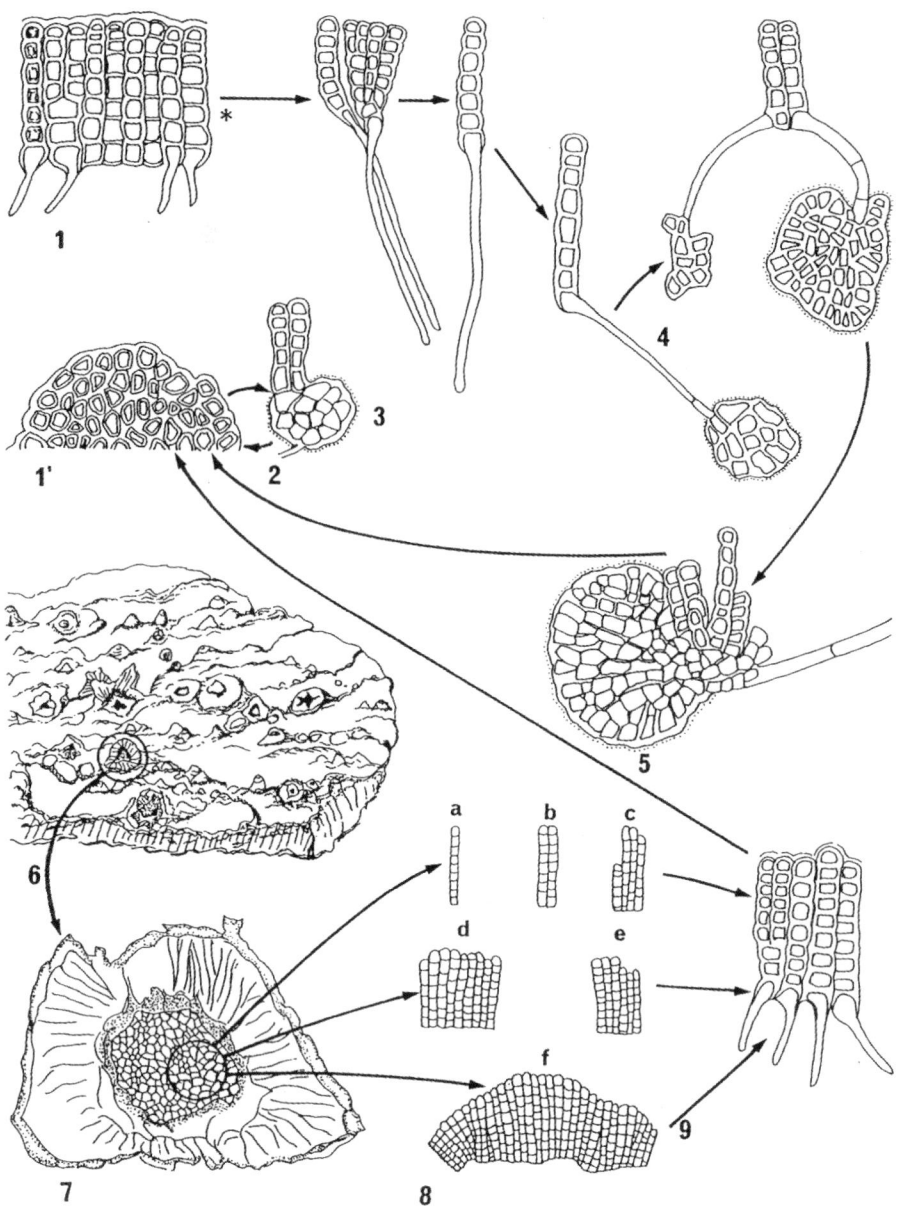

Plate 179
Hildenbrandia rivularis, figs 1–9 (Seto 1993b)
1. the side view of erect filaments with rhizoids, 1'. the surface view of a plant, 2. outgrowths of a stolon from the marginal portion of the mother plant, 3. the formation of stolonal filaments from erect filaments which were separated fragmentary from the mother plant, 4. the formation of prostrate system derived from a terminal cell of stolon, 5. erect filaments newly shooted upwardly from the marginal surface of a prostrate system, 6, the various stages in development of gemmae protuberances and crateriform hollows on the surface of a plant, 7. the surface view of a gammae protuberance in a mature stage, 8. the side view of mature gemmae which were separated from the protuberance, 9. the side view of erect filaments with rhizoids, * fragmentation of the erect filament. a–f; various erect systems consisting of from 1 to many erect filaments separated from gemmae.

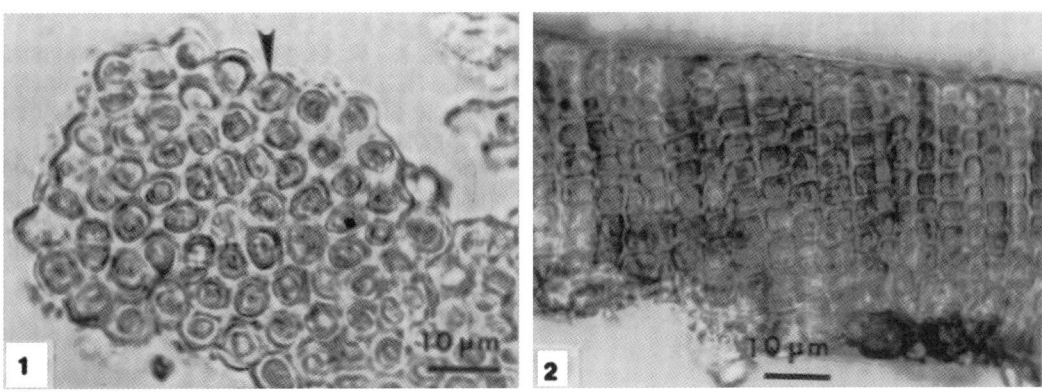

Plate 180
Hildenbrandia angolensis, figs 1–2 (Sheath & Cole 1996)
1. crimson-coloured crust in surface view with densely packed cells (arrowhead), 2. cross section of erect filament system with well-defined rows of cells.

Order **Balliales** Choi, Kraft & Saunders (2000, 278)

Type: family Balliaceae

Plants uniaxial, each axial cell of indeterminate axes bearing two determinate, opposite and distichous branchlets, which in turn usually bear two rows of unbranched or once-pinnate opposite distichous pinnules. Pit-plugs on both sides are usually dome-shaped at light microscopy levels. Sexual reproduction unknown.

Family **Balliaceae** Choi, Kraft & Saunders (2000, 278)

Type: genus *Ballia*

Reproduction by presumably mitotic, cruciate tetrasporangia. Other features as for the order.

Genus *Ballia* Harvey (1840:191) *emend*. Choi, Kraft & Saunders (2000, 278)

Type: *Ballia brunonia* Harvey = *Ballia callitricha* (C. Agardh) Montagne

Plants consisting of di-, trichotomously, or alternately branched main axes with each axial cell bearing opposite laterals; axis non-corticated or corticated by branched rhizoids, initially arising from the base of branchlets, extending downwards along axial cells; branchlets issued oppositely, each branchlet gradually tapering towards the apex; monosporangia spherical or obovoidal, lateral or terminal on branchlets. Sexual reproduction unknown in freshwater species.

Note: Molecular analysis of *B. callitricha*, the type of the genus, showed only a remote relationship to the Ceramiales (Choi *et al*. 2000). Distance and parsimony analyses based on nuclear small-subunit rDNA sequences of the type species and two additional marine species of *Ballia* revealed that they represent a distinct lineage sister to the Acrochaetiales, Batrachospermales, Nemaliales and Palmariales. As *B. callitricha* is the type species of the genus *Ballia* and is distinct in anatomical and molecular features from all recognized orders, a new order and family (Balliales and Balliaceae) were proposed. The relationship of marine species and

Balliales

freshwater species is uncertain (Kumano & Phang, 1990; Choi *et al.* 2000). It is clear that the two recognized freshwater species of the genus (*B. pinnulata* and *B. prieurii*) need to be re-examined. In fact, preliminary analysis of nuclear small-subunit rDNA sequences from one freshwater species, *B. prieurii*, has shown that it is related to the Batrachospermales rather than to the Balliales (Necchi, personal communication). Ultimately the freshwater species may form a new genus and family with affinities to the Balliales (Choi *et al.* 2000).

Genus *Ballia* Harvey (1840:191) in NCU–3 (1993:113)

Type: *Ballia brunonia* Harvey [= *Ballia callitricha* (C. Agardh) Montagne]

Plants consisting of di-, trichotomously, or alternately branched main axes with each axial cell bearing opposite laterals; axis non-corticated or corticated by branched rhizoids, initially arising from the base of branchlets, extending downwards along axial cells; branchlets issued oppositely, each branchlet gradually tapering towards the apex; monosporangia spherical or obovoidal, lateral or terminal on branchlets. Sexual reproduction unknown in freshwater species.

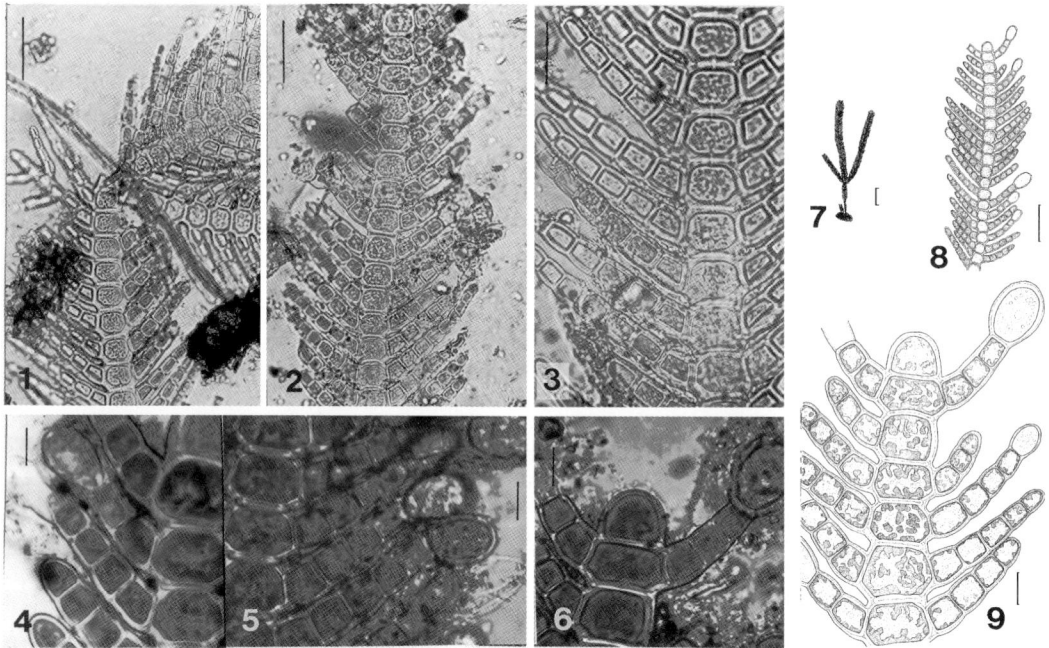

Plate 181 (a)
Ballia pinnulata, figs 1–6 (Kumano 1996u), figs 7–9 (original author, Kumano 1978)
1–3. axial cells and pinnate primary branchlets, 4–5, 8. an upper portion of a plant showing axial cells and monosporangia terminal on opposite pinnate primary branchlets, 7. the habit showing a trichotomously branched plant, 6,9. an apical portion of a plant showing a rounded apical cell and monosporangia terminal on pinnate opposite branchlets.
Scale bars = 100 µm (fig. 7); 50 µm (figs 1–2, 8);   20 µm (fig. 3); 10 µm (figs 4–6, 9).

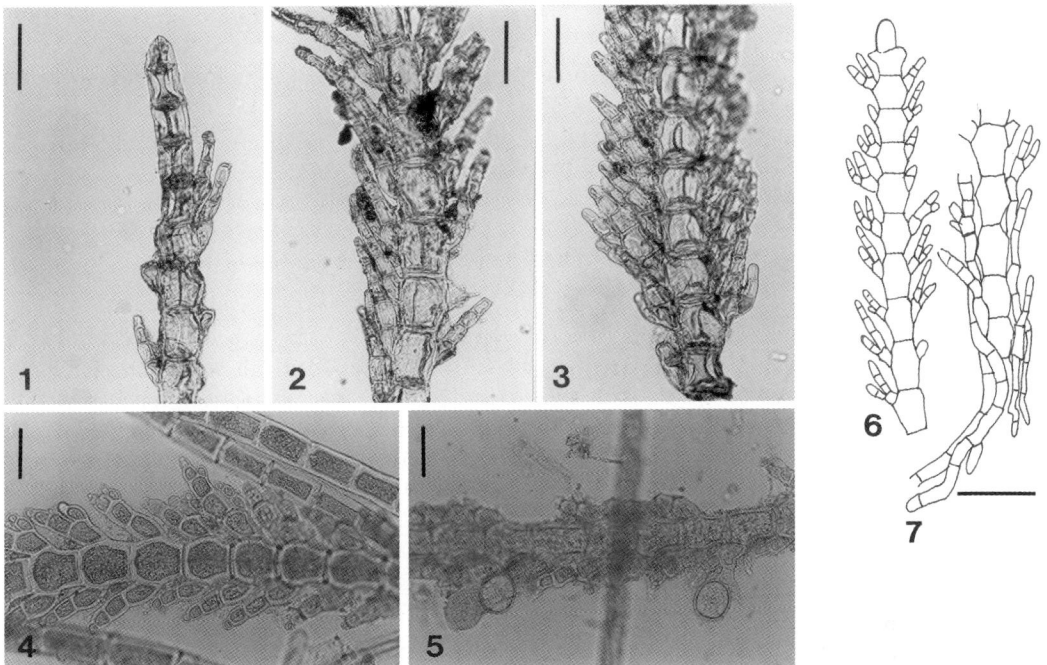

Plate 181 (b)
Ballia prieurii, figs 1–5 (Kumano 1997a), figs 6–7 (Kumano & Phang 1990)
1. an apical portion of an erect system showing a cylindrical apical cell with a rounded tip, 2–4, 6. an upper portion of an erect system showing a series of axial cells and primary branchlets opposite, unilaterally branched and bear secondary branchlets tapering gradually towards the apex, 5. monosporangia mostly lateral sometimes subterminal on lateral branchlets, 7. rhizoidal filaments arising from the basal cells of primary branchlets.
Scale bars = 100 µm (figs 6–7); 50 µm (figs 1–5).

Note: The freshwater species *B. pinnulata*, *B. prieurii* and *B. pygmaea* all differ from *B. callitricha* in bearing only abaxial, unbranched pinnulea, in producing unaggregated descending rhizoids, and in reproducing exclusively by monosporangia (not reported for *B. pygmaea*). Dome-shaped structure associated with pit-plugs have not been reported (Kumano 1978, Kumano & Phang 1990, Sheath *et al.* 1993). Choi *et al.* (2000) deferred judgment on the freshwater species until such time as molecular or ultrastructural data of taxonomic relevance become available. The comparative morphological simplicity of the marine *B. nana* and its superficial similarity to the freshwater species are additional reasons for maintaining the latter in the genus *Ballia* for the time being. Ultimately the freshwater species may form a new genus and family with affinities to the Balliales.

Couté & Sarthou (1990) studied the type specimens of *B. pygmaea* and *B. prieurii* and concluded that these two species should be merged, since there are no distinguishing features. Freshwater species have been reported only from tropical regions of Asia (Kumano 1978, Ratnasabapathy & Kumano 1982a, Kumano & Phang 1990), Central America (Sheath, Vis & Cole 1993b) and South America (Kützing 1847, Montagne 1850, Skuja 1944, Thérézien 1985, Couté & Sarthou 1990, Necchi & Zucchi 1995).

Balliales

Key to freshwater species of the genus *Ballia*

1. Branchlets oppositely and pinnately issued, unbranched...1) *B. pinnulata*
1. Branchlets oppositely and alternately issued, branched..2) *B. prieurii*

1) ***Ballia pinnulata*** Kumano (1978:98, fig. 1, A–C)

                  plate 181 (a), figs 1–6 (Kumano1996u)
                  figs 7–9 (original author, Kumano 1978)

Plants 2–3 mm high, ca. 100 µm in diameter, main axis unbranched or trichotomously branched; axis uncorticated, consisting of subhexagonal or octagonal cells, terminating in hemispherical cells; axial cells large, about 20 µm in diameter and about 14 µm long; branchlets oppositely issued in one plane, unbranched, 30–60 µm long, consisting of 3–9 cells; cells cylindrical or barrel-shaped, 9–11 µm in diameter and 9–10 µm long, each branchlet gradually tapering towards the apex, terminating in a subconical cell; monosporangia ellipsoidal or obovoidal, 10–14 µm in diameter, 16–20 µm long, terminal on branchlets.

    Type Locality: Field Studies Centre, University of Malaya along Sungai Gombak, Selangor, Malaysia in Southeast Asia.
    Holotype: Kobe University, Kumano, 30/IV 1971.
    Distribution: Known from the type locality in Southeast Asia and in South America (Couté & Sarthou 1990).

2) ***Ballia prieurii*** Kützing (1847:37)
Synonym: *Ballia pygmaea* Montagne (1850:291)

           plate 181 (b), figs 1–5 (Kumano 1997a), figs 6–7 (Kumano & Phang 1990)
plate 182, figs 1–5 (Kumano 1996a), figs 6–7 (Kumano & Phang 1990) as *Ballia pygmaea*,
                  plate 183, figs 1–4 (Necchi & Zucchi 1995)
                  plate 190, fig. 1 (Sheath, Vis & Cole 1993b)

Plants 5–7 mm high, 200–230 µm in diameter, main axis dichotomously branched uncorticated, consisting of subhexagonal or octagonal cells, terminating in broadly rounded tips; axial cells large, ca. 50 µm in diameter and ca. 60 µm long; rhizoids branched, initially arising from the base of branchlets, extending downwards along axial cells, consisting of 2–13 cells, cells cylindrical, 11–27 µm in diameter, 64–185 µm long; branchlets oppositely issued, unilaterally branched, 100–150 µm long, consisting of 3–6 cells; basal cells of branchlet pentagonal or barrel-shaped, 18–25 µm in diameter, 16–27 µm long; distal cells barrel-shaped or cylindrical, each branchlet gradually tapering towards the apex, terminating in dome-shaped cells of ca. 10 µm in diameter, 10–12 µm long; monosporangia spherical or obovoidal, 30–45 µm in diameter, 30–45 µm long, lateral or terminal on branchlets.

    Type Locality: Rivulet Gemeaus, Mahuri Mountains, French Guiana in South America.
    Holotype: L 938.92.104, Le Prieur no. 832.
    Distribution: Known from the type locality and in Sungai Sempanong, Sungai Jasin and Sungai Pelawar, Johore, Malaysia in Southeast Asia; in Belize and Costa Rica in Central America; in a stream in Adolpho Ducke Forest Reserve in Amazonas, Brazil in South America.

Ballia

Plate 182
Ballia prieurii as *B. pygmaea,* figs 1–5 (Kumano 1997a), figs 6–7 (Kumano & Phang 1990)
1–5, 7. an apical portion of an erect system showing conical apical cells and opposite branchlets, 6. a lower portion of an erect system showing rhizoidal filaments arising from the basal cell of primary branchlets. Scale bars = 100 μm (figs 1–3, 6–7); 50 μm (figs 4–5).

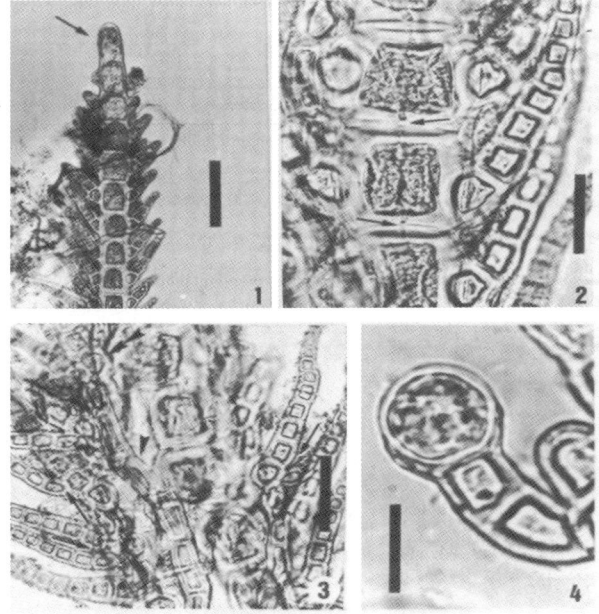

Plate 183
Ballia prieurii, figs 1–4 (Necchi & Zucchi 1995)
1. an apical portion of plant showing elongate apical cell (arrow), 2. detail o axial cells showing conspicuous pit connections (arrows), 3. detail of the basal portion of a mature plant showing branched pinnate (arrow) and rhizoidal filament (arrowhead) developing from basal cell (double arrowhead), 4. monosporangial branch with a spherical monosporangium.   Scale bars = 100 μm (fig. 1); 25 μm (figs 2–3); 10 μm (fig. 4).

Balliales

Genus *Ptilothamnion* Thuret in Le Jolis (1863:118)

Type: *Ptilothamnion pluma* (Dillwyn) Thuret

Plants alternately branched; axis uncorticated; rhizoids branched, initially arising from the base of branchlets, extending downwards along axial cells; branchlets alternately issued, each branchlet gradually tapering towards the apex.

1) *Ptilothamnion richardsii* Skuja (1944:44, pl. 8, figs 1–6)
Synonym: *Anfractutofilum umbracolens* Cribb (1965: 93), *Ptilothamnion umbracolens* (Cribb) Bourrelly (1970: 266) *nom. illeg.*

plate 184, figs 1–6 (original author, Skuja 1944)
plate 185, figs 1–10 (Entwisle & Foard 1999a)

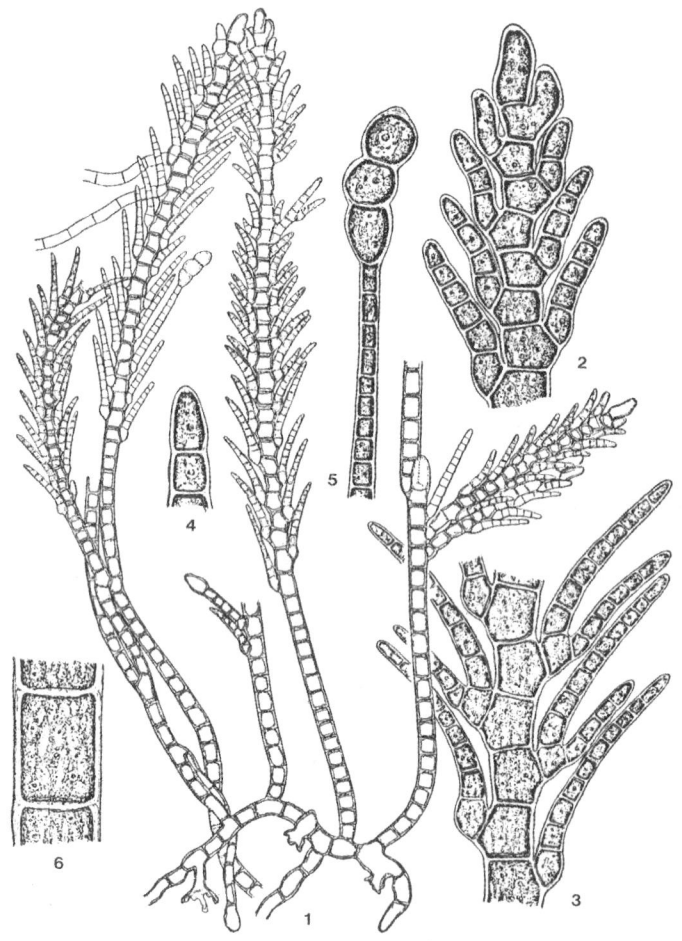

Plate 184
Ptilothamnion richardsii, figs 1–6 (original author, Skuja 1944)
1. many erect filaments arising from prostrate portion, 2. apical portion of main axis showing sympodial growth, 3. middle portion of main axis showing sympodial growth, 4. terminal portion of lateral branchlet, 5. sporangia terminal on lateral branchlet, 6. cells of lateral branchlet.

Plant cushion-like, consisting of prostrate filaments and erect filaments. Prostrate filaments irregularly and sparsely branched; cells mostly cylindrical, 8–28 μm in diameter, 23–45 μm long; holdfast arising from middle of cell, irregularly shaped, about 28 μm in diameter. Erect filaments with zigzag main axis by means of sympodial growth, consisting of discoidal to oblique-discoidal cells; cells (14–)17–29 μm in diameter, (14–)19–31 μm long; lateral branchlets alternately issued in one plane. Ultimate filaments up to 22 cells long, cells cylindrical to barrel-shaped, 6–15 μm in diameter, 8–18 μm long; apical cell sometimes elongate to 27 μm long and sometimes hyaline. Monosporangia on non-pinnately branched portions, 1 or 2 per subtending cell, sometimes clustered, obovoidal, 9–11 μm in diameter, 13–17 μm long.

Type Locality: first falls of River Essequibo, British Guiana in South America.

Holotype: NMF [NMW?] coll. 14/X, 1929, Richard.

Distribution: Known from the type locality at latitude ca. 6°N in South America; and Rangery Creek, a tributary of Cooper Creek, 25 km northeast of Lismore, New South Wales; dawn stream of Nagarigoon Fall, Nixons Creek, 3.9 km north of car parking, Lamington National Park, Queensland (Entwisle & Foard 1999); Cedar Creek Fall, Mt. Tamborine, (40 km north of Laminghton National Park), Cedar Creek National Park, Queensland (McLord 1975), at latitude ca. 28°S in the southernmost extent of tropical rain forest in Australia.

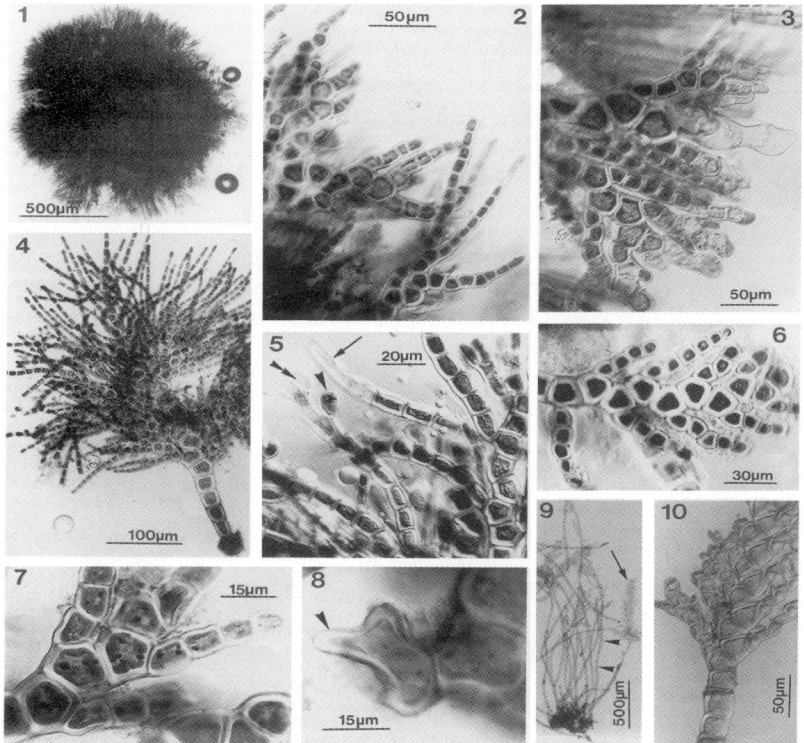

Plate 185
Ptilothamnion richardsii, figs 1–10 (Entwisle & Foard 1999a)
1. habit of colony, 2. edge of colony showing zigzag axes by means of sympodial growth, 3. edge of colony showing larger, flattened cells, 4. individual showing broad axial filament and elongate terminal filaments, 5. monosporangia with monospore (arrowhead) and empty (double arrowhead) and a hyaline apical cell (arrow) on individual illustrated in figs 4, 6. zigzag axis by means of sympodial growth and pinnate lateral branching, 7. detail of zigzag axis by means of sympodial growth showing chloroplasts and lateral branching, 8. holdfast (arrowhead), 9. holotype showing long, unbranched proximal portions (arrowheads) and pinnately branched portion (arrow), 10. holotype showing detail of pinnately branched portion.

Order **Ceramiales** Oltmanns (1904:700)

Type: family Ceramiaceae Dumortier

The axial cells are uniseriate, completely naked in some genera, or they may be covered by partial or complete cortication. Branching (to 3 or more orders) may be determinate or indeterminate, with branchlets being whorled, alternate, pinnate, or irregular in origin.

Family **Delesseriaceae** Bory (1828:181)

Type: genus *Delesseria* Lamouroux

Plants leafy segment, growing from an apical cell or a marginal meristem. In forms with an apical cell, four periaxial cells produced by proximal derivatives of apical cell. The two lateral periaxial cells usually first to be cut off followed by dorsal and ventral periaxial cells. Lateral periaxial cells continuing to divide, assuming the role of secondary initials and tertiary initials, continuing growth in the same plane and resulting in wings or membranous extensions from the primary axial row of cells.

Genus *Caloglossa* (Harvey) G. Martens (1869:232)

Type: *Caloglossa leprieurii* (Montagne) G. Martens

Plants dichotomously, rarely trichotomously branched, consisting of flat, articulate, narrow, leafy segments; midrib consisting of an axial cell and 4 periaxial cells, looking like 3 rows of cylindrical or barrel-shaped cells in surface view; two wing cell-rows arising from lateral periaxial cells, wing cell-row consisting of 5–20 cells; rhizoids and new leafy segments produced laterally from the constricted portion and from margins of leafy segments. Tetrasporangial sori formed on both side of midrib (Okamura 1908, King & Puttock 1993 etc.), Carpogonial branch is composed of a supporting cell and 4 segments including carpogonium. Spherical or hemispherical cystosarps form conspicuous projections upon midrib near tip of cystocarpic thallus (Papenfuss 1961, West 1991, Tanaka 1992, Tanaka & Kamiya 1993, etc.).

Tetrasporangia and cystocarps often unknown in freshwater habitat.

Note: Most authors such as King & Puttock (1994a) had attributed the genus to Agardh (1876) rather than Martens (1869) who did not provide a generic description of *Caloglossa*. However, Silva, Basson & Moe (1996) adopted Martens (1869) as the author of *Caloglossa* and *C. leprieurii* rather than J. Agardh (1876). The explanation is as follows. The generic name, *Caloglossa*, first appeared in the binomial *C. hookeri* Harvey, a name mentioned without a description in an article by Hooker & Harvey (1845a:270). Caloglossa was next mentioned in another article by Hooker & Harvey (1845b:542) as a generic name proposed by Harvey to accommodate *Delesseria leprieurii* Montagne, but subsequently abandoned. J. Agardh (1852 [1851–1863]: 680, 682) published *Caloglossa* as the epithet of the name of a taxon of unspecified rank within the genus *Delesseria*, but without giving Harvey explicit credit. Harvey (1853:98) specified the rank of the infrageneric taxon as subgenus and acredited it to himself. J. Agardh (1876:498) is traditionally accredited with elevating this intrageneric name to generic rank, but Martens (1869:232, 237) anticipated him, albeit in a cursory manner. While Martens did not explicitly state that he considered Harvey's subgenus to be worthy of generic rank, his use of *Caloglossa* as a generic name may be interpreted as a taxonomic statement.

Caloglossa

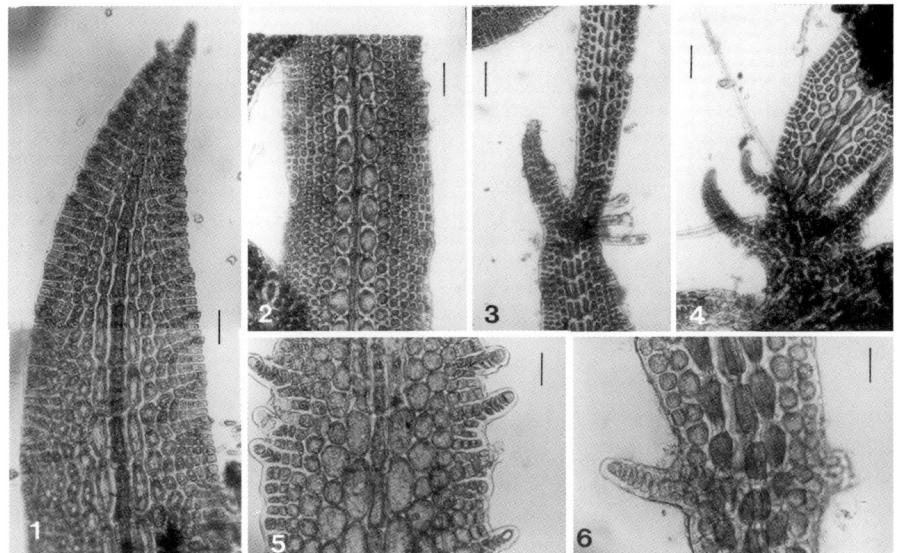

Plate 186 (a)
Caloglossa ogasawaraensis, figs 1–6 (Kumano 1996zq)
1. an apical portion of a leafy segment, 2. a middle portion of a leafy segment, 3. an adventitious branch and rhizoids developing from a nodal portion, 5. the development of adventitious branches from marginal wing cells, 4, 6. adventitious branches developed from the lateral periaxial cells around the node.
Scale bars = 200 μm (fig. 4); 100 μm (figs 1–3); 50 μm (figs 5–6) .

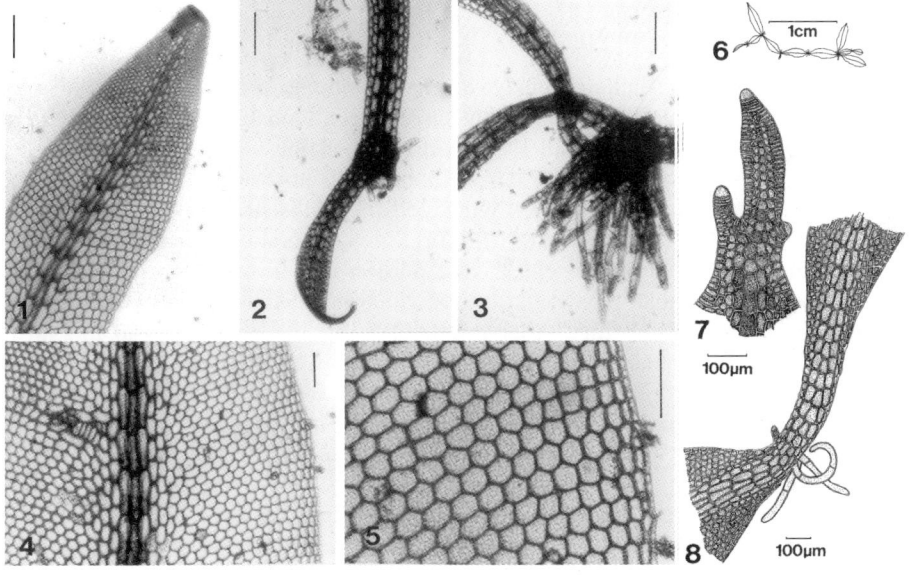

Plate 186 (b)
Caloglossa beccarii as *C. ogasawaraensis* var. *latifolia,* figs 1–5 (Kumano 1996s), figs 6–8 (original author, Kumano 1978)
1. an apical portion of a leafy segment, 2, 3, 8. new adventitious branches and rhizoids developing from a nodal portion, 4–5. a middle portion of a leafy segment, 6. the habit of a plant, 7. adventitious branches developed from the lateral periaxial cells around the nodal portion.
Scale bars = 1 cm (fig. 6); 100 μm (figs 1–3, 7–8); 50 μm (figs 4–5).

Ceramiales

Morphological and hybridization analysis of *Caloglossa* have been done by Kamiya *et al.* (1995, 1997, 1998). Recently, analysis of RuBisCo gene sequences used successively for phylogeny, evolutionary divergence and biogeographical studies at the ranks of species and populations of *Caloglossa,* such as *C. leprieurii and C. apomeitica* (Kamiya *et al.* 1998), *C. continua* (Kamiya *et al.* 1999), *C. intermedia, C. leprieurii and C. monosticha* (Kamiya *et al.* 2000).

Key to the species in freshwater habitat of the genus *Caloglossa*

1. Branching adventitious..2
1. Branching endogenous..3
2. Plant internodes mostly < 330 μm wide, 2–6 wing cell wide..................1) *C. ogasawaraensis*
2. Plant internodes > 700 μm wide, > 14 wing cell wide..2) *C. beccarii*
3. Rhizoids arising from transverse and lateral periaxial cells...............................3) *C. leprieurii*
3. Rhizoids arising from wing cells...4
4. Leafy segment 1–2 mm wide..4) *C. continua*
4. Leafy segment < 0.8 mm wide..5) *C. saigonensis*

1) ***Caloglossa ogasawaraensis*** Okamura (1897:13, figs A–D)
Synonym: *Delesseria zanzibariensis* K. I. Goebel (1898:65, figs 1–6); *Caloglossa zanzibariensis* (Goebel) De Toni (1900:731, 1924:357); *Caloglossa bombayensis* Børgesen (1933:127, figs 10–12)

plate 186 (a), figs 1–6 (Kumano 1996zq)
plate 187, figs 1–11 (Okamura 1908)
plate 190, figs 5–7 (Sheath, Vis & Cole 1993b)

Plants about 2 cm high, dichotomously, rarely trichotomously branched, consisting of flat, narrow, leafy segments, purple; segments linear-lanceolate, attenuated more narrowly towards the base, 150–600 μm in diameter, 1–7 mm long; midrib consisting of an axial cell and 4 periaxial cells, looking like 3 rows of cylindrical or barrel-shaped cells in surface view; two wing cell-rows arising from lateral periaxial cells, upper wing cell-row consisting of 5–7 cells, unilaterally branched, lower wing cell-row consisting of 5 or 6 cell, unbranched, rhizoids and new leafy segments produced laterally from the constricted portion and from margins of leafy segments. Tetrasporophytes are larger than plants, tetrasporangia are cut off from the lateral periaxial cells and the cells of the second and third order except 2–3 marginal cells, 35–40 μm in diameter. Spermatangial sori are produced on both side of a central axis of at the upper to middle portion of thallus. Spermatangia are cut off from almost all wing cells except a few marginal cells. Procarp consisting of a 4-celled carpogonial branch, a supporting cell and two mother cells of sterile groups (Tanaka & Kamiya 1993). A cystocarp developed on the central axis near the apex of a female thallus, hemispherical, 500–800 μm in diameter.

Reproductive organs often unknown in freshwater habitat.

Type Locality: Ogasawara Island, Bonin Islands, Japan in Eastern Asia. Known from freshwater habitat.

Lectotype: TI herb. Yendo, designated by Seto (1985).

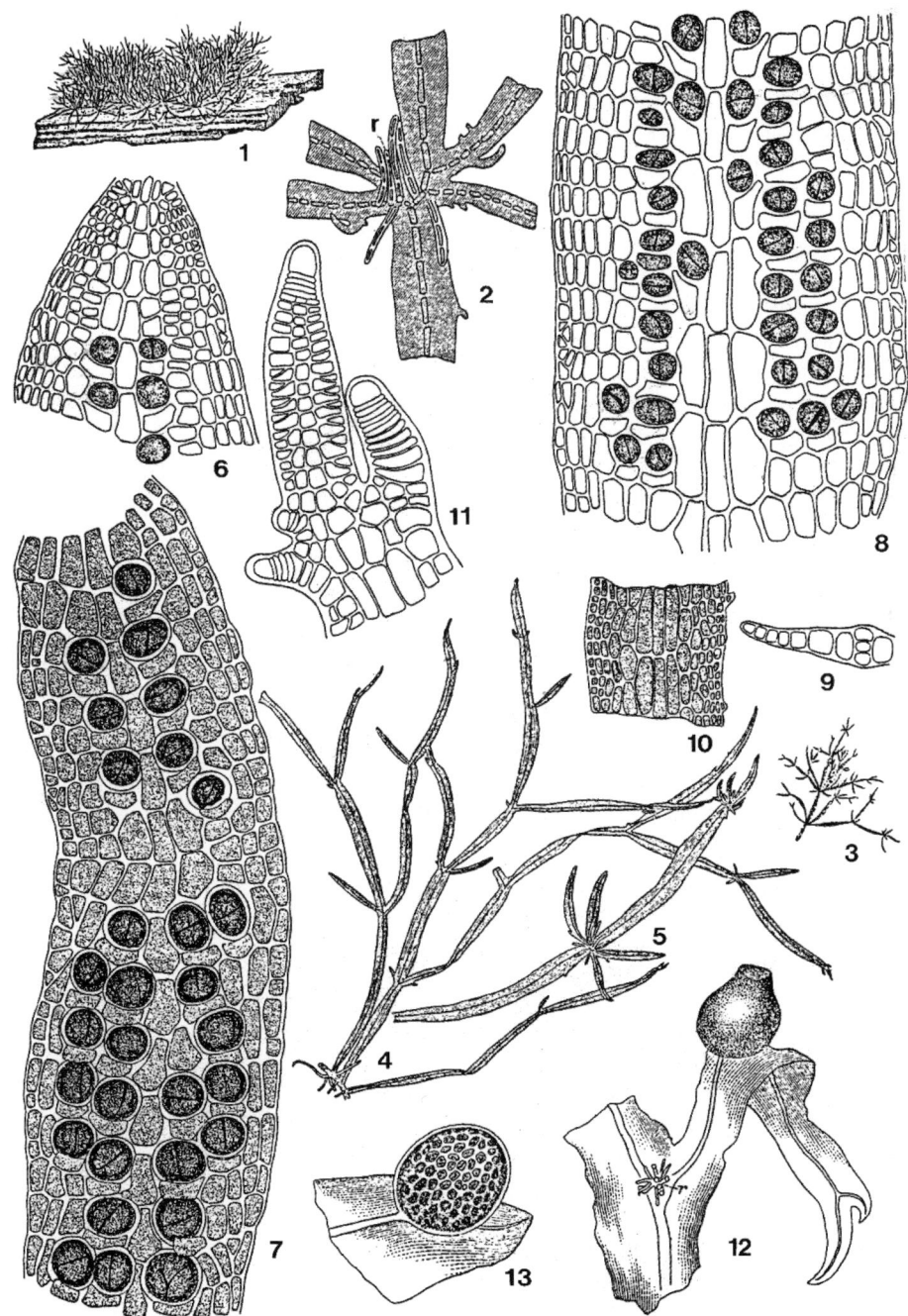

Plate 187
Caloglossa ogasawaraensis, figs 1–11 (Okamura 1908)
1. plant in natural habit, 2. mode of branching from constricted portion (r; rhizoid), 3. portion of plant detached, 4–5. marginal portion of plant, 6–8. upper, middle and lower portion of tetrasporangial sorus, 9. cross section of plant, 10. surface view of plant, 11. growing apices of plant.

Caloglossa continua as *C. leprieurii* auct. japon, figs 12–13 (Okamura 1908)
12–13. cystocarps.

Ceramiales

Distribution: Known from brackish habitats, in West Africa; western Pacific Ocean and eastern Australia; throughout Southeast Asia, Japan in Eastern Asia; Costa Rica in Central America. In Japan, Shikine-jima Island in Bonin Islands, Echizen Takada in Iwate, Ibaragi, Chiba, Tokyo, Mie, Fukui, Kyoto, Shimane, Osaka, Ehime, Kohchi, Nagasaki, Kagoshima. Epiphytic on stems of *Phragmites* and on trunks, roots, and pneumatophores of mangroves in areas with slow water flow and low turbility.

Also in freshwater habitats, in Asia, Eastern Asia, Southeast Asia. In Asia, Athirampally Water Fall, Sholayar River, Lerala State India. In Eastern Asia, Shiratori-no-ike, Kumanomiya-no-ike in Kumamoto, brachish Suga in Okinawa Island, Ohtaki in Ishigaki Island in Okinawa, Japan. In Southeast Asia, upstream of Sungai Air Dalam, Sungai Ayer Besar, Pulau Tioman, Malaysia.

2) *Caloglossa beccarii* (Zanardini) De Toni (1900:387)
Basionym: *Delesseria beccarii* Zanardini (1872b:140, pl. 5A)
Synonym: *Delesseria amboinensis* Karsten (1891:265, pl. 5); *Caloglossa amboinensis* (Karsten) De Toni (1900:731); *Caloglossa ogasawaraensis* var. *latifolia* Kumano (91:103, 1978)

plate 186 (b), figs 1–5 (Kumano 1996s)
figs 6–8 (original author, Kumano 1978) as *C. ogasawaraensis* var. *latifolia*

Plants about 3 cm high, dichotomously, rarely trichotomously branched, articulate, consisting of leafy segments, purple; segments lanceolate, ellipsoidal, attenuated more narrowly towards the base, more or less undulate at the margins, 600–1500 μm in diameter, 2–7 mm long; midrib consisting of an axial cell and 4 periaxial cells, looking like 3 rows of cylindrical or barrel-shaped cells in surface view; upper and lower wing cell-rows arising from lateral periaxial cells, upper wing cell-row consisting of 5–18 cells, unilaterally branched, lower wing cell-row consisting of 5 or 6 cells, unbranched; rhizoids and new leafy segments produced laterally from the constricted portion and from the margins of leafy segments.

Reproductive organs of this species unknown in freshwater habitat.

Type Locality: Epilithic on stones in a torrent, Sodomae near Gunung Poe (Pueh), western Sarawak, East Malaysia in Southeast Asia. Known from freshwater stream.

Lectotype: L, Beccari

Distribution: Restricted to India in Asia, Malaysia, Indonesia in Southeast Asia, western Pacific Ocean and Northern Australia. Epilithic on stones in freshwater coastal rivers such as Sungai Cherok, Perak, Malaysia in Southeast Asia. Epiphytic on mangrove trunks, roots and pneumatophores in areas with moderate water flow and high turbility.

Note: Post (1936) placed *Caloglossa amboinensis* [type locality: Ambon, Indonesia] in this synonymys of *C. beccarii*, and King & Puttock (1994a) included *C. ogasawaraensis* var. *latifolia* [type locality: Sungai Cherok, Perak, Malaysia; holotype: Kobe University, Kumano, 4/V 1971] to this species.

3) *Caloglossa leprieurii* (Montagne) G. Martens (1869:234, 237)
Basionym: *Delesseria leprieurii* Montagne (1840:193)
Synonym: *Hypoglossum leprieurii* (Montagne) Kützing (1849:875).

plate 190, figs 2–4 (Sheath, Vis & Cole 1993b)
plate 197, figs 3–4 (Entwisle & Kraft 1984)

Plant 0.5–2.0 cm long, forming dense mats on rocks or wood and consisting of leaf-like, lanceolate leafy segments linked by constricted nodes, red-brown to brown. Branching palmate to sub-dichotomous. Leaf-like segments 2–5 mm long, 350–800 μm in diameter, uniaxial, with 4 periaxial cells and monostromatic wings. New leafy segments often produced in rosettes at constricted nodes. Plants anchored by uniseriate, septate rhizoids, 15–20 m in diameter also arising at nodes. Tetrasporangial sori are 8–18 axial cells long, 7–9 row-cells wide and usually include the lateral periaxial cells and the submerged cells. Spermatangial sori developed on both face of terminal leafy segment. Spermatangia are cut off from the second and third-row of wing cells. Cystocarps are spherical, formed 1–2 on a leafy segment. Carpogonia and details see Papenfuss (1961). Plants often sterile in freshwater habitat. (Entwisle & Kraft 1984)

Type Locality: near Cayenne, French Guiana in South America.

Syntypes: PC Leprieur coll. 356 & 362.

Distribution: Widespread; commonly in brackish habitat such as mangroves and salt marshes, Atlantic Ocean; Indonesia, Southern China in Southeast Asia; in Nago River in Okinawa Island in Okinawa in Japan in Eastern Asia.

Also found in freshwater habitat on rocky coasts and occasionally above the tidal limit of coastal rivers such as Hopkins River, Hopkins Falls, Warrnambool on vertical walls of waterfall in Australia; in inland freshwater regions at altitude 400–500 m, about 12 km far from sea coast, Sierra de Luquillo, Puerto Rico in Caribbean Islands in Central America.

4) *Caloglossa continua* (Okamura) King et Puttock (1994:115)
Basionym: *C. leprieurii* var. *continua* Okamura (1903:129)
Synonym: *C. leprieurii* auct. japon (Okamura 1908:179 pl. 36, figs 1–15)

plate 187, figs 12–13 (Okamura 1908) as *C. leprieurii* auct. japon,
plate 188, figs 1–15 (Okamura 1908) as *C. leprieurii* auct. japon

Plant linear, dichotomously branched and consisting of leaf-like, lanceolate leafy segments linked by slightly constricted nodes. Leaf-like segments 1.3–3 mm long, 1–2 mm wide. New leafy segments often endogenously produced from axial cell at constricted nodes. Plants anchored by uniseriate, septate rhizoids also arising from back side at nodes. Tetrasporangial sori are produced on both side of a central axis in the upper portion of leafy segment. Tatrasporangia spherical, 40–55 μm. in diameter. Spermatangial sori are developed on both sides of a central axis at the upper and middle portion of a leafy segment. Procarp consisting of 4-celled carpogonial branch. A cystocarp formed upon the central axis near the apex of female leafy segment, hemispherical, 500–800 μm. in diameter (Tanaka 1992).

Reproductive organs often unknown in freshwater habitat.

Type Locality: Ko-yahagi-gawa River in Aichi, Japan in Eastern Asia.

Lectotype: SAP herb. Okamura Algae Japonicae Exsiccatae No. 67, vii 1902, designated by King & Puttock (1994).

Distribution: Known from *Phragmite* communities in intertidal brackish habitats, Korea, China, Japan in Eastern Asia; Indonesia in Southeast Asia; Australia. In Japan, north from Iwate, Chiba, Tokyo, Kanagawa, Aichi, Mie, Okayama, Ehime, Kochi, Fukuoka, Kumamoto, Kagoshima, Okinawa.

Ceramiales

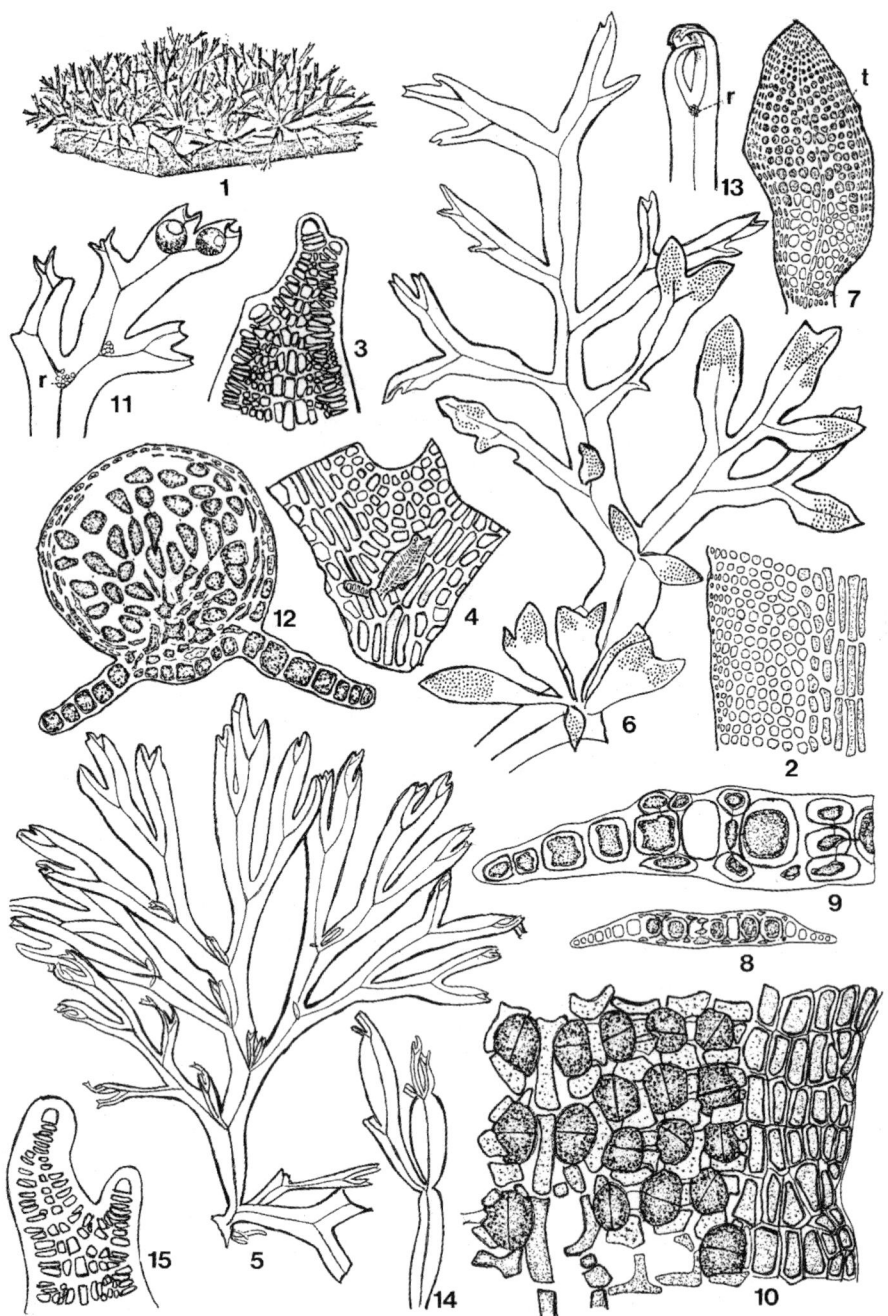

Plate 188
Caloglossa continua as *C. leprieurii* auct. japon, figs 1–15 (Okamura 1908)
1. plant in natural habit, 2. surface view of the half of plant, 3, 15. growing portion of plant, 4. young proliferation, 5. plant having more regular arrangement of segment, 6. plant having sub-alternate segments bearing tetrasporangial sori, 7. fertile segment with tetrasporangial sorus, 8–9. cross section of sorus, 10. surface view of a portion of sorus on the right node of midrib, showing tetrasporangia and cortical cells, 11. under surface of plant showing cystocarp, 12. longitudinal section of a cystocarp, 13, 14. terminal portion of plant bending toward the under surface (r; rhizoid).

Caloglossa

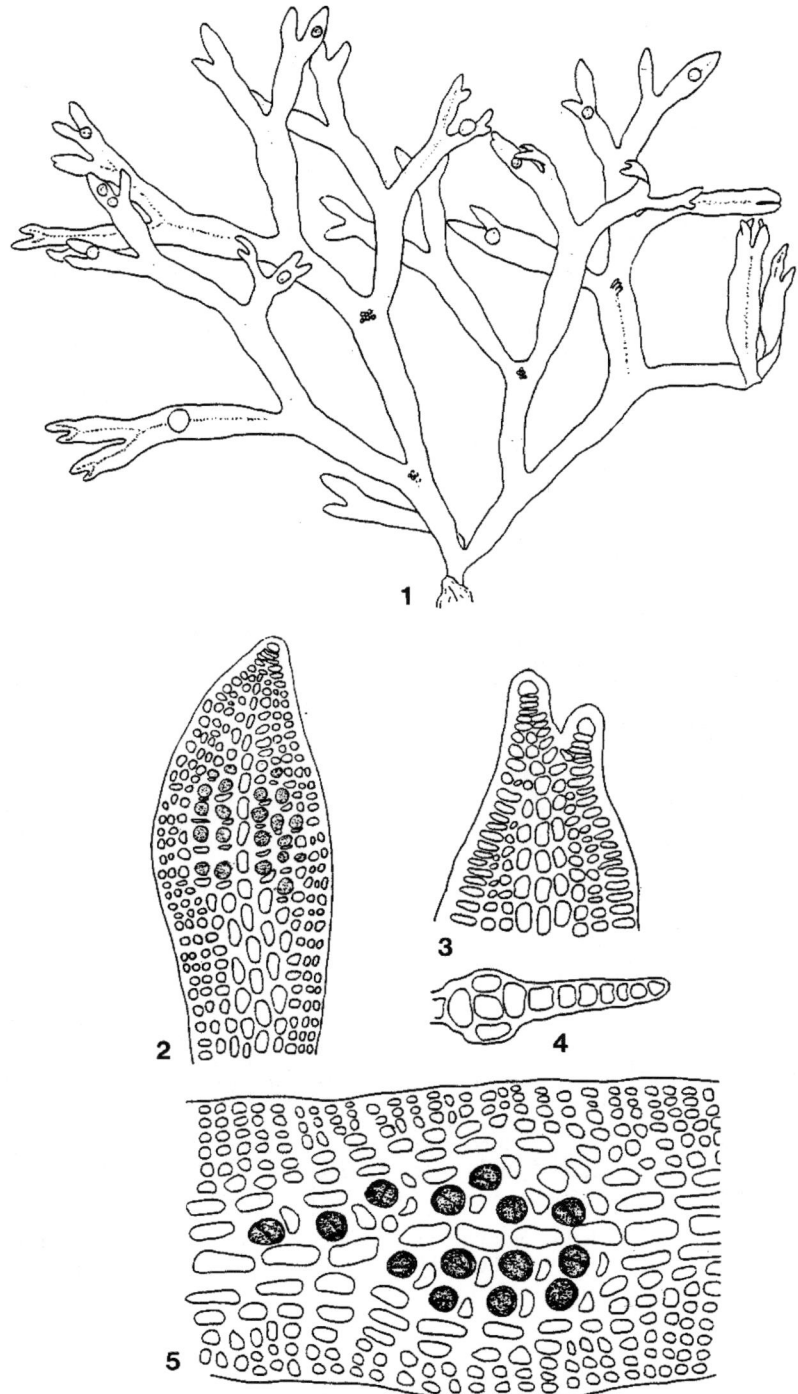

Plate 189
Caloglossa saigonensis, figs 1–5 (original authors, Tanaka & Pham-Hoàng Hô 1962)
1. habit of a cystocarpic plant, 2. apex of a tetrasporangial segment, 3. apex of a branch, 4. cross section of segment, 5. midrib portion of tetrasporangial segment.

Ceramiales

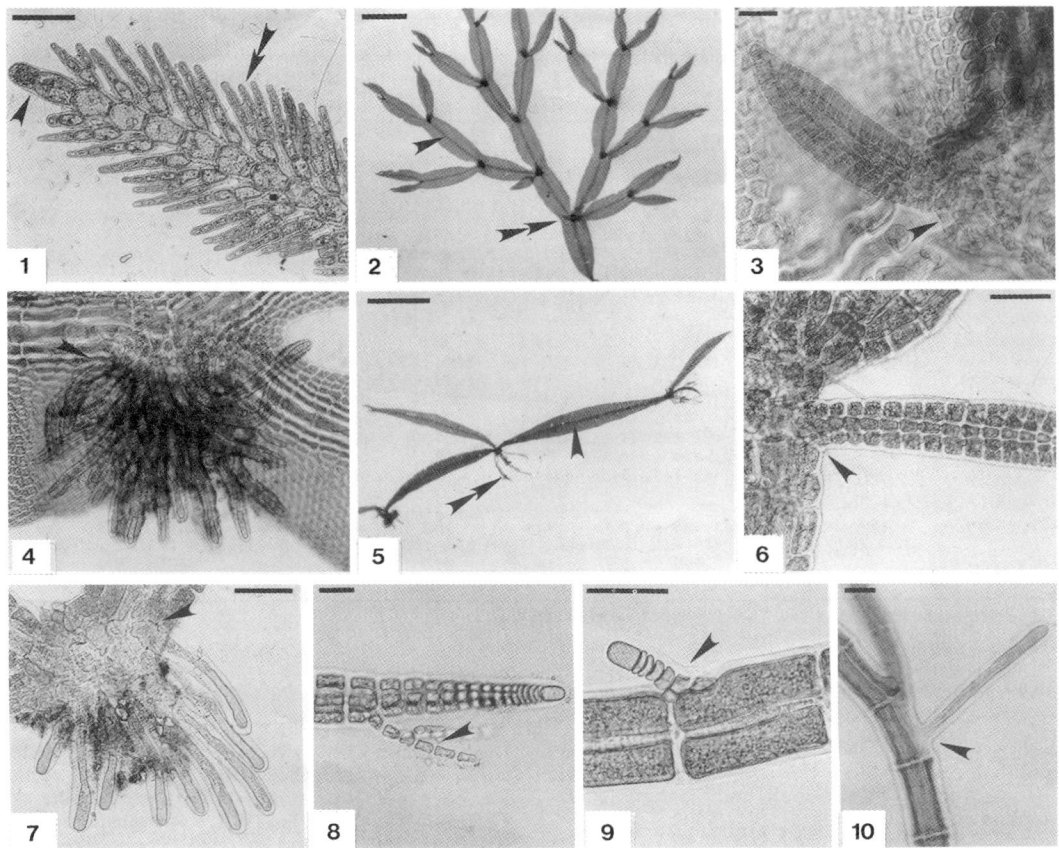

Plate 190
Ballia prieurii, fig. 1 (Sheath, Vis & Cole 1993b)
1. with elongate apical cell (arrowhead) and re-branched determinate branchlets (double arrowhead). Scale bar = 50 μm (fig. 1).

Caloglossa leprieurii, figs 2–4 (Sheath, Vis & Cole 1993b)
2. habit of plant showing sub-dichotomous ramification of broad blades, midrib (arrowhead) and only parietal constriction at nodes (double arrowhead). 3. branch initial (arrowhead) from periaxial cells of midrib, 4. rhizoids arising from midrib (arrowhead). Scale bars = 1500 μm (fig. 2); 50 μm (figs 3–4).

Caloglossa ogasawaraensis, figs 5–7 (Sheath, Vis & Cole 1993b)
5. habit of plant showing thin blades, which are constricted at the nodes, midrib (arrowhead), and rhizoids (double arrowhead), 6. blade initial (arrowhead) from blade margin, 7. rhizoids arising from blade margin (arrowhead). Scale bars = 1500 μm (fig. 5); 50 μm (figs 6–7).

Polysiphonia subtilissima, figs 8–10 (t by Sheath, Vis & Cole 1993b)
8. apices with trichoblast (arrowhead), 9. branch initial (arrowhead) from trichoblast scar, 10. rhizoid arising from periaxial cell (arrowhead). Scale bar = 50 μm.

5) ***Caloglossa saigonensis*** Tanaka et Pham-Hoàng Hô (1962:24, fig. 7, 8).
Synonym: *Caloglossa continua* subspecies *saigonensis* (Tanaka et Pham-Hoàng Hô) King et Puttock (1994:89); *Caloglossa leprieurii* var. *angusta* Jao (1941:274, pl. I, fig. 1)

plate 189, figs 1–5 (original authors, Tanaka et Pham-Hoàng Hô 1962) as *C. saigonensis*

Plants up to 2 cm long, blackish violet, when dried becoming purple, pulvinate-caespitose, stolon-shaped, dichotomously branched, slightly constricted or not constricted at the dichotomous portion, sometimes proliferous from both midribs and margins, attached by rhizoids arising from the base and the dichotomous portion of the plants, segments linear, regularly downwards arcuated, 3–7 mm long, 215–700 μm wide.

Reproductive organs often unknown in freshwater habitat.

Type Locality: Cau Chu Y, Cholon, Saigon, Vietnam in Southeast Asia.

Holotype: SAP 052172, Tanaka et Pham-Hoàng Hô, 2210, 24/IV 1961.

Distribution: Occurs epiphytic on mangrove populations in brackish rivers entering the South China Sea in Vietnam in Southeast Asia.

Also in freshwater habitat, an inland freshwater river, Chialing River near Pehpei, Szechwan, China in Eastern Asia.

Note: This taxon was described by Tanaka et Pham-Hoàng Hô from Vietnam in Southeast Asia. *Caloglossa leprieurii* var. *angusta* (Type Locality: Chialing River near Pehpei, Szechwan, China in Eastern Asia; holotype: SC 1105B) was placed in synonymy with *C. continua* subsp. *saigonensis* by King & Puttock, the thallus morphologies being virtually identical. Reproductive material is unknown. Seto & Jao (1984) described the leafy segments as being about 0.5–0.8 mm wide, but King & Puttock (1994) measured widths on the type material in the range of 0.15–0.45 mm.

This alga was treated at species rank by Wynne & Clerck (1999:212), but *C. monosticha* Kamiya in Kamiya *et al.* (1997:104) is better to be treated as separate species.

Family **Rhodomelaceae** Areschoug (1847:260) *nom. cons.*

Type: genus *Rhodomela* C. Agardh *nom. cons.*

Characterized by basic polysiphonous structure, as well as the usual production of laterals of two different types, ordinary branches and trichoblasts. The dome-shaped apical cell forming a single series of axial cells. The mode of formation of the periaxial cells distinctive to Rhodomelaceae. The number of periaxial cells varying between 4 and 20; constant in a given species, especially in taxa with few periaxial cells. Tetrasporangium-bearing structures often specially differentiated, constituting so-called stichidia.

Genus *Bostrychia* Montagne in Ramon de la Sagra (1842:39) *nom. cons.*

Type: *Bostrychia scorpioides* (Hudson) Montagne

Plants densely tufted, pseudodichotomously or dichotomously branched, corticated or uncorticated, polysiphonous axis and branches consisting of axial cells each bearing 2–6 whorls of 4–8 periaxial cells; flagellar haptera or hapterous branches present.

Tetrasporangial stichidia are produced on the upper portion of thallus. Tetrasporangia are developed by transformation of the terminal portion of polysiphonous branches with 4–5 periaxial cells around an axial cell (West & Calumpong 1988, Tanaka 1989, 1991, Kumano 1979, etc.). Spermatangia are developed on normal lateral branch or spermatangial stichidia (King & Puttock 1991, Tanaka 1989, etc.). In the species of *Bostrychia*, many authors reported 4-celled carpogonial branches that are standard in Ceramiales; for example in *B. scorpioides* (Falkenberg 1901), in *B. arbuscula* (Hommersand 1963), in *B. kelanensis* (Takana & Chihara 1984b), and in *B. pinnata* (King & Puttock 1988). However, 2–4-celled carpogonial branches are also reported in some species, for example, *B. tenella* (3–celled, Tanaka 1989, 2–3–4-celled, West & Calumpong 1988), *B. fragellifera* (3–4-celled, Kumano 1988), *B. harveyi, B. pinnata, B. moritziana, B. tenella*

Ceramiales

(3–4-celled, King & Puttock 1989), *B. simpliciuscula* (4–celled, rarely 3–celled, Kamiya *et al.* 1994). Ovoidal or spherical cystocarps are developed terminally on lateral branches.

Tetrasporangial stichidia and cystocarps often absent in freshwater species.

Zuccarello & West (1997) mentioned that *Bostrychia radicans* formed a clad distinct from *B. moritziana*, based on the sequences of Rubisco gene.

Key to the species in freshwater habitat of the genus ***Bostrychia***

1. Haptera (cladohaptera) representing transformed ordinary branch..2
1. Haptera (flagellar haptera) consisting of rhizoids issued from ventral
   and lateral periaxial cells..3
2. Ultimate branchlets all monosiphonous...1) *B. moritziana*
2. Ultimate branchlets polysiphonous...2) *B. radicans*
3. Non-corticated..4
3. Corticated...5
4. Plants pseudodichotomously or dichotomously branched.............................3) *B. simpliciuscula*
4. Plants alternately branched..4) *B. flagellifera*
5. Terminal portion of branchlets not spirally enrolled..5) *B. tenella*
5. Terminal portion of branchlets spirally enrolled...6) *B. scorpioides*

1) ***Bostrychia moritziana*** (Sonder) J. Agardh (1863: 862)
Basionym: *Polysiphonia moritziana* Sonder in Kützing (1849:838)
Synonym: *Bostrychia radicans* Montagne f. *moniliforme* Post (1987:437, figs 1–2).

plate 191, figs 1–9 (Kumano 1979b)
figs 10–11 (D'Lacoste & Ganesan 1987)
plate 192, figs 1–5 (Sheath, Vis & Cole 1993b)
plate 193, figs 1–5 (Kumano 1997j)
figs 6–7 (Kumano & Necchi 1987) as *B. radicans* f. *moniliforme*

Plants dull reddish purple, rather erect, repeatedly pinnate, uncorticated, two whorls of 7 or 8 periaxial cells to each axial cell, ultimate branchlets monosiphonous at apex, but polysiphonous at the base. Apical cells large, conspicuous; at tip of plants a lateral branch and a cladohapteron successively initiated from main axis. At the tip of the cladohapteron, the apical cell ceasing production of discoid segments posteriorly, the periaxial cells elongating along the axis, the cladohapteron thus formed bending over toward the substratum. Tetrasporangial stichidia developed below tips of lateral branches; terminal portion monosiphonous, each tetrasporangium accompanied by a supporting cell linked by pit connection with an axial cell and lower ones of two sets of two cover cells. Male plants unknown. Female plants rare, carpogonial branch 3–4-celled (King & Puttock 1989), trichogyne prominently projecting outside the fertile branch, each branchlet of the determinate branch bearing 1–3 swollen, spherical to ovoidal cystocarps, 540–680 μm in diameter, 500–720 μm long. Carposporangia elongate, 13–27 μm in diameter, 38–65 μm long.

Type Locality: St. Lucia, Lesser Antilles in Caribbean Sea in central America.
Syntype: MEL 672271.

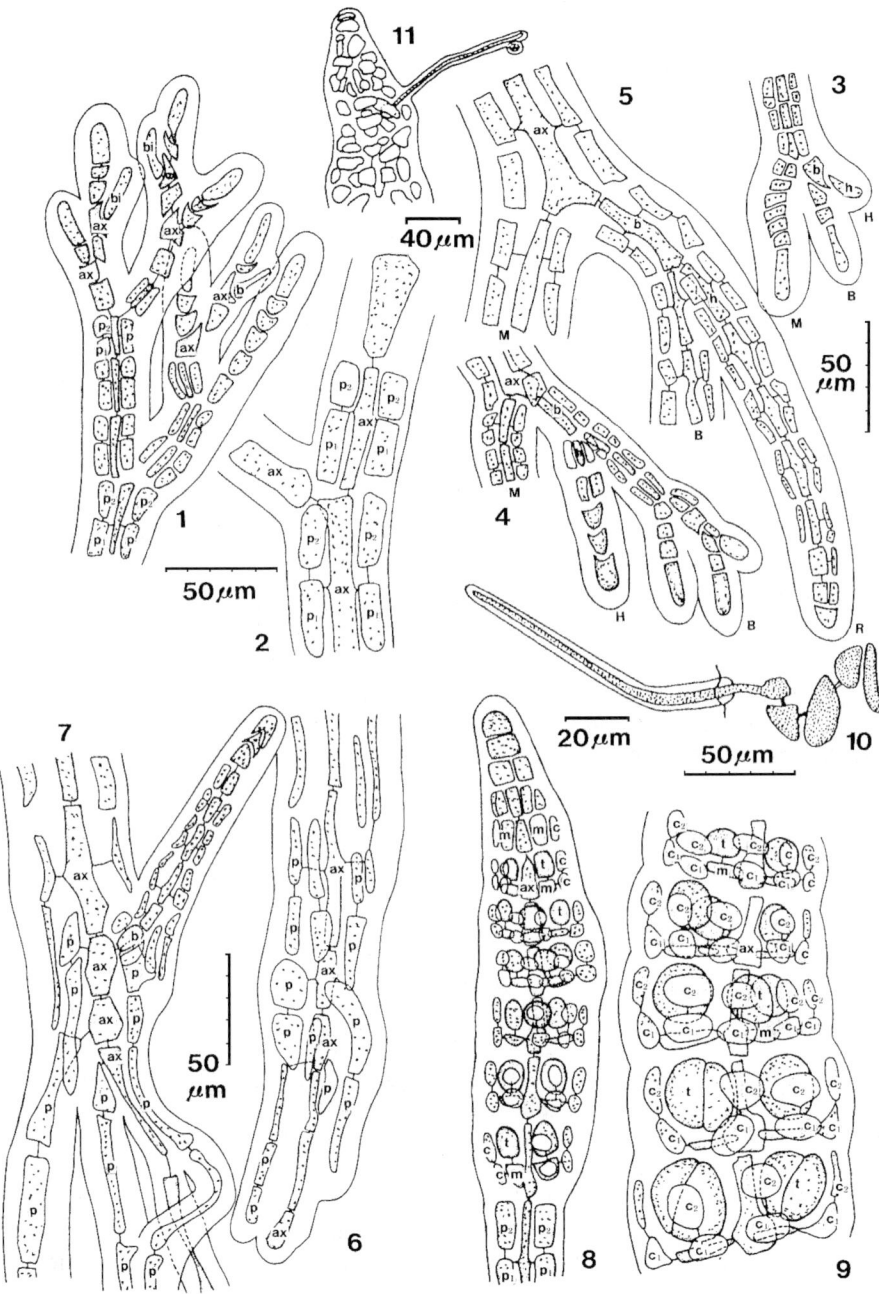

Plate 191
Bostrychia moritziana, figs 1–9 (Kumano 1979b, figs 10–11 by D'Lacoste & Ganesan 1987)
1–2. tip of monosiphonous ultimate branchlets, but polysiphonous at base, 3–5. cladohapteron formation, a lateral branch and a cladohapteron are successively initiated from main axis, 6–7. tip of caldo-hapteron and a new plant formation, 8–9. terminal tetrasporangial stichidia, each tetrasporangium is associated by two sets of two cover cells and a mother cell, 10. a procarp, 11. spermatium attached to the trichogyne. ax: axial cell, p: periaxial cell, h: basal segment of cladohapteron, M: main branch, B: lateral branch, H, R: cladohapteron, t: tetrasporangium, m: tetrasporangium mother cell, c: cover cell.
Scale bars = 20 μm (fig. 10); 40 μm (fig. 11).

Ceramiales

Distribution: (as *Bostrychia moritziana*): Known from mangrove populations in brackish habitat in Indian Ocean, in Hinai River, Iriomote Island in Okinawa, Japan in Eastern Asia; Indonesia in Southeast Asia; Micronesia in Pacific Ocean; Atlantic Ocean.

Also it found growing in the Caribbean Islands such as Guadeloupe, Dominica, Grenada, Martinique, Saint Lucia in Central America; Venezuela in South America.

As *Bostrychia radicans* Montagne f. *moniliforme,* known from mostly in marine and brackish habitats, but also in freshwater habitat such as a freshwater stream at the Serra de Baturité, Município Guaramiranga, in the northeastern State of Ceará in Brazil in South America. Known epilithically from a rivulet with clean running water, associated with a freshwater liverwort, *Lejeunea minutiloba* Evans and freshwater Simulidae (Diptera) larvae (Kumano & Necchi 1987).

Note: Female plants of this species are reported growing attached to stones forming dense masses at Rio El Pilar, Sucre State, in eastern Venezuela (D'Lacoste & Ganesan 1987).

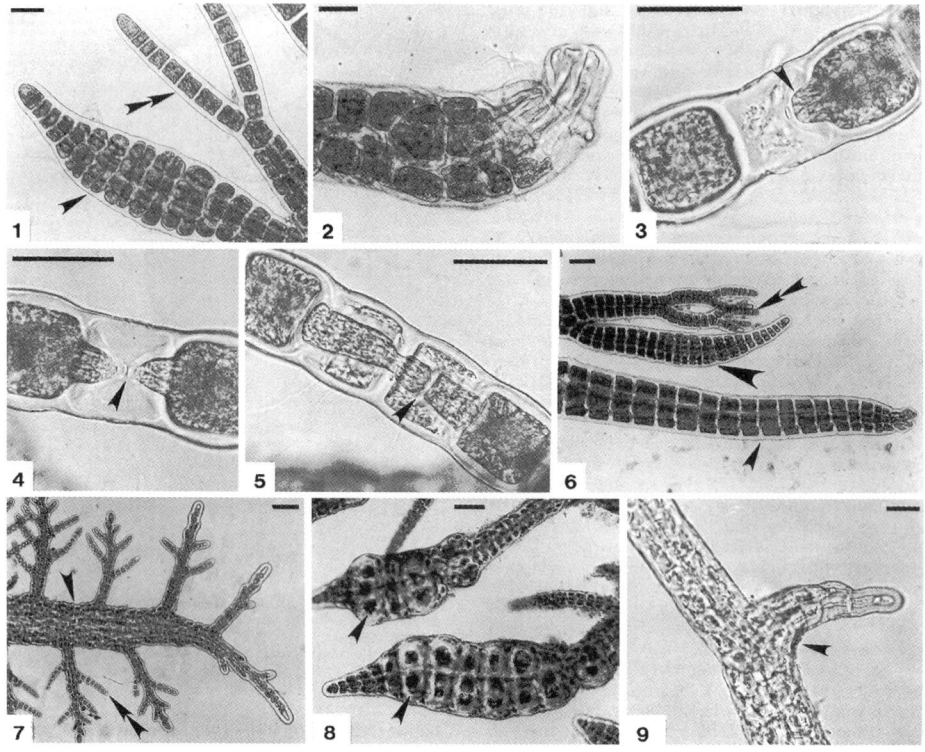

Plate 192

Bostrychia moritziana, figs 1–5 (Sheath, Vis & Cole 1993b)
1. stichidium (arrowhead) and monosiphonous ultimate branch (double arrowhead), 2. cladohapteron, 3. cell repair in monosiphonous branch with rhizoidal growth from super Janet cell (arrowhead), 4. cell repair with rhizoidal growth from super Janet and subjacent cells (arrowhead), 5. cell repair through two cells (arrowhead). Scale bar = 50 µm.

Bostrychia radicans, fig. 6 (Sheath, Vis & Cole 1993b)
6. with cladohapteron (arrowhead), polysiphonous ultimate branch (double arrowhead), and stichidium) large arrowhead). Scale bar = 50 µm.

Bostrychia tenella, figs 7–9 (Sheath, Vis & Cole 1993b)
7. corticated main axis (arrowhead) with opposite determinate branches that have monosiphonous tips (double arrowhead), 8. stichidia with tetrasporangia (arrowhead), 9. flagellar hapteron with rhizoids developing from periaxial cells (arrowhead). Scale bar = 50 µm.

2) ***Bostrychia radicans*** (Montagne) Montagne (1842:661)
Basionym: *Rhodomela radicans* Montagne (1840:193, pl. 5–6)

plate 192, fig. 6 (Sheath, Vis & Cole 1993b)
plate 194, figs 1–6 (Kumano 1979b)

Plant dull purplish, consisting of erect and prostrate portions, attaching with clado-haptera, pinnately branched, ecorticated and polysiphonous. Each axial cell with two whorls of periaxial cells. At tip of plants a lateral branch and a cladohapteron successively initiated from main axis. At the tip of the cladohapteron, the apical cell ceasing production of discoid segments posteriorly, the periaxial cells elongating along the axis, the cladohapteron thus formed bending over toward the substratum. Tetrasporangial stichidia developed below tips of lateral branches, each tetrasporangium accompanied by a supporting cell linked by pit connection with an axial cell and lower ones of two sets of two cover cells. Long conical spermatangial stichidia developed on lateral branches. Procarp consisting of a supporting cell and 4–celled carpogonial branch which includes a carpogonium and three hypogenous cells (Tanaka 1991). Cystocarp terminal on lateral branch, spherical, 500–650 µm in diameter.

Reproductive organs often unknown in freshwater habitat.
Type Locality: Sinnamary, French Guiana in South America.
Lectotype: MEL 672285, designated by King & Puttock 1989.
Distribution: Known from brackish habitat in Eastern Asia; Micronesia in Pacific Ocean; Caribbean Islands such as Dominican Republic, Guadeloupe, Dominica and Barbados in Central America. In Japan, brackish habitat in Urauchi-gawa River, Iriomote Island, Nozoko-gawa River, Nagura-gawa River, Ishigaki Island in Okinawa, Japan.

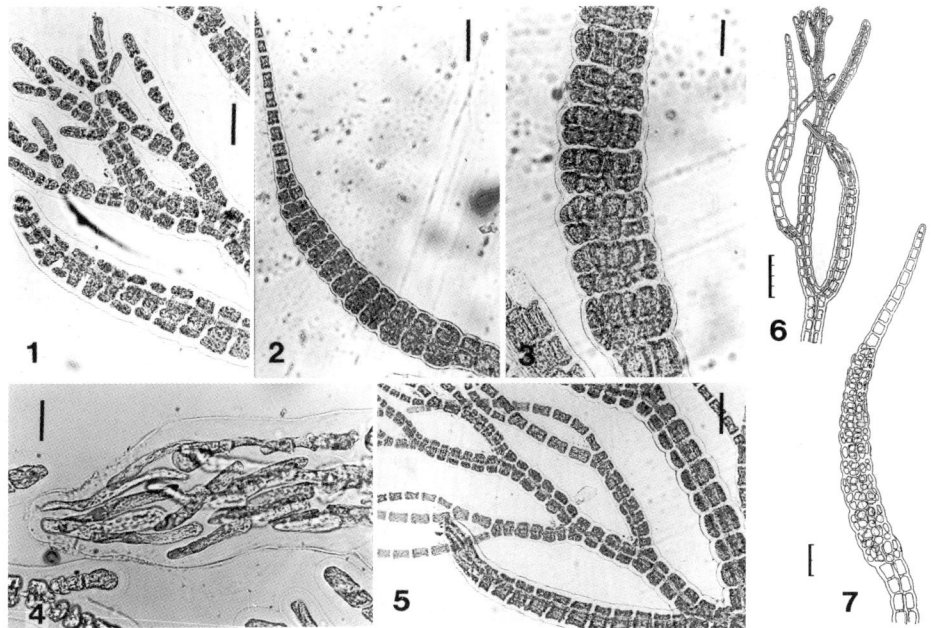

Plate 193
Bostrychia moritziana as *B. radicans* f. *moniliforme*, figs 1–5 (Kumano 1997j), figs 6–7 (Kumano & Necchi 1987)
1, 5, 6. an apical portion of a plant showing an erect branch and a cladohapteron are successively initiated from a main axis, 2, 3, 7. a terminal tetrasporangial stichidium, 4. an apical portion of a cladohapteron.
Scale bars = 200 µm (fig. 6); 100 µm (figs 2, 5, 7); 50 µm (figs 1, 3); 25 µm (fig. 4).

Ceramiales

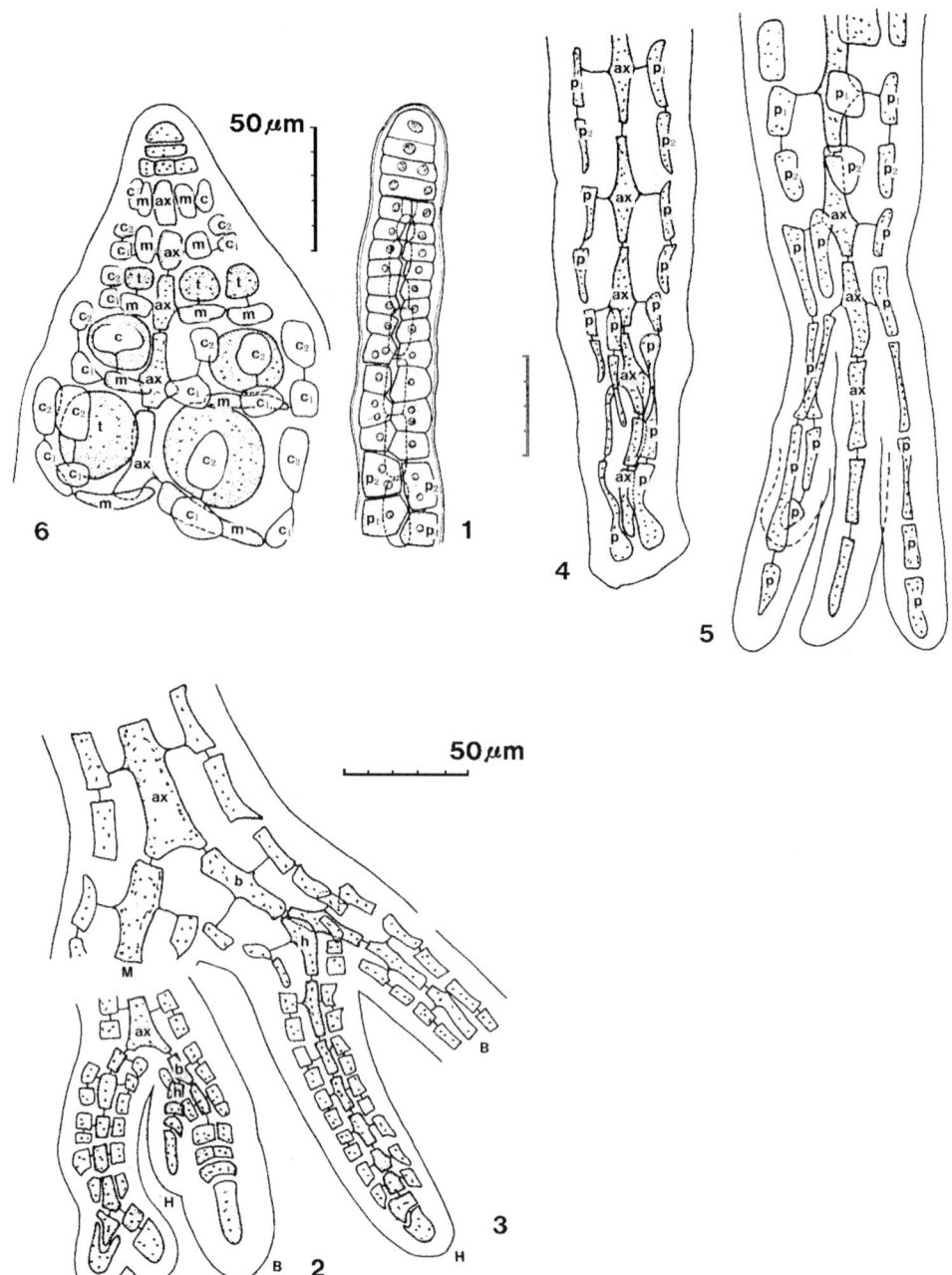

Plate 194
Bostrychia radicans, figs 1–6 (Kumano 1979b)
1. tip of plant, showing two whorls of periaxial cells with each axial cell, 2–3. cladohapteron formation, tip of plant, an erect branch and a cladohapteron are successively initiated from main axis, 4–5. tip of cladohapteron, 6. terminal tetrasporangial stichidium, each tetrasporangium is accompanied by two sets of two cover cells and a mother cell. ax: axial cell, p: periaxial cell, h: basal segment of cladohapteron, M: main branch, B: lateral branch, H: cladohapteron, t: tetrasporangium, m: tetrasporangium mother cell, c: cover cell.

Bostrychia

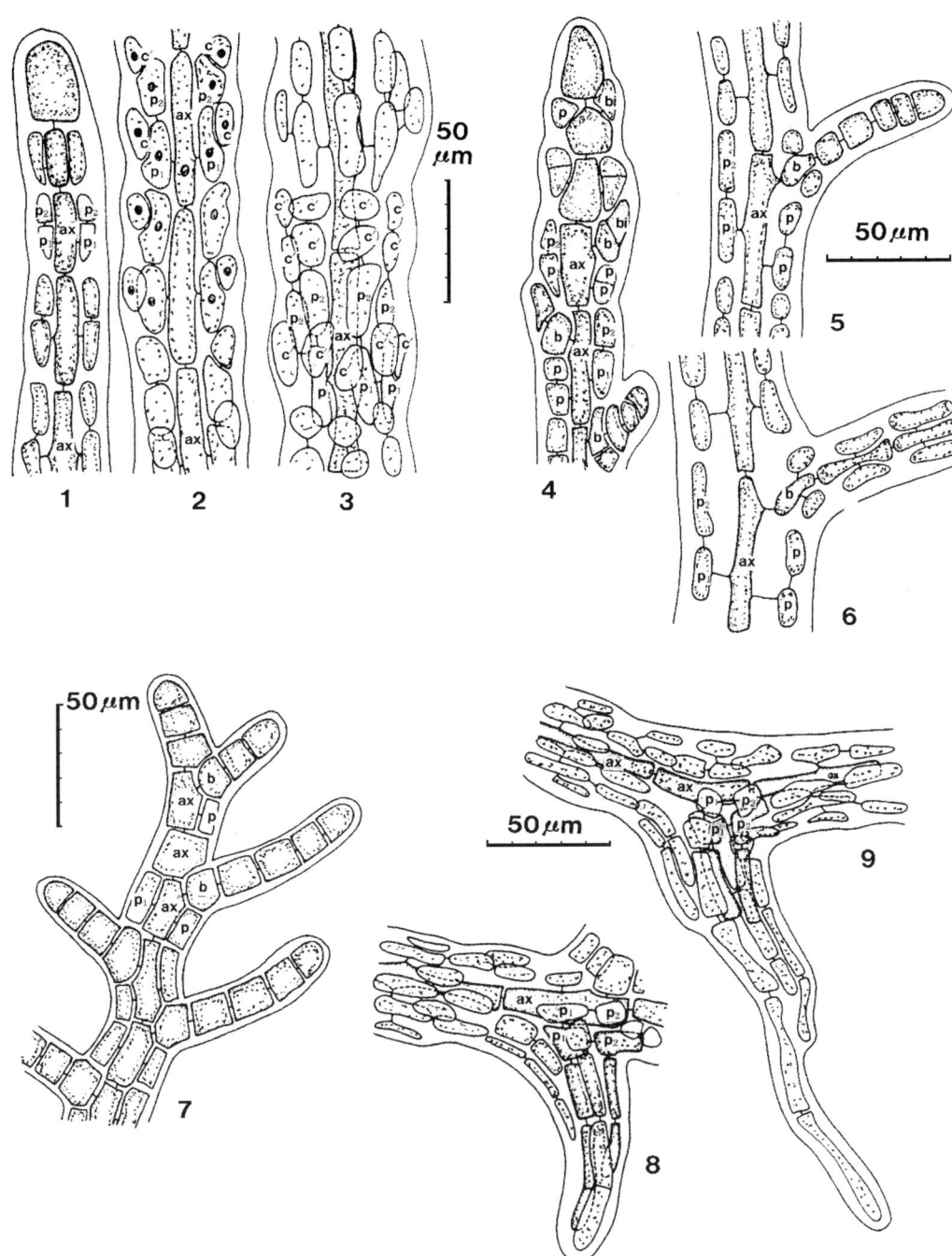

Plate 195
Bostrychia tenella, figs 1–9 (Kumano 1979b)
1. tip of plant showing a dome-shaped apical cell, axial cells and two whorls of periaxial cells, 2–3. cortical cell formation, 4–6. branch formation, 7. monosiphonous ultimate branchlets, 8–9. flagellar hapteron formation. ax: axial cell, p_1, p_2: periaxial cells, c: cortical cells, bi: branch initial, b: basal segment of branch.

Ceramiales

3) ***Bostrychia simpliciuscula*** Harvey ex J. Agardh (1863:354)
Synonym: *Bostrychia andoi* Okamura (1907: 102, pl. 22, figs 14–22); *Bostrychia tenuis* f. *simpliciuscula* (Harvey) Post (1936:6, 22–23); *Bostrychia hamana-tokidae* Post (1941:208)

plate 196 (a), figs 1–5 (Kumano 1996zp), figs 6–9 (Kumano 1979b) as *B. tenuis* f. *simpliciuscula*

Plants dark purplish brown, densely tufted, pseudodichotomously or dichotomously branched, non-corticated, ultimate branchlets polysiphonous at the base, monosiphonous at the apex; apical cell dome-shaped; each axial cell with two whorls of 4 or 5 periaxial cells; haptera flagellar outgrowths, originating from ventral and lateral periaxial cells, developing into flagellar haptera, consisting of bundles of 8–12 filaments, arising from ventral side of segments, then bending toward substrata. Tetrasporangial stichidia were sub-terminally produced on the upper portion of lateral branch. Each tetrasporangium accompanied by a supporting cell linked by pit-connection with an axial cell and lower ones of three sets of one or two cover cells. One or two spermatangial stichidia often developed on each lateral branch and some became bifurcate. Procarp consisting of a supporting cell and 4 celled (only one case 3-celled) carpogonial branch with long trichogyne (Kamiya *et al.* 1994). Cystocarp terminal on lateral branch, spherical, 130–480 μm in diameter. Carposporangia tear-drop shaped or lanceolate, 20–35 μm in diameter 40–70 μm long.

Reproductive organs often unknown in freshwater habitat.
Type Locality: Friendly Island, Tonga Island Kingdom in Pacific Ocean.
Lectotype: NSW, Sydney, designated by King & Puttock 1989.
Distribution: Known from the type locality and in marine and brackish habitats, Indian Ocean; Eastern Asia; Indonesia in Southeast Asia; Micronesia, Polynesia in Pacific Ocean. In Japan, brackish habitats in Chiba, Kagoshima, Su-ga in Okinawa, Japan.
Also found growing in a freshwater habitat such as Arakawa Fall in Ishigaki Island in Okinawa, Japan in Eastern Asia.

4) ***Bostrychia flagellifera*** Post (1936:1)

plate 196 (b), figs 1–5 (Kumano 1997i), figs 6–7 (Kumano 1988)

Plants dark brown, densely tufted, alternately branched, uncorticated, ultimate branchlets of polysiphonous at the base, monosiphonous at the apex; apical cell conspicuous; each axial cell with two whorls of periaxial cells; haptera flagellar outgrowths, originating from ventral and lateral periaxial cells, developing into flagellar haptera consisting of bundles of many filaments, arising from ventral side of segments, then bending toward substrata; tetrasporangial stichidia developed at tips of lateral branches, each tetrasporangium accompanied by a supporting cell linked by pit-connection with an axial cell and lower ones of two sets of two cover cells; procarps consisting of 3 or 4-celled carpogonial branches, mature trichogynes projecting obliquely toward tips of plants (Kumano 1988).

Type Locality: Paramatta River, Sydney in Australia.
Lectotype: MEL 672239, designated by King & Puttock 1989.
Distribution: Known from marine and brackish habitat in Australia, New Zealand, Indian Ocean. Also in freshwater habitat in Sonoyama-ike Pond, Sakura-jima Island in Kagoshima, Japan in Eastern Asia.

Bostrychia

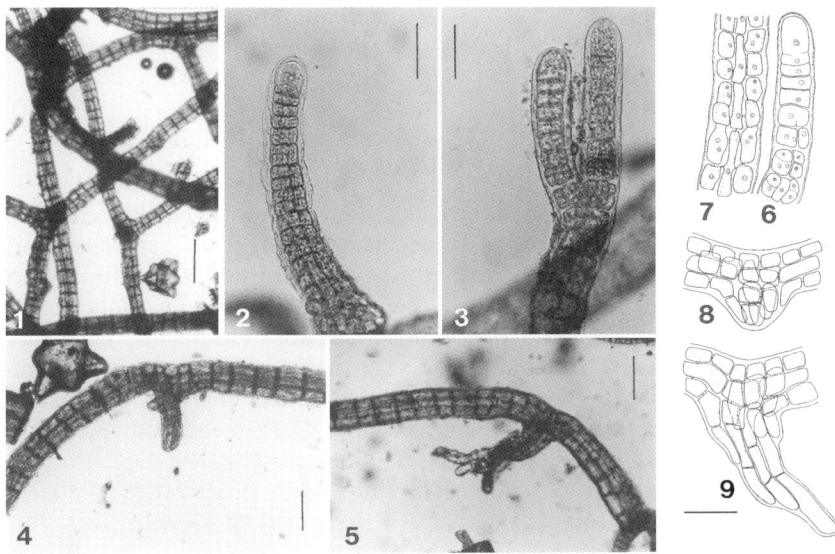

Plate 196 (a)

Bostrychia simpliciuscula as *B. tenuis* f. *simpliciuscula*, figs 1–5 (Kumano 1996zp), figs 7–9 (Kumano 1979b)
1. the habit of a plant showing pseudo-dichotomous ramification, 2, 3, 6. an apical portion of a plant showing a dome-shaped apical cell, discoid segments, an axial cell and periaxial cells, 4–5. a flagellar hapteron, 7. a terminal portion of a plant showing each axial cell with two whorls of periaxial cells, 8–9. the development of a flagellar hapteron.
Scale bars = 200 μm (fig. 1); 100 μm (figs 4–5); 50 μm (figs 2–3, 6–9).

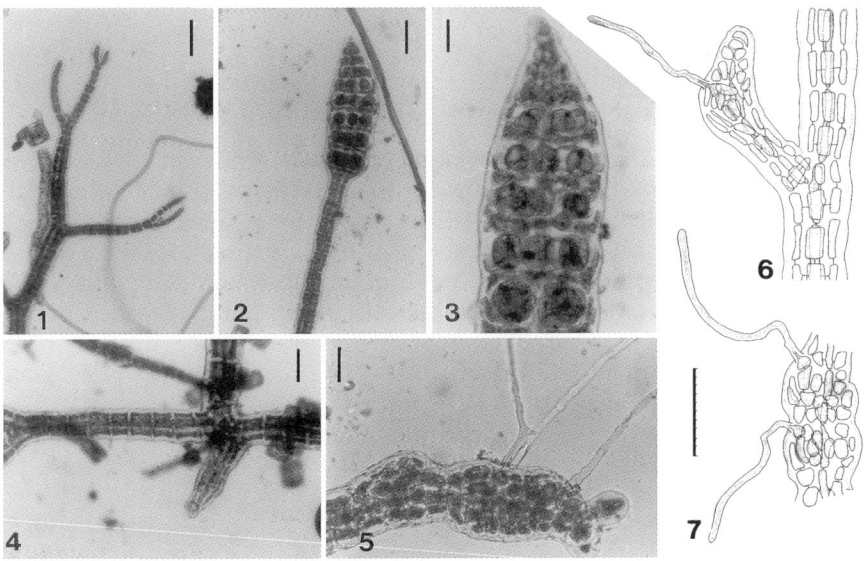

Plate 196 (b)

Bostrychia flagellifera, figs 1–5 (Kumano 1997i), figs 6–7 (Kumano 1988)
1. the habit of a plant shows dichotomous ramification, 2–3. a terminal tetrasporangial stichidium, 4. a middle portion of a plant showing a flagella hapteron, 5–7. procarps portion showing a long filamentous trichogyne and procarps. Scale bars = 200 μm (figs 1–2); 100 μm (figs 4, 6–7); 40 μm (figs 3, 5).

Ceramiales

5) ***Bostrychia tenella*** (Lamouroux) J. Agardh (1863:869)
Basionym: *Fucus tenellus* Vahl (1802:45) *nom. illeg.*; *Plocamium tenellum* Lamouroux (1813:21–47, 115–139, 267–293)

plate 192, figs 7–9 (Sheath, Vis & Cole 1993b)
plate 195, figs 1–9 (Kumano 1979b)

Plants dark brown, forming dense, large patches, alternately branched, corticated with 2–4 layers, ultimate branchlets, monosiphonous at the apex; apical cell conspicuous; each axial cell producing 6–8 periaxial cells, which divide into two whorls of periaxial cells; haptera flagellar outgrowths, originating from ventral and lateral periaxial cells, developing into flagellar haptera consisting bundles of many filaments, arising from ventral side of segments, then bending toward substrata. Tetrasporangial stichidia, observed by Sheath *et al.* (1993) in specimen from Barbados, developed in terminal portions of branchlets, linear or lanceolate often curved. Procarp consisting of 3–4-celled, (King & Puttock 1989), 3-celled (Tanaka 1989), 2–3–4-celled (West & Calumpong 1988) carpogonial branch.

Sexual reproductive organs often unknown for this species in freshwater habitat.

Type Locality: St. Croix, Virgin Island, Puerto Rico in Central America.

Lectotype: MEL 672309, designated by King & Puttock 1989.

Distribution: Known mostly from mangrove communities in brackish habitats, Sira-gawa River in Irimomte Island, Fukitohsi-gawa River in Ishigaki Island in Okinawa, Japan in Eastern Asia.

Also found growing in freshwater habitats such as on a rocky wall in a freshwater pool, Marbo Cave, Guam Island, Mariana Islands in Pacific Ocean; Barbados in the Caribbean Islands in Central America.

6) ***Bostrychia scorpioides*** (Hudson) Montagne (1842)
Synonym: *Fucus scorpioides* Hudson (1762:471)

plate 197, figs 5–6 (Entwisle & Kraft 1984)
plate 198, figs 1–5 (Bourrelly 1970)

Plants filamentous, 4–7 cm long, forming tangled mats similar in appearance to tree-fern roots. Filaments erect, attached at base and along short creeping stolons by rhizoidal holdfasts. Branching alternate to dichotomous, apices long and tapering, often spirally enrolled. Plants 100–180 µm in diameter, consisting of central row of axial cells each surrounded by 6 periaxial cells. Periaxial cells irregularly shaped, 5–10 µm in diameter, 22–30 µm long, dividing anticlinally to form cortex of irregularly spherical cells 10–18 µm in diameter. No reproductive organs observed.

Type Locality: Great Britain, UK in Europe.

Lectotype: Dillenius in Ray (1794: pl. 2: fig. 6)

Distribution: Known worldwide from marine habitats in UK, Belgium in Europe.

Also in freshwater habitats such as Bachamp Falls, Air River, Beech Forest, near Apollo Bay, Gordon River, Tasmania in Australia.

Plate 197
Audouinella hermannii, figs 1–2 (Entwisle & Kraft 1984)
1. habit of plant, 2. distal portion of plant. Scale bars = 2.5 mm (fig. 1); 10 μm (fig. 2).

Caloglossa leprieurii, figs 3–4 (Entwisle & Kraft 1984)
3. nodal portion of plant, 4. habit of plant.
Scale bars = 2.5 mm (fig. 4); 10 μm (fig. 3).

Bostrychia scorpioides, figs 5–6 (Entwisle & Kraft 1984)
5. habit of plant, 6. apices with characteristic curled tips.
Scale bars = 80 μm (fig. 1); 10 μm (fig. 2).

Ceramiales

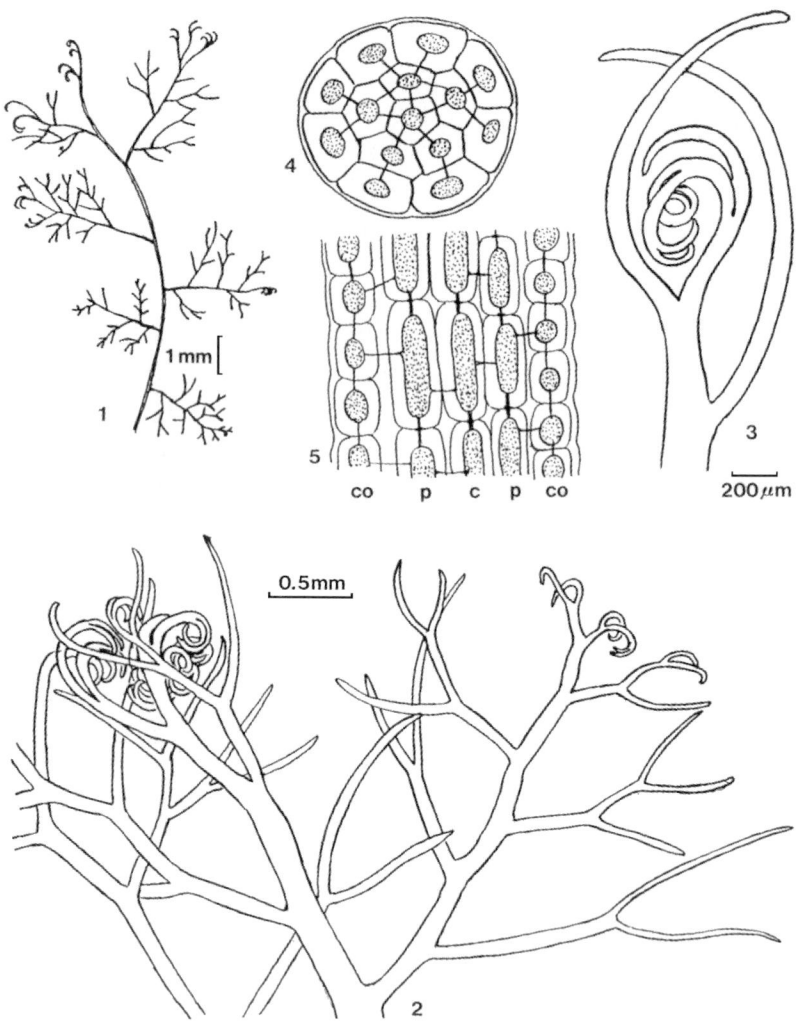

Plate 198
Bostrychia scorpioides, figs 1–5 (Bourrelly 1970)
1. general habit of plant, 2. apical portion of plant, 3. details of apical portion of plant, 4. cross section of juvenile plant, 5. longitudinal section of plant. co: cortex, p: periaxial cells, c: axial cells.

Genus *Polysiphonia* Greville (1823: pl. 90) *nom. cons.*

Type: *Polysiphonia urceolata* (Dillwyn) Greville *typ. cons.*

Plants radially organized, consisting of erect, polysiphonous axes arising from holdfast or from prostrate axes. The apical cell cutting off segments proximally which elongate and cut off periaxial cells the same length as the axial cell. Depending on the taxon, the number of periaxial cells is 4–24, and cortication may be added in some taxa. Trichoblasts colourless, simple or branched, arising exogenously in distal portion of plants. Reproductive organs unknown in species found in freshwater habitat.

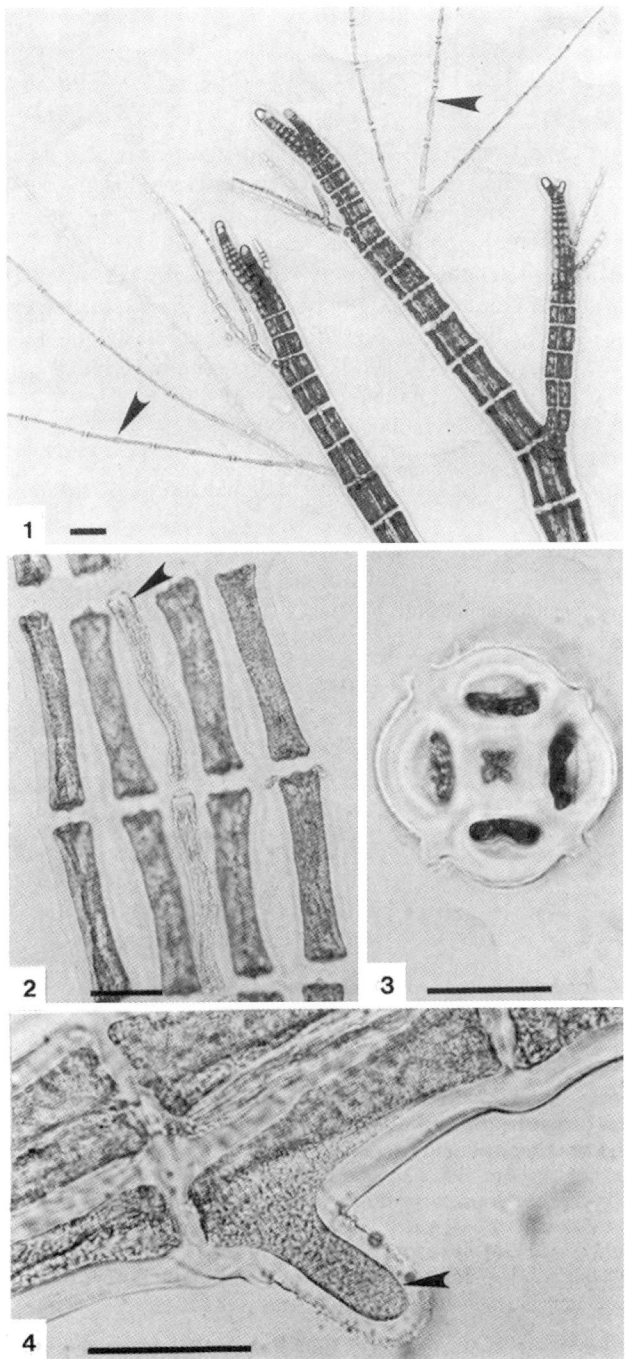

Plate 199
Polysiphonia subtilissima, figs 1–4 (Sheath & Cole 1990)
1. an apex of a plant with trichoblasts (arrowheads) showing sub-dichotomous branching, 2. a squashed branch showing four pigmented periaxial cells and colourless axial filaments (arrowhead), 3. the cross section of a branch with four periaxial cells, 4. rhizoid (arrowhead) originating as outgrowth of periaxial cells. Scale bars = 100 μm (figs 1–4).

Ceramiales

1) *Polysiphonia subtilissima* Montagne (1840:193, pl. 5–6)

> plate 190, figs 8–10 (Sheath, Vis & Cole 1993b)
> plate 199, figs 1–4 (Sheath & Cole 1990)

Plants 1.4–4.7 cm long. Branch 38–76 µm in diameter, periaxial cells 58–125 µm long. The presence of four periaxial cells, sub-dichotomous branching, apical trichoblasts and rhizoids arising from periaxial cells combined with a lack of cortication and reproductive cells is consistent with marine population of this species.

The maximum filament length, 4.7 cm is below the 15 cm noted by Taylor (1960) or 10–12 cm by Fralick and Mathieson (1975) but similar to the 4.0 cm given by Kapraun (1979, 1980). The range of branch diameters, 38–76 µm, compares to 50–80(–90) µm for marine filaments (Taylor 1960, Kapraun 1979, 1980). The maximum periaxial cell length, 125 µm is nearly equivalent to the 130 µm presented by Taylor (1960).

Type Locality: Cayenne, French Guiana in South America.

Holotype: PC herb. Montagne, Le Prieur coll. 682.

Distribution: Known from brackish habitats in Jamaica, Great Antilles in Caribbean Sea in Central America.

In North America, Florida, USA, Juniper Creek, Marion County, 170 km upstream from the drainage basin mouth and 140 km from tidal influence, represents the first finding of *Polysiphonia* in a truly freshwater habitat (Starmach 1977, Ott & Sommerfeld 1982, Sheath 1984, Bourrelly 1985).

Abbreviations used in nomenclature

ex: when author who validly publishes a name ascribes it to another person, the correct author citation is the name of the actual publishing author, but the name of the other person, followed by the connecting word 'ex', may be inserted before the name of the publishing author, if desired.

holotype: a holotype is the one specimen or other element used by the author or designated by him as the nomenclatural type.

in: when a name with a description or diagnosis (or reference to a description or diagnosis) supplied by one author is published in a work by another author, the word 'in' should be used to connect the names of the two authors. In such cases the name of the author who supplied the description or diagnosis is the most important and should be retained when it is desirable to abbreviate such a citation.

ined.: ineditus: unpublished.

isotype: an isotype is any duplicate (part of a single gathering made by a collector at one time) of the holotype; it is always a specimen.

lectotype: a lectotype is a specimen or other element selected from the original material to serve as a nomenclature type when no holotype was indicated at the time of publication or as long as it is missing.

nom. cons.: nomen conservandum: in order to avoid disadvantageous changes in the nomenclature of families, genera and species entailed by the strict application of the rules, and especially of the principle of priority in starting from the dates given in Art. 13, names that are conserved and must be retained as useful exceptions.

nom. illeg.: nomen illegitinum: an illegitimate name is one that is not in accordance with the rules, and one that is designated as such in Arts. 18.3 or 63–67.

nom. invalid.: nomen invalidum: the name invalidly published.

nom. nud.: nomen nudun: the name without Latin description, because in order to be validly published, a name of a new taxon of plants, algae (all fossils excepted), published on or after Jan. 1, 1935 must be accompanied by a Latin description or diagnosis or by reference to a previously and effectively published Latin description or diagnosis.

orth. var.: orthographia varia: orthographic variant: only the orthographic variant of any one name is treated as validly published, the form which appears in the original publication except as provided in Art. 73 (orthographic and typographic errors).

emend: pro parte: sensu, etc.: an alteration of the diagnostic characters or of the circumscription of a taxon without the exclusion of the type does not warrant the citation of the name of an author other than the one who first published its name.

Abbreviations of herbaria compiled in this book
based on Index Herbaria ver. 8, 1990

* Abbreviations indicate names of provinces in China

AKU:	Botany Department, University of Auckland, Auckland, New Zealand
B:	Botanischer Garten und Botanisches Museum Berlin-Dahlem, Germany
BA:	Museo Argentino de Ciências Naturales, Bernardino Rivadavia, e Instituto Nacional de Investigação de la Ciências Naturales Buenos Aires, Argentina
BH:	Liberty Hyde Bailey Herbarium, Cornel University, New York, USA
BM:	The Natural History Museum, London, UK
COL:	Herbario Nacional Colombiano, Instituto de Ciências Naturales, Museo de Historia Natural, Universidado Nacional de Colombia, Bogota, Colombia
DCR:	Doncaster Metropolitan Borough, Museum and Art Gallery, Doncaster, UK
G:	Conservatoire de Jardin Botanique de la Villa de Genève, Switzerland
FLOR:	Horto Botânico, Universidade Federal de Santa Catarina, Florianopolis, Brasil
GX*:	Guangxi (Kwangsi) Chuang Autonomous Region, China
HAS:	Fundação Zoobotânica do Rio Grande do Sul, Porto Alegre, Brasil
HBI:	Freshwater Algal Herbarium, Institute of Hydrobiology, Academia Sinica, Wuhan, China
HI:	Dr. Itono, H. Collection no. probably deposited in Faculty Science, Kagoshima University, Kagoshima, Japan.
HN*:	Hunan Province, China
HP*:	Hupeh (Hubei) Province, China
ICN:	Departamento de Botânica, Universidade Federal do Rio Grande do Sul, Porto Alegre, Brasil
INPA:	Departamento de Botânica, Instituto Nacional de Pesquisas da Amazônia, Manaus, Brasil
Kobe U:	Faculty of Science, Kobe University, Kobe, Japan
KRA:	Institute of Botany, Universitatis Jagellonicae Cracoviensis, Kraków, Poland
KSI*:	Kiangsi (Jiangxi) Province, China
L:	Rijksherbarium, Leiden, Netherlands
LD:	Botaniska Museum, Lund, Sweden
LISU:	Instituto Botânico, Faculdade de Ciências, Rua de Escole Politécnica, Lisboa, Portugal
MEL:	National Herbarium of Victoria, Royal Botanic Gardens, Melbourne, Australia
MELU:	Botany School, University of Melbourne, Melbourne, Australia
MELU TJE:	Botany School Herbarium, University of Melbourne, Melbourne, Australia
NSW:	National Herbarium of New South Wales, Royal Botanic Gardens, Sydney, Australia
NT:	Herbarium of the Northern Territory of Australia, Alice Springs, Australia
NY:	New York Botanical Garden, Bronx, New York, USA
OXF:	Fielding-Druce Herbarium, University of Oxford, Oxford, UK
PC:	Laboratoire de Cryptogamie, Muséum National d'Histoire Naturelle, Paris, France
PH:	Botany Department, Academy of Natural Sciences, Philadelphia, USA
R:	Departamento de Botânica, Museu Nacional, Universidade Federal do Rio de Janeiro, Rio de Janeiro, Brasil
RB:	Seção de Botânica Sistemática, Jardim Botânico de Rio de Janeiro, Rio de Janeiro, Brasil
REN:	Laboratoire de Botanique de la Faculté des Sciences, Rennes, France

RIG:	Botany and Ecology Department, P. Stuchka Latvian State University, Riga, Latvia
SAP:	Herbarium of Graduate School of Science, Hokkaido University, Sapporo, Japan
SAS*:	Shansi (Shanxi) Province, China
SC:	Biology Department, Salem College, Winston Salem, North Carolina, USA
SC*:	Szechwan (Sichuan) Province, China
SK*:	Name of locality in Szechwan Province, China
SP:	Herbario do Estado, "Maria Eneyda P. K. Figalgo", Instituto de Botanica, Saõ Paulo, Brasil
SXU:	Herbarium, Department of Biology, Shanxi University, China
TCD:	Herbarium, School of Botany, Trinity College, Dublin, Ireland
TI:	Botanical Gardens, University of Tokyo, Tokyo, Japan
TNS:	Botany Department, National Science Museum, Tokyo, Japan
UBC:	Department of Botany, University of British Colombia, Vancouver, Canada
UC:	University Herbarium, University of California, Berkeley, California, USA
UPNG:	Biological Department, Natural Sciences Resource Centre, University of Papua New Guinea, Port Moresby, Papua New Guinea
UPS:	Botanical Museum (Fytoteket), Uppsala University, Uppsala, Sweden
US:	United States National Herbarium, Botany Department, Smithsonian Institution, Washington DC, USA
WA:	Herbarium, Department of Plant Systematics and Geography, Institute of Botany, University of Warsaw, Aleje Ujazdowskie 4, PL–00–478 Warszawa, Poland
WELT:	Botany Department, National Museum of New Zealand (Te Papa Tongarewa), Wellington, New Zealand
XZTC:	Herbarium, Department of Biology, Xuzhou Teachers College, China
YN*:	Yunnan Province, China

Glossary

Phycological terms relating to the freshwater Rhodophyceae

alternation of generations : a life cycle in which both asexual and sexual plants occur, one giving rise to the other by the production of spores (carpospore from carposporophyte and monospore from chantransia phase), and by the fusion of gamete (spermatium and carpogonium in gametophyte) found in Rhodophyceae.

amoeboid : like an amoeba in locomotion in monospore of genus *Bangia*.

arcuate : arched, bow-shaped, sharply crescent-shaped, strongly curved as in drawn bow:

assimilatory filament(s) : [←photosynthetic filaments] : the filament(s) consisting of pigmented cells as found in family Thoreaceae.

audouinelloid, audouinella-like: a shape of filaments of fascicles found in some taxa of genus *Batrachospermum*.

axial cell(s) : [←central cell(s)]

axis (axes) : the central or median plane of plant.

basal cell of primary branchlets : [→pericentral cell(s)] : [→periaxial cell(s)].

bract filament(s) : [→involucral filament(s)].

branchlet(s) : [→fascicle(s)]

caespitose : clustered in fascicles, forming a mat or tangle.

carpogonial branch(es) : [→carpogonium-bearing branch(es)] :a specialized filament consisting of a definite number of modified cells subtending carpogonium found in the advanced members of the Rhodophyceae.

carpogonium (carpogonia) : a female reproductive cell, usually with a trichogyne found in the Rhodophyceae.

carpogonium-bearing branch(es) : [←carpogonial branch(es)] : the undifferentiated filament(s) consisting of indefinite number of unmodified cells bearing carpogonium found in the primitive members of the Rhodophyceae such as genus *Batrachospermum*.

carposporangium (carposporangia) : a cell producing carpospore.

carpospore(s) : spore that arises as a result of fertilized carpogonium found in the Rhodophyceae.

carposporophyte(s) : [←cystocarp(s)] : a phase produced following fertilization found in Nemaliophycidae, comprised of gonimoblast filaments bearing carposporangia.

central cell(s) : [→axial cell(s)].

chantransia-phase : [←chantransia-stage] : [←audouinella-phase] : a diploid phase producing gametophyte by means of meiotic division in vegetative cell.

cladohapteron (cladohaptera) : [←hapteronous branch(es)] : found in some species of genus *Bostrychia*, at a tip of plants a lateral branch and a cladohapteron successively initiated from main axis. At the tip of the cladohapteron, the apical cell stops to produce discoid segments posteriorly, the periaxial cells elongate along the axis, the cladohapteron thus formed bends over toward the substratum.

clavate, claviform : wedge-shaped.

contorted : irregularly twisted.

cortex : a layer of cells or filaments which invest or grow around an axial cell or filaments, forming an incising layer.

cortical cell(s) : cells which invest or grow around an axial cell or filaments, forming an incising layer.

cortical filament(s) : [←rhizoidal filaments] : filament(s) which invest or grow around an axial cell or filaments, forming an incising layer.

corymb : a flat-topped cluster.

costae (costa) : rib, ribbed.

cover cell(s) : cells that are cut off in association with tetrasporangia, serving as superficial, protective covers found in genus *Bostrychia*.

cushion-like : [→parenchymatous] : of a thallus composed of a mound of cells, two to many layers of cells.

cylindrical : elongate and round in cross section with parallel lateral margins.

determinate gonimoblast filament(s) : found in most taxa the sections of the genus *Batrachospermum*.

dichotomous : divided or forked into two parts, forking branches.

dioecious : plants issued male and female organs separately

disc, discoid : a circular, flat body, a plate.

distal : referring to the forward end, opposite from base.

18S rRNA gene: 18S rDNA: gene sequences coding 18 S ribosomal RNA, small subunit rRNA. Due to its relatively slow substitution rate, it has been frequently used for discussing the molecular phylogenetic relationship at higher taxonomic ranks (e. g. families, orders) in Rhodophyceae.

endospore(s) : a spore formed within a cell.

epilithic : living upon stones and rocks.

epiphytic : living upon a plant.

fascicle(s) : [←branchlets] : found in genus *Batrachospermum* and genus *Sirodotia*, densely branched laterals [fascicles] of limited growth, all of about same length arise from periaxial cells just below the septa separating the elongate cells of main axes, and from cortical cells developed from periaxial cells and creeping along axial cells.

female plant(s) : a plant producing female reproductive organs, carpogonia found in the Rhodophyceae.

filament : a linear arrangement of cells; thread of cells.

flagellar hapteron (haptera) : filaments modified to form an attaching organ found in some species of genus *Bostrychia*: flagellar outgrowths, originating from ventral and side periaxial cells, developing into flagellar haptera, consisting bundles of 8–12 filaments, arising from bentral side of segments, then bend toward substrata.

flagelliform : whip-shaped.

frond : a flat, leaf-like plant, a foliaceous thallus.

fused cell(s) : placenta : the cell produced by the union of the protoplasm and nuclei of two or more cells such as fertilized carpogonium, cells of carpogonium-bearing branch found in genus *Tuomeya*.

gametophyte(s) : the sexual, gamete-producing phase in the life cycle of a plant.

gonimoblast filament(s) : short filaments developing from the zygotes, and cut off carposporangia at their tip.

gonimoblast : [→carposporophyte] : a filament bearing carpospores of the entire collection of these filaments comprising the carposporophyte.

hapteron (haptera) : an anchoring organ found in genus *Bostrychia*.

hapteronous branch(es) : [→cladohapteron (cladohaptera)] : branches modified to form an attaching organ found in genus *Bostrychia*.

hemispherical protuberance(s) : found in carpogonia base of genus *Sirodotia*.

holdfast(s) : [←rhizoid(s)] : an anchoring organ: cells or filaments modified to form an attaching organ.

hypogenous cell(s) : the cell subtending, directly beneath, the carpogonium.

indeterminate gonimoblast filament(s): found in some taxa of section *Turfosa* of genus *Batrachospermum* and genus *Sirodotia*.

involucral filament(s) : [←bract filament(s)] : a sterile group of filaments forming an envelope around carpogonia found in genus *Batrachospermum*.
involucre : [←bract(s)] : a sterile group of cells or filaments forming an envelope around a reproductive structure(s).
laminate : plate-like.
lanceolate : lance-shaped, long and narrow with subparallel margins but tapered at the apex.
lateral and central periaxial cell(s) form flagellar hapteron found in genus *Bostrychia*.
leafy internode(s) : found in genus *Caloglossa*.
loosely agglomerated carposporophyte : found in the some species of section *Contorta* of genus *Batrachospermum*.
losette like involucral filament(s) : found in section *Aristata* of genus *Batrachospermum*.
male plant(s) : a plant producing male sexual reproductive organs, spermatangia.
medulla : central region of a thallus.
medullarly filament(s) : filaments composing central region of a thallus.
microsporangium (microsporangia): a sporangium producing microspores.
microspore(s): minute, asexual reproductive spores formed by some taxa of genus *Compsopogon*.
midrib(s) : a structure consisting of a bentral, a dorsal and two lateral periaxial cells around axial cell found in genus *Caloglossa*.
monoecious : of one household with both male and female sex organs on the same plant.
monopsore(s) : asexual spores forming on branches: a non-flagellate spore produced singly from a sporangium.
monosporangium (monosporangia) : a sporangium that produces a single spore.
mucilaginous : gelatinous.
multiaxial : an axis composed of many elongate filaments.
multipodial : with more than one axis; with more than one row of cells.
multiseriate : with more than one row of cells; with many filaments.
node : the site on an axis from which branches arise.
obconical : cone-shaped with the broader end foremost.
obovoidal : inversely ovoidal, with the broader end anterior or outer most.
one-celled gametophyte(s) : found in subgenus *Acarposporophytum* of genus *Batrachospermum*.
ovoidal : shaped like an egg.
parietal chloroplast(s) : chloroplast lying along the wall; peripheral in the cell.
parietal : lying along the wall, peripheral in the cell.
periaxial cell(s) : [←pericentral cell(s)] : [←basal cell of primary branchlets].
pericentral cell(s) : [→periaxial cell(s)] :a cell (one or several) enclosing a central cell.
pinnate: feather like.
pit connection : a discrete lens-shaped plug held within an aperture of the cross wall between two adjacent cells.
pit plug: structure of pit connection consisting of core, cap membrane enveloping core and cap layer, type 1: core only, type 2: core enveloped by cap membrane only, type 3: core enveloped by a cap layer, type 4: core enveloped by a cap layer and a cap membrane, type 5: core enveloped by inner and outer cap layers with intercarally cap membrane, type 6: core enveloped by inner and outer dome-shaped cap layers with intercarally cap membrane.
primary fascicle(s) : [←primary branchlet(s)] : found in genera *Batrachospermum* and *Sirodotia*, densely branched laterals (fascicles) of limited growth, all of about same length arise from periaxial cells just below the septa separating the elongate cells of main axes.
propagule(s) : zonately separated specialized carpospores formed on gonimoblast filaments found in *Batrachospermum breutelii*:
prostrate portion(s) : horizontally creeping portion.
pseudo-filament : a thread of cells incidentally arranged in a linear series, not a true filament.

pseudo-parenchymatous : resembling a mound of cells but actually constructed of closely grown filaments.
pyramidal: in the shape of a pyramid.
pyriform : pear-shaped, with narrow end foremost.
ray cell(s): a variation of periaxial cell(s) found in genus *Lemanea* and genus *Paralemanea*.
rhizoid(s) : [→holdhast] : an anchoring organ of elongate cells.
RuBisCo gene, *rbc*L and *rbc*S: gene sequences coding large and small sub-units of RuBisCo (ribulose-1,5-bis-phosphate carboxylase, oxigenase). Due to their relatively fast substitution rate, these regions have been successfully used for discussing the molecular phylogenetic and biogeographical relationships at lower taxonomic ranks (e. g. populations, species, genera) in Rhodophyceae.
saccate : like a sac; balloon-shaped.
secondary fascicle(s) : [←secondary branchlet(s)] : found in genus *Batrachospermum* and genus *Sirodotia*, densely branched laterals (fascicles) of limited growth, all of about same length arise from cortical cells developed from periaxial cells and creeping along axial cells.
short spinous branchlet(s) : found in genus *Compsopogon*.
simple ray cell(s) : a variation of periaxial cell found in genus *Paralemanea*.
spatulate : spatula-shaped.
spermatangial papillae (papilla) : found in family Lemaneaceae.
spermatangium (spermatangia) : male reproductive structure found in the Rhodophyceae, which produces a spermatangium.
spermatium (spermatia) : cells acting as male gamete; non-motile and colourless male cells released from sporangium found in the Rhodophyceae.
spermatiophore : formed at the tip of fascicles, spermatangia formed directly from cells of fascicles, most spermatangia occur at the apices of spermatiophores, which range from few-celled and unbranched complex found in *Batrachospermum spermatiophorum*.
sporophyte(s) : the diploid, usually, spore-producing plant or phase in a life cycle.
stellate : star-shaped.
stichidium (stichidia) : a specialized branch bearing tetrasporangia found in Nemaliophycidae.
supporting cell(s) : the cell bearing cover cells and tetrasporangium found in genus *Caloglossa*.
T- or L-shaped radial cell(s): a variation of periaxial cell(s) found in genus *Lemanea:*
terminal hair(s).
tetrasporangium (tetrasporangia) : a cell in which a diploid nucleus undergoes meiosis and four haploid spores, tetraspores, are produced in Nemaliophycidae.
tetraspore(s) : a spore produced within a tetrasporangium as a result of meiosis found in the Rhodophyceae.
tetrasporophyte(s) : a diploid phase producing tetrasporangia.
thallus (thalli) : a plant body in which there is little or no differentiation of cells to form tissues.
trichoblast : a simple or branched filaments, pigmented or colourless, arising exogenousely at the apices of plants found in family Rhodomelaceae.
trichogyne(s) : a receptive protuberance or elongation of a female gametangium (carpogonium) to which male gametes become attached.
truncate : flat at the top, flatly rounded.
unilateral : on one side, arising from one side only.
uniseriate : arranged in a single row or series.
whorl(s) : found in genus *Batrachospermum* and genus *Sirodotia*, each bead consisting of a whorl of densely branched laterals [→fascicles] of limited growth, all of about same length.
wing cell-row(s) : wing formation follows that of genus *Caloglossa*, being initiated from the lateral periaxial cells with production of secondary and tertiary rows.

Bibliography

References cited and related

References were seen by the author and the reviewer at University of California at Berkeley, except a few written in Japanese.

Agardh, C. A. 1810–12. Dispositio algarum Suecicae... Lund. pp. [1]–16 (1810), 17–26 (1811), 27–45 (1812).
Agardh, C. A. 1817. Synopsis algarum Scandinaviae... Lund. XL + 135 pp.
Agardh, C. A. 1820. Species algarum... Vol. 1, part 1. Lund. pp. [1]–168.
Agardh, C. A. 1822–1823. Species algarum... Vol. 1, part 2. Lund. pp. 169–398 (1822), 399–531 (1823).
Agardh, C. A. 1824. Systema algarum. Lund. XXXVIII + 312 pp.
Agardh, C. A. 1828. Species algarum... Vol. 2, sect. 1. Greifswald. LXXVI + 189 pp.
Agardh, J. G. 1841. In historiam algarum symbolae. Linnaea 15: 1–50, 443–457.
Agardh, J. G. 1842. Algae maris Mediterranei et Adriatici. Paris. X + 164 pp.
Agardh, J. G. 1847. Nya alger från Mexico. Öfvers. Förh. Kongl. Svenska Vetensk.-Akad. 4: 5–17.
Agardh, J. G. 1848. Species genera et ordines algarum... Vol. 1. Lund. VIII +363 pp.
Agardh, J. G. 1851–1863. Species genera et ordines algarum... Vol. 2. Lund. XII + 1291 pp. [Part 1, pp. [I]–XII + [1]–336 + 337–351 (Addenda and Index) (1851); part 2, fasc. 1, pp. 337–504 (1851); part 2, fasc. 2, pp. 505–700 + 701–720 (Addenda and Index) (1852); part 3, fasc. 1, pp. 701–786 (1852); part 3, fasc. 2, pp. 787–1291 (1139–1158 omitted) (1863).
Agardh, J. G. 1854. Nya algformer. Öfvers. Förh. Kongl. Svenska Vetensk.-Akad. 11: 107–111.
Agardh, J. G. 1876. Species genera et ordines algarum... Vol. 3, part 1. Leipzig. VII + 724 pp.
Agardh, J. G. 1883. Till algernes systematik. Nya bidrag. (Tredje afdelningen.) Lunds Univ. Års-Skr., Afd. Math. och Naturvidensk. 19(2). 177 pp.
Agardh, J. G. 1898. Species genera et ordines algarum... Vol. 3, part 3. Lund. 239 pp.
Aghajanian, J. G. & Hommersand, M. H. 1980. Growth and differentiation of axial and lateral filaments in *Batrachospermum sirodotii* (Rhodophyta). J. Phycol. 16: 15–28.
Akiyama, M. & Nishigami, K. 1959. Ecological studies on algal flora in Lake Shinji and Nakano-umi. I. Distribution of macroscopic algae. Sci. Rep. (Nat. Sci.) Shimane Univ. no. 9, 69–75.
Alexopoulos, C. J. & Bold, H. C. 1967. Algae and fungi. New York: Macmillan. vii + 135 pp.
Allen, M. B. 1959. Studies with *Cyanidium caldarium*, an anomalously pigmented Chlorophyte. Arch. Mikrobiol. 32: 270–277.
Anderson, F. W. & Kelsey, F. D. 1891. Common and conspicuous algae of Montana. Bull. Torrey Bot. Club 18: 137–146.
Ando, K., Haraguchi, K. & Kobayasi, H. 1971. Diatoms from Senjogaike, an irrigation pond, Saitama Pref. Bulletin of the Chichibu Musem of Natural History, 1971 (16): 57–79. (in Japanese)
Anton, A., Sato, H., Kumano, S. & Mohamed, M. 1999. *Batrachospermum gombakense* (Batrachospermaceae, Rhodophyta), new to Sabah, Malaysia. Nature and Human Activities No. 4, 1–8.
Arai, S., Sano, O & Waken, I. 1996. A mesothermic freshwater red alga *Compsopogon oishii* found in Lake Akan, Hokkaido, northernmost Japan. Marimo Res. 5: 12–15. (in Japanese)
Arasaki, S. 1937. Preliminary note on the life-history of *Thorea ramosissima*. Jap. J. Bot. 101: 715–721. (in Japanese)
Arcangeli, G. 1882. Sopra alcune specie di *Batrachospermum*. Nuovo Giorn. Bot. Ital. 14: 155–167.
Archer, W. 1876. On the minute structure and mode of growth of *Ballia callitricha*, Ag. (sensu latiori). Trans. Linn. Soc. London, Bot. 1: 211–232.
Areschoug, J. E. 1842. Algarum minus rite cognitarum pugillus primus. Linnaea 16: 225–236.
Areschoug, J. E. 1843. Algarum (Phycearum) minus rite cognitarum pugillus secundus. Linnaea 17: 257–269.
Areschoug, J. E. 1847. Phycearum, quae in maribus Scandinaviae crescunt, enumeratio. Sectio prior Fucaceas continens. Nova Acta Regiae Soc. Sci. Upsal. 13: 223–382.
Areschoug, J. E. 1850. Phycearum, quae in maribus Scandinaviae crescunt, enumeratio. Sectio posterior Ulvaceas continens. Nova Acta Regiae Soc. Sci. Upsal. 14: 385–454.

Areschoug, J. E. 1875. Observationes phycologicae. Particula tertia. De algis nonnullis scandinavicis et de conjunctione Phaeozoosporarum Dictyosiphonis hippuroidis. Nova Acta Regiae Soc. Sci. Upsal., ser. 3, 10: 1–36.

Areschoug, J. E. 1876. De algis nonnulis maris Baltici et Bahusiensis. Bot. Not. 1876: 33–37.

Atkinson 1889. Preliminary note on the synonymy of Entothrix grande Wolle. Bot. Gaz. 14: 292.

Atkinson, G. F. 1890. Monograph of the Lemaneaceae of the United States. Ann. Bot. (London) 4: 177–229.

Atkinson, G. F. 1904. A new *Lemanea* from Newfoundland. Torreya 4: 26.

Atkinson, G. F. 1931. Notes on the genus *Lemanea* in North America. Bot. Gaz. 92: 225–242.

Bachmann, H. 1921. Beiträge zur Algen3flora des Süsswassers von Westgrönland. Mitt. Naturf. Ges. Luzern 8: 1–181.

Bailey, F. M. 1895. Contributions to the Queensland flora. Queensland freshwater algae. Queensland Dept. Agric., Bot. Bull. 11. 69 pp.

Bailey, F. M. 1913. Comprehensive catalogue of Queensland plants both indigenous and naturalised... Brisbane. 879 pp.

Balakrishnan, M. S. & Chaugule, B. B. 1975. "Elimination cells" in the Batrachospermaceae. Curr. Sci. 44: 436–437.

Balakrishnan, M. S. & Chaugule, B. B. 1980a. Cytology and life history of *Batrachospermum mahabaleshwarensis* Balakrishnan et Chaugule. Cryptog. Algol. 1: 83–97.

Balakrishnan, M. S. & Chaugule, B. B. 1980b. Morphology and life history of *Batrachospermum kylinii* n. sp. J. Indian Bot. Soc. 59: 291–300.

Balakrishnan, M. S. & Chaugule, B. B. 1980c. Indian Batrachospermaceae. In: Desikachary, T. V. & V. N. Raja Rao (eds), *Taxonomy of algae...* Univ. Madras. pp. 223–248. [completed book did not appear until 1986]

Basson, P. W. 1979. Marine algae of the Arabian Gulf coast of Saudi Arabia. Bot. Mar. 22: 47–82.

Battiato, A., Cormaci, M., Furnari, G. & Lanfranco, E. 1979. The occurrence of *Compsopogon coeruleus* (Balbis) Montagne (Rhodophyta, Bangiophycideae) in Malta and of *Compsopogon chalybeus* Kützing in an aquarium at Catania (Sicily). Rev. Algol. N. S. 14: 11–16.

Beanland, W. R. & Woelkerling, W. J. 1982. Studies on Australian mangrove algae. II. Composition and geographic distribution of communities in Spencer Gulf, South Australia. Proc. Roy. Soc. Victoria 94: 89–106.

Bennett, J. L. 1888. Plants of Rhode Island... Providence. 128 pp.

Berkeley, M. J. 1857. Introduction to cryptogamic botany. London. viii + 604 pp.

Bischoff, H. W. 1965. *Thorea riekei* sp. nov. and related species. J. Phycol. 1: 111–117.

Blum, J. L. 1993a. *Lemanea* (Rhodophyceae) in Wisconsin. Trans. Wisconsin Acad. Sci. Arts & Lett. 81: 7–11.

Blum, J. L. 1993b. *Lemanea* (Rhodophyta, Florideophyceae) in Indiana. Proc. Indiana Acad. Sci. 102: 1–7.

Blum, J. L. 1997 ("1994"). *Paralemanea* species (Rhodophyceae) in California. Proc. Indiana Acad. Sci. 103: 1–24.

Bold, H. C. 1973. *Morphology of plants*. 3rd ed. New York: Harper & Row. xv + 668 pp.

Bold, H. C. & Wynne, M. J. 1978. *Introduction to the algae...* Englewood Cliffs, New Jersey: Prentice-Hall. xiv + 706 pp.

Borge, O. 1928. [Zellpflanzen Ostafrikas, gesammelt auf der Akademischen Studienfahrt 1910. Von Bruno Schröder. Teil VIII.] Süsswasseralgen. Hedwigia 68: 93–114.

Børgesen, F. 1911. The algal vegetation of the lagoons in the Danish West Indies. In: Rosenvinge, L. K. (ed.), *Biologist abrader tilegnede Eug. Warming...* Kobenhavn. pp. 41–56.

Børgesen, F. 1919. The marine algae of the Danish West Indies. Part III. Rhodophyceae (5). Dansk Bot. Ark. 3: 305–368.

Børgesen, F. 1933. Some Indian Rhodophyceae especially from the shores of the Presidency of Bombay. III. Bull. Misc. Inform. 1933: 113–142.

Børgesen, F. 1937. Contributions to a South Indian marine algal flora – II. J. Indian Bot. Soc. 16: 311–357.

Børgesen, F. 1945. Some marine algae from Mauritius. III. Rhodophyceae. Part 4. Ceramiales. Biol. Meddel. Kongel. Danske Vidensk. Selsk. 19(10): 1–68.

Bornemann, F. 1887. Beiträge zur Kenntniss der Lemaneaceen. Inaugural-Dissertation... Universität zu Freiburg. Berlin. 49 pp.

Bornet, E. & Thuret, G. 1866. Note sur la fécondation des Floridées. Mém. Soc. Imp. Sci. Nat. Cherbourg 12: 257–262.
Bornet, E. 1892. Les algues de P.-K.-A. Schousboe. Mém. Soc. Sci. Nat. Cherbourg 28: 165–376.
Bory de Saint-Vincent, J. B. 1797. Mémoire sur le genres *Conferva*. et *Byssun*, du chevalien O. Linné. Burdeaux (Louis Cavazza) 1979, 58 pp.
Bory de Saint-Vincent, J. B. 1804. Voyage dans les quatre principales îles des mers d'Afrique...Paris. Vol. 2. 431 pp.
Bory de Saint-Vincent, J. B. 1808a. Mémoire sur un genre nouveau de la cryptogamie aquatique, nommé *Thorea*. Ann. Mus. Hist. Nat. 12: 126–135.
Bory de Saint-Vincent, J. B. 1808b. Mémoire sur le genre *Lemanea* de la famille des Conferves. Ann. Mus. Hist. Nat. 12: 177–190.
Bory de Saint-Vincent, J. B. 1808c. Mémoire sur le genre *Batrachosperma*, de la famille des Conferves. Ann. Mus. Hist. Nat. 12: 310–332.
Bory de Saint-Vincent, J. B. 1809. Mémoire sur le genre *Lemanea*. Mag. Entdeck. Naturk. 3: 274–281. [German translation of 1808b]
Bory de Saint-Vincent, J. B. 1823a. *Auduinella*. Dict. Class. Hist. Nat. 3: 340–341.
Bory de Saint-Vincent, J. B. 1823b. Confervées. Dict. Class. Hist. Nat. 4: 392–394
Bory de Saint-Vincent, J. B. 1827–1829. Cryptogamie. In: Duperrey, L. I., ed. *Voyage autour du monde... La Coquille... Paris*. [pp. 1–96 (1827), 97–200 (1828), 201–301 (1829).]
Bourrelly, P. 1966. Quelques algues d'eau douce du Canada. Int. Rev. Gesammten Hydrobiol. 51: 45–126.
Bourrelly, P. 1970. *Les algues d'eau douce...* Paris: Boubée Vol. 3. 512 pp.
Bourrelly, P. 1985. *Les algues d'eau douce...* Ed. 2. Paris: Boubée. 606 pp.
Brand, F. 1910. Über die Süsswasserformen von *Chantransia* (DC) Schmitz, einschliesslich *Pseudochantransia* Brand. Hedwigia 49: 107–118.
Britton, M. E. 1944. *A catalog of Illinois Algae*. Evanston: Northwestern University. viii + 177 pp.
Brown, D. L. & Weier, T. E. 1968. Chloroplast development and ultrastructure in the freshwater red alga *Batrachospermum*. J. Phycol. 4: 199–206.
Brown, D. L. & Weier, T. E. 1970. Ultrastructure of the freshwater alga *Batrachospermum*. I. Thin-section and freeze-etch analysis of juvenile and photosynthetic filament vegetative cells. Phycologia 9: 217–235.
Brown, D. L. 1969. Ultrastructure of the freshwater red alga *Batrachospermum*. Unpublished Ph.D. Thesis. University of California, Davis. 154 pp.
Cassie, V. 1971. Contributions of Victor Lindauer (1888–1964) to New Zealand phycology. J. Roy. Soc. New Zealand 1: 89–98.
Cassie, V. 1984. *Revised checklist of the freshwater algae of New Zealand (excluding diatoms and Charophytes). Part I. Cyanophyta, Rhodophyta and Chlorophyta*. New Zealand, National Water and Soil Conservation Organisation, Water & Soil Techn. Publ. 25. lxiv + 116 pp.
Chapman, D. J. 1974. Taxonomic status of *Cyanidium caldarium*. The Porphyridiales and Goniotrichales. Nova Hedwigia 25: 673–682.
Chapman, V. J. 1962. *The algae*. London: Macmillan. viii + 472 pp.
Chapman, V. J. 1963. The marine algae of Jamaica. Part 2. Phaeophyceae and Rhodophyceae. Bull. Inst. Jamaica, Sci. Ser. 12(2): 1–201.
Chapman, V. J. & D. J. 1973. *The algae*. 2nd ed. London: Macmillan. xiv + 497 pp.
Chapman, V. J., Thompson, R. H. & Segar, E. C. M. 1957. Check list of the fresh-water algae of New Zealand. Trans. Roy. Soc. New Zealand 84: 695–747.
Charlton, S. E. D. & Hickman, M. 1988. Epilithic algal nitrogen fixation, standing crops and productivity in five rivers flowing through the oilsands region of Alberta/Canada. Arch. Hydrobiol. Suppl. 79: 109–143.
Chemin, E. 1940. Les Batrachospermes dans la région de Paris. Bull. Soc. Bot. France. 7: 231–241.
Chihara, M. 1976. *Compsopogonopsis japonica*, a new species of fresh water red algae. J. Jap. Bot. 51: 289–294.
Chihara, M. (ed.) 1997. *Biology of algal diversity*. Uchida Rokakuho, Tokyo, Japan. 386 pp. (in Japanese)
Chihara, M. (ed.) 1999. *Diversity and evolution of algae*. Shokabo, Tokyo, Japan. 346 pp. (in Japanese)

Chihara, M. & Nakamura, Tn. 1975. Occurrence of *Compsopogonopsis* (Rhodophyta) in Japan. Bull. Jap. Soc. Phycol. 23: 150–152. (in Japanese)

Chihara, M. & Nakamura, T. 1980. *Compsopogon corticrassus*, a new species of fresh water red algae (Compsopogonaceae, Rhodophyta). J. Jap. Bot. 55: 136–144.

Choi, H.-C., Kraft, G. T. & Saunders, G. W. 2000. Nuclear small-subunit rDNA sequences from *Ballia* spp. (Rhodophyta): proposal of the Balliales ord. nov., Balliaceae fam. nov., *Ballia nana* sp. nov. and *Inkyuleea* gen. nov. (Ceramiales). Phycologia 39: 272-287.

Christensen, T. 1980. Algae. *A taxonomic survey*. Fasc. 1. Odense: Aio Tryk. 216 pp.

Chung, J. 1970. A taxonomic study on the freshwater algae from Youngnam Area. Taeku. pp. 115.

Cole, K. M. & Sheath, R. G. (eds) 1990. *Biology of the red algae*. Cambridge: Cambridge University Press. 517 pp.

Collins, F. S. 1906. New species, etc., issued in the Phycotheca Boreali-Americana. Rhodora 8: 104–113.

Collins, F. S. & Hervey, A. B. 1917. *The algae of Bermuda*. Proc. Amer. Acad. Arts 53: 1–195.

Collins, F. S., Holden, I. & Setchell, W. A. 1895–1919. *Phycotheca boreali-americana*. Malden, Massachusetts. Fasc. I–XLVI + A–E, nos. 1–2300 + I–CXXV. [Exsiccata with printed labels.]

Colt, L. C. Jr. 1974. Some algae of the Connecticut River, New England, U.S.A. Nova Hedwigia 45: 195–209.

Compère, P. 1991a. Flore pratique des algues d'eau douce de Belgique. 3. Rhodophytes. Meise: Jardin Botanique National de Belgique. 55 pp.

Compère, P. 1991b. Taxonomic and nomenclatural notes on some taxa of the genus *Batrachospermum* (Rhodophyceae). Belg. J. Bot. 124: 21–26.

Cooke, M. C. 1882–1884. *British fresh-water algae*...London. viii + 329 pp. [pp. 1–110 (1882), 111–198 (1883), 199–329 (1884).]

Couté, A. & Sarthou, C. 1990. Révision des espèces d'eau douce du genre *Ballia* (Rhodophytes, Céramiales). Cryptog. Algol. 11: 265–279.

Cramer, C. 1891. Über *Caloglossa leprieurii* (Mont. Harv.) J. G. Agardh. Synon.: *Delesseria leprieurii.* Mont. *Hypoglossum leprieurii* (Mont.) Kg. *Delesseria* (subgen. *Caloglossa*) leprieurii (Mont.) Harvey. In: *Festschrift...von Nägeli ... und... von Kolliker...* Zürich. pp. 1–18.

Cribb, A. B. 1965. *Anfractutofilum umbracolens* gen. et sp. nov., a freshwater red alga from Queensland. Proc. Royal Soc. Queensland 76: 93–95.

Cribb, A. B. 1987. Some freshwater algae from the Jardine River area. Queensland Naturalist 28: 69–71.

Daily, W. A. 1943. First reports for the algae *Borzia, Aulosira* and *Asterocytis* in Indiana. Butler Univ. Bot. Stud. 6: 84–86.

Dallwitz, M. J. 1980. A general system for coding taxonomic descriptions. Taxon 29: 41–46.

Das, C. R. 1963. The Compsopogonales in India (a systematic account of the Indian representatives of the order). Proc. Natl. Inst. Sci. India, Pt. B, Biol. Sci. 13: 239–243.

Davey, A. & Woekerling, W. J. 1980. Studies on Australian mangrove algae. I. Victorian communities: composition and geographic distribution. Proc. Roy. Soc. Victoria 91: 53–66.

Davis, B. M. 1896. The fertilization of *Batrachospermum*. Ann. Bot. (London) 10: 49–76.

Dawson, E. Y. 1966. *Marine botany. An introduction*. New York: Holt, Rinehart and Winston. xii + 371 pp.

De Candolle, A. P. 1801. Extrait d'un rapport sur les Conferves, fait à la Société philomathique. Bull. Sci. Soc. Philom. Paris 3: 17–21.

De Candolle, A. P. 1802. Rapport sur les Conferves, fait à la Société philomathique. J. Phys. Chim. Hist. Nat. Arts 54: 421–441.

De Luca, P., Taddei, R. & Varano, L. 1978. "*Cyanidioschyzon merolae*": a new alga of thermal acidic environments. Webbia 33: 37–44.

De Toni, G. B. 1897. Sylloge algarum... Vol. IV. Florideae. Sectio I. Padova. pp. [I]–XX + [I]–LXI + [1]–386 + 387–388 (Index).

De Toni, G. B. 1900. Sylloge Algarum... Vol. IV. Florideae. Sectio II. Padova. pp. 387–774 + 775–776 (Index).

De Toni, G. B. 1924. Sylloge Algarum... Vol. VI. Florideae. Sectio V. Additamenta. Padova. XI + 767 pp.

Descy, J.-P. & Empain, A. 1974. *Thorea ramosissima* Bory (Rhodophyceae, Nemalionales) dans le bassin mosan belge. Bull. Soc. Roy. Bot. Belgique 107: 23–26.

Desvaux, N. A. 1808. Extrait des Annales du Muséum d'Histoire naturelle, sur les genres *Thorea* et *Lemanea*, de M. Bory de Saint-Vincent. J. Bot. (Desvaux) 1: 121–125.

Desvaux, N. A. 1818. Observations sur les plantes des environs d'Angers... Angers. 188 pp.

Dillard, G. E. 1966. The seasonal periodicity of *Batrachospermum macrosporum* Mont. and *Audouinella violacea* (Kütz.) Ham. in Turkey Creek, Moore County, North Carolina. J. Elisha Mitchell Sci. Soc. 82: 204–207.

Dillard, G. E. 1967. The fresh-water algae of South Carolina I. Previous work and recent additions. J. Elisha Mitchell Sci. Soc. 83: 128–131.

Dillenius, J. J. 1741. *Historia muscorum*... Oxford. xvi + 576 pp., LXXV pls.

Dillwyn, L. W. 1802. *British Confervae*. Fasc. 1. London. Pls. 1–12.

Dixon, P. S. 1958. The development of carpogonial branches and lateral branches of unlimited growth in *Batrachospermum vagum*. Bot. Not. 111: 645–649.

Dixon, P. S. 1964. On the concept of the "carpogonial branch" in the Florideae. Proc. Fourth Int. Seaweed Symp. [Biarritz, 1961], pp. 71–77.

Dixon, P. S. 1970. The Rhodophyta: some aspects of their biology. II. Oceanogr. Mar. Biol. Ann. Rev. 8: 307–352.

Dixon, P. S. 1973. *Biology of the Rhodophyta*. Edinburgh: Oliver & Boyd. xiii + 285 pp.

Dixon, P. S. & Irvine, L. M. 1977. *Seaweeds of the British Isles. Vol. I. Rhodophyta. Part 1. Introduction, Nemaliales, Gigartinales*. London: British Museum (Natural History). xi + 252 pp.

D'Lacoste, L. G. & Ganesan, E. K. 1972. A new freshwater species of *Rhodochorton* (Rhodophyta, Nemaliales) from Venezuela. Phycologia 11: 233–238.

D'Lacoste, L. G. & Ganesan, E. K. 1987. Notes on Venezuelan freshwater red algae– I. Nova Hedwigia 45: 263–281.

Drew, K. M. & Ross, R. 1965. Some generic names in the Bangiophycidae. Taxon 14: 93–99.

Drew, K. M. 1928. A revision of the genera *Chantransia*, *Rhodochorton*, and *Acrochaetium* with descriptions of the marine species of *Rhodochorton* (Näg.) gen. emend. on the Pacific coast of North America. Univ. Calif. Publ. Bot. 14: 139–224.

Drew, K. M. 1935. The life-history of *Rhodocorton violaceum* (Kütz.) comb. nov. (*Chantransia violacea* Kütz.). Ann. Bot. (London) 49: 439–450.

Drew, K. M. 1936. *Rhodochorton violaceum* (Kütz.) Drew and *Chantransia boweri* Murray et Barton. Ann. Bot. (London) 50: 419–421.

Drew, K. M. 1946. Anatomical observations on a new species of *Batrachospermum*. Ann. Bot. (London), ser. 2, 10: 339–352.

Drouet, F. 1933. Algal vegetation of the large Ozark springs. Trans. Amer. Microscop. Soc. 52: 83–100.

Duby, J. E. 1830. Aug. Pyrami de Candolle Botanicon gallicum... Pars secunda plantas cellulares continens. Paris. pp. 545–1068 + i–lviii.

Ducker, S. C. 1990. History of Australian marine phycology. In: Clayton, M. N. & King, R. J. (eds), *Biology of marine plants*. Melbourne: Longman Cheshire. pp. 415–430.

Dumortier, B. C. 1822. Commentationes botanicae. Observations botaniques... Tournay. 116 pp.

Duthie, H. C. & Ostrofsky, M. L. 1975. Freshwater algae from western Labrador. III. Cyanophyta and Rhodophyta. Nova Hedwigia 26: 555–559.

Duthie, H. C. & Socha, R. 1976. A checklist of the freshwater algae of Ontario. exclusive of the Great Lakes. Naturaliste Canad. 103: 83–109.

El-Gamal, Ahmed D. & Salah El-Din, Rawheya A. 1999. New species of the genus *Compsopogon* (*C. helwanii*) from Egypt. Phycos 38: 37–42.

Emoto, Y. & Hirose, H. 1942. Studien über die Thermalflora von Japan. XXI. Thermal Bakterien und Algen aus den heissen Quellen von Shiobara in Totigi Prefektur (2). Onsen Kagaku 2: 79–85. (in Japanese)

Engler, A. 1892. Syllabus der Vorlesungen über spezielle und medizinische-pharmaceutische Botanik. Berlin. XXIII + 184 pp.

Engler, H. G. A. & Prantl, K. A. E. 1891–1897. Die natürlich Pflanzenfamilien. 1 (1, 2). Englmann, Leipzig. (Schmitz, F. & Hauptfleisch, P. : Rhodophyceae. Wille, N. : Conjugatae und Chlorophyceae).

Entwisle, T. J. 1989a. Macroalgae in the Yarra River basin: flora and distribution. Proc. Roy. Soc. Victoria 101: 1–76.

Entwisle, T. J. 1989b. *Psilosiphon scoparium* gen. et sp. nov. (Lemaneaceae), a new red alga from south-eastern Australian streams. Phycologia 28: 469–475.

Entwisle, T. J. 1990. The lean legacy of freshwater phycology in Victoria. In: Short, P. S. (ed.), *History of systematic botany in Australia*. South Yarra, Victoria: Australian Systematic Botany Society. 239–246, 1 fig., I table.

Entwisle, T. J. 1992. The setaceous species of *Batrachospermum* (Rhodophyta): a re-evaluation of *B. atrum* (Hudson) Harvey and *B. puiggarianum* Grunow including the description of *B. diatyches* sp. nov. from Tasmania, Australia. Muelleria 7: 425–445.

Entwisle, T. J. 1993. The discovery of batrachospermalean taxa (Rhodophyta) in Australia and New Zealand. Muelleria 8: 5–16.

Entwisle, T. J. 1995. *Batrachospermum antipodites*, sp. nov. (Batrachospermaceae): a widespread freshwater red alga in eastern Australia and New Zealand. Muelleria 8: 291–298.

Entwisle, T. J. 1997. Following the trail of the Little Water Besom: discovery of *Psilosiphon scoparium* in New Zealand. Australian Society of Phycology and Aquatic Botany Newsletter 16: 11–12.

Entwisle, T. J. 1998. Batrachospermaceae (Rhodophyta) in France: 200 years of study. Cryptogamie: Algologie 19: 149–159.

Entwisle, T. J. & Foard, H. J. 1997. *Batrachospermum* (Batrachospermales, Rhodophyta) in Australia and New Zealand: new taxa and emended circumscription in sections Aristata, Batrachospermum, Turfosa and Virescentia. Austral. Syst. Bot. 10: 331–380.

Entwisle, T. J., & Foard, H. J. 1998. *Batrachospermum latericium* sp. nov. (Batrachospermales, Rhodophyta) from Tasmania, Australia, with new observations on *B. atrum* and a discussion of their relationships. Muelleria 11: 27–40.

Entwisle, T. J., & Foard, H. J. 1999a. Freshwater Rhodophyta in Australia: *Ptilothamnion richardsii* (Ceramiales) and *Thorea conturba* sp. nov. (Batrachospermales). Phycologia 38: 47–53.

Entwisle, T. J., & Foard, H. J. 1999b. *Sirodotia* (Batrachospermales, Rhodophyta) in Australia and New Zealand. Austral. Syst. Bot. 12: 605–613.

Entwisle, T. J., & Foard, H. J. 1999c. *Batrachospermum* (Batrachospermales, Rhodophyta) in Australia and New Zealand: New taxa and records in Sections *Contorta* and *Hybrida*. Austral. Syst. Bot. 12: 615–633.

Entwisle, T. J. & Kraft, G. T. 1984. Survey of freshwater red algae (Rhodophyta) of south-eastern Australia. Austral. J. Mar. Freshwater Res. 35: 213–259.

Entwisle, T. J., & Necchi, O. Jr. 1992. Phylogenetic systematics of the freshwater red algal order Batrachospermales. Jap. J. Phycol. 40: 1–12.

Entwisle, T. J., Vis, M. L. & Foard, H. J. 2000. Biogeography of *Psilosiphon* (Batrachospermales, Rhodophyta) im Australia and New Zealand. Cryptogamie, Algol. 21: 133–148.

Falkenberg, P. 1901. Die Rhodomelaceen des Golfes von Neapel und der angrenzenden Meeres-Abschnitte. Fauna und Flora des Golfes von Neapel, Monogr. 26. XVI + 754 pp.

Fan, K. C. 1952. The structure, methods of branching and tetrasporangia formation of *Caloglossa*. Taiwan Fish. Res. Inst., Lab. Hydrobiol. Rep. no. 4, 16 pp.

Farr, E. R., Leussink, J. A. & Stafleu, F. A. 1979. Index nominum genericorum (plantarum). Vol. 1. Regnum Veg. 100. pp. I–XXVI + 1–630.

Feldmann, J. 1953. L'évolution des organes femelles chez les Floridées. Proc. First Int. Seaweed Symp. [Edinburgh, 1952], pp. 11–12.

Feldmann, J. 1962. The Rhodophyta order Acrochaetiales and its classification. Proc. Ninth Pacific Sci. Congr. [Bangkok, 1957] 4: 219–221.

Fischer, H. 1984. Turgor regulation in *Caloglossa leprieurii* (Montagne) J. Agardh (Delesseriaceae: Rhodophyta). J. Exp. Mar. Biol. Ecol. 81: 235–239.

Flint, L. H. 1947. Studies of freshwater red algae. Amer. J. Bot. 34: 125–131.

Flint, L. H. 1948. Studies of fresh-water red algae. Amer. J. Bot. 35: 428–433.

Flint, L. H. 1949. Studies of freshwater red algae. Amer. J. Bot. 36: 549–552.

Flint, L. H. 1950. Studies of freshwater red algae. Amer. J. Bot. 37: 754–757.

Flint, L. H. 1951. Some winter red algae of Louisiana. Proc. Louisiana Acad. Sci. 14: 34–36.

Flint, L. H. 1953a. Two new species of *Batrachospermum*. Proc. Louisiana Acad. Sci. 16: 10–15.

Flint, L. H. 1953b. *Kyliniella* in America. Phytomorphology 3: 76–80.

Flint, L. H. 1954a. *Sirodotia* in Louisiana. Proc. Louisiana Acad. Sci. 17: 59–65.
Flint, L. H. 1954b. *Nemalionopsis* in America. Phytomorphology 4: 76–79.
Flint, L. H. 1955. *Hildenbrandia* in America. Phytomorphology 5: 185–189.
Flint, L. H. 1957a. Notes on algae of Quebec. I. Mont Tremblant Provincial Park. Naturaliste Canad. 84: 157–160.
Flint, L. H. 1957b. Notes on algae of Quebec. II. Laurentide Park. Naturaliste Canad. 84: 179–181.
Flint, L. H. 1970. *Freshwater red algae of North America. An introduction.* New York: Vantage Press. 110 pp.
Forest, H. S. 1954. *Handbook of algae with special reference to Tennessee and the southeastern United States.* Knoxville: Univ. Tennessee Press. 467 pp.
Forti, A. 1907. Myxophyceae. In: De Toni, G. B. (ed.), *Sylloge algarum... Vol. V.* Padova. 761 pp.
Fralick, R. A. & Mathieson, A. C. 1975. Physiological ecology of four *Polysiphonia* species (Rhodophyta, Ceramiales). Mar. Biol. (Berl.) 29: 29–36.
Friedrich, G. 1967. *Compsopogon hookeri* Montagne (Rhodophyceae, Bangioideae), neue für Deutschland. Nova Hedwigia 12;399–403.
Fries, E. M. 1825. *Systema orbis vegetabilis... Pars I. Plantae Homonemeae.* Lund. VIII + 374 pp.
Fritsch, F. E. 1945. *The structure and reproduction of the algae.* Vol. 2. Cambridge: Cambridge Univ. Press. xiv + 939 pp.
Galdieri, A. 1899. Su di un'alga che cresce intorno alle fumarole della Solfatera. Rendiconti Reale Accad. Sci. Fis. e Mat. [Napoli], ser. 3, 5: 160–164.
Gantt, E. & Conti, S. F. 1965. Ultrastructure of *Porphyridium cruentum*. J. Cell Biol. 26: 365–375.
Gantt, E., Edwards, M. R. & Conti, S. F. 1968. Ultrastructure of *Porphyridium aerugineum*, a blue-green colored rhodophytan. J. Phycol. 4: 65–71.
Garbary, D. J. 1987. The Acrochaetiaceae (Rhodophyta): an annotated bibliography. Biblioth. Phycol. 77: 1–267.
Garbary, D. J. & Gabrielson, P. W. 1990. Taxonomy and evolution. In: Cole, K. M. & Sheath, R. G. (eds), *Biology of the red algae.* Cambridge: Cambridge University Press. pp. 477–498.
Garbary, D. J., Hansen, G. I. & Scagel, R. F. 1980. A revised classification of the Bangiophyceae (Rhodophyta). Nova Hedwigia 33: 145–166.
Geesink, R. 1973. Experimental investigation on marine and freshwater *Bangia* (Rhodophyta) in the Netherlands. J. Exp. Mar. Biol. Ecol. 11: 239–247.
Geitler, L. & Ruttner, F. 1936. Die Cyanophyceen der Deutschen Limnologischen Sunda-Expedition, ihre Morphologie, Systematik und Ökologie. Zweiter Teil. Arch. Hydrobiol. Suppl. 14: 371–483.
Geitler, L. 1933. Diagnosen neuer Blaualgen von den Sunda-Inseln. Arch. Hydrobiol. Suppl. 12: 622–634.
Geitler, L. 1944. Ein neues einheimisches *Batrachospermum* sowie Beobachtungen an anderen einheimischen *Batrachospermum*-Arten. Wiener Bot. Z. 93: 127–137.
Gobi, Ch. 1879. Forschungen im Finnischen Meerbunsen. Arb. St. Petersb. Nat., Ges. 10: 83–92.
Goebel, K. 1897. Morphologische und biologische Bemerkungen. 6. Über einige Süsswasserflorideen aus Britisch-Guyana. Flora 83: 436–444.
Goebel, K. 1898. Morphologische und biologische Bemerkungen. 8. Eine Süsswasserfloridee aus Ostafrika. Flora 85: 65–68.
Goff, L. J. & Coleman, A. W. 1990. DNA: microspectrofluorometric studies. In: Cole, K. M. & Sheath, R. G. (eds), *Biology of the red algae.* Cambridge: Cambridge University Press. pp. 43–71.
Greuter, W., Burmitt, R. K., Farr, E. Killian, N, Kirk, P. M. & Silva, P. C. (eds) 1993. *Names in Current Use for Extant Plant Genera-3. The International Association for Plant Taxonomy.* Koeltz Scientific Books, Königstein, Germany. xxviii+1464 pp.
Greuter, W. *et al.* (eds) 1994. *International Code of Botanical Nomenclature* (Tokyo Code).... Regnum Veg. 131. Königstein: Koeltz Scientific Books. viii + 389 pp.
Greville, R. K. 1823. *Scottish cryptogamic flora...* vol. 2. Edinburgh. pls. 61–90 (with text).
Guiry, M. D. 1978. The importance of sporangia in the classification of the Florideophyceae. In: Irvine, D. E. G. & Price, J. H. (eds), *Modern approaches to the taxonomy of red and brown algae.* London: Academic Press. pp. 111–144.
Guiry, M. D. 1990. The life history of *Liagora harveyana* (Nemaliales, Rhodophyta) from south-eastern Australia. Brit. Phycol. J. 25: 353–362.

Habeeb, H. & Drouet, F. 1948. A list of freshwater algae from New Brunswick. Rhodora 50: 67–71.
Hamel, G. 1925a. Floridées de France. III. Rev. Algol. 2: 39–67. Ibid. IV. Rev. Algol. 2: 280–309.
Hamel, G. 1925b. Sur la synonymie des Chantransiées. C. R. Congr. Soc. Savantes 1925 [Paris]. pp. 239–241.
Hansgirg, A. 1885. Ein Beitrag zur Kenntniss von der Verbreitung der Chromatophoren und Zellkerne bei den Schizophyceen (Phycochromaceen). Ber. Deutsch. Bot. Ges. 3: 14–22.
Hara, Y. 1993. *Porphyridium purpureum* (Bory) Drew et Ross. In: Hori, T. (ed.), *An illustrated atlas of the life history of algae. Vol. 2. Brown and red algae.* Tokyo: Uchida Rokakuho. pp. 182–183. (in Japanese)
Hanyuda, T., Kumano, S., Arai, S., Suzawa, Y., Iima, M.& Ueda, K. 2001. The molecular phylogenetic analyses of the Thoreaceae. Jpn. J. Phycol. (Sorui) 49: 114 (in Japanese)
Hara, Y. & Chihara, M. 1974. Comparative studies on the chloroplast ultrastructure in the Rhodophyta with special reference to their taxonomic significance. Sci. Rep. Tokyo Kyoiku Daigaku, Sect. B, 15: 209–235.
Hara, Y., Ikawa, T. & Chihara, M. 1989. A taxonomic study of *Porphyridium* and related algae (Porphyridiales, Rhodophyta). Abstr. Korea-Japan Phycol. Symp: 48. (in Japanese)
Hara, Y., Ono, J. & Chihara, M. 1988. Systematic position of the unicellular rhodophycean genus *Porphyridium*. Jap. J. Phycol. 36: 108. (in Japanese).
Haraguchi, K. & Kobayasi, H. 1969. On the differentiation of the thallus of *Batrachospermum moniliforme* Roth directly from the prostrate thallus of the chantransia-stage without forming erect parts. Bull. Jap. Soc. Phycol. 17: 61–65. (in Japanese)
Hardy, A. D. 1906. The fresh-water algae of Victoria. Part III. Victoria Naturalist 23: 18–22, 33–42.
Harvey, W. H. 1836. *Algae*. In: Mackay, J. T., Flora hibernica... Dublin. Parts 2 & 3. pp. 157–254.
Harvey, W. H. 1840. Description of *Ballia*, a new genus of algae. J. Bot. (Hooker) 2: 190–193.
Harvey, W. H. 1841. *A manual of the British Algae...* London. lvii + 229 pp.
Harvey, W. H. 1848. *Phycologia britannica...* [fasc. 36]. London. Pls. CCXI–CCXVI.
Harvey, W. H. 1851. *Nereis boreali-americana... Part I. Melanospermeae.* Smithsonian Contr. Knowl. 3(4). 150 pp.
Harvey, W. H. 1853. *Nereis boreali-americana... Part II. Rhodospermeae.* Smithsonian Contr. Knowl. 5(5). 258 pp.
Harvey, W. H. 1858. *Nereis borealis-americana... Part III. Chlorospermeae.* Smithsonian Contr. Knowl. 10(2). 140 pp.
Harvey, W. H. 1863. *Phycologia australica... Vol. 5.* London. Pls. CCXLI–CCC (with text).
Hassall, A. H. 1845. *A history of the British freshwater algae...* London. Vol. I. viii + 462 pp. Vol. II. CIII pls., 24 pp. explanation.
Hedgcock, G. G. & Hunter, A. A. 1900. Notes on *Thorea*. Bot. Gaz. 28: 425–429.
Herndon, W. R. 1964. *Boldia*: a new rhodophycean genus. Amer. J. Bot. 51: 575–581.
Hinton, G. C. F. & Maulood, B. K. 1980. Freshwater red algae [:] a new addition to the Iraqi flora. Nova Hedwigia 33: 487–497.
Hirose, H. 1950. Studies on a thermal alga, *Cyanidium caldarium*. Bot. Mag. Tokyo 63: 745–746.
Hirose, H. 1958. Rearrangement of the systematic position of a thermal alga, *Cyanidium caldarium*. Bot. Mag. Tokyo 71: 347–352.
Hirose, H. 1959. *General Phycology*. Tokyo: Uchida Rokakuho. 506 pp. (in Japanese)
Hirose, H., Kumano, S. & Madono, K. 1969. Spectroscopic studies on the phycoerythrins from cyanophyceaen and rhodophycean algae with special reference to their phylogenetical relations. Bot. Mag. Tokyo 82: 197–203.
Hirose, H. & Kumano, S. 1966. Spectroscopic studies on the phycoerythrins from rhodophycean algae with special reference to their phylogenetical relations. Bot. Mag. Tokyo 79: 105–113.
Hirose, H. & Seto, R. 1959. Some new knowledge of the chantransia stage of *Batrachospermum moniliforme* Roth. Bull. Jap. Soc. Phycol. 7: 52–58. (in Japanese)
Hirose, H. & Yamagishi, T. (eds) 1977. *Illustrations of the Japanese freshwater algae*. Tokyo: Uchida Rokakuho. 933 pp. (in Japanese)
Hirsch, A. & Palmer, C. M. 1958. Some algae from the Ohio River drainage basin. Ohio J. Sci. 58: 375–382.

Hollenberg, G. J. 1963. A new species of *Malaconema* (Rhodophyta) from the Marshall Islands. Phycologia 2: 169–172.

Holmgren, P. K., Holmgren, N. H. & Barnett, L. C. (eds) 1990. Index herbariorum. Part I: The herbaria of the world. 8th. ed. Regnum Veg. 120. In: *The Herbaria of the World*, 8th edition. New York: New York Botanical Garden. x + 693 pp.

Hooker, J. D. & Harvey, W. H. 1845a. Algae Antarctica, being characters and descriptions of the hitherto unpublished species of algae, discovered in Lord Auckland's Group, Cambell's Island, Kerguelen's Land, Falkland Islands, Cape Horn and other southern circumpolar regions, during the voyage of H. M. Discovery ships "Erebus" and "Terror". Lond. J. Bot. 4: 249–276, 293–298.

Hooker, J. D. & Harvey, W. H. 1845b. Algae Novae-Zelandiae; being a catalogue of all the species of algae yet recorded as inhabiting the shores of New Zealand, with charaters and briet descriptions of the new species discovered during the voyage of H.M. Discovery ships "Erebus" and "Terror", and of others communicated to Sir W. Hooker by Dr. Sinclair, the Rev. Colenso, and M. Raoul. Lond. J. Bot. 4: 521–551.

Hori, T. (ed.) 1993. *An illustrated atlas of the life history of algae. Vol. 2. Brown and red algae*. Tokyo: Uchida Rokakuho. xix + 345 + 51 pp. (in Japanese)

Horn af Rantzien, H. 1950. *Tristicha, Najas*, and *Sirodotia* in Liberia. Meddeland. Göteborgs Bot. Trädg. 18: 185–197.

Hornemann, J. W. 1818. *Icones plantarum ... Florae danicae*. Vol. 9, fasc. 27. Copenhagen. pp. [1]–11, pls. MDLXI–MDCXX.

Howard, R. V. & Parker, B. C. 1979. *Nemalionopsis shawii* forma *caroliniana* (forma nov.) (Rhodophyta: Nemalionales) from the southeastern United States. Phycologia 18: 330–337.

Howard, R. V., & Parker, B. C. 1980. Revision of *Boldia erythrosiphon* Herndon (Rhodophyta, Bangiales). Amer. J. Bot. 67: 413–422.

Howe, M. A. 1902. *Caloglossa leprieurii* in mountain streams. Torreya 2: 149–152.

Hua, D. & Shi, Z.-X. 1996. A new species of *Batrachospermum* from Jiangsu, China. Acta Phytotax. Sini. 34: 324–326. (in Chinese)

Hudson, H. 1762. *Flora anglica...* London. viii [xvi] + 506 [+ 22] pp.

Hudson, H. 1778. *Flora anglica...* Editio altera. London. xxxviii + 690 pp.

Hughes, E. O. 1949. Fresh-water algae of the Maritime Provinces. Proc. Nova Scotian Inst. Sci. 22: 1–63.

Hurdelbrink, L. & Schwantes, H. O. 1972. Sur le cycle de développement de *Batrachospermum*. Mém. Soc. Bot. France 1972: 269–274.

Hylander, C. J. 1928. *The algae of Connecticut*. Connecticut State Geol. & Nat. Hist. Surv. Bull. no. 42. 245 pp.

Hymes, B. J. & Cole, K. M. 1984a ('1983'). The cytology of *Audouinella hermannii* (Rhodophyta, Florideophyceae). I. Vegetative and hair cells. Canad. J. Bot. 61: 3366–3376.

Hymes, B. J. & Cole, K. M. 1984b ('1983'). The cytology of *Audouinella hermannii* (Rhodophyta, Florideophyceae). II. Monosporogenesis. Canad. J. Bot. 61: 3377–3385.

Ichimura, S. & Kobayasi, H. 1964. Primary production in Tokyo Bay. Information Bulletin on Planktology in Japan, 1964 (11): 6–8. (in Japanese).

Index Herbaria ver. 8, 1990. See Holmgren, P. K., Holmgren, N. H. & Barnett, L. C. (eds) 1990.

Iriki, Y. & Tsuchiya, Y. 1963. Constituents of the cell wall of *Batrachospermum virgatum* from Shiga Heights 1. Bull. Inst. Biol. Shiga Heights, Shinshu Univ. 2: 1–8. (in Japanese)

Israelson [Israelsson], G. 1938. Svenska batrachospermaceer i J. G. Agardhs algherbarium. Bot. Not. 1938: 34–36.

Israelson, G. 1942. The freshwater Florideae of Sweden. Studies on their taxonomy, ecology, and distribution. Symb. Bot. Upsal. 6(1). 134 pp.

Jaasund, E. 1965. Aspects of the marine algal vegetation of North Norway. Bot. Gothob. 4. 174 pp.

Jao, C. C. 1940. Studies on the freshwater algae of China. VIII. A preliminary account of the Chinese freshwater Rhodophyceae. Sinensia 12: 245–290.

Jao, C. C. 1941. Studies on the freshwater algae of China. IV. Subaerial and aquatic algae from NanYoh, Huna Part II. Sinensia 11: 241–361.

John, D. M., Price, J. H., Maggs, C. A. & Lawson, G. W. 1979. Seaweeds of the western coast of tropical Africa and adjacent islands: a critical assessment. III. Rhodophyta (Bangiophyceae). Bull. Brit. Mus. (Nat. Hist.), Bot. 7: 69–82.

Johnstone, I. M., Mukiu, J., Nagari, T., Pokihian, M. & Rau, M. 1980. *Batrachospermum*, first freshwater red algal record for New Guinea. Sci. in New Guinea 7: 1–5.

Jose, L. & Patel, R. J. 1990. *Caloglossa ogasawaraensis* (Rhodophyta, Delesseriaceae), a fresh water Rhodophyceae new to India. Cryptog. Algol. 11: 225–228.

Kaczmarczyk, D. & Sheath, R. G. 1991. The effect of light regime on the photosynthetic apparatus of the freshwater red alga *Batrachospermum boryanum*. Cryptog. Algol. 12: 249–263.

Kaczmarczyk, D. & Sheath, R. G. 1992. Pigment content and carbon to nitrogen ratios of freshwater red algae growing at different light levels. Jap. J. Phycol. 40: 279–282.

Kaczmarczyk, D., Sheath, R. G., & Cole, K. M. 1993 ("1992"). Distribution and systematics of the freshwater genus *Tuomeya* (Rhodophyta, Batrachospermaceae). J. Phycol. 28: 850–855.

Kamiya, M., Tanaka, J. & Hara, Y. 1995. A morphological study and hybridization analysis of *Caloglossa leprieurii* (Delesseriaceae, Rhodophyta) from Japan, Singapore and Australia. Phycol, Res. 43: 81–91.

Kamiya, M., Tanaka, J. & Hara, Y. 1997. Comparative morphology, crossability and taxonomy within *Caloglossa continua* (Delesseriaceae, Rhodophyta) complex from the western Pacific. J. Phycol. 33: 97–105.

Kamiya, M., Tanaka, J., King, R. J., West, J. A., Zuccarello, G. C. & Kawai, H. 1999. Reproductive and genetic distribution between broad and narrow entities of *Caloglossa continua* (Delesseriaceae, Rhodophyta). Phycologia 38: 356–367.

Kamiya, M., West, J. A. & Hara, Y. 1994. Reproductive structures of *Bostrychia simpliciuscula* (Ceramiales, Rhodophyceae) in the field and in culture. Jpn. J. Phycol. 42: 165–174.

Kamiya, M., West, J. A., King, R. J., Zuccarello, G. C., Tanaka, J. & Hara, Y. 1998. Evolutionary divergence in the red algae *Caloglossa leprieurii* and *C. apomeiotica*. J. Phycol. 34: 361–370.

Kamiya, M., West, J. A., Zuccarello, G. C. & Kawai, H. 2000. *Caloglossa intermedia*, sp, nov. (Rhodophyta) from the western Atlantic coast: Molecular and morphological analyses with special reference to *C. lprieurii* and *C. monosticha*. J. Phycol. 36: 411–420.

Kamiya, T. 1953. Ecological observation of winter freshwater algae in a waterway. Hokuriku J. Bot. 2: 37–39. (in Japanese)

Kamiya, T. 1955. The new locality of *Compsopogon oishii* Okamura and a consideration of its distribution in Japan. Hokuriku J. Bot. 4: 18–20. (in Japanese)

Kapraun, D. F. 1979. The genus *Polysiphonia* (Ceramiales, Rhodophyta) in the vicinity of Port Aransas, Texas. Contr. Mar. Sci. 22: 105–120.

Kapraun, D. F. 1980. *An illustrated guide to the benthic marine algae of coastal North America 1. Rhodophyta*. Univ. North Carolina Press, Chapel Hill, 206 pp.

Karsten, G. 1891. *Delesseria* (*Caloglossa* Harv.) *amboinensis*. Eine neue Süsswasser-Floridee. Bot. Zeit. 41: 265–271.

Karsten, U., West, J. A., Mostaert, A. S., King, R. J., Barrow, K. D. & Kirst, G. O. 1992. Mannitol in the red algal genus *Caloglossa* (Harvey) J. Agardh. J. Plant Physiol. 140: 292–297.

Kawasaki, Y. 1935. On a freshwater red alga, *Audouinella* sp. Kishu Do-Shokubutsu 2: 21–23. (in Japanese)

Khan, M. 1970. On two fresh water red algae from Dehradun. Hydrobiologia 35: 249–253.

Khan, M. 1973. On edible *Lemanea* Bory de St Vincent – a fresh water red alga from India. Hydrobiologia 43: 171–175.

Khan, M. 1979 ("1978"). On *Thorea* Bory (Nemalionales, Rhodophyta), a freshwater red alga new to India. Phykos 17: 55–58.

Kim, C. S & Chang, Y. K. 1958. A study on the freshwater red alga of Korea (Preliminary report). J. Biol. Sci. 3: 14–16. (in Korean).

King, R. J. 1981. Mangroves and saltmarsh plants. In: Clayton, M. N., & King, R. J. (eds), *Marine botany: an Australian perspective*. Melbourne: Longman Cheshire. pp. 308–328.

King, R. J. & Puttock, C. F. 1989. Morphology and taxonomy of *Bostrychia* and *Stictosiphonia* (Rhodomelaceae/Rhodophyta) Austral. Syst. Bot. 2: 1–73.

King, R. J. & Puttock, C. F. 1994a. Morphology and taxonomy of *Caloglossa* (Delesseriaceae, Rhodophyta). Austral. Syst. Bot. 7: 89–124.

King, R. J. & Puttock, C. F. 1994b. Macroalgae associated with mangroves in Australia: Rhodophyta. Bot. Mar. 37: 181–191.

King, R. J. & Wheeler, M. D. 1985. Composition and geographic distribution of mangrove macroalgal communities in New South Wales. Proc. Linn. Soc. New South Wales 108: 97–117.

Klas, Z. 1935. Eine neue *Thorea* aus Jugoslawien, *Thorea brodensis* Klas sp. n. Hedwigia 75: 273–284.

Kobayasi, H. 1962. Diatoms from River Arakawa (1). Bulletin of the Chichibu Museum of Natural History, 1962 (11): 33–40. (in Japanese)

Kobayasi, H. 1964. Diatoms from River Arakawa (2). Bulletin of the Chichibu Museum of Natural History, 1964 (12): 65–77. (in Japanese)

Kobayasi, H. & Haraguchi, K. 1969. Diatom-association from spring pools in the vicinity of Kawagoe City, Saitama Pref. Bulletin of the Chichibu Museum of Natural History, 1969 (15): 27–54. (in Japanese)

Kobayasi, H. & Kobayashi, H. 1988. A study of *Epithemia amphicephala* (Østr.) comb. et stat. nov. and *E. reticulata* Kütz., with special reference to the areolar occlusion. In: F.E. Round (ed.), *Proceedings of the 9th International Diatom Symposium*. Biopress, Bristol, pp. 459–467.

Kondo, T., Yokoyama, A., Ohhashi, H. & Hara, K. 1998. Morphology and systematics of cyanidian algae. Jap. J. Phycol. 46: 83. (in Japanese)

Korch, J. E. & Sheath, R. G. 1989. The phenology of *Audouinella violacea* (Acrochaetiaceae, Rhodophyta) in a Rhode Island stream, USA. Phycologia 28: 228–236.

Koster, J. T. 1969. Type collection of algae. Taxon 18: 549–559.

Krishnamurthy, V. 1961a. A note on *Compsopogon leptoclados* Montagne. Rev. Algol., ser. 2, 5: 260–265.

Krishnamurthy, V. 1961b. A *Compsopogon* occurring in the Reddish Canal, near Manchester. Brit. Phycol. Bull. 2: 87–88.

Krishnamurthy, V. 1962a. The morphology and taxonomy of the genus *Compsopogon* Montagne. J. Linn. Soc. London, Bot. 58: 208–222.

Krishnamurthy, V. 1962b. The formation of "microaplanospores" in *Compsopogon coeruleus* (Balbis) Montagne. Curr. Sci. 31: 99–100.

Kumano, S. 1977. Rhodophyta. In: Hirose, H, & Yamagishi, T. (eds), *Illustrations of the Japanese freshwater algae. Tokyo*: Uchida-Rokakuho Shinsya. pp. 157–175. (in Japanese).

Kumano, S. 1978a. Notes on freshwater red algae from West Malaysia. Bot. Mag. Tokyo 91: 97–107.

Kumano, S. 1978b. Occurrence of a new freshwater species of the genus *Acrochaetium*, Rhodophyta, in Japan. Jap. J. Phycol. 26: 105–108.

Kumano, S. 1979. Studies on the taxonomy and the phylogenetic relationships of the Batrachospermaceae of Japan and Malaysia. (Doctorate Thesis; Hokkaido University).

Kumano, S. 1980. On the distribution of some freshwater red algae in Japan and Southeast Asia. Proceedings of the 1st Workshop for the Promotion of Limnology in Developing Countries, Kyoto. pp. 3–6.

Kumano, S. 1982a. Two taxa of the section *Contorta* of the genus *Batrachospermum* (Rhodophyta, Nemalionales) from Iriomote Jima and Ishigaki Jima, subtropical Japan. Jap. J. Phycol. 30: 181–187.

Kumano, S. 1982b. Four taxa of the sections *Moniliformia, Hybrida* and *Setacea* of the genus *Batrachospermum* (Rhodophyta, Nemalionales) from temperate Japan. Jap. J. Phycol. 30: 289–296.

Kumano, S. 1982c. Development of carpogonium and taxonomy of six species of the genus *Sirodotia*, Rhodophyta, from Japan and West Malaysia. Bot. Mag. Tokyo 95: 125–137.

Kumano, S. 1983. Studies on freshwater Rhodophyta of Papua New Guinea. II. *Batrachospermum woitapense,* sp. nov. from the Papuan Highlands. Jap. J. Phycol. 31: 76–80.

Kumano, S. 1984a. Studies on freshwater red algae of Malaysia. V. Early development of carposporophytes of *Batrachospermum cylindrocellulare* Kumano and *B. tortuosum* Kumano. Jap. J. Phycol. 32: 24–28.

Kumano, S. 1984b. Some observations on *Batrachospermum intortum* Jao and *B. sinense* Jao (Rhodophyta, Nemalionales) from Szechwan in China. Jap. J. Phycol. 32: 221–226.

Kumano, S. 1986. Studies on freshwater red algae of Malaysia. VI. Morphology of *Batrachospermum gibberosum* (Kumano), comb. nov. Jap. J. Phycol. 34: 19–24.

Kumano, S. 1989. Freshwater species of the genus *Ballia* (Rhodophyta). Program and Abstracts. Korea–Japan Phycological Symposium. p. 47, 1989.

Kumano, S. 1990. Carpogonium and carposporophytes of Montagne's taxa of *Batrachospermum* (Rhodophyta) from French Guiana. Cryptog. Algol. 11: 281–292.

Kumano, S. 1993a. Taxonomy of the family Batrachospermaceae (Batrachospermales, Rhodophyta). Jap. J. Phycol. 41: 253–272.

Kumano, S. 1993b. *Batrachospermum brasiliense* Necchi. In: Hori, T. (ed.), *An illustrated atlas of the life history of algae. Vol. 2. Brown and red algae*. Tokyo: Uchida Rokakuho. pp. 220–221. (in Japanese)

Kumano, S. 1993c. *Sirodotia delicatula* Skuja. In: Hori, T. (ed.), *An illustrated atlas of the life history of algae. Vol. 2. Brown and red algae*. Tokyo: Uchida Rokakuho. pp. 224–225. (in Japanese)

Kumano, S. 1996a. *Ballia pygmaea* Montagne. In: Yamagishi, T. & Akiyama, M. (eds), *Photomicrographs of freshwater algae*. Vol. 16. Tokyo: Uchida Rokakuho. p. 7. (in Japanese and English)

Kumano, S. 1996b. *Batrachospermum ambiguum*. In: Yamagishi, T. & Akiyama, M. (eds), *Photomicrographs of freshwater algae*. Vol. 16. Tokyo: Uchida Rokakuho. p. 8. (in Japanese and English)

Kumano, S. 1996c. *Batrachospermum bakarense*. In: Yamagishi, T. & Akiyama, M. (eds), *Photomicrographs of freshwater algae*. Vol. 16. Tokyo: Uchida Rokakuho. p. 9. (in Japanese and English)

Kumano, S. 1996d. *Batrachospermum beraense*. In: Yamagishi, T. & Akiyama, M. (eds), *Photomicrographs of freshwater algae*. Vol. 16. Tokyo: Uchida Rokakuho. p. 10. (in Japanese and English)

Kumano, S. 1996e. *Batrachospermum boryanum*. In: Yamagishi, T. & Akiyama, M. (eds), *Photomicrographs of freshwater algae*. Vol. 16. Tokyo: Uchida Rokakuho. p. 11. (in Japanese and English)

Kumano, S. 1996f. *Batrachospermum brasiliense*. In: Yamagishi, T. & Akiyama, M. (eds), *Photomicrographs of freshwater algae*. Vol. 16. Tokyo: Uchida Rokakuho. p. 12. (in Japanese and English)

Kumano, S. 1996g. *Batrachospermum cayennense*. In: Yamagishi, T. & Akiyama, M. (eds), *Photomicrographs of freshwater algae*. Vol. 16. Tokyo: Uchida Rokakuho. p. 13. (in Japanese and English)

Kumano, S. 1996h. *Batrachospermum cipoense*. In: Yamagishi, T. & Akiyama, M. (eds), *Photomicrographs of freshwater algae*. Vol. 16. Tokyo: Uchida Rokakuho. p. 14. (in Japanese and English)

Kumano, S. 1996i. *Batrachospermum confusum*. In: Yamagishi, T. & Akiyama, M. (eds), *Photomicrographs of freshwater algae*. Vol. 16. Tokyo: Uchida Rokakuho. p. 15. (in Japanese and English)

Kumano, S. 1996j. *Batrachospermum crispatum*. In: Yamagishi, T. & Akiyama, M. (eds), *Photomicrographs of freshwater algae*. Vol. 16. Tokyo: Uchida Rokakuho. p. 16. (in Japanese and English)

Kumano, S. 1996k. *Batrachospermum cylindrocellulare*. In: Yamagishi, T. & Akiyama, M. (eds), *Photomicrographs of freshwater algae*. Vol. 16. Tokyo: Uchida Rokakuho. p. 17. (in Japanese and English)

Kumano, S. 1996l. *Batrachospermum doboense*. In: Yamagishi, T. & Akiyama, M. (eds), *Photomicrographs of freshwater algae*. Vol. 16. Tokyo: Uchida Rokakuho. p. 18. (in Japanese and English)

Kumano, S. 1996m. *Batrachospermum equisetoideum*. In: Yamagishi, T. & Akiyama, M. (eds), *Photomicrographs of freshwater algae*. Vol. 16. Tokyo: Uchida Rokakuho. p. 19. (in Japanese and English)

Kumano, S. 1996n. *Batrachospermum faroense*. In: Yamagishi, T. & Akiyama, M. (eds), *Photomicrographs of freshwater algae*. Vol. 16. Tokyo: Uchida Rokakuho. p. 20. (in Japanese and English)

Kumano, S. 1996o. *Batrachospermum gibberosum*. In: Yamagishi, T. & Akiyama, M. (eds), *Photomicrographs of freshwater algae*. Vol. 16. Tokyo: Uchida Rokakuho. p. 21. (in Japanese and English)

Kumano, S. 1996p. *Batrachospermum gombakense*. In: Yamagishi, T. & Akiyama, M. (eds), *Photomicrographs of freshwater algae*. Vol. 16. Tokyo: Uchida Rokakuho. p. 22. (in Japanese and English)

Kumano, S. 1996q. *Batrachospermum guyanense*. In: Yamagishi, T. & Akiyama, M. (eds), *Photomicrographs of freshwater algae*. Vol. 16. Tokyo: Uchida Rokakuho. p. 23. (in Japanese and English)

Kumano, S. 1996r. *Batrachospermum hirosei*. In: Yamagishi, T. & Akiyama, M. (eds), *Photomicrographs of freshwater algae*. Vol. 16. Tokyo: Uchida Rokakuho. p. 24. (in Japanese and English)

Kumano, S. 1996s. *Caloglossa ogasawaraensis* var. *latifolia*. In: Yamagishi, T. & Akiyama, M. (eds), *Photomicrographs of freshwater algae*. Vol. 16. Tokyo: Uchida Rokakuho. p. 26. (in Japanese and English)

Kumano, S. 1996t. *Sirodotia segawae*. In: Yamagishi, T. & Akiyama, M. (eds), *Photomicrographs of freshwater algae*. Vol. 16. Tokyo: Uchida Rokakuho. p. 66. (in Japanese and English)

Kumano, S. 1996u. *Ballia pinnulata* Kumano. In: Yamagishi, T. & Akiyama, M. (eds), *Photomicrographs of freshwater algae*. Vol. 17. Tokyo: Uchida Rokakuho. p. 4. (in Japanese and English)

Kumano, S. 1996v. *Batrachospermum hypogynum*. In: Yamagishi, T. & Akiyama, M. (eds), *Photomicrographs of freshwater algae*. Vol. 17. Tokyo: Uchida Rokakuho. p. 5. (in Japanese and English)

Kumano, S. 1996w. *Batrachospermum intortum*. In: Yamagishi, T. & Akiyama, M. (eds), *Photomicrographs of freshwater algae*. Vol. 17. Tokyo: Uchida Rokakuho. p. 6. (in Japanese and English)

Kumano, S. 1996x. *Batrachospermum iriomotense*. In: Yamagishi, T. & Akiyama, M. (eds), *Photomicrographs of freshwater algae*. Vol. 17. Tokyo: Uchida Rokakuho. p. 7. (in Japanese and English)

Kumano, S. 1996y. *Batrachospermum kushiroense*. In: Yamagishi, T. & Akiyama, M. (eds), *Photomicrographs of freshwater algae*. Vol. 17. Tokyo: Uchida Rokakuho. p. 8. (in Japanese and English)

Kumano, S. 1996z. *Batrachospermum macrosporum*. In: Yamagishi, T. & Akiyama, M. (eds), *Photomicrographs of freshwater algae*. Vol. 17. Tokyo: Uchida Rokakuho. p. 9. (in Japanese and English)

Kumano, S. 1996za. *Batrachospermum mahalacense*. In: Yamagishi, T. & Akiyama, M. (eds), *Photomicrographs of freshwater algae*. Vol. 17. Tokyo: Uchida Rokakuho. p. 10. (in Japanese and English)

Kumano, S. 1996zb. *Batrachospermum nechochoense*. In: Yamagishi, T. & Akiyama, M. (eds), *Photomicrographs of freshwater algae*. Vol. 17. Tokyo: Uchida Rokakuho. p. 11. (in Japanese and English)

Kumano, S. 1996zc. *Batrachospermum nodiflorum*. In: Yamagishi, T. & Akiyama, M. (eds), *Photomicrographs of freshwater algae*. Vol. 17. Tokyo: Uchida Rokakuho. p. 12. (in Japanese and English)

Kumano, S. 1996zd. *Batrachospermum nonocense*. In: Yamagishi, T. & Akiyama, M. (eds), *Photomicrographs of freshwater algae*. Vol. 17. Tokyo: Uchida Rokakuho. p. 13. (in Japanese and English)

Kumano, S. 1996ze. *Batrachospermum nova-guineense*. In: Yamagishi, T. & Akiyama, M. (eds), *Photomicrographs of freshwater algae*. Vol. 17. Tokyo: Uchida Rokakuho. p. 14. (in Japanese and English)

Kumano, S. 1996zf. *Batrachospermum onobodense*. In: Yamagishi, T. & Akiyama, M. (eds), *Photomicrographs of freshwater algae*. Vol. 17. Tokyo: Uchida Rokakuho. p. 15. (in Japanese and English)

Kumano, S. 1996zg. *Batrachospermum tabagatense*. In: Yamagishi, T. & Akiyama, M. (eds), *Photomicrographs of freshwater algae*. Vol. 17. Tokyo: Uchida Rokakuho. p. 16. (in Japanese and English)

Kumano, S. 1996zh. *Batrachospermum tapirense*. In: Yamagishi, T. & Akiyama, M. (eds), *Photomicrographs of freshwater algae*. Vol. 17. Tokyo: Uchida Rokakuho. p. 17. (in Japanese and English)

Kumano, S. 1996zi. *Batrachospermum tiomanense*. In: Yamagishi, T. & Akiyama, M. (eds), *Photomicrographs of freshwater algae*. Vol. 17. Tokyo: Uchida Rokakuho. p. 18. (in Japanese and English)

Kumano, S. 1996zj. *Batrachospermum torridum*. In: Yamagishi, T. & Akiyama, M. (eds), *Photomicrographs of freshwater algae*. Vol. 17. Tokyo: Uchida Rokakuho. p. 19. (in Japanese and English)

Kumano, S. 1996zk. *Batrachospermum tortuosum* var. *majus*. In: Yamagishi, T. & Akiyama, M. (eds), *Photomicrographs of freshwater algae*. Vol. 17. Tokyo: Uchida Rokakuho. p. 20. (in Japanese and English)

Kumano, S. 1996zl. *Batrachospermum tortuosum* var. *tortuosum*. In: Yamagishi, T. & Akiyama, M. (eds), *Photomicrographs of freshwater algae*. Vol. 17. Tokyo: Uchida Rokakuho. p. 21. (in Japanese and English)

Kumano, S. 1996zm. *Batrachospermum turgidum*. In: Yamagishi, T. & Akiyama, M. (eds), *Photomicrographs of freshwater algae*. Vol. 17. Tokyo: Uchida Rokakuho. p. 22. (in Japanese and English)

Kumano, S. 1996zn. *Batrachospermum virgato-decaisneanum*. In: Yamagishi, T. & Akiyama, M. (eds), *Photomicrographs of freshwater algae*. Vol. 17. Tokyo: Uchida Rokakuho. p. 23. (in Japanese and English)

Kumano, S. 1996zo. *Batrachospermum woitapense*. In: Yamagishi, T. & Akiyama, M. (eds), *Photomicrographs of freshwater algae*. Vol. 17. Tokyo: Uchida Rokakuho. p. 24. (in Japanese and English)

Kumano, S. 1996zp. *Bostrychia tenius* f. *simpliciusculae*. In: Yamagishi, T. & Akiyama, M. (eds), *Photomicrographs of freshwater algae*. Vol. 17. Tokyo: Uchida Rokakuho. p. 25. (in Japanese and English)

Kumano, S. 1996zq. *Caloglossa ogasawaraensis* var. *ogasawaraensis*. In: Yamagishi, T. & Akiyama, M. (eds), *Photomicrographs of freshwater algae*. Vol. 17. Tokyo: Uchida Rokakuho. p. 26. (in Japanese and English)

Kumano, S. 1996zr. *Sirodotia sinica*. In: Yamagishi, T. & Akiyama, M. (eds), *Photomicrographs of freshwater algae*. Vol. 17. Tokyo: Uchida Rokakuho. p. 74. (in Japanese and English)

Kumano, S. 1996zs. *Sirodotia yutakae*. In: Yamagishi, T. & Akiyama, M. (eds), *Photomicrographs of freshwater algae*. Vol. 17. Tokyo: Uchida Rokakuho. p. 75. (in Japanese and English)

Kumano, S. 1997a. *Ballia prieurii* Montagne. In: Yamagishi, T. & Akiyama, M. (eds), *Photomicrographs of freshwater algae*. Vol. 18. Tokyo: Uchida Rokakuho. p. 7. (in Japanese and English)

Kumano, S. 1997b. *Batrachospermum atrum*. In: Yamagishi, T. & Akiyama, M. (eds), *Photomicrographs of freshwater algae*. Vol. 18. Tokyo: Uchida Rokakuho. p. 8. (in Japanese and English)

Kumano, S. 1997c. *Batrachospermum breutelii*. In: Yamagishi, T. & Akiyama, M. (eds), *Photomicrographs of freshwater algae*. Vol. 18. Tokyo: Uchida Rokakuho. p. 9. (in Japanese and English)

Kumano, S. 1997d. *Batrachospermum gelatinosum* var. *obtrullatum*. In: Yamagishi, T. & Akiyama, M. (eds), *Photomicrographs of freshwater algae*. Vol. 18. Tokyo: Uchida Rokakuho. p. 10. (in Japanese and English)

Kumano, S. 1997e. *Batrachospermum heteromorphum*. In: Yamagishi, T. & Akiyama, M. (eds), *Photomicrographs of freshwater algae*. Vol. 18. Tokyo: Uchida Rokakuho. p. 11. (in Japanese and English)

Kumano, S. 1997f. *Batrachospermum sirodotii*. In: Yamagishi, T. & Akiyama, M. (eds), *Photomicrographs of freshwater algae*. Vol. 18. Tokyo: Uchida Rokakuho. p. 12. (in Japanese and English)

Kumano, S. 1997g. *Batrachospermum turfosum* var. *turfosum*. In: Yamagishi, T. & Akiyama, M. (eds), *Photomicrographs of freshwater algae*. Vol. 18. Tokyo: Uchida Rokakuho. p. 13. (in Japanese and English)

Kumano, S. 1997h. *Batrachospermum turfosum* var. *undulate-pedicellatum*. In: Yamagishi, T. & Akiyama, M. (eds), *Photomicrographs of freshwater algae*. Vol. 18. Tokyo: Uchida Rokakuho. p. 14. (in Japanese and English)

Kumano, S. 1997i. *Bostrychia flagellifera*. In: Yamagishi, T. & Akiyama, M. (eds), *Photomicrographs of freshwater algae*. Vol. 18. Tokyo: Uchida Rokakuho. p. 15. (in Japanese and English)

Kumano, S. 1997j. *Bostrychia radicans* f. *moniliforme*. In: Yamagishi, T. & Akiyama, M. (eds), *Photomicrographs of freshwater algae*. Vol. 18. Tokyo: Uchida Rokakuho. p. 16. (in Japanese and English)

Kumano, S. 1997k. *Compsopogon prolificus*. In: Yamagishi, T. & Akiyama, M. (eds), *Photomicrographs of freshwater algae*. Vol. 18. Tokyo: Uchida Rokakuho. p. 22. (in Japanese and English)

Kumano, S. 1997l. *Sirodotia huillensis*. In: Yamagishi, T. & Akiyama, M. (eds), *Photomicrographs of freshwater algae*. Vol. 18. Tokyo: Uchida Rokakuho. p. 84. (in Japanese and English)

Kumano, S. 1997m. *Sirodotia suecica*. In: Yamagishi, T. & Akiyama, M. (eds), *Photomicrographs of freshwater algae*. Vol. 18. Tokyo: Uchida Rokakuho. p. 85. (in Japanese and English)

Kumano, S. 1998. *Bangia atropurpurea* (Roth) Agardh. In: Yamagishi, T. & Akiyama, M. (eds), *Photomicrographs of freshwater algae*. Vol. 20. Tokyo: Uchida Rokakuho. p. 9. (in Japanese and English)

Kumano, S. 2000. *Taxonomy of Freshwater Rhodophyta*. (Japanese Version) Uchida Rokakuho Tokyo. xiv+395 pp. (in Japanese)

Kumano, S. & Bowden-Kerby, W. A. 1986. Studies on the freshwater Rhodophyta of Micronesia. I. Six new species of *Batrachospermum* Roth. Jap. J. Phycol. 34: 107–128.

Kumano, S. & Johnstone, I. M. 1983. Studies on the freshwater Rhodophyta of Papua New Guinea. I. *Batrachospermum nova-guineense* sp. nov. from the Papuan Lowlands. Jap. J. Phycol. 31: 65–70.

Kumano, S., Hirose, H. & Seto, R. 1962. On the variation of three species of *Batrachospermum*. Bot. Mag. Tokyo 75: 199–204.

Kumano, S. & Liao, L. M. 1987. A new species of the section *Contorta* of the genus *Batrachospermum* (Rhodophyta, Nemalionales) from Nonoc Island, the Philippines. Jap. J. Phycol. 35: 99–105.

Kumano, S. & Necchi, O. Jr. 1985. Studies on the freshwater Rhodophyta of Brazil. II. Two new species of *Batrachospermum* from states of Amazonas and Minas Gerais. Jap. J. Phycol. 33: 181–189.

Kumano, S. & Necchi, O. Jr. 1990. *Batrachospermum macrosporum* Montagne from South America. Jap. J. Phycol. 38: 119–123.

Kumano, S. & Ohsaki, M. 1983. *Batrachospermum kushiroense*, sp. nov. (Rhodophyta, Nemalionales) from Kushiro Moor in cool temperate Japan. Jap. J. Phycol. 31: 156–160.

Kumano, S. & Phang, S. M. 1987. Studies on freshwater red algae of Malaysia. VII. *Batrachospermum tapirense* sp. nov. from Sungai Tapir, Johor, Peninsular Malaysia. Jap. J. Phycol. 35: 259–264.

Kumano, S. & Phang, S. M. 1990. *Ballia leprieurii* Kützing and the related species (Ceramiales, Rhodophyta). Jap. J. Phycol. 38: 125–134.

Kumano, S. & Ratnasabapathy, M. 1982. Studies on freshwater red algae of Malaysia. III. Development of carposporophytes of *Batrachospermum cayennense* Montagne, *B. beraense* Kumano and *B. hypogynum* Kumano et Ratnasabapathy. Bot. Mag. Tokyo 95: 219–228.

Kumano, S. & Ratnasabapathy, M. 1984. Studies on freshwater red algae of Malaysia. IV. *Batrachospermum bakarense*, sp. nov. from Sungai Bakar, Kelantan, West Malaysia. Jap. J. Phycol. 32: 19–23.

Kumano, S., Seto, R. & Hirose, H. 1970. On the development of the carposporophytes in several species of the Batrachospermaceae with special reference to their phylogenetical relations. Bull. Jap. Soc. Phycol. 18: 116–120. (in Japanese)

Kumano, S. & Watanabe, M. 1983. Two new varieties of *Batrachospermum* (Rhodophyta) from Mt. Albert Edward, Papua New Guinea. Bull. Natl. Sci. Mus. [Tokyo], Ser. B, 9: 85–94.

Kützing, F. T. 1843. *Phycologia generalis...* Leipzig. XXXII + 458 pp.

Kützing, F. T. 1845. *Phycologia germanica...* Nordhausen. X + 340 pp.

Kützing, F. T. 1847. Diagnosen und Bemerkungen zu neuen oder kritischen Algen. Bot. Zeit. 5: 1–5, 22–25, 33–38, 52–55, 164–167, 177–180, 193–198, 219–223.

Kützing, F. T. 1849. *Species algarum*. Leipzig. VI + 922 pp.

Kützing, F. T. 1857. *Tabulae phycologicae... Vol. 7*. Nordhausen. II + 40 pp., 100 pls.

Kützing, F. T. 1862. *Tabulae phycologicae... Vol. 12*. Nordhausen. IV + 30 pp., 100 pls.

Kylin, H. 1912. Studien über die schwedischen Arten der Gattungen *Batrachospermum* Roth und *Sirodotia* nov. gen. Nova Acta Regiae Soc. Sci. Upsal., ser. 4, 3(3). 40 pp.

Kylin, H. 1914. Studien über die Entwicklungsgeschichte von *Rhodomela virgata* Kjellm. Svensk. Bot. Tidskr. 8: 33–69.

Kylin, H. 1916. Über die Befruchtung und Reproduktionsteilung bei *Nemalion multifidum*. Ber. Deutsch. Bot. Ges. 34: 257–271.

Kylin, H. 1917. Über die Entwicklungsgeschichte von *Batrachospermum moniliforme*. Ber. Deutsch. Bot. Ges. 35: 155–164.

Kylin, H. 1923. Studien über die Entwicklungsgeschichte der Florideen. Kongl. Svenska Vetenskapsakad. Handl., [ser. 4], 63(11). 139 pp.

Kylin, H. 1934. Über den Aufbau der Prokarpien bei den Rhodomelaceen nebst einigen Worten über *Odonthalia dentata*. Förh. Kongl. Fysiogr. Sällsk. Lund 4(9): 69–90 (1–22 as separate).

Kylin, H. 1937. Über eine marine *Porphyridium*-Art. Förh. Kongl. Fysiogr. Sällsk. Lund 7(10): 119–123 (1–5 as separate).

Kylin, H. 1944. Die Rhodophyceen der schwedischen Westküste. Lunds Univ. Årsskr., N. F., Avd. 2, 40(2). 104 pp.

Kylin, H. 1956. *Die Gattungen der Rhodophyceen.* Lund: CWK Gleerup. XV + 673 pp.

Lamouroux, J. V. F. 1813. Essai sur les genres de la famille des thalassiophytes non articulées. Ann. Mus. Hist. Nat. [Paris] 20: 21–47, 115–139, 267–293.

Le Jolis, A. 1863. Liste des algues marines de Cherbourg. Soc. Imp. Sci. Nat. Cherbourg, Mém. 10: 1–168.

Lee, R. E. 1989. *Phycology.* 2nd ed. Cambridge: Cambridge University Press. xv + 645 pp.

Lee, Y. P. & Lee. I. K. 1988. Contribution to the generic classification of the Rhodochortaceae (Rhodophyta, Nemaliales). Bot. Mar. 31: 119–131.

Lee, Y. P. & Yoshida, T. 1997. The Acrochaetiaceae (Acrochaetiales, Rhodophyta) in Hokkaido. Sci. Rep. Inst. Algol. Res., Fac. Sci., Hokkaido Univ. 9: 159–229.

Lee, Y. P. 1980. *Taxonomic study on the Acrochaetiaceae (Rhodophyta).* Dr. Sci. thesis, Hokkaido Univ. Sapporo. 302 pp.

Lichtle, C. & Giraud, G. 1970. Aspects ultrastructuraux particuliers au plaste du *Batrachospermum virgatum* (Sirodot) – Rhodophycée – Némalionale. J. Phycol. 6: 281–289.

Liebmann, F. 1839. Om et not *Erythroclathrus* af Algernes familie Kröy. Nat. Tidskr., 169–175.

Linnaeus, C. 1753. *Species plantarum...* Vol. 2. Stockholm. pp. 561–1200.

Linnaeus, C. 1755. *Flora suecica... Editio secunda...* Stockholm. XXXII + 464 pp.

Lobban, C. S. & Wynne, M. J. (eds) 1981. *The biology of seaweeds.* Berkeley & Los Angeles: University of California Press. [Botanical Monographs 17.] xi + 786 pp.

Lowe, C. W. 1923. *Freshwater algae and freshwater diatoms. Report of the Canadian Arctic Expedition 1913–18.* Vol. IV: Botany. Part A. Ottawa. 53 pp.

Lowe, C. W. 1927. Some freshwater algae of southern Quebec. Trans. Roy. Soc. Canada, ser. 3, 20: 291–316.

Lyngbye, H. C. 1819. *Tentamen hydrophytologiae danicae...* Copenhagen. XXXII + 248 pp.

Magne, F. 1961. Sur le cycle cytologique de *Nemalion helmintoides* (Velley) Batters. C. R. Acad. Sci. [Paris] 252: 157–159.

Magne, F. 1967a. Sur l'existence, chez les *Lemanea* (Rhodophycées, Némalionales), d'une type de cycle de développement encore inconnu chez les algues rouges. C. R. Acad. Sci. [Paris], Sér. D, 264: 2632–2633.

Magne, F. 1967b. Sur le déroulement et le lieu de la meiose chez les Lémanéacées (Rhodophycees, Némalionales). C. R. Acad. Sci. [Paris], Sér. D, 265: 670–673.

Magne, F. 1972, Le cycle de développement des Rhodophycées et son évolution. Mém. Soc. Bot. France 1972: 247–267.

Mann, F. D. & Steinke, T. D. 1988. Photosynthetic and respiratory responses of the mangrove-associated red algae, *Bostrychia radicans* and *Caloglossa leprieurii.* S. African J. Bot. 54: 203–207.

Martens, G. von 1869. Beiträg zur Algen-Flora Indiens. Flora 52: 233–238.

Matsudaira, Y., Adachi, R., Endo, T., Fuse, S., Hirota, R., Ichimura, S., Iizuka, S., Irie, H., Iwasaki, H., Kajiwara, T., Kawamura, T., Kobayasi, H., Matsudaira, C., Minota, T., Ogawa, T., Saijyo, Y., Sakamoto, I., Sakamoto, T., Takeshita, K., Tanaka, O., Tomiyama, T., Tsuda, T., Yamaji, I. & Yone, Y. 1964. The report of the comprehensive study on primary production in Japanese coastal area. Information Bulletin on Planktology in Japan, 1964 (11): 24–73.

Meneghini, G. 1841. Lettera del Prof. Giuseppe Meneghini al Dott. Corinaldi. Giorn. Tosc. Sci. Med. 1: 186–189.

Merola, A., Castaldo, R., De Luca, P., Gambardella, R., Musacchio, A. & Taddei, R. 1982. Revision of *Cyanidium caldarium.* Three species of acidophilic algae. Giorn. Bot. Ital. 115: 189–195.

Migita, S. 1986. Culture of freshwater red alga, *Nemalionopsis tortuosa.* Bull. Dept. Fish. Nagasaki Univ. 59: 23–28. (in Japanese)

Migita, S.. 1993. *Nemalionopsis tortuosa* Yoneda et Yagi. In: Hori, T. (ed.), *An illustrated atlas of the life history of algae. Vol. 2. Brown and red algae.* Tokyo: Uchida Rokakuho. pp.228–229. (in Japanese)

Migita, S. & Takahashi, M. 1991. *Thorea okadai,* new to Amaki city, Kumamoto Pref. Japan. Bull. Dept. Fish. Nagasaki Univ. 69: 1–5. (in Japanese)

Mitomo, K. & Kobayasi, H. 1994. Diatoms found in the digestive organs of two sweetfishes (*Plecoglossus altivelis* Tem. & Schl.) collected from Ara-kawa (Ara River). Bulletin of the Saitama Museum of Natural History, 1994 (12): 65–72. (in Japanese)

Möbius, M. 1888. Über einige in Portorico gesammelte Süsswasser- und Luft-algen. Hedwigia 9 u 10: 1–29.

Möbius, M. 1892a. Beiträg zur Kenntniss der Gattung *Thorea*. Ber. Deutsch. Bot. Ges. 9: 333–344.

Möbius, M. 1892b. Über einige brasilianische Algen. Ber. Deutsch. Bot. Ges. 10: 17–26.

Möbius, M. 1892c. Australische Süsswasseralgen. Flora 75: 421–450.

Möbius, M. 1892d. Bemerkungen über die systematische Stellung von *Thorea* Bory. Ber. Deutsch. Bot. Ges. 10: 266–270.

Möbius, M. 1894. Australische Süsswasseralgen. II. Abh. Senckenberg. Naturf. Ges. 18: 309–350.

Moi, Phang Siew & Leong, P. 1987. Freshwater algae from the Ulu Endou area, Johore, Malaysia. Malayan Nat. J. 41: 145–157.

Montagne, C. 1839. Sertum patagonicum. Cryptogames de la Patagonie. In: D'Orbigny, A., *Voyage dans l'Amerique meridionale...* Vol. 7, part 1. Paris. 19 pp.

Montagne, C. 1840. Seconde centurie de plantes cellulaires exotiques nouvelles. Décades I et II. Ann. Sci. Nat. Bot., sér. 2, 13: 193–207.

Montagne, C. 1842a. Botanique. Plantes cellulaires. In: Sagra, R. de la, *Histoire physique, politique et naturelle de l'île de Cuba*. Vol. 11. Paris. x + 549 pp.

Montagne, C. 1842b. *Bostrychia*. Dict. Univ. Hist. Nat. [Orbigny] 2: 660–661.

Montagne, C. 1845. Plantes cellulaires. In: Hombron, J. B. & Jacquinot, H., eds, *Voyage au Pole Sud et dans l'Océanie sur les corvettes 'Astrolabe et la Zélée... pendant les années 1837–1838–1839–1840, sous le commandement de M.J. Dumont-d'Urville*. Botanique. Paris. Vol. 1. XIV + 349 pp.

Montagne, C. 1846. Ordo I. Phyceae. In: Durieu de Maisonneuve, M. C., *Exploration scientifique de l'Algérie pendant les années 1840, 1841, 1842...* Sciences physiques. Botanique. Cryptogamie. Paris. pp. [1] – 197.

Montagne, C. 1850. Cryptogamia guyanensis, seu plantarum cellularium in Guyana gallica annis 1835–1849 a cl. Leprieur collectarum enumeratio universalis. Ann. Sci. Nat. Bot., sér. 3, 14: 283–309.

Mori. M. 1953. Communities of brackish water algae in the lower part of the River Ohno, Kumamoto Pref. Bull. Soc. Plant Ecol. 1: 130–132. (in Japanese)

Mori, M. 1961. Studies on the difference between the vegetative and tetrasporic thallus of *Caloglossa* in Japan. Bull. Jap. Soc. Phycol. 9: 57–62.

Mori, M. 1970. Taxonomical and ecological discussions on *Batrachospermum ectocarpum* Sirodot. Bull. Jap. Soc. Phycol. 18: 1–8.

Mori, M. 1975. Studies on the genus *Batrachospermum* in Japan. Jap. J. Bot. 20: 461–485.

Mori, M. 1977. Studies on primordia in the genus *Batrachospermum*. Bull. Jap. Soc. Phycol. 25 (suppl.): 189–193. (in Japanese)

Mori, M. 1984a. *Batrachospermum bruziense*. In: Yamagishi, T. & Akiyama, M. (eds), *Photomicrographs of freshwater algae*. Vol. 1. Tokyo: Uchida Rokakuho. p. 7. (in Japanese and English)

Mori, M. 1984b. *Batrachospermum coerurescens*. In: Yamagishi, T. & Akiyama, M. (eds), *Photomicrographs of freshwater algae*. Vol. 1. Tokyo: Uchida Rokakuho. p. 8. (in Japanese and English)

Mori, M. 1984c. *Batrachospermum ectocarpum*. In: Yamagishi, T. & Akiyama, M. (eds), *Photomicrographs of freshwater algae*. Vol. 1. Tokyo: Uchida Rokakuho. p. 9. (in Japanese and English)

Mori, M. 1984d. *Batrachospermum radicans*. In: Yamagishi, T. & Akiyama, M. (eds), *Photomicrographs of freshwater algae*. Vol. 1. Tokyo: Uchida Rokakuho. p. 10. (in Japanese and English)

Mori, M. 1985a. *Batrachospermum decaisneanum*. In: Yamagishi, T. & Akiyama, M. (eds), *Photomicrographs of freshwater algae*. Vol. 3. Tokyo: Uchida Rokakuho. p. 7. (in Japanese and English)

Mori, M. 1985b. *Batrachospermum godronianum*. In: Yamagishi, T. & Akiyama, M. (eds), *Photomicrographs of freshwater algae*. Vol. 3. Tokyo: Uchida Rokakuho. p. 8. (in Japanese and English)

Mori, M. 1985c. *Batrachospermum japonicum*. In: Yamagishi, T. & Akiyama, M. (eds), *Photomicrographs of freshwater algae*. Vol. 4. Tokyo: Uchida Rokakuho. p. 6. (in Japanese and English)

Mori, M. 1985d. *Batrachospermum rubescens*. In: Yamagishi, T. & Akiyama, M. (eds), *Photomicrographs of freshwater algae*. Vol. 4. Tokyo: Uchida Rokakuho. p. 7. (in Japanese and English)

Mori, M. & Ikeda, H. 1961. Studies on the freshwater *Caloglossa* of Japan. Jap. J. Limnol. 22: 225–229. (in Japanese)

Mostaert, A. S. & King, R. J. 1993. The cell wall of the halotolerant red alga *Caloglossa leprieurii* (Montagne) J. Agardh (Ceramiales, Rhodophyta) from freshwater and marine habitats: effect of changing salinity. Cryptog. Bot. 4: 40–46.

Moul, E. T. & Buell, H. F. 1979. Algae of the Pine Barrens. In: Forman, R. T. T. (ed.), *Pine Barrens; ecosystem and landscape.* New York: Academic Press. pp. 425–440.

Mullahy, J. H. 1952. The morphology and cytology of *Lemanea australis* Atk. Bull. Torrey Bot. Club 79: 393–406, 471–484.

Müller, K. M., Gutell, R. R. & Sheath, R. G. 1998. A preliminary analysis of the group I introns in the 18S rRNA gene of *Bangia* (Bangiales) Rhodophyta. J. Phycol. 34 (suppl.): 102–103. (abstract)

Müller, K. M., Sheath, R. G., Vis, M. L., Crease, T. J. & Cole, K. M. 1998. Biogeography and systematics of *Bangia* (Bangiales, Rhodophyta) based on the RuBisCo spacer, *rbc*L gene and 18S rRNA gene sequences and morphometric analyses. I. North America. Phycologia 37: 195–207.

Müller, K. M., Vis, M. L., Chiasson, W. B., Whittick, A. & Sheath, R. G. 1997. Phenology of a *Batrachospermum* population in a boreal pond and its implications for the systematics of section *Turfosa* (Batrachospermales, Rhodophyta). Phycologia 36: 68–75.

Nagai, M. 1941. Marine algae of the Kurile Islands. II. J. Fac. Agric. Hokkaido Imp. Univ. 46: 139–310.

Nagashima, H. 1993a. *Cyanidioschyzon merolae* De Luca, Taddei & Varano. In: Hori, T. (ed.), *An illustrated atlas of the life history of algae. Vol. 2. Brown and red algae.* Tokyo: Uchida Rokakuho. pp. 188–189. (in Japanese)

Nagashima, H. 1993b. *Galdieria sulphuraria* (Galdieri) Merola. In: Hori, T. (ed.), *An illustrated atlas of the life history of algae. Vol. 2. Brown and red algae.* Tokyo: Uchida Rokakuho. pp. 186–187. (in Japanese)

Nagashima, H. 1993c. *Cyanidium caldarium* (Tilden) Geitler. In: Hori, T. (ed.), *An illustrated atlas of the life history of algae. Vol. 2. Brown and red algae.* Tokyo: Uchida Rokakuho. pp.184–185. (in Japanese)

Nagashima, H. & Fukuda, I. 1981. Morphological properties of *Cyanidium caldarium* and related algae in Japan. Jap. J. Phycol. 29: 237–242.

Nägeli, C. 1849. *Gattungen einzelliger Algen...* Neue Denkschr. Allgem. Schweiz. Gesammten Naturwiss. 10[7]. VIII + 139 pp.

Nagumo, T. & Kobayasi, H. 1979. Raphe structure of *Pinnularia sundaensis* Hust. (Studies on the fine structure of diatoms obtained by clonal culture 1.). Bulletin of Nippon Dental University, General Education, 1979 (8): 151–161. (in Japanese)

Nakamura, T. 1980. Some remarks on *Thorea* found in the Kanto district, Japan. Jap. J. Phycol. 28: 249–254. (in Japanese)

Nakamura, T. 1984a. *Compsopogon aeruginosus.* In: Yamagishi, T. & Akiyama, M. (eds), *Photomicrographs of freshwater algae. Vol. 1.* Tokyo: Uchida Rokakuho. p. 24. (in Japanese and English)

Nakamura, T. 1984b. *Compsopogon corticrassus.* In: Yamagishi, T. & Akiyama, M. (eds), *Photomicrographs of freshwater algae. Vol. 1.* Tokyo: Uchida Rokakuho. p. 25. (in Japanese and English)

Nakamura, T. 1984c. *Compsopogon hookeri.* In: Yamagishi, T. & Akiyama, M. (eds), *Photomicrographs of freshwater algae. Vol. 1.* Tokyo: Uchida Rokakuho. p. 26. (in Japanese and English)

Nakamura, T. 1984d. *Compsopogon oishii.* In: Yamagishi, T. & Akiyama, M. (eds), *Photomicrographs of freshwater algae. Vol. 1.* Tokyo: Uchida Rokakuho. p. 27. (in Japanese and English)

Nakamura, T. 1984e. *Compsopogonopsis japonica.* In: Yamagishi, T. & Akiyama, M. (eds), *Photomicrographs of freshwater algae. Vol. 1.* Tokyo: Uchida Rokakuho. p. 28. (in Japanese and English)

Nakamura, T. 1984f. *Thorea okadai.* In: Yamagishi, T. & Akiyama, M. (eds), *Photomicrographs of freshwater algae. Vol. 1.* Tokyo: Uchida Rokakuho. p. 96. (in Japanese and English)

Nakamura, T. 1993a. *Compsopogonopsis japonica* Chihara. In: Hori, T. (ed.), *An illustrated atlas of the life history of algae. Vol. 2. Brown and red algae.* Tokyo: Uchida Rokakuho. pp. 172–173. (in Japanese)

Nakamura, T. 1993b. *Compsopogon hookeri* Montagne. In: Hori, T. (ed.), *An illustrated atlas of the life history of algae. Vol. 2. Brown and red algae.* Tokyo: Uchida Rokakuho. pp. 174–175. (in Japanese)

Nakamura, T. & Chihara, M. 1977a. Occurrence of *Thorea* species (Rhodophyta) in Kanto district, central Japan. Bull. Jap. Soc. Phycol. 25: 159–162. (in Japanese)

Nakamura, T. & Chihara, M. 1977b. Life history of *Compsopogonopsis japonica*, a fresh water red alga. Bull. Jap. Soc. Phycol. 25 (suppl.): 195–201. (in Japanese)

Nakamura, T. & Chihara, M. 1983. *Compsopogon aeruginosus* and *C. hookeri* (Compsopogonaceae, Rhodophyta) newly found in Japan. J. Jap. Bot. 58: 54–61. (in Japanese)

Nakamura, Y. 1941. The species of *Rhodochorton* from Japan. I. Sci. Pap. Inst. Algol. Res. Fac. Sci. Hokkaido Imp. Univ. 2: 273–291.

Nakamura, Y. 1944. The species of *Rhodochorton* from Japan. II. Sci. Pap. Inst. Algol. Res. Fac. Sci. Hokkaido Imp. Univ. 3: 99–119.

Nardo, G. D. 1834. De novo genere algarum cui nomen est *Hildenbrandtia prototypus*. Isis (Oken) 1834: 675–676.

Nasr, A. H. 1947. Synopsis of the marine algae of the Egyptian Red Sea coast. Bull. Fac. Sci. Fouad I Univ. 25: 1–155.

NCU–3. See Greuter, W., Burmitt, R. K., Farr, E. Killian, N, Kirk, P. M. & Silva, P. C. (eds) 1993.

Necchi, O. Jr. 1986. Studies on the freshwater Rhodophyta of Brazil – 4. Four new species of *Batrachospermum* (section *Contorta*) from the southern state of São Paulo. Revista Brasil. Biol. 46: 517–525.

Necchi, O. Jr. 1987a. Studies on freshwater Rhodophyta of Brazil – 3. *Batrachospermum brasiliense* sp. nov. from the state of São Paulo, southern Brazil. Revista Brasil. Biol. 47: 441–446.

Necchi, O. Jr. 1987b. Sexual reproduction in *Thorea* Bory (Rhodophyta, Thoreaceae). Jap. J. Phycol. 35: 106–112.

Necchi, O. Jr. 1988. Revisão de género *Batrachospermum* Roth (Rhodophyta, Batrachospermales) no Brasil. Ph.D. Thesis. Universidade Estadual Paulista, Julio de Masquita Filho, Rio Caro.

Necchi, O. Jr. 1989. Rhodophyta de água doce do estado de São Paulo: levantamento taxónomico. Bol. Bot. Univ. São Paulo 11: 11–69.

Necchi, O. Jr. 1990a. Revision of the genus *Batrachospermum* Roth (Rhodophyta, Batrachospermales) in Brazil. Biblioth. Phycol. 84. iii + 201 pp.

Necchi, O. Jr. 1990b. Geographic distribution of the genus *Batrachospermum* (Rhodophyta, Batrachospermales) in Brazil. Revista Brasil. Biol. 49: 663–669.

Necchi, O. Jr. 1991a. Evaluation of numeric taxonomic characters in Brazilian species of *Batrachospermum* (Rhodophyta, Batrachospermales). Revista Brasil. Biol. 50: 627–641.

Necchi, O. Jr. 1991b. The section *Sirodotia* of *Batrachospermum* (Rhodophyta, Batrachospermales) in Brazil. Arch. Hydrobiol. Suppl. 89 (Algol. Stud. 62): 17–30.

Necchi, O. Jr. 1997. Microhabitat and plant structure of *Batrachospermum* (Batrachospermales, Rhodophyta) populations in four streams of São Paulo State, southeastern Brazil. Phycol. Res. 45: 39–45.

Necchi, O. Jr. & Bicudo, D. de C. 1993. Criptógamos do Parque Estadual das Fontes do Ipiranga, São Paulo, SP. Algas, 3: Rhodophyceae. Hoehnea 19: 89–92.

Necchi, O. Jr., Braga, M. R. A., & Moulton, T. P. 1998. Survey and distribution of freshwater Rhodophyta from Cardoso Island, São Paulo State, southeastern Brazil. Arch. Hydrobiol. Suppl. 123 (Algol. Stud. 88): 111–124.

Necchi, O. Jr. & Branco C. C. Z. 1999. Phenology of a dioecious population of *Batrachospermum delicatulum* (Batrachospermales, Rhodophyta) in a stream from southeastern Brazil. Phycol. Res. 47: 251–256.

Necchi, O. Jr., Branco, C. C. Z. & Branco, L. H. Z. 1999. Distribution of Rhodophyta in streams from São Paulo State, southeastern Brazil. Archiv für Hydrobiologie, Stuttgart, 147: 73–89.

Necchi, O. Jr., Branco, C. C. Z. & Branco, L. H. Z. 2000. Distribution of stream macroalgae in São Paulo State, southeastern Brazil. Archiv für Hydrobiologie 132, Algological Studies Supplement, Stuttgart, 97: 43–57.

Necchi, O. Jr., Branco, C. C. Z. & Gomes, R. R. V. 1999. Microhabitat and plant structure of *Compsopogon coeruleus* (Comsopogonaceae, Rhodophyta) populations in streams from São Paulo State, southeastern Brazil. Cryptogamie, Algologie, Paris, 20: 75–87.

Necchi, O. Jr. & Dip, M. R. 1992. The family Compsopogonaceae (Rhodophyta) in Brazil. Arch. Hydrobiol. Suppl. 94 (Algol. Stud. 66): 105–118.

Necchi, O. Jr. & Entwisle, T. J. 1990. A reappraisal of generic and subgeneric classification in the Batrachospermaceae (Rhodophyta). Phycologia 29: 478–488.

Necchi, O. Jr., Goes, R. M., & Dip, M. R. 1990. Phenology of *Compsopogon coeruleus* (Balbis) Montagne (Compsopogonaceae, Rhodophyta) and evaluation of taxonomic characters of the genus. Jap. J. Phycol. 38: 1–10.

Necchi, O. Jr. & Kumano, S. 1984. Studies on the freshwater Rhodophyta of Brazil. 1. Three taxa of *Batrachospermum* Roth from the northeastern state of Serigipe. Jap. J. Phycol. 32: 348–353.

Necchi, O. Jr. & Pascoaloto, D. 1995. Morphometry of *Compsopogon coeruleus* (Compsopogonaceae, Rhodophyta) populations in a tropical river basin of southeastern Brazil. Arch. Hydrobiol. Suppl. 106 (Algol. Stud. 76): 61–73.

Necchi, O. Jr. & Sheath, R. G. 1992. Karyology of Brazilian species of *Batrachospermum* (Rhodophyta, Batrachospermales). Brit. Phycol. J. 27: 423–427.

Necchi, O. Jr., Sheath, R. G. & Cole, K. M. 1993a. Distribution and systematics of the freshwater genus *Sirodotia* (Batrachospermales, Rhodophyta) in North America. J. Phycol. 29: 236–243.

Necchi, O. Jr., Sheath, R. G. & Cole, K. M. 1993b. Systematics of freshwater *Audouinella* (Acrochaetiaceae, Rhodophyta) in North America. 1. The reddish species. Arch. Hydrobiol. Suppl. 98 (Algol. Stud. 70): 11–28.

Necchi, O. Jr., Sheath, R. G. & Cole, K. M. 1993c. Systematics of freshwater *Audouinella* (Acrochaetiaceae, Rhodophyta) in North America. 2. The bluish species. Arch. Hydrobiol. Suppl. 100 (Algol. Stud. 71): 13–21.

Necchi, O. Jr. & Zucchi, M. R. 1995a. Systematics and distribution of freshwater *Audouinella* (Acrochaetiaceae, Rhodophyta) in Brazil. Eur. J. Phycol. 30: 209–218.

Necchi, O. Jr. & Zucchi, M. R. 1995b. Record of *Paralemanea* (Lemaneaceae, Rhodophyta) in South America. Arch. Hydrobiol. Suppl. 109 (Algol. Stud. 78): 33–38.

Necchi, O. Jr. & Zucchi, M. R. 1996 ("1995"). Occurrence of the genus *Ballia* (Ceramiaceae, Rhodophyta) in freshwater in Brazil. Hoehnea 22: 229–235.

Necchi, O. Jr. & Zucchi, M. R. 1997a. Taxonomy and distribution of *Thorea* (Thoreaceae, Rhodophyta) in Brazil. Arch. Hydrobiol. Suppl. 118 (Algol. Stud. 84): 83–90.

Necchi, O. Jr. & Zucchi, M. R. 1997b. *Audouinella macrospora* (Acrochaetiaceae, Rhodophyta) is the chantransia stage of *Batrachospermum* (Batrachospermaceae). Phycologia 36: 220–224.

Negoro, K. 1943. Über die Algenvegetation der Thermen von Kusatu, Gunma Präfektur, Japan. Bot. Mag. (Tokyo) 57: 302–312. (in Japanese)

Nichols, H. W. 1964b. Developmental morphology and cytology of *Boldia erythrosiphon*. Amer. J. Bot. 51: 653–659.

Nichols, H. W. 1964a. Culture and developmental morphology of *Compsopogon coeruleus*. Amer. J. Bot. 51: 180–188.

Nichols, H. W. 1965. Culture and development of *Hildenbrandia rivularis* from Denmark and North America. Amer. J. Bot. 52: 9–15.

Noda, M. 1963. Freshwater algae of north-eastern China. V. Rhodophyta. p. 174–177. Niigata Univ. Niigata, Japan.

Nohda, M. 1982. A new record of *Compsopogon oishii* collected from River Ayagawa, Kagawa Prefecture. Kagawa Seibutsu 10: 109–110. (in Japanese)

Notoya, M. & Kikuchi, N. 1993. *Bangia atropurpurea* (Roth) C. Agardh. In: Hori, T. (ed.), *An illustrated atlas of the life history of algae. Vol. 2. Brown and red algae*. Tokyo: Uchida Rokakuho. pp. 198–199. (in Japanese)

Ohishi, Y. 1901. *Batrachospermum gallaei*. Shinsen Nihon Shokubutsu Zusestu (Matsumura & Miyoshi eds), Kato Inka-rui-bu 2(4): 79. (in Japanese)

Ohno, Y. 1899. *Batrachospermum moniliforme*. Shinsen Nihon Shokubutsu Zusestu (Matsumura & Miyoshi eds), Kato Inka-rui-bu 1(2): 10. (in Japanese)

Ohno, Y. 1901. *Porphyridium purpureum*. Shinsen Nihon Shokubutsu Zusestu (Matsumura & Miyoshi eds), Kato Inka-rui-bu 2(4): 79. (in Japanese)

Okada, Y. 1939a. *Rhodophyceae. Inka Shokubutsu Zukan*. (Asahina, ed.). Sanseido, Tokyo (in Japanese).

Okada, Y. 1939b. Report on the flora of the "Suga" or the salt-river in Okinawa Island, Ryukyu. J. Jap. Bot. 15: 48–53. (in Japanese)

Okada, Y. 1944. Notes on *Bangia atropurpurea* (Roth) Ag. found in Japan. J. Jap. Bot. 20: 201–204. (in Japanese)

Okada, Y. 1950a. A contribution to our knowledge of *Thorea okadai* Yamada. J. Jap. Bot. 25: 145–147.
Okada, Y. 1950b. A contribution to our knowledge of *Thorea okadai* Yamada. II. J. Kagoshima Fish. Coll. 1: 148–150. (in Japanese)
Okada, Y. & Koba, K. 1933. Outline of the northern Kuriles. Bull. Biogeogr. Soc. Japan 4: 1–46. (in Japanese)
Okada, Y. & Migita, S. 1956. On life history of *Nemalionopsis tortuosa*. Bull. Dept. Fish. Nagasaki Univ. 4: 1–6. (in Japanese).
Okamura, K. 1897. On the algae from Ogasawara-jima (Bonin Islands). Bot. Mag. Tokyo 11: 11–17.
Okamura, K. 1902. *Nihon sorui meii*. Tokyo: Keigyosha. VII + 276 pp. (in Japanese)
Okamura, K. 1903a. *Algae japonicae exsiccatae*. Tokyo. Fasc. 2. Nos. 51–100.
Okamura, K. 1903b. Contents of the "Algae japonicae exsiccatae" Fasciculus II. Bot. Mag. [Tokyo] 17: 129–132.
Okamura, K. 1907–1909. *Icones of Japanese algae. Vol. 1*. Tokyo. 258 pp., pls. I–L.
Okamura, K. 1913–1915. *Icones of Japanese algae. Vol. 3*. Tokyo. 218 pp., pls. CI–CL.
Okamura, K. 1916. *Nihon sorui meii. Ed. 2*. Tokyo: Seibi do. 362 pp. (in Japanese)
Okamura, K. 1936. *Nihon kaiso shi*. Tokyo: Uchida Rokaku ho. 964 pp. (in Japanese)
Oltmanns, F. 1904. *Morphologie und Biologie der Algen. Vol. 1*. Jena: Gustav Fischer. VI + 733 pp.
Ortega, M. M. 1984. *Cátlogo de algas continentales recientes de México*. México: Univ. Nac. Auton. Mexico. 566 pp.
Osterhout, W. J. V. 1900. Befruchtung bei *Batrachospermum*. Flora 87: 109–115.
Ott, F. D. 1973. A review of the synonyms and the taxonomic positions of the algal genus *Porphyridium* Nägeli 1849. Nova Hedwigia 23: 237–289.
Ott, F. D. & Seckbach, J. 1994. New classification for the genus *Cyanidium* Geitler 1933. In: Seckbach, J. (ed.). *Evolutionary pathways and enigmatic algae:* Cyanidium caldarium *(Rhodophyta) and related cells*. Dordrecht etc.: Kluwer Academic Publishers. pp. 145–152.
Ott, F. D. & Sommerfelt, M. R. 1982. Freshwater Rhodophyceae: Introduction and bibliography. In: Rosowski, J. R. & Parker, B. C. (eds) *Selected papers in Phycology II*. Phycol. Soc. America, Lawrence, Kansas, pp. 671–681.
Palmer, C. M. 1941. A study of *Lemanea* in Indiana with notes on its distribution in North America. Butler Univ. Bot. Stud. 5: 1–26.
Palmer, C. M. 1942. *Lemanea* herbarium packets containing more than one species. Butler Univ. Bot. Stud. 5: 222–223.
Palmer, C. M. 1945. A preliminary study of *Sacheria* in western North America. Butler Univ. Bot. Stud. 7: 176–181.
Papenfuss, G. F. 1945. Review of the *Acrochaetium–Rhodochorton* complex of the red algae. Univ. Calif. Publ. Bot. 18: 229–334.
Papenfuss, G. F. 1946. Proposed names for the phyla of algae. Bull. Torrey Bot. Club 73: 217–218.
Papenfuss, G. F. 1947. Further contributions toward an understanding of the *Acrochaetium–Rhodochorton* complex of the red algae. Univ. Calif. Publ. Bot. 18: 433–447.
Papenfuss, G. F. 1955. Classification of algae. In: *A century of progress in the natural sciences, 1853–1953*. San Francisco: California Acad. Sci. pp. 115–224.
Papenfuss, G. F. 1958. Notes on algal nomenclature. IV. Taxon 7: 104–109.
Papenfuss, G. F. 1961. The structure and reproduction of *Caloglossa leprieurii*. Phycologia 1: 8–31.
Papenfuss, G. F. 1966. A review of the present system of classification of the Florideophycidae. Phycologia 5: 247–255.
Parker, B. C., Samsel, G. L. Jr. & Prescott, G. W. 1973. Comparison of microhabitats of macroscopic subalpine stream macroalgae. Amer. Midl. Naturalist 90: 143–153.
Paschér, A. & Schiller, J. 1925. Rhodophyta (Rhodophyceen). In: Pascher, A. (ed.). *Die Süsswasser-Flora Deutschlands, Österreichs und der Schweiz*. Fasc. 11. Jena: Gustav Fischer. pp. 134–206.
Patel, R. J. 1970. New freshwater species of *Acrochaetium* from Gujarat, India (*Acrochaetium godwardense* sp. nov.). Rev. Algol., ser. 2, 10: 30–36.
Patel, P. J. & Francis, M. A. 1968. On *Batrachospermum* from Gujarat. Proc. Indian Acad. Sci., Sect. B, 67: 230–232.

Patel, P. J. & Francis, M. A. 1970. Some interesting observations on *Compsopogon aeruginosus* (J. Ag.) Kützing, a species new to India. Phykos 8: 46–51.

Patel, R. J. 1965. *Compsopogon iyengarii* Krishnamurthy from Gujarat. Curr. Sci. 34: 644–645.

Post, E. 1936. Systematische und pflanzengeographische Notizen zur *Bostrychia*–*Caloglossa*–Assoziation. Rev. Algol. 9: 1–84.

Post, E. 1939. *Bostrychia tangatensis* spec. nov., eine neue *Bostrychia* der ostafrikanischen Mangrove. Arch. Protistenk. 92: 152–156.

Post, E. 1941. *Bostrychia hamana-tokidai* spec. nov., eine neue südjapanische *Bostrychia*. Beih. Bot. Centralbl., Abt. B, 61: 208–210.

Post, E. 1943. Zur Morphologie und Ökologie von *Caloglossa*. Ergebnisse der Sunda-Expedition der Notgemeinschaft der deutschen Wissenschaft 1929/30. Arch. Protistenk. 96: 123–220.

Post, E. 1957. Fruktifikationen und Keimlinge bei *Caloglossa*. Hydrobiologia 9: 105–125.

Post, E. 1963. Zur Verbreitung und Ökologie der *Bostrychia*–*Caloglossa*-Assoziation. Int. Rev. Gesamten Hydrobiol. 48: 47–152.

Post, E. 1965. *Caloglossa beccarii* im Golf von Mexico. Hydrobiologia 26: 184–188.

Post, E. 1966a. *Caloglossa stipitata* in Florida. Hydrobiologia 27: 109–112.

Post, E. 1966b. Neues zur Verbreitungsökologie neuseeländischer und mittel-amerikanischer *Bostrychia*–*Caloglossa*-Assoziation. Rev. Algol., ser. 2, 8: 127–150.

Post, E. 1966c. *Caloglossa ogasawaraensis* in Westafrika. Hydrobiologia 27: 317–322.

Post, E. 1968. Zur Verbreitungs-Ökologie des Bostrychietum. Hydrobiologia 31: 241–316.

Prescott, G. W. 1951. History of phycology. In: Smith, G. M. (ed.), *Manual of Phycology*. Waltham, Massachusetts: Chronica Botanica. pp. 1–11.

Prescott, G. W. 1962. *Algae of the Western Great Lakes area*...2nd ed. Dubuque, Iowa: Wm. C. Brown. xiii + 977 pp.

Prescott, G. W. 1978. *How to know the freshwater algae*. 3rd ed. Dubuque, Iowa: Wm. C. Brown. x + 293 pp.

Price, S. R. 1914. Notes on *Batrachospermum*. New Phytol. 13: 276–279.

Pueschel, C. M. & Cole, K. M. 1982. Rhodophycean pit plugs: an ultrastructural survey with taxonomic implications. Amer. J. Bot. 69: 703–720.

Pueschel, C. M. 1990. Cell structure. In: Cole, K. M. & Sheath, R.G. (eds), *Biology of the red algae*. Cambridge: Cambridge University Press. pp. 7–41.

Pujals, C. 1967. Presencia en la Argentina del género "*Thorea*" (Rhodophycophyta, Florideae). Comun. Mus. Argent. Ci. Nat. "Bernardino Rivadavia", Hidrobiol. 1: 55–64.

Rabenhorst, L. 1854. *Batrachospermum kuhnianum* Rabenh. Hedwigia 1: 42, pl. 7, fig. 1.

Rabenhorst, L. 1855. Beitrag zur Kryptogamen-Flora Süd-Afrikas. Pilze und Algen. Allg. Deutsche Naturhist. Zeitung, ser. 2, 1: 280–283.

Rabenhorst, L. 1859. *Algen Sachsens...* Decades 83/84, nos. 821–840. (exsiccata)

Rabenhorst, L. 1863. *Kryptogamen-Flora von Sachsen...* Vol. 1. Leipzig. XX + 653 pp.

Rabenhorst, L. 1868. *Flora europaea algarum aquae dulcis et submarinae*. Vol. 3. Leipzig. XX + 461 pp.

Raikwar, S. K. S. 1962. A new freshwater species of *Acrochaetium* (*A. indica* sp. nov.). Rev. Algol., ser. 2, 6: 98–104.

Ratnasabapathy, M. 1972. Algae from Gunong Jerai (Kedah Peak), Malaysia. Gard. Bull. Singapore 26: 95–110.

Ratnasabapathy, M. 1977 ('1975'). Preliminary observations on Gombak river algae at the Field Studies Centre, University of Malaya. Phykos 14: 15–23.

Ratnasabapathy, M. 1978. *Freshwater biology of Pulau Tioman*. Kuala Lumpur: Univ. Press. 150 pp.

Ratnasabapathy, M. & Kumano, S. 1982a. Studies on freshwater red algae of Malaysia. I. Some taxa of the genera *Batrachospermum, Ballia* and *Caloglossa* from Pulau Tioman, West Malaysia. Jap. J. Phycol. 30: 15–22.

Ratnasabapathy, M. & Kumano, S. 1982b. Studies on freshwater red algae of Malaysia. II. Three species of *Batrachospermum* from Sungai Gombak and Sungai Pusu, Selangor, West Malaysia. Jap. J. Phycol. 30: 119–124.

Ratnasabapathy, M. & Seto, R. 1981. *Thorea prowsei* sp. nov. and *Thorea clavata* sp. nov. (Rhodophyta, Nemaliales) from West Malaysia. Jap. J. Phycol. 29: 243–250.

Raven, P. H., Evert, R. F. & Curtis, H. 1981. *Biology of plants*. 3rd ed. New York: Worth Publishers. xvi + 686 pp.

Reinsch, P. F. 1875. *Contributiones ad algologiam et fungologiam*. Nurnberg. XII + 103 pp.

Reis, M. P. dos. 1954. Contribuição para o conhecimento das espécies de *Batrachospermum* Roth da flora portuguesa. Ciências [Madrid] 19: 349–357.

Reis, M. P. dos. 1955. Sur le *Batrachospermum corbula* Sirodot var. *alcoense* P. Reis n. var. Compt. Rend. VIII Congr. Int. Bot. [Paris] 17: 70–71.

Reis, M. P. dos. 1958. Subsídios para o conhecimento das Rodofíceas de água doce de Portugal – I. Bol. Soc. Brot., ser. 2, 32: 101–126 [127=148 = expl. pls.].

Reis, M. P. dos. 1959. Variabilidade do tricogínio no género *Batrachospermum* Roth. Proc. IX Int. Bot. Congr. [Montreal] 2: 307.

Reis, M. P. dos. 1960a. Variabilité du trichogyne chez le genre *Batrachospermum* Roth. Bol. Soc. Brot., ser. 2, 34: 29–36.

Reis, M. P. dos 1960b. Revisão dos espécimes de *Batrachospermum* Roth e *Sirodotia* Kylin dos herbários dos institutos botanicos de Coimbra e Lisboa. Bol. Soc. Brot., ser. 2, 34: 37–55.

Reis, M. P. dos. 1961a. Sobre a identificação de *Chantransia violacea* Kütz. Bol. Soc. Brot., ser. 2, 35: 141–147 [149–154 = expl. pls.]..

Reis, M. P. dos. 1961b. Subsídios para o conhecimento das Rodofíceas de água doce de Portugal. II. Bol. Soc. Brot., ser. 2, 35: 163–178 [179–184 = expl. pls.].

Reis, M. P. dos. 1962a. Subsídios para o conhecimento das Rodofíceas de água doce de Portugal. III. Mem. Soc. Brot. 15: 57–71.

Reis, M. P. dos. 1962b. Una nova espécie de *Lemanea* Bory encontrada em Portugal. Bol. Soc. Brot., ser. 2, 36: 175–176 [177–178 = expl. pl.].

Reis, M. P. dos. 1963. Subsídios para o conhecimento das Rodofíceas de água doce de Portugal. IV. Bol. Soc. Brot., ser. 2, 37: 115–126.

Reis, M. P. dos. 1964. On a new species of *Batrachospermum* of the section *Hybrida* from Portugal. Abstr. Tenth Int. Bot. Congr. [Edinburgh], p. 449.

Reis, M. P. dos. 1965a. Subsídios para o conhecimento das Rodofíceas de água doce de Portugal. V. Bol. Soc. Brot., ser. 2, 39: 137–156.

Reis, M. P. dos. 1965b. *Batrachospermum gulbenkianum*, sp. nov.. Anuário Soc. Brot. 31: 31–44.

Reis, M. P. dos. 1966. Subsídios para o conhecimento das Rodofíceas de água doce de Portugal. VI. Anuário. Soc. Brot. 32: 33–47.

Reis, M. P. dos. 1967. Duas espécies novas de *Batrachospermum* Roth: *B. azeredoi* e *B. ferreri*. Bol. Soc. Brot., ser. 2, 41: 167–189.

Reis, M. P. dos. 1969. Subsídios para o conhecimento das Rodofíceas de água doce de Portugal. VII. Bol. Soc. Brot., ser. 2, 43: 183–192.

Reis, M. P. dos. 1971. Rhodophyceae novae. Mem. Soc. Brot. 21: 23–31.

Reis, M. P. dos. 1972a. *Batrachospermum henriquesianum*, sp. nov.. Bol. Soc. Brot., ser. 2, 46: 181–190.

Reis, M. P. dos. 1972b. Estudo comparativo de *Batrachospermum helminthosum* Bory, *B. coerulescens* Sirod. e *B. azeredoi* P. Reis e descrição de una variedade nova. Bol. Soc. Brot., ser.. 2, 46: 191–217.

Reis, M. P. dos. 1973. Subsídios para o conhecimento das Rodofíceas de água doce de Portugal. VIII. Bol. Soc. Brot. 47, ser. 2: 139–156.

Reis, M. P. dos. 1974. Chaves para a identifição das espécies portuguesas de *Batrachospermum* Roth. Anuário. Soc. Brot. 40: 37–129.

Reis, M. P. dos. 1981. Sobre as Rodofíceas da Ria de Aveiro. Bol. Soc. Brot., ser.2, 53: 1407–1436.

Reis, M. P. dos. 1982. Novidades ficológicas da Ria de Aveiro. [II]. Bol. Soc. Brot., ser. 2, 55: 117–119.

Rhodes, R. G. & Terzis, A. J. 1970. Some algae of the upper Cuyahoga River system in Ohio. Ohio J. Sci. 70: 295–299.

Rider, D. E. & Wagner, R. H. 1972. The relationship of light, temperature, and current to the seasonal distribution of *Batrachospermum* (Rhodophyta). J. Phycol. 8: 323–331.

Rieth, A. 1979a. Ein *Batrachospermum* der Sektion *Contorta* Skuja aus Kuba. Kulturpflanze 27: 265–281.

Rieth, A. 1979b. Ein Standort der epiphytischen Süsswasser-Rotalge *Balbiania investiens* (Lenormand) Sirodot 1976 in Mitteleuropa. Arch. Protistenk. 121: 401–416.

Rintoul, T., Sheath, R. G. & Vis, M. L. 1998. Systematics and biogeography of the Compsopogonales (Rhodophyta) with emphasis on freshwater genera in North America. J. Phycol. 34(suppl.): 50. (abstract)

Rintoul, T., Sheath, R. G. & Vis, M. L. 1999. Systematics and biogeography of the Compsopogonales (Rhodophyta) with emphasis on freshwater genera in North America. Phycologia. 38: 517–527.

Roeder, D. R. 1977. A floridean red alga new to Iowa: *Audouinella violacea* (Kütz.) Hamel (Acrochaetiaceae, Rhodophyta). Proc. Iowa Acad. Sci. 84: 139–143.

Roeder, D. R. & Peck, J. H. 1977. *Batrachospermum* Roth (Rhodophyta), a genus of red algae new to Iowa. Proc. Iowa Acad. Sci. 84: 133–138.

Roscoe, M. V. 1931. The algae of St. Paul Island. Rhodora 33: 127–131.

Rosenberg, M. 1935. On the germination of *Lemanea torulosa* in culture. Ann. Bot. (London) 49: 621–622.

Rosenvinge, L. K. 1918 ('1917'). The marine algae of Denmark... Part II. Rhodophyceae II (Cryptonemiales). Kongel. Danske Vidensk. Selsk. Skr., 7 Raekke, Naturvidensk. og Math. Afd. 7: 153–284.

Rosenvinge, L. K. 1924 ('1923'). The marine algae of Denmark... III. Rhodophyceae III. (Ceramiales). Kongel. Danske Vidensk. Selsk. Skr., 7 Raekke, Naturvidensk. og Math. Afd. 7: 285–487.

Roth, A. W. 1797a. Bemerkungen über das Studium der cryptogamischen Wassergewächse. Hannover. 109 pp.

Roth, A. W. 1797b. *Catalecta botanica..* Fasc. 1. Leipzig. VIII + 244 pp.

Roth, A. W. 1800. *Tentamen florae germanicae*. Vol. 3, part 1. Leipzig. vii + 578 pp.

Roth, A. W. 1806. *Catalecta botanica..* Fasc. 2. Leipzig.

Round, F. E. 1965. *The biology of the algae*. London: Edward Arnold. vii + 269 pp.

Round, F. E. 1981. *The ecology of algae*. Cambridge: Cambridge University Press. Cambridge. vii + 653 pp.

Ruprecht, F. J. 1851. Tange des Ochotskischen Meeres. In: Middendorff, A. Th. v., *Reise in den äussersten Norden und Osten Sibiriens*...Vol. 1, part. 2(1): 191–435. St. Petersburg.

Ruttner, F. 1960. Über die Kohlenstoffaufnahme bei Algen aus der Rhodophyceen-Gattung *Batrachospermum*. Schweiz. Z. Hydrol. 22: 280–291.

Ryan, B. F., Joiner, B. L. & Ryan, T. A. Jr. 1985. *MINITAB handbook*. 2nd ed. Boston: Duxbury Press. ix + 374 pp.

Saenger, P., Ducker, S. C. & Rowan, K. S. 1971. Two species of Ceramiales from Australia and New Zealand. Phycologia 10: 105–111.

Saida, K. 1887. Development of *Batrachospermum*. Bot. Mag. Tokyo. 1: 51–53. (in Japanese)

Saida, K. 1910. Index of native and foreign plants. Part. Crypt.. Tokyo. (in Japanese)

Sanchez Rodriguez, M. E. & Huerta M., L. 1969. Una nueva especie de *Lemanea* (Rhodoph., Florid.), para la flora dulceacuícola méxicana. Ciência [México] 27: 27–30.

Sanchez Rodriguez, M. E. 1974. Rodofíceas dulceacuícolas de México. Bol. Soc. Bot. México. 33: 31–37.

Sankaran, V. 1984. *Batrachospermum desikacharyi* sp. nov. (Rhodophyta) from Valparai, Anamalais, Tamil Nadu. Phykos 23: 163–170.

Sato, H., Anton, A. & Kumano, S. 1999. Freshwater algae of Tabin Wildlife Reserve, Sabah. In: *Tabin Scientific Expedition* (Mohamaed, M., Andau, M., Dalimin, M. N. & Malim, T. P. eds), Universiti Malaysia Sabah, Malaysia. pp.19–32.

Saunders, De A. 1901. Papers from the Harriman Alaska Expedition. XXV. The algae. Proc. Wash. Acad. Sci. 3: 391–486.

Saunders, G. W. & Bailey, J. C. 1999. Molecular systematic analyses indicate that the enigmatic Apophlaea is a member of the Hildenbrandiales (Rhodophyta, Florideophycidae). J. Phycol. 35: 171-175.

Schmidle, W. 1899a. Algologische Notizen [VIII. IX.]. Allg. Bot. Zeit. Syst. 5: 2–4.

Schmidle, W. 1899b. Einiges über die Befruchtung, Keimung und Haarinsertion von *Batrachospermum*. Bot. Zeitung 57: 125–135.

Schmitz, F. 1883. Untersuchungen über die Befuchung der Florideen Sitzungsber. K. Preuss. Akad. Wiss. Berlin 1883(1): 215–258.

Schmitz, F. 1892a. [6. Klasse Rhodophyceae] 2. Unterklasse Florideae. In: Engler, A., *Syllabus der Vorlesungen über specielle und medicinisch-pharmaceutische Botanik...* Grosse Ausgabe. Berlin. pp. 16–23.

Schmitz, F. 1892b. Die systematische Stellung der Gattung *Thorea* Bory. Ber. Deutsch. Bot. Ges. 10: 115–142.

Schmitz, F. 1893. Die Gatung *Lophothalia* J. Ag. Be. Deutsch Bot. Ges. 11: 212–232.

Schmitz, F. 1896a. Bangiaceae. In: Engler, A. & Prantl, K. (eds), *Die natürlichen Pflanzenfamilien*...1(2). Leipzig. pp. 307–316.

Schmitz, F. 1896b. Compsopogonaceae. In: Engler, A. & Prantl, K. (eds), *Die natürlichen Pflanzenfamilien*...1(2). Leipzig. pp. 318–320.

Schneider, C. W. & Searles, R. B. 1991. *Seaweeds of the southeastern United States; Cape Hatteras to Cape Canaveral.* Durham, North Carolina: Duke University Press. xiv + 553 pp.

Schumacher, G. J. & Whitford, L. A. 1961. Additions to the fresh-water algae in North Carolina. V. J. Elisha Mitchell Sci. Soc. 77: 274–280.

Scott, J. 1983. Mitosis in the freshwater red alga *Batrachospermum ectocarpum*. Protoplasma 118: 56–70.

Segawa, S. 1939. Two species of *Sirodotia* found in Japan. Bot. & Zool. 7: 2033–2036. (in Japanese)

Segawa, S. & Kamura, S. 1960. *Marine flora of Ryukyu Islands.* Biol. Inst. Ryukyus. 72 pp. (in Japanese)

Setchell, W. A. 1890. Concerning the structure and development of *Tuomeya fluviatilis*. Proc. Amer. Acad. Arts & Sci. 25: 53–68.

Seto, R. 1977. On the vegetative propagation of a fresh water red alga, *Hildenbrandia rivularis* (Liebm.) J. Ag. Jap. J. Phycol. 25: 129–136. (in Japanese).

Seto, R. 1979. Comparative study of *Thorea gaudichaudii* (Rhodophyta) from Guam and Okinawa. Micronesica 15: 35–40.

Seto, R. 1982. Notes on the family Compsopogonaceae (Rhodophyta, Bangiales) in Okinawa Prefecture, Japan. Jap. J. Phycol. 30: 57–62.

Seto, R. 1985. Typification of *Caloglossa ogasawaraensis* Okamura (Ceramiales, Rhodophyta). Jap. J. Phycol. 33: 317–319.

Seto, R. 1987. Study of a freshwater red alga, *Compsopogonopsis fruticosa* (Jao) Seto comb. nov. (Compsopogonales, Rhodophyta) from China. Jap. J. Phycol. 35: 265–267.

Seto, R. 1993a. *Compsopogon oishii* Okamura. In: Hori, T. (ed.), *An illustrated atlas of the life history of algae. Vol. 2. Brown and red algae.* Tokyo: Uchida Rokakuho. pp. 170–171. (in Japanese)

Seto, R. 1993b. *Hildenbrandia rivularis* (Liebman) J. Agardh. In: Hori, T. (ed.), *An illustrated atlas of the life history of algae. Vol. 2. Brown and red algae.* Tokyo: Uchida Rokakuho. pp. 270–271. (in Japanese)

Seto, R. 1998a. *Thorea clavata.* In: Yamagishi, T. & Akiyama, M. (eds), *Photomicrographs of freshwater algae.* Vol. 20. Tokyo: Uchida Rokakuho. p. 82. (in Japanese and English)

Seto, R. 1998b. *Thorea prowsei.* In: Yamagishi, T. & Akiyama, M. (eds), *Photomicrographs of freshwater algae.* Vol. 20. Tokyo: Uchida Rokakuho. p. 83. (in Japanese and English)

Seto, R., Hirose, H. & Kumano, S. 1974. On the growth of a fresh water red alga, *Hildenbrandia rivularis*. Bull. Jap. Soc. Phycol. 22: 10–16. (in Japanese)

Seto, R. & Jao, C. C. 1984. Morphological study of the freshwater red alga, *Caloglossa leprieurii* (Mont.) J. Ag. var. *angusta* Jao (Rhodophyta, Ceramiales) from China. Jap. J. Phycol. 32: 216–220.

Seto, R. & Kumano, S. 1993. Reappraisal of some taxa of the genera *Compsopogon* and *Compsopogonopsis* (Compsopogonaceae, Rhodophyta). Jap. J. Phycol. 41: 333–340.

Seto, R., Migita, S., Madono, K. & Kumano, S. 1993. A freshwater red alga, *Thorea okadai* Yamada from the Yasumuro-river in Hyogo Prefecture and the geographical distribution of the species of *Thorea* in Japan. Jap. J. Phycol. 41: 355–357. (in Japanese)

Seto, R., Yadava, R. N. & Kumano, S. 1991. Development of short spinous branchlets of *Compsopogon aeruginosus* var. *catenatum* (Compsopogonaceae, Rhodophyta). Jap. J. Phycol. 39: 369–373.

Sheath, R. G. 1984. The biology of freshwater red algae. In: Round, F. E. & Chapman, D.J., (eds), *Progress in Phycological Research*, Vol. 3: 89–157, Biopress, Bristol, U.K.

Sheath *et al.* 1986a. See Sheath, Morison, Cole & Vanalstyne

Sheath *et al.* 1986b. See Sheath, Morison, Korch, Kacznarczyk & Cole

Sheath *et al.* 1989. See Sheath, Hamilton, Hambrook & Cole

Sheath *et al.* 1996a. See Sheath, Müller, Colbo & Cole

Sheath *et al.* 1996b. See Sheath, Müller, Larson & Cole

Sheath *et al.* 1996c. See Sheath, Müller, Vis & Entwisle

Sheath *et al.* 1996d. See Sheath, Müller, Whittick & Cole

Sheath *et al.* 1996e. See Sheath, Vis, Hambrook & Cole

Sheath, R. G. & Burkholder, J. M. 1983. Morphometry of *Batrachospermum* populations intermediate between *B. boryanum* and *B. ectocarpum* (Rhodophyta). J. Phycol. 19: 324–331.

Sheath, R. G. & Burkholder, J. M. 1985. Characteristics of softwater streams in Rhode Island. II. Composition and seasonal dynamics of macroalgal communities. Hydrobiologia 128: 109–118.

Sheath, R. G. & Cole, K. M. 1980. Distribution and salinity adaptations of *Bangia atropurpurea* (Rhodophyta), a putative migrant into the Laurentian Great Lakes. J. Phycol. 16: 412–420.

Sheath, R. G. & Cole, K. M. 1984a. Systematics of *Bangia* (Rhodophyta) in North America. I. Biogeographic trends in morphology. Phycologia 23: 383–396.

Sheath, R. G. & Cole, K. M. 1984b. Computerized image analysis of *Batrachospermum* (Rhodophyta): biogeographic trends in morphometry. J. Phycol. 20 (suppl.): 7. (abstract)

Sheath, R. G. & Cole, K. M. 1990a. Differential alcian blue staining in freshwater Rhodophyta. Brit. Phycol. J. 25: 281–285.

Sheath, R. G. & Cole, K. M. 1990b. *Batrachospermum heterocorticum* sp. nov. and *Polysiphonia subtilissima* (Rhodophyta) from Florida spring-fed streams. J. Phycol. 26: 563–568.

Sheath, R. G. & Cole, K. M. 1992. Biogeography of stream macroalgae in North America. J. Phycol. 28: 448–460.

Sheath, R. G. & Cole, K. M. 1993. Distribution and systematics of *Batrachospermum* (Batrachospermales, Rhodophyta) in North America. 2. Chromosome numbers. Phycologia 32: 304–306.

Sheath, R. G. & Cole, K. M. 1996. Stream macroalgae of the Fiji Islands: a preliminary study. Pacific Sci. 50: 46–54.

Sheath, R. G. & Hambrook, J. A. 1988. Mechanical adaptations to flow in freshwater red algae. J. Phycol. 24: 107–111.

Sheath, R. G. & Hambrook, J. A. 1990. Freshwater ecology. In: Cole, K. M. & Sheath, R. G. (eds), *Biology of the red algae*. Cambridge: Cambridge University Press. pp. 423–453.

Sheath, R. G. & Hymes, B. J. 1980. A preliminary investigation of the freshwater red algae in streams of southern Ontario, Canada. Canad. J. Bot. 58: 1295–1318.

Sheath, R. G. & Morison, M. O. 1982. Epiphytes on Cladophora glomerata in the Great Lakes and St. Lawrence Seaway with particular reference to the red alga *Chroodactylon ramosum* (= *Asterocytis smargdina*). J. Phycol. 18: 385–391.

Sheath, R. G. & Müller, K. M. 1997. Ultrastructure of carpogonia and carpogonial branches of *Batrachospermum helminthosum* and *Batrachospermum involutum* (Batrachospermales, Rhodophyta). Phycol. Res. 45: 177–181.

Sheath, R. G. & Müller, K. M. 1998. A proposal for a new red algal order, the Balbianiales. J. Phycol. 34 (suppl.): 54. (abstract)

Sheath, R. G. & Müller, K. M. 1999. Systematic status and phylogenetic relationships of the freshwater genus *Balbiania* (Rhodophyta). J. Phycol. 35: 855–864.

Sheath, R. G. & Vis, M. L. 1995. Distribution and systematics of *Batrachospermum* (Batrachospermales, Rhodophyta) in North America. 7. Section *Hybrida*. Phycologia 34: 431–438.

Sheath, R. G. & Whittick, A. 1995. The unique gonimoblast propagules of *Batrachospermum breutelii* (Batrachospermales, Rhodophyta). Phycologia 34: 33–38.

Sheath, R. G., Hellebust, J. A. & Sawa, T. 1979a. Effects of low light and darkness on structural transformations in plastids of the Rhodophyta. Phycologia 18: 1–12.

Sheath, R. G., Hellebust, J. A. & Sawa, T. 1979b. Floridean starch metabolism of *Porphyridium purpureum* (Rhodophyta). I. Changes during aging of batch cultures. Phycologia 18: 149–163.

Sheath, R. G., Hellebust, J. A. & Sawa, T. 1979c. Floridean starch metabolism of *Porphyridium purpureum* (Rhodophyta). II. Changes during the cell cycle. Phycologia 18: 185–190.

Sheath, R. G., Hellebust, J. A. & Sawa, T. 1981. Floridean starch metabolism of *Porphyridium purpureum* (Rhodophyta). III. Effects of darkness and metabolic inhibitors. Phycologia 20: 22–31.

Sheath, R. G., Müller, K. M. & Sherwood, A. R. 2000. A proposal for a new red algal order, the Thoreales. J. Phycol. 36, Suppl. 62.

Sheath, R. G., Vis, M. L. & Cole, K. M. 1992. Distribution and systematics of *Batrachospermum* (Batrachospermales, Rhodophyta) in North America. 1. Section *Contorta*. J. Phycol. 28: 237–246.

Sheath, R. G., Vis, M. L. & Cole, K. M. 1993a. Distribution and systematics of the freshwater red algal family Thoreaceae in North America. Eur. J. Phycol. 28: 231–241.

Sheath, R. G., Vis, M. L. & Cole, K. M. 1993b. Distribution and systematics of freshwater Ceramiales (Rhodophyta) in North America. J. Phycol. 29: 108–117.
Sheath, R. G., Vis, M. L. & Cole, K. M. 1993c. Distribution and systematics of *Batrachospermum* (Batrachospermales, Rhodophyta) in North America. 3. Section *Setacea*. J. Phycol. 29: 719–725.
Sheath, R. G., Vis, M. L. & Cole, K. M. 1994a. Distribution and systematics of Batrachospermum (Batrachospermales, Rhodophyta) in North America. 4. Section *Virescentia*. J. Phycol. 30: 108–117.
Sheath, R. G., Vis, M. L. & Cole, K. M. 1994b. Distribution and systematics of *Batrachospermum* (Batrachospermales, Rhodophyta) in North America. 5. Section *Aristata*. Phycologia 33: 404–414.
Sheath, R. G., Vis, M. L. & Cole, K. M. 1994c. Distribution and systematics of *Batrachospermum* (Batrachospermales, Rhodophyta) in North America. 6. Section *Turfosa*. J. Phycol. 30: 872–884.
Sheath, R. G., Whittick, A. & Cole, K. M. 1994. *Rhododraparnaldia oregonica*, a new freshwater red algal genus and species intermediate between the Acrochaetiales and the Batrachospermales. Phycologia 33: 1–7.
Sheath, R. G., Hamilton, P. B., Hambrook, J. A. & Cole, K. M. 1989. Stream macroalgae of the eastern boreal forest region of North America. Canad. J. Bot. 67: 3553–3562.
Sheath, R. G., Morison, M. O., Cole, K. M. & Vanalstyne, K. L. 1986. A new species of freshwater Rhodophyta, *Batrachospermum carpocontortum*. Phycologia 25: 321–330.
Sheath, R. G., Müller, K. M., Colbo, M. H. & Cole, K. M. 1996. Incorporation of freshwater Rhodophyta into the case of chironomid larvae (Chironomidae, Diptera) from North America. J. Phycol. 32: 949–952.
Sheath, R. G., Müller, K. M., Larson, D. J. & Cole, K. M. 1996 ('1995'). Incorporation of freshwater Rhodophyta into the cases of caddisflies (Trichoptera) from North America. J. Phycol. 31: 889–896.
Sheath, R. G., Müller, K. M., Vis, M. L. & Entwisle, T. J. 1996. A re-examination of the morphology, ultrastructure and classification of genera in the Lemaneaceae (Batrachospermales, Rhodophyta). Phycol. Res. 44: 233–246.
Sheath, R. G., Müller, K. M., Whittick, A. & Entwisle, T. J. 1996. A re-examination of the morphology and reproduction of *Nothocladus lindaueri* (Batrachospermales, Rhodophyta). Phycol. Res. 44: 1–10.
Sheath, R. G., Vis, M. L., Hambrook, J. A. & Cole, K. M. 1996. Tundra stream macroalgae of North America: composition, distribution and physiological adaptations. In: Kristiansen, J. (ed.), *Biogeography of freshwater algae*. Dordrecht: Kluwer Academic Publishers. pp. 67–82. (Hydrobiologia 336: 67–82)
Sheath, R. G., Morison, M. O., Korch, J. E., Kaczmarczyk, D. & Cole, K. M. 1986. Distribution of stream macroalgae in south-central Alaska. Hydrobiologia 135: 259–269.
Sherwood, A. R. & Sheath, R. G. 1998. A comparison of freshwater and marine *Hildenbrandia* (Rhodophyta) in North America. J. Phycol. 34 (suppl.): 54–55. (abstract)
Sherwood, A. R. & Sheath, R. G. 1999. Biogeography and systematics of *Hildenbrandia* (Rhodophyta, Hildenbrandiales) in North America: inference from morphometrics and *rbc*L and 18S rRNA gene sequence analyses. Eur. J. Phycol. 34: 523–532.
Sherwood, A. R. & Sheath, R. G. 2000. Biogeography and systematics of *Hildenbrandia* (Rhodophyta, Hildenbrandiales) in Europe: inference from morphometrics and *rbc*L and 18S rRNA gene sequence analyses. Eur. J. Phycol. 35: 143–152.
Shi, Z. 1994. Two new species of the genus *Batrachospermum* (Rhodophyta) in China. Acta Phytotax. Sin. 32: 275–280.
Shi, Z., Hu, Z. & Kumano, S. 1993. *Batrachospermum heteromorphum*, sp. nov. (Rhodophyta) from Hubei Province, China. Jap. J. Phycol. 41: 295–302.
Shi, Z., Wie, Y. & Li, Y. 1994. A preliminary investigation on Rhodophyta, Cryptophyta, Pyrrophyta, Chrysophyta, Xanthophyta, Chloromomadophyceae and Charophyta from the Wuling mountain region, China, p. 218–226. In: *Compilation of Reports on the Survey of Algal Resources in South-Western China* (Shi, Z. *et al.* eds), Science Press, Beijing, China.
Shubert, L. E. (ed.) 1984. *Algae as ecological indicators*. London etc.: Academic Press. xii + 434 pp.
Shyam, R. & Sarma, Y. S. R. K. 1980. Cultural observstions on the morphology, reproduction and cytology of a freshwater alga *Compsopogon* Mont. from India. Nova Hedwigia 32: 745–765.
Silva, P. C. 1952. A review of nomenclatural conservation in the algae from point of view of the type method. Univ. Calif. Publ. Bot. 25: 241–323.

Silva, P. C. 1959. Remarks on algal nomenclature. II. Taxon 8: 60–64.
Silva, P. C., Basson, P. W. & Moe, R. L. 1996. *Catalogue of the Benthic Marine Algae of the Indian Ocean*. Univ. California Press, Berkeley. xiv+1259pp.
Silva, P. C. & Cleary, A. P. 1954. The structure and reproduction of the red alga *Platysiphonia*. Amer. J. Bot. 41: 251–260.
Singh, D. 1960. Occurrence of *Thorea ramosissima* Bory in India. Curr. Sci. 29: 490.
Singh, M. 1964. Morphology and reproduction of a form of *Compsopogon hookeri* Mont. from Delhi, India. Phykos 3: 37–40.
Singh, N. B. & Pandey, D. C. 1986. A new form of *Compsopogon aeruginosus* (J. Ag.) Kützing. Phykos 25: 84–87.
Sirodot, S. 1870. Organes et phénomènes de la fécondation dans le genre *Lemanea*. Compt. Rend. Acad. Sci. [Paris] 70: 691–694.
Sirodot, S. 1872. Étude anatonique, organogénique et physiologique sur les algues d'eau doce de la familie des Lémanéacées. Ann. Sci. Nat. Bot., ser. 5, 16: 5–95.
Sirodot, S. 1873a. Nouvelle classification des algues d'eau douce du genre *Batrachospermum*; devéloppement; générations alternantes. Compt. Rend. Acad. Sci. [Paris] 76: 1216–1220.
Sirodot, S. 1873b. Développement des algues d'eau douce du genre *Batrachospermum*; générations alternantes. Compt. Rend. Acad. Sci. [Paris] 76: 1335–1339.
Sirodot, S. 1874. Observations sur les phénomènes essentiels de la fécondation chez les algues d'eau douce du genre *Batrachospermum*. Compt. Rend. Acad. Sci. [Paris] 79: 1366–1369.
Sirodot, S. 1875. Observations sur le développement des algues d'eau douce composant le genere *Batrachospermum*. Bull. Soc. Bot. France 22: 128–145.
Sirodot, S. 1876. Le *Balbiania investiens*. Étude organogénique et physiologique. Ann. Sci. Nat. Bot., ser. 6, 3: 146–174.
Sirodot, S. 1877. Rapports morphologiques entre les anthéridies et les sporules développées dans la ramification verticillé d'une forme particulière du *Batrachospermum moniliforme*. Compt. Rend. Acad. Sci. [Paris] 84: 683–684.
Sirodot, S. 1880. Transformation d'une ramification fructifère issue de fécondation, et une végétation prothalliforme. Compt. Rend. Acad. Sci. [Paris] 91: 862–864.
Sirodot, S. 1884. *Les Batrachospermes: organisation, fonctions, développement, classification*. Paris. 299 pp., 50 pls.
Skuja, H. 1926. Eine neue Süsswasserbangiacee *Kyliniella latvica* n. g., n. sp. Acta Horti Bot. Univ. Latv. 1: 1–6.
Skuja, H. 1928. Vorarbeiten zu einer Algenflora von Lettland. IV. Acta Horti Bot. Univ. Latv. 3: 103–218.
Skuja, H. 1931a. Einiges zur Kenntnis der brasilianischen Batrachospermen. Hedwigia 71: 78–87.
Skuja, H. 1931b. Untersuchungen über die Rhodophyceen des Süsswassers. [I, II]. Arch. Protistenk. 74: 297–309.
Skuja, H. 1933. Untersuchungen über die Rhodophyceen des Süsswassers. III. *Batrachospermum breutelii* Rabenhorst und seine Brutkörper. Arch. Protistenk. 80: 357–366.
Skuja, H. 1934. Untersuchungen über die Rhodophyceen des Süsswassers. [IV–VI]. Beih. Bot. Centralb., Abt. B, 52: 173–192.
Skuja, H. 1938a. Die Süsswasserrhodophyceen der Deutschen Limnologischen Sunda-Expedition. Arch. Hydrobiol. Suppl. 15: 603–637.
Skuja, H. 1938b. Comments on fresh-water Rhodophyceae. Bot. Rev. 4: 665–676.
Skuja, H. 1939. Versuch einer systematischen Einteilung der Bangioideen oder Protoflorideen. Acta Horti Bot. Univ. Latv. 11/12: 23–38.
Skuja, H. 1944. Untersuchungen über die Rhodophyceen des Süsswassers. [VII–XII]. Acta Horti Bot. Univ. [Latv.] 14: 3–64.
Skuja, H. 1964. Weiteres zur Kenntnis der Süsswasserrhodophyceen der Gattung *Nothocladus*. Rev. Algol., ser. 2, 7: 304–314.
Skuja, H. 1969. Eigentümliche morphologische Anpassung eines *Batrachospermum* gegen mechanishe Schädigung in fliessenden Wasser. Österr. Bot. Z. 116: 55–64.
Smith, G. M. 1933. *The fresh-water algae of the United States*. New York: McGraw-Hill. xi + 716 pp.
Smith, G. M. 1938. *Cryptogamic botany. Vol. 1. Algae and fungi*. New York: McGraw-Hill. viii + 545 pp.

Smith, G. M. 1950. *The fresh-water algae of the United States*. 2nd ed. New York: McGraw-Hill. vii + 719 pp.

Smith, J. E. & Sowerby, J. 1800. *English botany*... Vol. 10. London. Pls. 649–720.

Soto, R. 1982. *Caloglossa ogasawaraensis* ['ogassuarensis'] Skuja (Rhodophyta, Ceramiales, Delesseriaceae) en Costa Rica. Brenesia 19/20: 251–253.

South, G. R. & Whittick, A. 1987. *Introduction to phycology*. Oxford etc.: Blackwell Scientific Publications. viii + 341 pp.

Squires, L .E., Rushforth, S. R. & Brotherson, J. D. 1979. Algal response to a thermal effluent: study of a power station on the Provo River, Utah, USA. Hydrobiologia 63: 17–32.

Staker, R. D. 1976. Algal composition and seasonal distribution in the Dunk River system, Prince Edward Island. Nova Hedwigia 27: 731–745.

Starmach, K. 1952. The reproduction of the fresh water Rhodophyceae *Hildenbrandia rivularis* (Liebm.) J. Ag.. Acta Soc. Bot. Polon. 21: 447–474. (in Polish)

Starmach, K. 1969a. Growth of thalli and reproduction of the red alga *Hildenbrandtia rivularis* (Liebm.) J. Ag.. Acta Soc. Bot. Polon. 38: 523–533.

Starmach, K. 1969b. *Hildenbrandtia rivularis* and its associated algae in the stream Cedronka near Wejherowo (Gdansk voivode). Fragm. Florist. et Geobot. 15: 387–398. (in Polish)

Starmach, K. 1969c. *Hildenbrandtia rivularis* (Liebm.) J. Ag. and *Chamaesiphon fuscoviolaceus* n. sp., and accompanying algae in the stream Lubogoszcz in the Beskid Wyspowy (Polish Western Carpathians). Fragm. FLorist. et Geobot. 15: 487–501. (in Polish)

Starmach, K. 1975. Algae from montane streams on the island of Mahé, in the Seychelles. Acta Hydrobiol. 17: 201–209.

Starmach, K. 1977. Phaeophyta – brunatnice. Rhodophyta – krasnorosty. In: Starmach, K. & Sieminska, J. (eds), *Flora slodkowodna polski*. Vol. 14. Warsaw-Kraków: Panstwowe Wydawnictwo Naukowe. 445 pp.

Starmach, K. 1978. *Compsopogon aeruginosus, Pithophora varia* und epiphytische Cyanophyceen im Bassin des Gewachshauses im Botanischen Garten in Krakow. Fragm. Florist. et Geobot. 24: 157–164.

Starmach, K. 1980. *Batrachospermum vagum* (Roth) Ag. und epiphytische Blaualgen im See Silm bei Ilawa. Fragm. Florist. et Geobot. 26: 165–174.

Starmach, K. 1984a. Red algae in the Kryniczanla stream. Fragm. Florist. et Geobot. 28: 257–293.

Starmach, K. 1984b. New species to Poland and new localities of some species of *Batrachospermum* Roth (Rhodophyceae). Fragm. Floirst. et Geobot. 28: 303–309. (in Polish)

Starmach, K. 1985. *Chantransia hermannii* (Roth) Desvaux and the systematic position of the genera *Chantransia, Pseudochantransia* and *Audouinella*. Acta Soc. Bot. Polon. 54: 273–284.

Starmach, K. 1988. Some taxa of freshwater red algae (Rhodophyta) from Cuba. Fragm. Florist. et Geobot. 31/32: 473–494.

Stein, J. R. & Borden, C. A. 1980. Checklist of freshwater algae of British Columbia. Syesis 12: 3–39.

Steinman. A. D. & Lamberti, G. A. 1988. Lotic algal communities in the Mt. St. Helens region six years following the eruption. J. Phycol. 24: 482–489.

Swale, E. M. F. 1963. Notes on the morphology and anatomy of *Thorea ramosissima* Bory. J. Linn. Soc. Bot. 58: 429–434.

Swale, E. M. F. & Belcher, J. H. 1963. Morphological observations on wild and cultured material of *Rhodochorton investiens* (Lenormand) nov. comb. (*Balbiania investiens* (Lenorm.) Sirodot.). Ann. Bot. (London), ser. 2, 27: 281–290.

Symoens, J. J. 1975. Une nouvelle station d'*Hildenbrandia rivularis* en Belgique. Bull. Soc. Roy. Bot. Belgique 108: 328–329.

Taft, C. E. & Taft, C. W. 1971. The algae of western Lake Erie. Bull. Ohio Biol. Surv., ser. 2, 4(1). 185 pp.

Tanaka, J. 1987. Taxonomic studies on Japanese mangrove macroalgae III. Two taxa from the Amami Islands. Bull. Nat. Sci. Mus. [Tokyo], Ser. B (Bot.), 13: 17–23.

Tanaka, J. 1989a. Taxonomy of the genus *Bostrychia* (Rhodophyceae). Misc. Jpn. Bot. Tax. 8: 15–23. (in Japanese).

Tanaka, J. 1989b. Morphology of *Bostrychia tenella* (Rhodophyceae) in Indonesia. Bull. Nat. Sci. Mus. [Tokyo], Ser. B (Bot.), 15: 115–124.

Tanaka, J. 1991. Morphology of *Bostrychia radicans* (Montagne) Montagne (Rhodophyceae) in Indonesia. Bull. Nat. Sci. Mus. [Tokyo], Ser. B (Bot.), 17: 5–13.

Tanaka, J. 1991a. Macroalgae in mangrove forests at Amatory Bay of Iriomote Island. Bull. Inst. Oceanic Res. & Develop., Tokai Univ. 11/12: 113–119. (in Japanese).

Tanaka, J. 1992. Reproductive structure of *Caloglossa leprieurii* f. *continua* (Ceramiales, Rhodophyceae) in Japan. Bull. Nat. Sci. Mus. [Tokyo], Ser. B (Bot.), 18: 139–147.

Tanaka, J. & Chihara, M. 1984a. Taxonomic studies of Japanese mangrove macroalgae. I. Genus *Bostrychia* (Ceramiales, Rhodophyceae).(1). Bull. Natl. Sci. Mus. [Tokyo], Ser. B (Bot.), 10: 115–126.

Tanaka, J. & Chihara, M. 1984b. Taxonomic studies of Japanese mangrove macroalgae. I. Genus *Bostrychia* (Ceramiales, Rhodophyceae) (2). Bull. Natl. Sci. Mus. [Tokyo], Ser. B (Bot.), 10: 169–176.

Tanaka, J. & Chihara, M. 1985a. Taxonomic studies of Japanese mangrove macroalgae. II. Two taxa of *Caloglossa* (Ceramiales, Rhodophyta). Bull. Natl. Sci. Mus. [Tokyo], Ser. B (Bot.), 11: 41–50.

Tanaka, J. & Chihara, M. 1985b. Species composition and vertical distribution of macroalgae in brackish waters of Japanese mangrove forests. Bull. Natl. Sci. Mus. [Tokyo], Ser. B (Bot.), 13: 141–150.

Tanaka, J. & Chihara, M. 1988a. Macroalgae in Indonesian mangrove forests. Bull. Natl. Sci. Mus. [Tokyo], ser. B (Bot.), 14: 93–106.

Tanaka, J. & Chihara, M. 1988b. Macroalgal flora in mangrove brackish areas of East Indonesia. In: Ogino, K. & Chihara, M. (eds), *Biological system of mangroves. A report of East Indonesian Mangrove Expedition*. Ehime: Ehime University. pp. 21–34.

Tanaka, J. & Kamiya, M. 1993. Reproductive structure of *Caloglossa ogasawaraensis* Okamura (Ceramiales, Rhodophyceae) in nature and culture. Jap. J. Phycol. 41: 113–121.

Tanaka, J. & Murakami, H. 1996. Marine and brackish water algae of Yokohama City. The organisms of river and sea in Yokohama, vol. 7. Sea. Report from the Environmental Protection Bureau of Yokohama City, No. 183. pp. 219–230. (in Japanese).

Tanaka, J., Kobayashi, A. & Kashimura, N. 1999. Marine and brackish water algae of Yokohama City. The organisms of river and sea in Yokohama, vol. 8. Sea. Report from the Environmental Protection Bureau of Yokohama City, No. 188. pp. 125–152. (in Japanese).

Tanaka, J. & Shameel, M. 1992. Macroalgae in mangrove forests of Pakistan. Nakaike & Malik (eds), *Cryptogamic flora of Pakistan*, National Science Museum, Tokyo. 1: 75–85.

Tanaka, T. 1950. On the species of *Bangia* from Japan. Bot. Mag. Tokyo 63: 163–169.

Tanaka, T. & Pham-Hoàng Hô, 1962. Notes on some marine algae from Viet-Nam. I. Mem. Fac. Fish. Kagoshima Univ. 11: 24–40.

Taylor, W. R. 1957. *Marine algae of the northeastern coast of North America*. 2nd ed. Ann Arbor: University of Michigan Press. ix + 509 pp.

Taylor, W. R. 1960. Marine algae of the eastern tropical and subtropical coast of the Americas. University of Michigan Press, Ann Arbor, 870 pp.

Tell, G. 1985. Catálogo de las algas de agua dulce de la Rep'blica Argentina. Biblioth. Phycol. 70. iii + 283 pp.

Thore, J. 1800. Sur la *Conferva hispida*. Mag. Encycl., An 5, 6: 398–403.

Thérézien, Y. 1985. Contribution à l'étude des algues d'eau douce de la Guyane Française à l'exclusion des diatomées. Biblioth. Phycol. 72. 275 pp.

Tiffany, L. H. & Britton, M. E. 1952. *The algae of Illinois*. Chicago: Univ. Chicago Press. xiv + 407 pp.

Tilden, J. E. 1898. Observations on some West American thermal algae. Bot. Gaz. 25: 89–105.

Tilden, J. E. 1935. *The algae and their life relations*. Minneapolis: Univ. Minnesota Press. xii + 550 pp.

Tokida, J. 1938. Phycological observations. IV. Trans. Sapporo Nat. Hist. Soc. 15: 212–222.

Tokida, J. 1939. On some little known marine algae of Japan, with special reference to the species *Bostrychia*. Bot. & Zool. 7: 522–530. (in Japanese)

Tokida, J. 1941. On some little known marine algae of Japan (2) . Bot. & Zool. 9: 49–56. (in Japanese)

Tomas, X. 1981. *Thorea ramosissima* en un canal del littoral valenciano. Folia Bot. Misc. 2: 71–74.

Umezaki, I. 1960. On *Sirodotia delicatula* Skuja from Japan. Acta Phytotax. et Geobot.. 18: 208–214.

Umezaki, I. 1967. The tetrasporophyte of *Nemalion vermiculare* Suringar. Rev. Algol., ser. 2, 9: 19–24.

Vahl, M. 1802. Endeel kryptogamiske Planter fra St.-Croix. Skr. Naturhist.-Selsk. 5(2): 29–47.

Van Meel, L. I. J. 1939. Sur la répartition du genre *Batrachospermum* en Belgique. Bull. Jard. Bot. État 15: 319–331.

Van Meel, L. 1943. Sur la répartition du genre *Batrachospermum* en Belgique. II. Bull. Jard. Bot. État 17: 49–53.

Vis, M. L., Carlson, T. A., & Sheath, R.G. 1991. Phenology of *Lemanea fucina* (Rhodophyta) in a Rhode Island river, USA. Hydrobiologia 222: 141–146.

Vis, M. L., Saunders, G. W., Sheath, R. G., Dunse, K. & Entwisle, T. J. 1998. Phylogeny of the Batrachospermales (Rhodophyta) inferred from rbcL and 18S ribosomal RNA gene sequences. J. Phycol. 34: 341–350.

Vis, M. L. & Sheath, R. G. 1992. Systematics of the freshwater red algal family Lemaneaceae in North America. Phycologia 31: 164–179.

Vis, M. L. & Sheath, R. G. 1993. Distribution and systematics of *Chroodactylon* and *Kyliniella* (Porphyridiales, Rhodophyta) from North American streams. Jpn. J. Phycol. 41: 237–241.

Vis, M. L. & Sheath, R. G. 1996. Distribution and systematics of *Batrachospermum* (Batrachospermales, Rhodophyta) in North America. 9. Section *Batrachospermum*: description of five new species. Phycologia 35: 124–134.

Vis, M. L. & Sheath, R. G. 1997. Biogeography of *Batrachospermum gelatinosum* (Batrachospermales, Rhodophyta) in North America based on molecular and morphological data. J. Phycol. 33: 520–526.

Vis, M. L. & Sheath, R. G. 1998. A molecular investigation of the systematic relationship among *Sirodotia* species (Batrachospermales, Rhodophyta) in North America. J. Phycol. 34(suppl.): 61. (abstract)

Vis, M. L. & Sheath, R. G. 1999. A molecular investigation of the systematic relationship among *Sirodotia* species (Batrachospermales, Rhodophyta) in North America. Phycologia 38: 261–266.

Vis, M. L., Sheath, R. G. & Cole, K. M. 1992. Systematics of the freshwater red algal family Compsopogonaceae in North America, Phycologia 31: 564–575.

Vis, M. L., Sheath, R. G. & Cole, K. M. 1996a. Distribution and systematics of *Batrachospermum* (Batrachospermales, Rhodophyta) in North America. 8a. Section *Batrachospermum*: *Batrachospermum gelatinosum*. Eur. J. Phycol. 31: 31–40.

Vis, M. L., Sheath, R. G. & Cole, K. M. 1996b. Distribution and systematics of *Batrachospermum* (Batrachospermales, Rhodophyta) in North America. 8b. Section *Batrachospermum*: previously described species excluding *Batrachospermum gelatinosum*. Eur. J. Phycol. 31: 189–199.

Vis, M. L., Sheath, R. G. & Entwisle, T. J. 1995. Morphometric analysis of *Batrachospermum* section *Batrachospermum* (Batrachospermales, Rhodophyta) type specimens. Eur. J. Phycol. 30: 35–55.

Vis, M. L., Sheath, R. G., Hambrook, J. A. & Cole, K. M.. 1994. Stream macroalgae of the Hawaiian Islands: a preliminary study. Pacific Science 48: 175–187.

von Stosch, H. A. & Theil, G. 1979. A new mode of life history in the freshwater red algal genus *Batrachospermum*. Amer. J. Bot. 66: 105–107.

Watanabe, M. 1994. *Chroodactylon ramosum*. In: Yamagishi, T. & Akiyama, M. (eds), *Photomicrographs of freshwater algae.* Vol. 12. Tokyo: Uchida Rokakuho. p. 5. (in Japanese)

Webber, E. E. 1963. The ecology of some attached algae of Worcester County. Massachusetts. Amer. Midl. Naturalist 70: 175–186.

Weber-van Bosse, A. 1896. Notes on *Sarcomenia miniata* Ag. J. Bot. 34: 281–285.

Weber-van Bosse, A. 1921. Liste des algues du Siboga. II. Rhodophyceae. Première partie. Protoflorideae, Nemalionales, Cryptonemiales. Siboga-Exped. Monogr. 59: 187–310.

Webster, R. N. 1958. The life history of the freshwater red alga *Tuomeya fluviatilis* Harvey. Butler Univ. Bot. Stud. 13: 141–159.

West, J. A. 1972. Environmental control of hair and sporangial formation in the marine red alga *Acrochaetium proskaueri* sp. nov. Proc. 7th Int. Seaweed Symp. [Sapporo]. Tokyo. pp. 377–384.

West, W. & West G. S. 1897. Welwitsch's African freshwater algae. J. Bot. 35: 1–7, 33–42, 77–89, 113–122, 172–183, 235–243, 264–272, 287–304.

West, G. S. & Fritsch, F. E. 1927. *A treatise on the British freshwater algae.* 2nd ed. Cambridge. xviii + 534 pp.

Whitford, L. A. & Schumacher, G. J. 1963. Communities of algae in North Carolina streams and their seasonal relations. Hydrobiologia 22: 133–196.

Whitford, L. A. & Schumacher, G. J. 1968. Notes on the ecology of some species of fresh-water algae. Hydrobiologia 32: 225–236.

Whitford, L. A. 1956. The communities of algae in the springs and spring streams of Florida. Ecology 37: 433–442.
Whittick, A. 1992. Culture, cytology and reproductive biology of *Callithamnion corymbosum* (Rhodophyceae, Celamiaceae) from a western Newfoundland fjord. Canad. J. Bot. 70: 1154–1156.
Wittmann, W. 1965. Aceto-iron-haematoxylin-chloral hydrate for chromosomes staining. Stain Technol. 40: 161–164.
Wittrock, V. & Nordstedt, O. 1883. Algae aquae dulcis exsiccatae... fasc. 11 (501–55); 12(551–600). Holmiae. 25.v.1883. Bot. Not. 1883: 145–152.
Woelkerling, W. J. 1971. Morphology and taxonomy of the *Audouinella* complex (Rhodophyta) in southern Australia. Austral. J. Bot. Suppl. 1. 91 pp.
Woelkerling, W. J. 1975. Observations on *Batrachospermum* (Rhodophyta) in southeastern Wisconsin streams. Rhodora 77: 467–477
Wolle, F. 1882. Fresh-water algae. VI. Bull. Torrey Bot. Club. 9: 25–30.
Wolle, F. 1887. *Fresh-water algae of the United States...* Bethlehem, Pennsylvania. pp. i–xx, 21–364.
Womersley, H. B. S. & Shepley, E. A. 1959. Studies on the *Sarcomenia* group of the Rhodophyta. Austral. J. Bot. 7: 168–223.
Wood, H. C. Jr. 1873 ("1872"). A contribution to the history of the fresh-water algae of North America. Smithsonian Contr. Knowl. 19(3). ix + 262 pp.
Woodson, B. R. & Afzal, M. 1976. The taxonomy and ecology of algae in the Appomattox River, Chesterfield County, Virginia. Virginia J. Sci. 27: 5–9.
Woodson, B. R. Jr. & Wilson, W. Jr. 1973. A systematic and ecological survey of algae in two streams of Isle of Wight County, Virginia. Castanea 38: 1–18.
Woolcott, G. W. & King, R. J. 1998. *Porphyra* and *Bangia* (Bangiaceae, Rhodophyta) in warm temperate waters of eastern Australia: Morphology and molecular analyses. Phycol. Res. 46: 111–123.
Wynne, M. J. & De Clerck, O. 1999. Taxonomic notes on *Caloglossa monosticha* and *C. saigonensis* (Delesseriaceae, Rhodophyta). Contribution from the University of Michigan Herbarium 22: 209–213.
Xie, S. L. & Ling, Y. J. 1998a. Two new species of *Compsopogon* (Rhodophyta) from Shanxi and Guangxi, China. Acta Phytotax. Sin. 36: 81–83.
Xie, S. L. & Ling, Y. J. 1998b. A new species of fresh-water red algae – *Audouinella heterospora* from China. Acta Phytotax. Sin. 36: 281–283. (in Chinese)
Yabu, H. & Tokida, J. 1966. Application of aceto-iron-haematoxylin-chloral hydrate method to chromosomes staining in marine algae. Bot. Mag. Tokyo 79: 381.
Yadava, R. N. & Kumano, S. 1985. *Compsopogon prolificus* sp. nov. (Compsopogonaceae, Rhodophyta) from Allahabad, Uttar Pradesh in India. Jap. J. Phycol. 33: 13–20.
Yadava, R. N. & Pandey, D. C. 1977. An interesting observation on a *Compsopogon* growing at Allahabad. Curr. Sci. 46: 713–714.
Yadava, R. N. & Pandey, D. C. 1980. Observations on a new variety of *Compsopogon* (*C. aeruginosus* (J. Ag.) Kützing var. *catenatum* var. nov.). Phykos 19: 15–22.
Yagi, S. & Yoneda, Y. 1940. A new species of freshwater Rhodophyceae, *Nemalionopsis turtuosa* nov. sp.. Acta Phytotax. et Geobot. 9: 82–86. (in Japanese)
Yamada, Y. 1943. On *Thorea ramosissima* Bory from Nagasaki Prefecture. J. Jap. Bot. 19: 136–138. (in Japanese).
Yamada, Y. 1949. On the species of *Thorea* from the Far Eastern Asia. J. Jap. Bot. 24: 155–158.
Yamagishi, T. 1999. *Introduction to the Freshwater Algae*. Uchida Rokakuho. Tokyo. 49 + 646 pp. (in Japanese)
Yamagishi, T. & Akiyama, M. (eds), 1984–1998. *Photomicrographs of the Freshwater Algae*. Vol. 1–20. Tokyo: Uchida Rokakuho. (in Japanese and English)
Yamagishi, T. & Kobayasi, H. 1971. Algae from *Sphagnum*-bogs of Mt. Omine. General Education Review, College of Agriculture and Veterinary Medicine, Nihon University, 1971 (7): 25–51. (in Japanese)
Yokoyama, A., Kondo, T., Yokoyama, J., Ohhashi, H. & Hara, K. 1998. Molecular systematics of cyanidian algae (Rhodophyta). Jap. J. Phycol. 46: 83. (in Japanese)
Yoneda, Y. 1949. Notes on the freshwater algae of Kikusui-sen, a rheocrene at Yoro-mura in Province Mino. J. Jap. Bot. 24: 169–175. (in Japanese)

Yoshida, T. 1959. Life-cycle of a species of *Batrachospermum* found in northern Kyushu, Japan. Jap. J. Bot. 17: 29–42.

Yoshida, T. 1993. *Batrachospermum* sp. In: Hori, T. (ed.), *An illustrated atlas of the life history of algae. Vol. 2. Brown and red algae.* Tokyo: Uchida Rokakuho. pp. 218–219. (in Japanese)

Yoshida, T. 1998. *Marine Algae of Japan.* Uchida Rokakuho. Tokyo. 25+1222 pp. (in Japanese)

Yoshizaki, M. 1986. The morphology and reproduction of *Thorea okadai* (Rhodophyta). Phycologia 25: 476–481.

Yoshizaki, M. 1993a. *Thorea okadai* Yamada. In: Hori, T. (ed.), *An illustrated atlas of the life history of algae. Vol. 2. Brown and red algae.* Tokyo: Uchida Rokakuho. pp. 226–227. (in Japanese)

Yoshizaki, M. 1993b. *Caloglossa ogasawaraensis* Okamura. In: Hori, T. (ed.), *An illustrated atlas of the life history of algae. Vol. 2. Brown and red algae.* Tokyo: Uchida Rokakuho. pp. 328–329. (in Japanese)

Yoshizaki, M., Fujita, T. & Iura, H. 1985. On *Caloglossa leprieurii* and *C. ogasawaraensis* from Kujukuri in Chiba Prefecture, Japan. Chiba Seibutsu Shi 34: 49–54. (in Japanese).

Yoshizaki, M., Fujita, T., Hatogai, T. & Iura, H. 1986. Seasonality on *Caloglossa reprieurii* and *C. ogasawaraensis* and *Bostrychia simpliciuscula* from Kujukuri in Chiba Prefecture, Japan. Chiba Seibutsu Shi 35: 64–70. (in Japanese).

Yoshizaki, M., Iura, K., Miyaji, K. & Kasaki, H. 1983. Two species of *Caloglossa* from Kujukuri hama, Chiba Prefecture. Nanki Seibutsu 25: 191–192. (in Japanese)

Yoshizaki, M & Kamiya, M. 1999. Division Rhodophyta. In: *Diversity and evolution of algae.* Chihara, M. (ed.) Shokabo, Tokyo, Japan. pp. 177–189. (in Japanese)

Yung, Y.-K., Stokes, P. & Gorham, E. 1986. Algae of selected continental and maritime bogs in North America. Canad. J. Bot. 64: 1825–1833.

Zanardini, G. 1872a. Nota intorno ad un viaggio a Borneo recentemente intrapreso dal botanico fiorentino O. Beccari. Atti Reale Ist. Veneto Sci. Lett. ed Arti, ser. 4, 1: 379–388.

Zanardini, G. 1872b. Phycearum indicarum pugillus a Cl. Eduardo Beccari ad Borneum, Sincapoore et Ceylanum annis MDCCCLXV–VI–VII collectarum... Mem. Reale Ist. Veneto Sci. Lett. ed Arti. 17: 129–170.

Zuccarello, G. C. & West, J. A. 1997. Hybridization studies in *Bostrychia*; 2. Correlation of crossing data and plastid DNA sequence data within *B. radicans* and *B. moritziana* (Ceramiales, Rhodophyta). Phycologia 36: 293–304.

(*the following references have not been seen by the author and the reviewer)

* Chessman, B. C. 1982. La Trobe Valley Water Resources Biological Studies. vol. 3. Algae and Functional Ecology. La Trobe Valley Water and Sewerage Board. Traralgon.
* Kalbe, L. 1974. Die rotalge *Thorea ramosissima* Bory in Wasser der Warnow. Limnologica 9: 239–241.
* Kerner, A. 1882. Schedae ad Floram exsiccatam Austro-Hungaricam a Museo botanico Universitatis Vindobonensis anno 1881 editam, fasc. II 8. Bot. Centralb. 10: 360–362. [This is a review by Freyn. We need to see the original, which is lacking at UC.]
*Bailey, F. M. 1893. Contributions to the Queensland flora. Queensland freshwater algae. Queensland Dept. Agric., Bot. Bull. 6.
*Dillenius, J. J. 1719. Conferva fontinalis geniculata, lubrica, major. Catta. Giss. p.199. Ephem. Nat. Cur. Cent., V & VI. appendice, p. 60, tab. 13, fig. 3.
*Hara, Y., Nagumo, T. & Katou, S. 1989. Microalgal flora of mangrove forests in Japan. In: *Recent advances in microbial ecology. publisher*. pp. 292–296.
*Hoffmann, L. 1987. Réparition et écologie d'*Hildenbrandia rivularis* (Liebm.) J. Agardh (Rhodophyceae) en Belgique et au Grand-Duche de Luxembourg. Dumortiera 38: 9–11, fig. 1.
*La Rivers, I. 1978. Algae of the Western Great Basin. Reno: Bioresources Center, University of Nevada. 390 pp.
*Meyer, R. L. 1969. The freshwater algae of Arkansas. I. Introduction and recent additions. Proc. Arkansas Acad. Sci. 23: 145–156.
*Nakatsu, C. H. 1983. The algal flora of three brown water systems. M. Sc. Thesis, University of Toronto.
*Reed, E. W. 1972. Studies on the ecology of *Lemanea fucina* Bory with notes on *Audouinella violacea* Kütz. Ph.D. Thesis, Washington State University.
*Sabater, S., Aboal, M. & Cambra. 1989. Neuvas observationes de rodofíceas en aguas epicontinentales del NE y SE de Espanã. Limnetica. 5: 93–100.

Species Index

Acarposporophytum 69
Acrochaetiaceae 39
Acrochaetiales 39
Acrochaetium amahatanum 40, 42
Acrochaetium godwardense 58
Acrochaetium godwardense 58
Acrochaetium indica 59
Acrochaetium indica 60
Acrochaetium sarmae 59
Acrochaetium sarmae 60
Ambigua 203
Anfractutofilum umbracolens 300
Aristata 162, 163
Asterocystis 15
Asterocystis ramosa 15
Audouinellaceae 39
Audouinella 39
Audouinella amahatana 40
Audouinella amahatana 42
Audouinella chalybea 51
Audouinella chalybea 54
Audouinella cylindrica 56
Audouinella cylindrica 57
Audouinella eugenea 41, 46
Audouinella eugenea 43, 52
Audouinella eugenea var. *secundata* 46
Audouinella glomerata 56
Audouinella glomerata 57
Audouinella hermannii 47
Audouinella hermannii 44, 45, 52, 53, 321
Audouinella investiens 62, 63
Audouinella lanosa 48
Audouinella lanosa 57
Audouinella leibleinii 50, 54
Audouinella macrospora 49, 53
Audouinella macrospora 46, 48, 51, 53
Audouinella meiospora 53, 61
Audouinella miniata 47
Audouinella pygmaea 50
Audouinella pygmaea 51, 53, 55
Audouinella serpens 49
Audouinella serpens 46
Audouinella sinensis 52
Audouinella sinensis 55
Audouinella subtilis 50
Audouinella tenella 41
Audouinella tenella 43, 52, 53

Baileya americana 238
Balbiania 61
Balbiania invcstiens 62
Balbiania investiens 63
Balbiania meiospora 61
Balbiania meiospora 62

Balbianiaceae 61
Balbianiales 61
Ballia 295, 296
Ballia brunonia 295, 296
Ballia callitricha 295, 296
Ballia pinnulata 298
Ballia pinnulata 296
Ballia prieurii 298
Ballia prieurii 297, 299, 310
Ballia pygmaea 298, 299
Balliaceae 295
Balliales 295
Bangia 35
Bangia atropurpurea 35
Bangia atropurpurea 36
Bangia fuscopurpurea 35
Bangiaceae 35
Bangiales 35
Batrachospermaceae 67
Batrachospermales 66
Batrachospermum 67, 70
Batrachospermum abilii 158
Batrachospermum alpestre 74
Batrachospermum ambiguum 221
Batrachospermum ambiguum 223
Batrachospermum anatinum 88
Batrachospermum anatinum 90, 91
Batrachospermum androinvolucrum 124
Batrachospermum androinvolucrum 125
Batrachospermum angolense 129
Batrachospermum antipodites 109
Batrachospermum antipodites 110, 111
Batrachospermum antiquum 116
Batrachospermum antiquum 117
Batrachospermum arcuatoideum 88
Batrachospermum arcuatum 93, 102
Batrachospermum arcuatum 91, 93
Batrachospermum aristatum 165
Batrachospermum atrum 129
Batrachospermum atrum 129, 130, 132
Batrachospermum atrum var. *puiggarianum* 131
Batrachospermum australe 162, 168
Batrachospermum australicum 209
Batrachospermum australicum 210
Batrachospermum azeredoi 148
Batrachospermum bakarense 147
Batrachospermum bakarense 148
Batrachospermum basilare 221
Batrachospermum beraense 168
Batrachospermum beraense 169
Batrachospermum bicudoi 221
Batrachospermum boryanum 97
Batrachospermum boryanum 97, 98, 99
Batrachospermum brasiliense 69

Batrachospermum brasiliense 70
Batrachospermum breutelii 156
Batrachospermum breutelii 156, 157
Batrachospermum breviarticulatum 194
Batrachospermum breviarticulatum 194
Batrachospermum bruziense 152
Batrachospermum campyloclonum 107
Batrachospermum campyloclonum 108
Batrachospermum capense 201
Batrachospermum capense 202
Batrachospermum capense
  var. *breviarticulatum* 194
Batrachospermum capensis 202
Batrachospermum carpocontortum 82
Batrachospermum carpocontortum 83
Batrachospermum carpoinvolucrum 96
Batrachospermum carpoinvolucrum 96
Batrachospermum cayennense 123, 163
Batrachospermum cayennense 162, 163, 164, 165
Batrachospermum cipoense 186, 197
Batrachospermum coerulescens 143, 152
Batrachospermum confusum 74
Batrachospermum confusum 76, 77, 98
Batrachospermum confusum.......................
  f. *spermatogloberatum* 74
Batrachospermum corbula 88
Batrachospermum corbula var. *alcoense* 88
Batrachospermum crispatum 143
Batrachospermum crispatum 144
Batrachospermum crouanianum 74, 76, 77
Batrachospermum curvatum 182
Batrachospermum curvatum 185
Batrachospermum cylindrocellulare 85
Batrachospermum cylindrocellulare 86
Batrachospermum dasyphillum 211
Batrachospermum dasyphillum 215
Batrachospermum debilis 105
Batrachospermum debilis 106
Batrachospermum decaisneanum 88
Batrachospermum deminutum 204
Batrachospermum deminutum 206
Batrachospermum densum 87
Batrachospermum desikacharyi 154
Batrachospermum desikacharyi 155
Batrachospermum diatyches 127
Batrachospermum diatyches 125, 128
Batrachospermum dillenii 123, 129
Batrachospermum dimorphum 156
Batrachospermum discorum 111
Batrachospermum discorum 102, 112
Batrachospermum distensum 77
Batrachospermum doboense 181, 184
Batrachospermum ectocarpoideum 92
Batrachospermum ectocarpum 88, 90, 109
Batrachospermum elegans 150
Batrachospermum elegans 151, 152
Batrachospermum equisetifolium 171

Batrachospermum equisetifolium 172
Batrachospermum equisetoideum 187
Batrachospermum equisetoideum 186
Batrachospermum excelsum 168
Batrachospermum exsertum 221
Batrachospermum faroense 181
Batrachospermum faroense 183, 184
Batrachospermum ferreri 149
Batrachospermum flagelliforme 145
Batrachospermum fluitans 92
Batrachospermum fruticulosum 74
Batrachospermum gallaei 129, 132
Batrachospermum gelatinosum 86, 102, 107
Batrachospermum gelatinosum 87, 89
Batrachospermum gibberosum 203
Batrachospermum gibberosum 204
Batrachospermum globosporum 197
Batrachospermum globosporum .. 186, 198, 199
Batrachospermum godronianum 88
Batrachospermum gombakense 144
Batrachospermum gombakense 145
Batrachospermum gracillimum 212
Batrachospermum gracillimum 216
Batrachospermum graibussoniense 152
Batrachospermum gulbenkianum 147
Batrachospermum guyanense 196
Batrachospermum guyanense 196
Batrachospermum helminthosum 152
Batrachospermum helminthosum ... 74, 153, 154
Batrachospermum henriquesianum 178
Batrachospermum heterocorticum 98
Batrachospermum heterocorticum 78, 100
Batrachospermum heteromorphum 82
Batrachospermum heteromorphum 84
Batrachospermum hirosei 207
Batrachospermum hirosei 205
Batrachospermum huillense 234
Batrachospermum hybridum 86
Batrachospermum hypogynum 171
Batrachospermum hypogynum 172
Batrachospermum intortum 174
Batrachospermum intortum 174, 175
Batrachospermum involutum 74
Batrachospermum involutum 75
Batrachospermum iriomotense 192
Batrachospermum iriomotense 193
Batrachospermum iyengarii 214
Batrachospermum iyengarii 219
Batrachospermum japonicum 88
Batrachospermum jolyi 197
Batrachospermum keratophytum 121, 137
Batrachospermum keratophytum 138, 139
Batrachospermum kraftii 113
Batrachospermum kraftii 114
Batrachospermum kushiroense 190
Batrachospermum kushiroense 192
Batrachospermum kylinii 217

Batrachospermum kylinii 222
Batrachospermum latericiuum 124
Batrachospermum latericiuum 126
Batrachospermum lindaueri 243
Batrachospermum lochmodes 73
Batrachospermum lochmodes 73
Batrachospermum longiarticulatum 166
Batrachospermum longiarticulatum 165, 167
Batrachospermum longipedicellatum 94
Batrachospermum longipedicellatum 95
Batrachospermum louisianae 193
Batrachospermum louisianae 182, 183
Batrachosperma ludibunda [alpha] *confusa* .. 74
Batrachospermum ludibundum
    var. *caerulescens* 86
Batrachospermum ludibundum var. *stagnale* . 86
Batrachospermum lusitanicum 177
Batrachospermum macrosporum 168
Batrachospermum macrosporum 169, 170
Batrachospermum macrosporum
    var. *excelsum* 168, 170
Batrachospermum macrosporum
    var. *oxycladum* 168, 170
Batrachospermum mahabaleshwarensis 218
Batrachospermum mahabaleshwarensis 215
Batrachospermum mahlacense 209
Batrachospermum mahlacense 211
Batrachospermum mikrogyne 159
Batrachospermum moniliforme 87
Batrachospermum moniliforme f. *lipsiensis* ... 86
Batrachospermum moniliforme typicum 86
Batrachospermum moniliforme
    var. *chlorosum* 86
Batrachospermum moniliforme
    var *helminthoideum* 88
Batrachospermum moniliforme
    var. *isoeticola* 88
Batrachospermum moniliforme
    var. *obtrullatum* 88
Batrachospermum moniliform
    var. *pisanum* 86
Batrachospermum moniliforme
    var. *rubescens* 86
Batrachospermum moniliforme var. *scopula* . 88
Batrachospermum moniliforme
    var. *trullatum* 88
Batrachospermum moniliforme var. *vagum* . 141
Batrachospermum nechochoense 199
Batrachospermum nechochoense 200
Batrachospermum nigrescens 131
Batrachospermum nodiflorum 212
Batrachospermum nodiflorum 213
Batrachospermum nodusum 246
Batrachospermum nonocense 196
Batrachospermum nonocense 197
Batrachospermum nothogeae 124, 127
Batrachospermum nova-guineense 92

Batrachospermum nova-guineense 94
Batrachospermum omobodense 220
Batrachospermum omobodense 221
Batrachospermum orthostichum 136
Batrachospermum orthostichum 136
Batrachospermum oxycladum 162, 168
Batrachospermum periplocum 139
Batrachospermum periplocum 140, 141
Batrachospermum procarpum 88
Batrachospermum procarpum
    var. *americanum* 188
Batrachospermum procarpum
    var. *americanum* 189
Batrachospermum procarpum
    var. *procarpum* 187
Batrachospermum procarpum
    var. *procarpum* 188
Batrachospermum prominens 118
Batrachospermum prominens 119
Batrachospermum pseudocarpum 175
Batrachospermum puiggarianum 131
Batrachospermum puiggarianum 132, 133
Batrachospermum pulchrum 76
Batrachospermum pulchrum 78
Batrachospermum pygmaeum 88
Batrachospermum pyramidale 88
Batrachospermum radians 88
Batrachospermum ranuliferum 115
Batrachospermum ranuliferum 115
Batrachospermum reginense 88
Batrachospermum rubrum 62
Batrachospermum schwacheanum 131
Batrachospermum setigenum 74
Batrachospermum sinense 85
Batrachospermum sinense 87
Batrachospermum sirodotii 152, 154
Batrachospermum skujae 72
Batrachospermum skujae 71, 73
Batrachospermum skujanum 201
Batrachospermum skujanum 200
Batrachospermum spermatiophorum 190
Batrachospermum spermatiophorum 191
Batrachospermum spermatoinvolucrum 79
Batrachospermum spermatoinvolucrum 80
Batrachospermum sporulans 73
Batrachospermum stagnale 86
Batrachospermum suevorum 137
Batrachospermum szechwanense 82
Batrachospermum szechwanense 84
Batrachospermum tabagatense 195
Batrachospermum tabagatense 195
Batrachospermum tapirense 135
Batrachospermum tapirense 135
Batrachospermum tenuissimum 129
Batrachospermum terawhiticum 118
Batrachospermum terawhiticum 120
Batrachospermum testale 152

Batrachospermum theaquum 103
Batrachospermum theaquum 104
Batrachospermum tiomanense 214
Batrachospermum tiomanense 220
Batrachospermum torridum 180
Batrachospermum torridum 181
Batrachospermum torsivum 213
Batrachospermum torsivum 218
Batrachospermum tortuosum* var. *majus ... 179
Batrachospermum tortuosum var. majus 180
***Batrachospermum tortuosum*
var. *tortuosum*** 178
Batrachospermum tortuosum var. tortuosum 179
Batrachospermum transtaganum 150
Batrachospermum transtaganum 159
Batrachospermum trichocontortum 79
Batrachospermum trichocontortum 81
Batrachospermum trichofurcatum 79
Batrachospermum trichofurcatum 80
Batrachospermum turfosum 141
Batrachospermum turfosum 134, 142, 215
Batrachospermum turgidum 166
Batrachospermum turgidum 167
Batrachospermum vagum 137, 141, 142, 215
Batrachospermum vagum f. tennuissimum .. 210
Batrachospermum vagum
  var. flagelliforme 145, 146
Batrachospermum vagum var. guyanense 196
Batrachospermum vagum
  var. keratophytum 137, 138
Batrachospermum vagum var. nodiflorum ... 212
Batrachospermum vagum
  var. periplocum 139, 140, 141
Batrachospermum vagum var. torridum 180
Batrachospermum vagum
  var. undulato-pedicellatum 141
Batrachospermum virgato-decaisneanum .. 158
Batrachospermum virgato-
  decaisneanum 159, 160, 161, 162
Batrachospermum virgato-decaisneanum
  var. cochleophilum 160
Batrachospermum virgatum 152
Batrachospermum vittatum 207
Batrachospermum vittatum 208
Batrachospermum vogesiacum 145
Batrachospermum vogesiacum 146
Batrachospermum wattsii 121
Batrachospermum wattsii 122
Batrachospermum woitapense 176
Batrachospermum woitapense 176
Batrachospermum zeylanicum 216
Batrachospermum zeylanicum 219
Boldia 37
Boldia angustata 37
Boldia erythrosiphon 37
Boldia erythrosiphon 38
Boldiaceae 37

Bostrychia 311
Bostrychia andoi 318
Bostrychia flagellifera 318
Bostrychia flagellifera 319
Bostrychia hamana-tokidae 318
Bostrychia moritziana 312
Bostrychia moritziana 313, 314, 315
Bostrychia radicans 315
Bostrychia radicans 314, 316
Bostrychia radicans f. moniliforme 312, 315
Bostrychia scorpioides 320
Bostrychia scorpioides 311, 321, 322
Bostrychia simpliciuscula 318
Bostrychia simpliciuscula 319
Bostrychia tenella 320
Bostrychia tenella 314, 317
Bostrychia tenuis f. simpliciuscula 318, 319

Caloglossa 302
Caloglossa amboinensis 306
Caloglossa beccarii 306
Caloglossa beccarii 303
Caloglossa bombayensis 304
Caloglossa continua 307
Caloglossa continua 305, 308
Caloglossa continua
  subspecies saigonensis 310
Caloglossa leprieurii 306
Caloglossa leprieurii 305, 310, 321
Caloglossa leprieurii auct. Japon 307, 308
Caloglossa leprieurii var. angusta 310
Caloglossa leprieurii var. continua 307
Caloglossa ogasawaraensis 304
Caloglossa ogasawaraensis 303, 305, 310
Caloglossa ogasawaraensis
  var. latifolia 303, 306
Caloglossa saigonensis 310
Caloglossa saigonensis 309
Caloglossa zanzibariensis 304
Ceramiaceae 295
Ceramiales 295, 302
Chantransia chalybea 51, 54
Chantransia eugenea 41, 43
Chantransia hermannii 44, 47
Chantransia investiens 62
Chantransia leibleinii 50, 54
Chantransia macrospora 46, 49
Chantransia pygmaea 50, 55
Chantransia subtilis 50
Chantransia tenella 41, 43
Chantransia violacea 47
Chara gelatinosa var. vaga 141
Chroodactylon 15
Chroodactylon ornatum 16
Chroodactylon ornatum 17
Chroodactylon ramosum 15

Chroodactylon ramosum 16
Chroodactylon wolleanum 15
Compsopogon 23
Compsopogon aeruginosus 23, 24
Compsopogon aeruginosus 25
Compsopogon aeruginosus var. *catenatum* ... 26
Compsopogon chalybeus 27
Compsopogon chalybeus 30
Compsopogon coeruleus 33
Compsopogon coeruleus 30
Compsopogon corinaldii 27, 30
Compsopogon corticrassus 27
Compsopogon corticrassus 28
Compsopogon fruticosus 21, 22
Compsopogon helwanii 33
Compsopogon hookeri 31
Compsopogon hookeri 21, 29
Compsopogon leptoclados 20, 22
Compsopogon minutus 28
Compsopogon minutus 31
Compsopogon oishii 33
Compsopogon prolificus 24
Compsopogon prolificus 25
Compsopogon sparsus 29
Compsopogon sparsus 32
Compsopogon tenellus 27
Compsopogon tenellus 32
Compsopogonaceae 19
Compsopogonales 19
Compsopogonopsis 19
Compsopogonopsis fruticosa 22
Compsopogonopsis fruticosa 21
Compsopogonopsis japonica 22
Compsopogonopsis japonica 20
Compsopogonopsis leptoclados 22
Compsopogonopsis leptoclados 20
Conferva atra 129
Conferva atropurpurea 35
Conferva chalybea 51
Conferva coerulea 33
Conferva flexuosa 280
Conferva fluviatilis 256
Conferva gelatinosa 86, 107
Conferva hermannii 47
Conferva hispida 280
Conferva ornatum 17
Contorta 173
Cyanidiaceae 10
Cyanidioschyzon 11
Cyanidioschyzon merolae 11
Cyanidioschyzon merolae 12
Cyanidium 10
Cyanidium caldarium 11
Cyanidium caldarium 11
Cyanidium sulphuraria 14

Delesseria amboinensis 306
Delesseria beccarii 306
Delesseria leprieurii 306
Delesseria zanzibariensis 304
Delesseriaceae 302

Entothrix grandis 269
Erythroclathrus rivularis 293
Eulemanea 264

Fucus scorpioides 320
Fucus tenellus 320

Galdieria 14
Galdieria sulphuraria 14
Galdieria sulphuraria 13
Gonimopropagulum 156
Goniotrichiaceae 15
Goniotrichiales 15

Hildenbrandia 292
Hildenbrandia angolensis 293
Hildenbrandia angolensis 295
Hildenbrandia prototypus 292
Hildenbrandia rivularis 293
Hildenbrandia rivularis 294
Hildenbrandia rubra 292
Hildenbrandiaceae 292
Hildenbrandiales 292
Hybrida 158
Hybride 158
Hypoglossum leprieurii 306

Intorta 173

Kushiroense 189
Kyliniella 18
Kyliniella latvica 18
Kyliniella latvica 17

Lemanea 252, 264
Lemanea annulata 265, 267
Lemanea annulata var. *franciscana* 271
Lemanea australis 269, 270
Lemanea borealis 253
Lemanea catenata 265, 266
Lemanea ciliata 260
Lemanea ciliata 260
Lemanea condensata 254
Lemanea corallina 252

Lemanea corinaldii 27
Lemanea daldinii 256
Lemanea feldmannii 265
Lemanea fluviatilis 256
Lemanea fluviatilis 257
Lemanea fucina 263
Lemanea fucina 258, 259
Lemanea fucina var. *mamillosa* 259, 263
Lemanea fucina var. **parva** 264
Lemanea fucina var. *rigida* 261
Lemanea grandis 269, 270
Lemanea kalchenbenneri 256
Lemanea mamillosa 263
Lemanea mamillosa 255, 259
Lemanea mamillosa var. *fucina* 263
Lemanea mexicana 265
Lemanea nodosa 265, 267
Lemanea pleocarpa 265
Lemanea rigida 261
Lemanea rigida 262, 270
Lemanea simplex 254
Lemanea simplex 255
Lemanea sinica 261
Lemanea sinica 262
Lemanea subgenus *Paralemanea* 264
Lemanea subgenus *Sacheria* 252
Lemanea sudetica 256
Lemanea sudetica 260
Lemanea torulosa 261, 270
Lemaneaceae 251

Macrospora 168
Moniliformes 70
Moniliformia 70, 123
Moniliformia subsection *Capillacea* 123
Moniliformia subsection *Setacea* 123

Nemalionopsis 290
Nemalionopsis shawii 292
Nemalionopsis shawii 281, 290
Nemalionopsis shawii f. *caroliniana* 285
Nemalionopsis tortuosa 290
Nemalionopsis tortuosa 281, 285, 291
Nothocladus 243
Nothocladus afroaustralis 243
Nothocladus afroaustralis 244
Nothocladus lindaueri 243
Nothocladus lindaueri 244, 245
Nothocladus nodosus 246
Nothocladus nodosus 247, 248
Nothocladus tasmanicus 248

Paralemanea 264
Paralemanea annulata 265

Paralemanea annulata 267, 268
Paralemanea brandegeei 275
Paralemanea brandegeei 274, 276
Paralemanea californica 275
Paralemanea californica 273, 276
Paralemanea catenata 265
Paralemanea catenata 266, 267
Paralemanea deamii 271
Paralemanea deamii 272
Paralemanea gardneri 271
Paralemanea gardneri 273, 276
Paralemanea grandis 269
Paralemanea grandis 270
Paralemanea mexicana 265
Paralemanea parishii 271
Paralemanea parishii 274, 276
Paralemanea tulensis 275
Paralemanea tulensis 274
Pericystis aeruginosa 23
Phragmonemataceae 18
Pleurococcus sulphurarius 14
Plocamium tenellum 320
Polysiphonia 322
Polysiphonia moritziana 312
Polysiphonia subtilissima 324
Polysiphonia subtilissima 310, 323
Polysiphonia urceolata 322
Porphyridiaceae 10
Porphyridiales 10
Porphyridium 10
Porphyridium cruentum 10
Porphyridium purpureum 10
Porphyridium purpureum 12
Procarpa 184
Protococcus botryoides f. *caldaria* 11
Pseudochantransia chalybea 51
Pseudochantransia leibleinii 50
Pseudochantransia pygmaea 50
Pseudochantransia serpens 46, 49
Psilosiphon 249
Psilosiphon scoparium 249
Psilosiphon scoparium 250
Psilosiphonaceae 249
Ptilothamnion 300
Ptilothamnion pluma 300
Ptilothamnion richardsii 300
Ptilothamnion richardsii 300, 301
Ptilothamnion umbracolens 300

Rhodochortaceae 39
Rhodochorton investiens 62
Rhodochorton violaceum 45, 47
Rhodococcus 11
Rhodococcus caldarius 11
Rhododraparnaldia 64
Rhododraparnaldia oregonica 64

Rhododraparnaldia oregonica 65
Rhodomela radicans 315
Rhodomelaceae 311
Rhodomela 311
Rhodophyceae 9

Sacheria 252
Sacheria ciliata 260
Sacheria fluviatilis 256, 257
Sacheria fucina 263
Sacheria mamillosa 255, 263
Sacheria rigida 261, 262
Setacea 68, 118, 123, 124, 134, 137
Setaces 123
Sirodotia 224
Sirodotia acuminata 230
Sirodotia angolensis 129
Sirodotia ateleia 234
Sirodotia cirrhosa 234
Sirodotia delicatula 236
Sirodotia delicatula 237
Sirodotia fennica 230, 233
Sirodotia gardneri 234
Sirodotia gardneri 236
Sirodotia geobelii 227
Sirodotia goebelii 229
Sirodotia huillensis 234
Sirodotia huillensis 224, 234, 235, 236
Sirodotia nigrescens 131
Sirodotia segawae 225
Sirodotia segawae 226, 232
Sirodotia sinica 227
Sirodotia sinica 228
Sirodotia sp. 224, 225, 232
Sirodotia suecica 230
Sirodotia suecica 224, 231, 232, 233
Sirodotia tenuissima 230, 231
Sirodotia yutakae 224
Sirodotia yutakae 226, 232

Thorea 278
Thorea andina 280, 282
Thorea bachmannii 284
Thorea bachmannii 288, 289
Thorea brodensis 289
Thorea clavata 279
Thorea clavata 278, 279, 281
Thorea conturba 282
Thorea conturba 283
Thorea gaudichaudii 280, 286
Thorea hispida 280
Thorea hispida 278, 281
Thorea lehmannii 280
Thorea nemalionopsis 277
Thorea okadae 282

Thorea okadae 280, 284, 287
Thorea prowsei 286
Thorea prowsei 290
Thorea ramosissima 278, 280
Thorea riekei 284
Thorea riekei 285
Thorea trailii 187
Thorea violacea 280
Thorea violacea 281, 285, 286, 289
Thorea zollingeri 280
Thorea zollingeri 281
Thoreaceae 66, 277, 278
Thoreales 277
Torrida 177
Trentepohlia aeruginosa 51
Trentepohlia pulchella f. *chalybea* 51
Tuomeya 238
Tuomeya americana 238
Tuomeya americana 239, 240, 241, 242
Tuomeya fluviatilis 238, 239, 240
Tuomeya gibberosa 203, 205
Tuomeya grandis 269
Turficola 134
Turficoles 134
Turfosa 134
Turfosa 134

Verts 143
Virescentia 143
Viridia 143

Addendum

Order **Bangiales**

Gargiolo *et al.* (2001) report that an alternation between a macroscopic gametophytic phase and a microscopic filamentous chonchocelis phase was found, and spermatia and carpogonia are described for the first time in a freshwater population of *Bangia atropurpurea* in the field and in culture.

Mueller *et al.* (2001) propose a new genus and species within the order Bangiales, *Pseudobangia kaycoleia* (ined.) for a freshwater specimen from the Virgin Island with numerous plastids in all cells, all of which contain a pyrenoid.

Gargiolo, G. M., Genovese, G., Morabito, M., Culoso, F. & de Masi, F. 2001. Sexual and asexual reproduction in a freshwater population of *Bangia atropurpurea* (Bangiales, Rhodophyta) from eastern Sicily (Italy). Phycologia 40:88–96.

Mueller, K. M., Thompson, S. L., Cannone, J. J. & Sheath, R. G. 2001. A molecular phylogenetic analysis of the Bangiales (Rhodophyta) and description of a new genus, *Pseudobangia*. Phycologia 40: Suppl. 21.

Order **Batrachospermales**

Vis *et al.* (2001) state that *B. helminthosum* is a morphologically plastic species with some genetic variation among samples from throughout its North American distribution. Both types of molecular and morphological data suggest that the biogeography of this taxon in North America is complex.

Vis, M. L., Miller, E. J. & Hall, M. M. 2001. Biogeographical analyses of *Batrachospermum helminthosum* (Batrachospermales, Rhodophyta) in North America using molecular and morphological data. Phycologia 40:2–9.

Order **Gigartinales**

Based on the morphological differences from *Sterrocladia*, Sheath *et al.* (2001) propose a new gigartinalean genus *Pseudosterrocladia* (ined.) for a Belizean specimen, which was positioned in the Gigartinales by phylogenetic analysis using the rbc gene and the 18S rRNA gene.

Sheath, R. G., Sherwood, A. R. & Necchi Jr, O. 2001. Freshwater members of the Gigartinales (Rhodophyta). Phycologia 40: Suppl. 22.